THE NUCLEAR COOPER PAIR

This monograph presents a unified theory of nuclear structure and nuclear reactions in the language of quantum electrodynamics-Feynman diagrams. It describes how two-nucleon transfer reaction processes can be used as a quantitative tool to interpret experimental findings with the help of computer codes and nuclear field theory. Making use of Cooper pair transfer processes, the theory is applied to the study of pair correlations in both stable and unstable exotic nuclei. Special attention is given to unstable, exotic halo systems, which lie at the forefront of the nuclear physics research being carried out at major laboratories around the world. This volume is distinctive in dealing in both nuclear structure and reactions and benefits from comparing the nuclear field theory with experimental observables, making it a valuable resource for incoming and experienced researchers who are working in nuclear pairing and using transfer reactions to probe them.

GRÉGORY POTEL AGUILAR obtained his PhD at the University of Sevilla, Spain, in 2005. His work concerning the reaction mechanism of Cooper pair transfer in the study of pairing in atomic nuclei led him to earn a FRIB Theory Fellowship at the Facility for Rare Isotopes Beams (FRIB), Michigan. He is currently a staff scientist at the Lawrence Livermore National Laboratory.

RICARDO A. BROGLIA earned his PhD at the Institute J. Balseiro, University of Cuyo, Argentina, in 1965. Following positions at the University of Buenos Aires, the Niels Bohr Institute, and the University of Minnesota, he joined the staff of the Niels Bohr Institute in 1970, where he is now an affiliated professor. In 1985 he was called to occupy the Chair of Nuclear Structure at the University of Milan, a position at which he stayed until 2009. He is known for helping to develop the nuclear field theory based on quantum electrodynamics-Feynman diagrams and the theory of heavy ion reactions. His previous books include *Oscillations in Finite Quantum Systems* with George Bertsch and *Nuclear Superfluidity* with David Brink, which can be considered companions to the present volume.

THE NUCLEAR COOPER PAIR
STRUCTURE AND REACTIONS

GRÉGORY POTEL AGUILAR

Lawrence Livermore National Laboratory

RICARDO A. BROGLIA

Niels Bohr Institute

CAMBRIDGE
UNIVERSITY PRESS

CAMBRIDGE
UNIVERSITY PRESS

University Printing House, Cambridge CB2 8BS, United Kingdom

One Liberty Plaza, 20th Floor, New York, NY 10006, USA

477 Williamstown Road, Port Melbourne, VIC 3207, Australia

314–321, 3rd Floor, Plot 3, Splendor Forum, Jasola District Centre, New Delhi – 110025, India

103 Penang Road, #05–06/07, Visioncrest Commercial, Singapore 238467

Cambridge University Press is part of the University of Cambridge.

It furthers the University's mission by disseminating knowledge in the pursuit of education, learning, and research at the highest international levels of excellence.

www.cambridge.org
Information on this title: www.cambridge.org/9781108843546
DOI: 10.1017/9781108919036

First published 2022

A catalogue record for this publication is available from the British Library.

Library of Congress Cataloging-in-Publication Data
Names: Aguilar, Grégory Potel, 1972- author. | Broglia, R. A., author.
Title: The nuclear Cooper pair : structure and reactions / Grégory Potel Aguilar, Ricardo A. Broglia.
Description: Cambridge ; New York, NY : Cambridge University Press, 2022. | Includes bibliographical references and index.
Identifiers: LCCN 2021024756 (print) | LCCN 2021024757 (ebook) | ISBN 9781108843546 (hardback) | ISBN 9781108919036 (ebook)
Subjects: LCSH: Pairing correlations (Nuclear physics) | Nucleon-nucleon interactions. | Superconductivity. | Superfluidity. | Unified nuclear models. | Unified field theories. | BISAC: SCIENCE / Physics / Nuclear
Classification: LCC QC794 .A37 2022 (print) | LCC QC794 (ebook) | DDC 539.7/5–dc23
LC record available at https://lccn.loc.gov/2021024756
LC ebook record available at https://lccn.loc.gov/2021024757

ISBN 978-1-108-84354-6 Hardback

Additional resources for this publication at www.cambridge.org/nuclearcooperpair

Contents

Preface

The elementary modes of nuclear excitation are vibrations and rotations, single-particle motion, and pairing vibrations and rotations. The reactions that specifically probe them are inelastic scattering plus Coulomb excitation and single- and two-particle transfer processes, respectively.

The interweaving of the elementary modes of excitation leads to the renormalization of energy, radial wavefunction, and particle content of the single-particles. It also leads to the renormalization of energy, width, and collectivity of vibrations and rotations. This implies renormalization of transition densities and formfactors, as well as of deformation (order) parameters, both in 3D and in gauge space. A consequence is the emergence of a variety of properties, such as generalized rigidity and long-range correlations in connection with collective pairing states, implying, for example, that pair transfer is dominated by the successive tunneling of entangled nucleons and, consequently, the need to go beyond lowest-order distorted wave Born approximation, that is, to second-order DWBA.

Within this context one can posit that nuclear structure (bound) and reactions (continuum) are but two aspects of the same physics. Even more concerning is the study of light exotic halo nuclei, in which case the distinction between bound and continuum states is, to a large extent, blurred. This is also the reason why these two aspects of nuclear physics are treated in the present monograph on equal footing, within the framework of a unified nuclear field theory of structure and reactions $(NFT)_{(s+r)}$.

This theory provides the (graphical) rules to diagonalize in a compact and economic way the nuclear Hamiltonian for both bound and continuum states. It does so in terms of Feynman diagrams, which describe the coupling of elementary modes of excitation, correcting for the overcompleteness of the basis (structure) and for the nonorthogonality of the scattering states (reaction), as well as for Pauli principle violation. The outcome connects directly with observables: absolute reaction cross sections and decay probabilities.

(NFT)$_{(s+r)}$ focuses on the scattering amplitudes that determine the absolute cross sections for the variety of physical processes, involving also those in which (quasi) bosons and fermions are created or annihilated, connecting such processes to form-factors and transition densities, processes where one set of particles with given energies, momenta, angular momenta, etc. go in and another group (or the same) comes out.

Pairing vibrations and rotations, closely connected with nuclear superfluidity, are paradigms of quantal nuclear phenomena. They play an important role within the field of nuclear structure. It is natural that two-nucleon transfer plays a similar role concerning the probing of the structure of the nucleus.

At the basis of fermionic pairing phenomena[1] one finds Cooper pairs[2] – weakly bound, very extended, strongly overlapping (quasi-) bosonic entities made out of pairs of nucleons dressed by collective vibrations and interacting through the exchange of these vibrations as well as through the bare NN-interaction. Cooper pairs change, under certain conditions,[3] the statistics of the nuclear stuff around the Fermi surface and, condensing, the properties of nuclei close to their ground state. They also display a rather remarkable mechanism of tunneling between target and projectile in direct two-nucleon transfer reactions.

Cooper pair partners, being weakly bound ($\ll \epsilon_F$, Fermi energy), are correlated over distances (correlation length ξ) much larger than nuclear dimensions ($\gg R$, nuclear radius). On the other hand, Cooper pairs – building blocks of the so-called abnormal (pair) density – are forced to be confined within regions in which normal, single-particle density is present, that is, within nuclear dimensions. Said differently, the mean field acts on Cooper pairs as a strong external field, distorting their spatial structure.

The correlation length paradigm comes into evidence, for example, when two nuclei are set into weak coupling in a direct nuclear reaction with distance of closest approach $D_0 \lesssim \xi$. In such a case, the partner nucleons of a Cooper pair have a finite probability to be confined each within the mean field of a different nucleus, equally well pairing correlated than when both nucleons are in the same nucleus. It is then natural that a Cooper pair can also tunnel between target and projectile, equally well correlated, through simultaneous rather than through successive transfer processes. Because of the weak binding of the Cooper pair, let alone the fact that tunneling probability falls off exponentially with increasing mass, successive is the dominant transfer process.

---

[1] Bardeen et al. (1957a,b).

[2] Cooper (1956).

[3] Value of the intrinsic excitation energy, rotational frequency, and distance of closest approach (D_0) in Cooper pair tunneling between superfluid nuclei in nuclear reactions, smaller than the correlation length (ξ).

Although one does not expect supercurrents in nuclei, one can study long-range pairing correlations in terms of individual quantal states and of the tunneling of single Cooper pairs. Such (weak coupling) Cooper pair transfer reminds the tunneling mechanism of electronic Cooper pairs across a barrier (e.g., a dioxide layer of dimensions much smaller than the correlation length) separating two low-temperature, metallic superconductors and known as a Josephson junction.[4]

In the nuclear time-dependent junction transiently established in direct two-nucleon transfer process, only one or sometimes none of the two weakly interacting nuclei is superfluid. On the other hand in nuclei, a paradigmatic example of finite quantum many-body systems (FQMBS), zero-point fluctuations (ZPF) in general, and those associated with pair addition and pair subtraction modes known as pairing vibrations in particular, are, as a rule, much stronger than in condensed matter. Thus, pairing correlations based on even a single Cooper pair can lead to distinct effects in two-nucleon transfer processes.

Nucleonic Cooper pair tunneling has played and is playing a central role in the probing of these subtle quantal phenomena, both in the case of light exotic nuclei and for medium and heavy nuclei lying along the stability valley. They have been instrumental in shedding light on the subject of pairing in nuclei at large, and on nuclear superfluidity in particular. Consequently, and as already said, the subject of two-nucleon transfer reactions occupies a central place in the present monograph, both concerning the conceptual and the computational aspects of the description of nuclear pairing, as well as regarding the quantitative confrontation of the theoretical results with the experimental findings, in terms of absolute differential cross sections.

Concerning exotic nuclei, experimental studies carried out at TRIUMF, Vancouver (Canada),[5] have provided the basis for what can be considered a nuclear embodiment[6] of the Cooper pair model: a pair of fermions (nucleons N) moving in time reversal states on top of the Fermi surface and interacting through the short-range, bare NN- and the long range, induced-pairing interaction.[7] This last one results from the exchange of a long-wavelength dipole vibration (quasi-boson), leading to an extended, weakly bound system.

Regarding medium heavy nuclei lying along the stability valley, studies of heavy ion reactions between superfluid nuclei carried out at energies around and below the Coulomb barrier at the National Laboratory of Legnaro (Italy)[8] have

---

[4] Josephson (1962); Anderson (1964b).
[5] Tanihata et al. (2008).
[6] Barranco et al. (2001); Potel et al. (2010).
[7] Fröhlich (1952); Bardeen and Pines (1955).
[8] Montanari et al. (2014).

provided a measure of the neutron Cooper pair size (mean square radius or correlation length[9]).

In the present monograph, interdisciplinarity[10] is employed as a tool to attack concrete nuclear problems and, making use of the unique laboratory provided by the atomic nucleus,[11] to shed light on condensed matter results in terms of analogies involving individual, quantal states and tunneling of single Cooper pairs.

Because of the central role the interweaving of the variety of elementary modes of nuclear excitation plays in nuclear superfluidity, the study of Cooper pair tunneling in nuclei requires a consistent description of nuclear structure in terms of dressed quasiparticles and, making use of the resulting renormalized wavefunctions (formfactors), of single-nucleon transfer processes.[12] This is similar to the situation encountered in superconductors, in connection with strongly renormalized systems,[13] studied through one-electron tunneling experiments.[14]

In the present monograph the general physical arguments and technical computational details concerning the calculation of absolute one-and two-nucleon transfer differential cross sections within the framework of DWBA, making use of state-of-the-art NFT structure input, are discussed in detail.

As a result of this approach, it is expected that both theoretical and experimental nuclear practitioners can use the present monograph at profit. To help this use, the basic nuclear structure formalism, in particular that associated with single-particle and collective motion in both normal and superfluid nuclei, is economically introduced through general physical arguments. This is also in keeping with the availability in the current literature of detailed discussions of the corresponding material. Within this context, the monographs *Nuclear Superfluidity* by Brink and Broglia and *Oscillations in Finite Quantum Systems* by Bertsch and Broglia, published also by Cambridge University Press, can be considered companion volumes to the present one. This volume shares with those a similar aim: to provide a broad physical view of central issues in the study of finite quantal many-body nuclear systems accessible to motivated students and practitioners. However, neither the present one nor the other two are introductory texts. In particular, in the present one, an attempt at unifying structure and reactions is made. On the other hand, unifying discrete (mainly structure) and continuum (mainly reactions) configuration spaces implies that one will be dealing with those structure results that can be tested

---

[9] Potel et al. (2021).

[10] Quoting de Gennes (1974): "what a theorist can and should systematically introduce is comparison with other fields."

[11] Broglia (2020) (overview).

[12] In other words, one recognizes the difficulties of extracting spectroscopic factors from experiment, in terms of single-particle transfer cross sections calculated making use of mean field wavefunctions.

[13] Eliashberg (1960).

[14] Giaever (1973).

by means of experiment, a fact that makes the subject of the present monograph a chapter of quantum mechanics, and thus open to a wide range of practitioners.[15]

Concerning the notation, we have divided each chapter into sections. Each section may, in turn, be broken down into subsections. Equations and figures are identified by the number of the chapter and that of the section. Thus (5.1.33) labels the thirty-third equation of Section 1 of Chapter 5. Similarly, "Fig. 5.1.2" labels the second figure of Section 1 of Chapter 5. Concerning the appendices, they are labeled by the chapter number and by a Latin letter in alphabetical order, e.g., App. 2.A, App. 2.B. Concerning equations and figures, a sequential number is added. Thus (2.B.2) labels the second equation of Appendix 2.B of Chapter 2, while "Fig. 2.B.1" labels the first figure of Appendix 2.B of Chapter 2. References are called in terms of the author's surname and publication year and are found in alphabetic order in the bibliography at the end of the monograph.

A methodological approach used in the present monograph concerns a certain degree of repetition. Similar but not the same issues are dealt with more than once using different but equatable terminologies. This approach reflects the fact that useful concepts like reaction channels or correlation length, let alone elementary modes of excitation, are easy to understand but difficult to define.[16] This is because their validity is not exhausted by a single perspective. But even more important, it is because their power in helping at connecting[17] seemingly unrelated results and phenomena is difficult to fully appreciate the first time around, spontaneous symmetry breaking and associated emergent properties providing an example of this fact.

---

[15] Within this context let us mention the intimately correlated subjects of Random Phase Approximation (RPA) and Particle Vibration Coupling (PVC) not found in a fourth-year curriculum. They are explained and referred to in a number of places throughout the present monograph, starting from a pedestrian level and for both surface (particle-hole) and pairing (particle-particle and hole-hole) vibrations (Sect. 1.2, Fig. 1.2.3 (inset) and Sect. 1.3), and then extended to include further details and facets (see Sects. 1.5, 1.6, and 2.3, Fig. (2.3.1), Sect. 2.3.1 (Fig. 2.3.9), App. 2.B, and Sect. 3.1 (Fig. 3.1.6)). Furthermore, in the case of RPA of pairing vibrations around the closed shell system ^{208}Pb, one provides in Sect. 3.5 detailed documentation of the numerical calculation of the associated wavefunctions (X- and Y-amplitudes) at the level of an exercise in a fourth-year course. A similar situation is encountered in connection with the subject of the Distorted Wave Born Approximation (DWBA), again a subject not found in fourth-year curricula. It is treated at the pedestrian level in Sects. 5.6 and 6.4 in connection with one-particle and two-particle transfer reactions, respectively. And, once more in full detail, without eschewing complexities, but again in the style of an example exercise in Sects. 5.1 and 6.2 (within this context we quote, "Sometimes one has to say difficult things, but one ought to say them as simply as one knows how" (G. H. Hardy, *A mathematician's apology*, Cambridge University Press, Cambridge (1969)).

[16] "Gentagelsen den er virkeligheden, og Tilværelsens Alvor" (Repetition is the reality and the seriousness of life: S. Kierkegaard Gjentagelsen (1843)).

[17] "The concepts and propositions get 'meaning' viz. 'content,' only through their connection with sense-experiences. ... The degree of certainty with which this connection, viz., intuitive combination, can be undertaken, and nothing else, differentiates empty fantasy from scientific 'truth.'... A correct proposition borrows its 'truth' from the truth-content of the system to which it belongs" (A. Einstein, Autobiographical notes, in *Albert Einstein*, Ed. P. A. Schilpp, Vol I, Harper, New York (1951) p. 13.).

Throughout, a number of footnotes are found. This is in keeping with the fact that footnotes can play a special role within the framework of an elaborated presentation. In particular, they are useful to emphasize relevant issues in an economic way. Being outside the main text, they give the possibility of stating eventual important results, without the need of elaborating on the proof, but referring to the corresponding sources. Within this context, and keeping the natural distances, one can mention that in the paper in which Born[18] introduces the probabilistic interpretation of Schrödinger's wavefunction, the fact that this probability is connected with its [modulus] squared and not with the wavefunction itself is only referred to in a footnote.

A large fraction of the material contained in this monograph has been the subject of the lectures of the fourth-year course Nuclear Structure Theory, which RAB delivered throughout the years at the Department of Physics of the University of Milan, as well as at the Niels Bohr Institute (Copenhagen) and at Stony Brook (State University of New York). Part of it was also presented by the authors in the course Nuclear Reactions held at the PhD School of Physics of the University of Milan.

GP wants to thank the tutoring of Ben Bayman concerning specific aspects of two-particle transfer reactions. Discussions with Ian Thompson and Filomena Nunes on a variety of reaction subjects are gratefully acknowledged. RAB acknowledges the essential role the collaboration with Francisco Barranco and Enrico Vigezzi has played concerning the variety of nuclear structure aspects of the present monograph. His debt to the late Aage Winther regarding the reaction aspects of it is difficult to express in words. The overall contributions of Daniel Bès, Ben Bayman and the late Pier Francesco Bortignon are only too explicitly evident throughout the text and constitute a daily source of inspiration. GP and RAB have received important suggestions and comments regarding concrete points and the overall presentation of the material discussed below from Ben Bayman, Pier Francesco Bortignon, Willem Dickhoff, Vladimir Zelevinsky, and the late David Brink, who are here gratefully acknowledged. We are especially beholden to Elena Litvinova and Horst Lenske for much constructive criticism and many suggestions.

---

[18] Born (1926). Within this context, it is of note that the extension of Born probabilistic interpretation to the case of many-particle systems is also found in a footnote (Pauli (1927)), footnote on p. 83 of this reference; see Pais (1986)).

1

Introduction

Toute théorie physique est fondée sur l'analogie qu'on établi entre des choses
malconnues et des choses simples.

Every physical theory is based on the analogy which one establishes between
things not well known and things that are simple.

Simone Weil

1.1 Views of the Nucleus

In the atom, the nucleus provides the Coulomb field in which negatively charged
electrons $(-e)$ move independently of each other in single-particle orbitals. The
filling of these orbitals explains Mendeleev's periodic table. Thus the valence of the
chemical elements as well as the particular stability of the noble gases associated
with the closing of shells (2(He), 10(Ne), 18(Ar), 36(Kr), 54(Xe), 86(Ra)). The
dimension of the atom is measured in angstroms ($Å=10^{-8}$cm) and typical energies
in eV, the electron mass being $m_e \approx 0.511$ MeV (MeV$=10^6$eV).

The atomic nucleus is made out of positively charged protons $(+e)$ and of
(uncharged) neutrons, nucleons, of mass $\approx 10^3$ MeV ($m_p = 938.3$ MeV, $m_n =
939.6$ MeV). Nuclear dimensions are of the order of a few fermis (fm$= 10^{-13}$ cm).
The stability of the atom is provided by a source external to the electrons, namely,
the atomic nucleus. On the other hand, this system is self-bound as a result of the
strong interaction of range $a_0 \approx 0.9$ fm and strength $v_0 \approx -100$ MeV that acts
among nucleons.

1.1.1 The Liquid Drop and the Shell Models

While most of the atom is empty space, the density of the atomic nucleus is
conspicuous ($\rho = 0.17$ nucleon/fm^3). The "closed packed" nature of this system
implies, a priori, a short mean free path λ as compared to nuclear dimensions.
This can be estimated from classical kinetic theory $\lambda \approx (\rho\sigma)^{-1} \approx 1$ fm, where

$\sigma \approx 2\pi a_0^2$ is the nucleon–nucleon cross section. It seems, then, natural to liken the atomic nucleus to a liquid drop.[1] This picture of the nucleus provided the framework to describe the basic features of the fission process.[2]

The leptodermic properties of the atomic nucleus are closely connected with the semi-empirical mass formula:[3]

$$m(N, Z) = (Nm_n + Zm_p) - \frac{1}{c^2} B(N, Z),\qquad(1.1.1)$$

the binding energy being

$$B(N, Z) = \left(b_{vol} A - b_{surf} A^{2/3} - \frac{1}{2} b_{sym} \frac{(N - Z)^2}{A} - \frac{3}{5} \frac{Z^2 e^2}{R_c} \right),\qquad(1.1.2)$$

where $A = N + Z$ is the mass number, sum of the number of neutrons (N) and of protons (Z). The first term is the volume energy representing the binding energy in the limit of large A, for $N = Z$ and in the absence of the Coulomb interaction ($b_{vol} \approx 15.6$ MeV) . The second term represents the surface energy, where

$$b_{surf} = 4\pi r_0^2 \gamma.\qquad(1.1.3)$$

The nuclear radius is written as $R = r_0 A^{1/3}$, with $r_0 = 1.2$ fm, the surface tension energy being $\gamma \approx 0.95$ MeV/fm^2. The third term in (1.1.2) is the symmetry term, which reflects the tendency toward stability for $N = Z$, with $b_{sym} \approx 50$ MeV. The symmetry energy can be divided into a kinetic and a potential energy part. A simple estimate of the kinetic energy part can be obtained by making use of the Fermi gas model, which gives $(b_{sym})_{kin} \approx (2/3)\epsilon_F \approx 25$ MeV ($\epsilon_F \approx 36$ MeV). Consequently,

$$V_1 = (b_{sym})_{pot} = b_{sym} - (b_{sym})_{kin} \approx 25 \text{ MeV}.\qquad(1.1.4)$$

The last term of (1.1.2) is the Coulomb energy corresponding to a uniformly charged sphere of radius $R_c = 1.25 A^{1/3}$ fm.

When, in a heavy ion reaction, two nuclei come within the range of the nuclear forces, the Coulomb trajectory of relative motion will be changed by the attraction that will act between the nuclear surfaces. This surface interaction is a fundamental quantity in heavy ion reactions. Assuming two spherical nuclei at a relative distance $r_{aA} = R_a + R_A$, where R_a and R_A are the corresponding half-density radii, the (maximum) force acting between the two surfaces is

$$\left(\frac{\partial U_{aA}^N}{\partial r} \right)_{r_{aA}} = 4\pi \gamma \frac{R_a R_A}{R_a + R_A}.\qquad(1.1.5)$$

---

[1] Bohr and Kalckar (1937).
[2] Meitner and Frisch (1939); Bohr and Wheeler (1939).
[3] Weizsäcker (1935).

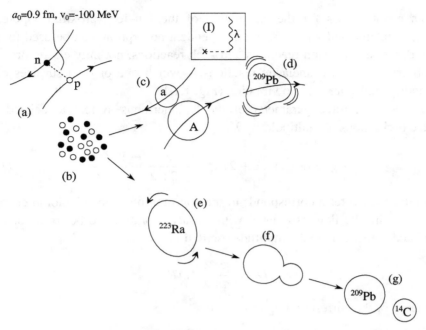

Figure 1.1.1 (a) Nucleon–nucleon (NN) interaction in a scattering experiment; emergent properties (collective nuclear modes). (b) Assembly of nucleons condensing into drops of nuclear matter displaying emergent properties, examples of which are shown in (c) and (e). (c) Anelastic heavy ion reaction $a + A \rightarrow a + A^*$ setting the nucleus A into an octupole surface oscillations (d). In inset (I) the time-dependent nuclear plus Coulomb field associated with the reaction (c) is represented by a cross followed by a dashed line, while the wavy line labeled λ describes the propagation of the $\lambda^\pi = 3^-$ surface vibration schematically shown in (d), time running upward. (e) The (weakly) quadrupole deformed nucleus ^{223}Ra can rotate as a whole with a moment of inertia considerably smaller than the rigid moment of inertia, a fact intimately connected with the role played by pairing in nuclei. The role becomes overwhelming in the phenomenon of exotic decay displayed in (f) in which the nuclear surface zero-point fluctuations (quadrupole ($\lambda = 2$), octupole ($\lambda = 3$), etc.) can get, with a small but finite probability ($P \approx 10^{-10}$), spontaneously in phase and produce a neck-in (saddle conformation), leading eventually to the (exotic) decay mode ^{223}Ra$\rightarrow ^{209}$Pb$+^{14}$C, as experimentally observed (g) (Rose and Jones (1984); see Brink and Broglia (2005), chapter 7 and refs. therein). *As correctly explained in Matsuyanagi et al. (2013) for vibrations in general, and valid also in the case of the ZPF leading to the saddle (neck-in) configuration, such fluctuations are associated with genuine quantum vibrations (where superfluidity and shell structure play a central role), and thus are essentially different in character from surface oscillations of a classical liquid drop. The intimate connection between pairing and collective vibrations reveals itself through the inertial masses governing the collective kinetic energies.*

This result allows for the calculation of the ion–ion (proximity) potential, which, supplemented with a position-dependent absorption, can be used to accurately describe heavy ion reactions.[4] In such reactions, not only elastic processes are observed, but also anelastic reactions in which one or both surfaces of the interacting nuclei are set into vibration (Fig. 1.1.1).

The restoring force parameter of the leptodermous system associated with surface oscillations of multipolarity λ is

$$C_\lambda = (\lambda - 1)(\lambda + 2)R_0^2\gamma - \frac{3}{2\pi}\frac{\lambda - 1}{2\lambda + 1}\frac{Z^2e^2}{R_c}, \qquad (1.1.6)$$

where the second term corresponds to the contribution of the Coulomb energy to C_λ. Assuming the flow associated with surface vibrations to be irrotational, the associated inertia for small amplitude oscillations is

$$D_\lambda = \frac{3}{4\pi}\frac{1}{\lambda}AmR^2, \qquad (1.1.7)$$

the energy of the corresponding mode being

$$\hbar\omega_\lambda = \hbar\sqrt{\frac{C_\lambda}{D_\lambda}}. \qquad (1.1.8)$$

The label λ stands for the angular momentum of the vibrational mode. Furthermore, the vibrations can be characterized by the parity quantum number $\pi = (-1)^\lambda$ and the third component of λ, denoted μ. Aside from λ, μ, surface vibrations can also be characterized by an integer $n (= 1, 2, \ldots)$, an ordering number indicating increasing energy. For simplicity, a single common label α will also be used.

A picture apparently antithetic to that of the liquid drop, the shell model, emerged from the study of experimental data, plotting them against either the number of protons (atomic number) or the number of neutrons in nuclei, rather than against the mass number. One of the main nuclear features that led to the development of the shell model was the study of the stability and abundance of nuclear species and the discovery of what are usually called magic numbers.[5] What makes a number magic is that a configuration of a magic number of neutrons, or of protons, is unusually stable whatever the associated number of other nucleons is.[6]

The strong binding of a magic number of nucleons and weak binding for one more reminds one of the results concerning the atomic stability of rare gases. In the nuclear case, the spin–orbit coupling plays an important role, as can be seen

[4] See, e.g., Broglia and Winther (2004) p. 110, and refs. therein.
[5] Elsasser (1933); Mayer (1948); Haxel et al. (1949).
[6] Mayer (1949); Mayer and Teller (1949).

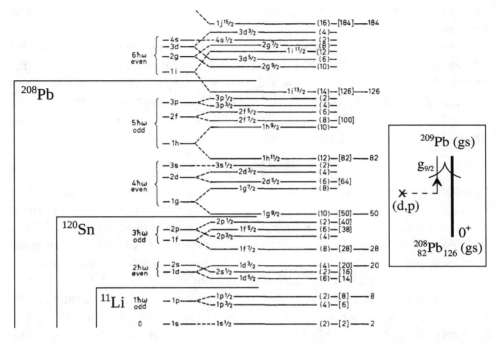

Figure 1.1.2 Sequence of levels of the harmonic oscillator potential labeled with the principal oscillator quantum number ($N(\hbar\omega) = 0(\hbar\omega), 1(\hbar\omega), 2(\hbar\omega), \ldots$, the parity being $\pi = (-1)^N$). The next column shows the splitting of major shell degeneracies obtained using a more realistic potential (Woods–Saxon), the quantum number being the number of radial nodes of the associated single-particle s, p, d, etc. states. The levels shown at the center result when a spin-orbit term is also considered, the quantum numbers nlj characterizing the states of degeneracy $(2j + 1)$ $(j = l \pm 1/2)$. To the left we schematically (in particular in the case of Li, which displays non-Meyer and Jensen sequence) indicate the Fermi energy associated with a light (exotic), medium, and heavy nucleus, namely, $^{11}_3$Li, $^{120}_{50}$Sn, and $^{208}_{82}$Pb. In the inset, a schematic graphical representation of the reaction ^{208}Pb$(d, p)^{209}$Pb(gs) is shown. A cross followed by a horizontal dashed line represents, in the present case, the (d, p) field, while a single arrowed line describes the odd nucleon moving in the $g_{9/2}$ orbital above the $N = 126$ shell closure drawn as a bold line labeled 0^+. After Mayer and Jensen (1955).

from the level scheme shown in Fig. 1.1.2, obtained by assuming that nucleons move independently of each other in an average potential of spherical symmetry.

A closed shell, or a filled level, has angular momentum zero, in keeping with the fact that, in such a case, there is a single way to arrange the fermions. Thus, nuclei with one nucleon outside (missing from the) closed shell should have the spin and parity of the orbital associated with the odd nucleon (nucleon hole), a prediction confirmed by the data (available at that time) throughout the mass table.

Such a picture implies that the nucleon mean free path is large compared to nuclear dimensions.

Systematic studies of the binding energies leading to the shell model found also that the relation (1.1.2) had to be supplemented to take into account the fact that nuclei with A odd, that is, with an odd number of either protons or neutrons, are energetically unfavored compared with the neighboring even-even ones by a quantity of the order of $\delta \approx 33$ MeV$/A^{3/4}$, called the pairing energy[7] and found at the basis of the odd-even staggering effect.

1.1.2 Nuclear Excitations

In addition to the quantum numbers λ, μ, and π, one can characterize nuclear excitations by additional quantum numbers, such as isospin τ and spin σ. Furthermore, one can assign a particle (baryon or transfer) quantum number[8] β. For a nucleon moving above the Fermi surface, one has $\beta = +1$, while for a hole in the Fermi sea, $\beta = -1$. For (quasi-) bosonic excitations, $\beta = 0$ for a mode associated with, for example, surface oscillations, which can also be viewed as a correlated particle-hole (p-h) excitation (within this context, see Fig. 1.2.3). In particular, the low-lying quadrupole and octupole vibrations of even-even nuclei (see Fig. 1.1.3) have quantum numbers $\beta = 0$, $\lambda^{\pi} = 2^{+}, 3^{-}, \tau = 0$ (protons and neutrons oscillate in phase), and $\sigma = 0$ (no spin-flip in the excitation).

For modes that involve the addition or substraction of two correlated nucleons to the nucleus, $\beta = +2$ (Fig. 1.3.1) and $\beta = -2$, respectively. The excitation that, around closed shells, connects the ground state of an even nucleus to the ground state of the next even nucleus, which is a monopole pairing vibration ($\lambda^{\pi} = 0^{+}$, $\beta = +2$), is of this type (Fig. 1.3.2). Multipole pairing vibrations with quantum numbers $\beta = \pm 2$ and[9] $\lambda^{\pi} = 2^{+}, 4^{+} \ldots$, have also been observed throughout the mass table.[10]

The low-lying excited state of closed shell nuclei can be interpreted as a rule, as a harmonic quadrupole, or as an octupole collective surface vibration (Fig. 1.1.3) described by the collective Hamiltonian[11]

$$H_{coll} = \sum_{\lambda\mu} \left(\frac{1}{2D_{\lambda}} |\hat{\Pi}_{\lambda\mu}|^2 + \frac{C_{\lambda}}{2} |\hat{\alpha}_{\lambda\mu}|^2 \right). \qquad (1.1.9)$$

---

[7] Mayer and Jensen (1955) p. 9. Connecting with further developments associated with the BCS theory of superconductivity (Bardeen et al. (1957a,b)) and its extension to the atomic nucleus (Bohr et al. (1958)), the quantity δ is identified with the pairing gap Δ parameterized according to $\Delta = 12$MeV$/\sqrt{A}$ (Bohr and Mottelson (1969)). It is of note that for typical superfluid nuclei like ^{120}Sn, the expression of δ leads to a numerical value that can be parameterized as $\delta \approx 33$ MeV$/(A^{1/4} \times A^{1/2}) \approx 10$ MeV$/\sqrt{A}$.

[8] Bohr (1964).

[9] It is of note that the quantum numbers of pairing vibrations are $\beta = \pm 2$ and $\pi = (-1)^{\lambda}$ (see App. 7.D).

[10] See, e.g., Flynn et al. (1971, 1972a); Brink and Broglia (2005) chapter 5. See also footnotes 42, 43, and 44 of chapter 7, and references therein.

[11] Classically, $\Pi_{\lambda\mu} = D_{\lambda}\dot{\alpha}_{\lambda\mu}$.

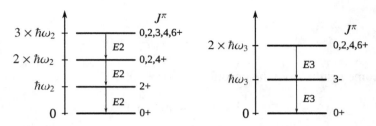

Figure 1.1.3 Schematic representation of harmonic quadrupole and octupole liquid drop collective surface vibrational modes.

Following Dirac (1930), one can describe the oscillatory motion introducing boson creation (annihilation) operator $\Gamma^\dagger_{\lambda\mu}$ ($\Gamma_{\lambda\mu}$) obeying the commutation relation

$$\left[\Gamma_\alpha, \Gamma^\dagger_{\alpha'}\right] = \delta(\alpha, \alpha') \tag{1.1.10}$$

and leading to

$$\hat{\alpha}_{\lambda\mu} = \sqrt{\frac{\hbar\omega_\lambda}{2C_\lambda}} \left(\Gamma^\dagger_{\lambda\mu} + (-1)^\mu \Gamma_{\lambda-\mu}\right). \tag{1.1.11}$$

A similar expression is valid for the conjugate momentum variable $\hat{\Pi}_{\lambda\mu}$, resulting in

$$\hat{H}_{coll} = \sum_{\lambda\mu} \hbar\omega_\lambda \left((-1)^\mu \Gamma^\dagger_{\lambda\mu}\Gamma_{\lambda-\mu} + 1/2\right). \tag{1.1.12}$$

The frequency of the mode is $\omega_\lambda = (C_\lambda/D_\lambda)^{1/2}$, while $(\hbar\omega_\lambda/2C_\lambda)^{1/2}$ is the amplitude of the zero-point fluctuation of the bosonic vacuum state $|0\rangle_B$, $|n_{\lambda\mu} = 1\rangle = \Gamma^\dagger_{\lambda\mu}|0\rangle_B$ being the one-phonon state. To simplify the notation, in many cases, one writes $|n_\alpha = 1\rangle$.

The ground and low-lying states of nuclei with one nucleon outside a closed shell can be described by the single-particle Hamiltonian

$$H_{sp} = \sum_\nu \epsilon_\nu a^\dagger_\nu a_\nu, \tag{1.1.13}$$

where $a^\dagger_\nu$ (a_ν) is the single-particle creation (annihilation) operator,

$$|\nu\rangle = a^\dagger_\nu|0\rangle_F \tag{1.1.14}$$

being the single-particle state of quantum numbers $\nu(\equiv nljm$, namely, number of nodes, orbital and total angular momentum, and its third component) and energy ϵ_ν, $|0\rangle_F$ being the Fermion vacuum. It is of note that

$$\left[H_{coll}, \Gamma^\dagger_{\lambda'\mu'}\right] = \hbar\omega_{\lambda'}\Gamma^\dagger_{\lambda'\mu'} \tag{1.1.15}$$

and

$$\left[H_{sp}, a_{\nu'}^{\dagger}\right] = \epsilon_{\nu'} a_{\nu'}^{\dagger}. \tag{1.1.16}$$

This outcome results from the bosonic

$$\left[\Gamma_{\alpha}, \Gamma_{\alpha'}^{\dagger}\right] = \delta(\alpha, \alpha') \tag{1.1.17}$$

and fermionic

$$\left\{a_{\nu}, a_{\nu'}^{\dagger}\right\} = \delta(\nu, \nu') \tag{1.1.18}$$

commutation (anti-commutation) relations.

The existence of drops of nuclear matter displaying both collective surface vibrations and independent-particle motion are emergent properties not contained in the particles forming the system, nor in the forces acting among them.

Expressed differently, generalized rigidity closely connected to the inertial parameter D_{λ} implies that acting on a nucleus with an external $\beta = 0$, time-dependent (nuclear/Coulomb) field, the system reacts as a whole (collective vibrations; also rotations see Sect. 1.4), while acting with fields that change particle number by one ($\beta = \pm 1$; e.g. (d, p) and (p, d) reactions), the system reacts in terms of independent particle motion, feeling the pushings and pullings of the other nucleons only when trying to leave the nucleus. Such a behavior can hardly be inferred from the study of the NN-forces in free space, being truly emergent properties of the finite, quantum many-body nuclear system.

Collective surface vibrations and independent particle motion are examples of what are called elementary modes of excitation in finite many-body physics and collective variables in soft-matter physics.

1.2 The Particle–Vibration Coupling

The oscillation of the nucleus under the influence of the surface tension implies that the potential $U(r, R)$ in which nucleons move independently of each other changes with time. For low-energy collective vibrations, this change is slow as compared with single–particle motion. Within this scenario the nuclear radius can be written as

$$R = R_0 \left(1 + \sum_{\lambda \mu} \alpha_{\lambda \mu} Y_{\lambda \mu}^{*}(\hat{r})\right). \tag{1.2.1}$$

Assuming small-amplitude motion,

$$U(r, R) = U(r, R_0) + \delta U(r), \tag{1.2.2}$$

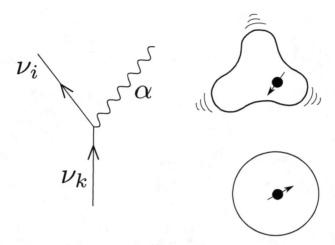

Figure 1.2.1 Graphical representation of a process by which a nucleon, bouncing inelastically off the nuclear surface, sets it into vibration. Particles are represented by an arrowed line pointing upward, which is also the direction of time, while the vibration is represented by a wavy line. In the cartoon to the right, the black dot represents a nucleon moving in a spherical mean field of which it excites, through the PVC vertex, an octupole vibration after bouncing inelastically off the surface.

where

$$\delta U = \kappa \hat{\alpha} \hat{F} = \Lambda_\alpha \left(\Gamma_{\lambda\mu}^\dagger + (-1)^\mu \Gamma_{\lambda-\mu} \right) \hat{F} = H_c, \tag{1.2.3}$$

with

$$\Lambda_\alpha = \kappa \sqrt{\frac{\hbar\omega_\lambda}{2C_\lambda}}, \tag{1.2.4}$$

is the particle–vibration coupling (PVC) strength (Fig. 1.2.1), product of the dynamic deformation

$$\beta_\lambda = \sqrt{2\lambda + 1} \sqrt{\frac{\hbar\omega_\lambda}{2C_\lambda}} \tag{1.2.5}$$

and of the strength κ, while

$$\hat{F} = \sum_{\nu_1\nu_2} \langle \nu_1 | F | \nu_2 \rangle a_{\nu_1}^\dagger a_{\nu_2} \tag{1.2.6}$$

is a single-particle field with (dimensionless) formfactor

$$F = -\frac{R_0}{\kappa} \frac{\partial U}{\partial r} Y_{\lambda\mu}^*(\hat{r}). \tag{1.2.7}$$

An estimate of κ is given below (Eq. (1.2.13)).

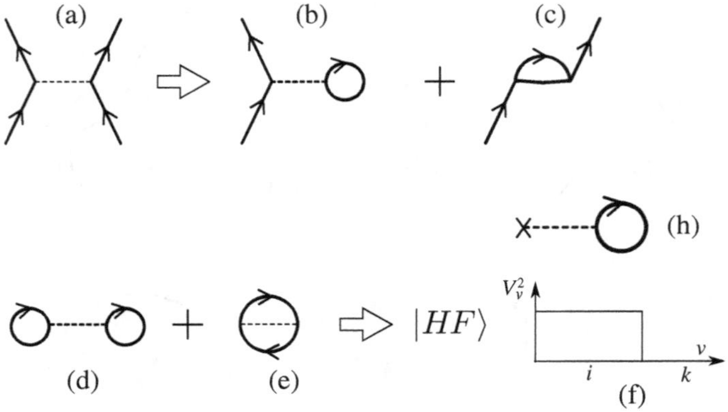

$$\alpha_\nu |HF\rangle = 0; \quad \alpha_\nu = \begin{cases} a_k & (\epsilon_k > \epsilon_F) \\ b_i = (-1)^{phase} a_{\tilde{i}}^\dagger & (\epsilon_i \leq \epsilon_F) \end{cases}$$

(g)

Figure 1.2.2 Schematic representation of the processes characterizing the Hartree–Fock ground state (single-particle vacuum), in terms of Feynman diagrams. (a) Nucleon–nucleon interaction through the bare (instantaneous) NN-potential. (b) Hartree mean field contribution. (c) Fock mean field contribution. (d,e) ground state correlations (ZPF) associated with the Hartree and Fock fields. (f) There is, in HF (mean field) theory, a complete decoupling between occupied and empty states, labeled i and k, respectively, and thus a sharp discontinuity at the Fermi energy of the occupation probability, from the value of 1 to 0. (g) This decoupling allows for the definition of two annihilation operators: $a_k(b_i)$ particle (hole) annihilation operators, implying the existence of hole (antiparticle) states ($b_i^\dagger |HF\rangle$) with quantum numbers time reversed to that of particle states (for details, see, e.g., Brink and Broglia (2005), App. A). In other words, the $|HF\rangle$ ground (vacuum) state is filled up to the Fermi energy (ϵ_F) with N nucleons. The system with ($N - 1$) nucleons can, within the language of (Feynman's) field theory, be described in terms of the degrees of freedom of that of the missing nucleon (hole-, antiparticle state). Such a description is considerably more economic than that corresponding to an antisymmetric wavefunction with ($N - 1$) spatial and spin coordinates (r_i, σ_i). Within the above scenario, a stripping reaction $N(d, p)(N + 1)$ can be viewed as the creation of a particle state ($a_k^\dagger |HF\rangle = |k\rangle$) and that of a pickup reaction $N(p, d)(N - 1)$ as that of a hole state ($b_i^\dagger |HF\rangle \equiv |\tilde{i}\rangle$). (h) Hartree, mean field contribution to the nuclear density, the density operator being represented by a cross followed by a dashed horizontal line (see also Fig. 1.8.1).

Diagonalizing δU making use of the graphical (Feynman) rules of nuclear field theory (NFT) to be discussed in the following chapter, one obtains structure results that can be used in the calculation of absolute transition probabilities and differential reaction cross sections, quantities that can be compared with the experimental findings.

Within the framework of NFT, single particles are to be calculated as the Hartree–Fock solution of the NN-interaction $v(|\mathbf{r} - \mathbf{r}'|)$ (Fig. 1.2.2, diagrams (a)–(c)), for example, a regularized NN-bare interaction in terms of renormalization group methods or alternative techniques (v_{low-k}), taking eventually also $3N$ terms into account[12] – leading, in particular, to

$$U(r) = \int d\mathbf{r}' \rho(r') v \left(|\mathbf{r} - \mathbf{r}'| \right). \tag{1.2.8}$$

that is, the Hartree field[13] expressing the self-consistency between density ρ and potential U (Fig. 1.2.2 (b)). Collective (p-h) vibrations are to be calculated in the random phase approximation[14] (RPA), making use of the same interaction (Fig. 1.2.3), extending the self-consistency to fluctuations $\delta\rho$ of the density and δU of the mean field, that is,

$$\delta U(r) = \int d\mathbf{r}' \delta\rho(r') v \left(|\mathbf{r} - \mathbf{r}'| \right). \tag{1.2.9}$$

Making use of the solution to this relation, one obtains the transition density $\delta\rho$. The matrix elements $\langle n_\lambda = 1, \nu_i | \delta U | \nu_k \rangle$ provide the particle–vibration coupling matrix elements to work out the variety of coupling processes between single-particle and collective (p-h) vibrations[15] (Fig. 1.2.1), that is, the matrix elements of the PVC Hamiltonian[16] H_c. However, the role of the NN-interaction v is not exhausted either by H_{HF} or by $H_{RPA} + H_c$. Diagonalizing

$$H = H_{HF} + H_{RPA} + H_c + v \tag{1.2.10}$$

by applying, in the basis of single-particle and collective modes, that is, solutions of H_{HF} and of H_{RPA}, respectively, the graphical NFT rules (see the next chapter), one obtains an exact solution of the total Hamiltonian, to the order of $1/\Omega$ of the Feynman diagrams calculated. The quantity Ω is the effective degeneracy in which the nucleonic excitations are allowed to correlate through v to give rise to the collective modes, $1/\Omega$ being the small parameter of the NFT diagrams[17]

---

[12] See Bogner et al. (2010) and refs. therein.

[13] To this potential one has to add the Fock potential resulting from the fact that nucleons are fermions (Fig. 1.2.2 (c), Eq. (2.3.4)).

[14] Bohm and Pines (1951, 1953). The sum of the so-called bubble (ring) diagrams is taken into account in RPA to infinite order. This is the reason why bubble contributions in the diagonalization of Eq. (1.2.10) are not allowed in NFT, being already contained in the basis states (see next chapter, Sect. 2.5).

[15] It is of note that the so-called scattering vertex shown in Fig. 1.2.1 is not operative in RPA. Being this a harmonic approximation, either two fermion lines (particle-hole) enter a vertex and a (quasi) boson line comes out (forward going process), or two fermion lines and a boson line come in or go out from the vertex (backward going process). See inset Fig. 1.2.3, graphs (b) and (c), respectively, as well as Fig. 1.2.4 (a).

[16] Mottelson (1967, 1968); Hamamoto (1969, 1970a,b, 1977); Bès and Broglia (1971a); Broglia et al. (1971a,b); Flynn et al. (1971), and refs. therein.

[17] According to *renormalized* NFT, v is the NN-interaction which, eventually combined with a k-mass, is used to calculate the bare single-particle states (HF-approximation), collective vibrations (RPA), and

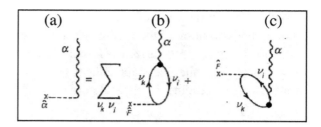

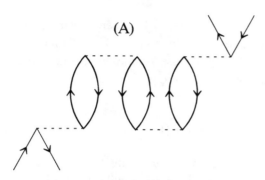

Figure 1.2.3 (**A**) Typical Feynman diagram diagonalizing, in the harmonic approximation (RPA), the NN-interaction $v(|r - r'|)$ (horizontal dashed line) in a particle-hole basis provided by the Hartree–Fock solution of v. Bubbles going forward in time (inset (b); lead to the amplitudes $X^\alpha_{ki} = \frac{\Lambda_\alpha \langle \tilde{i}|F|k \rangle}{(\epsilon_k - \epsilon_i) - \hbar\omega_\alpha}$) are associated with configuration mixing of particle-hole states ($|\tilde{i}\rangle = \tau|i\rangle$), τ being the time-reversal operator). It is of note that $\epsilon_k > \epsilon_F$ and $\epsilon_i \lesssim \epsilon_F$, and $\epsilon_j = \epsilon_{\tilde{j}}$ (Kramers degeneracy). Bubbles going backward in time (inset (c), leading to the amplitudes $Y^\alpha_{ki} = -\frac{\Lambda_\alpha \langle \tilde{i}|F|k \rangle}{(\epsilon_k - \epsilon_i) + \hbar\omega_\alpha}$) are associated with zero-point fluctuations (ZPF) of the ground state (term $(1/2)\hbar\omega_\alpha$ for each degree of freedom in Eq. (1.1.12)). The self-consistent solutions of (A), eigenstates of the dispersion relation $\sum_{ki} \frac{2(\epsilon_k - \epsilon_i)|\langle \tilde{i}|F|k \rangle|^2}{(\epsilon_k - \epsilon_i)^2 - (\hbar\omega_\alpha)^2} = 1/\kappa_\alpha$, are represented by a wavy line (inset), that is, a collective mode which can be viewed as a correlated particle- (arrowed line going upward) hole (arrowed line going downward) excitation. The quantity $\Lambda_\alpha = \kappa_\alpha \sqrt{\frac{\hbar\omega_\alpha}{2C_\alpha}}$ is the particle–vibration coupling vertex (normalization constant of the RPA amplitudes $\sum_{ki} \left(X^\alpha_{ki}\right)^2 - \left(Y^\alpha_{ki}\right)^2 = 1$). See Bohm and Pines (1951, 1953) and also Brink and Broglia (2005), Sect. 8.3.

particle–vibration coupling vertices so that once the corresponding renormalization diagrams (including also four-point vertices, i.e., v) have been worked out, the resulting dressed elementary modes of excitation reproduce the experimental findings. In the case in which one is only interested in the collective vibrations for the purpose to dress the single-particle degrees of freedom, one can take them from experiment (empirical renormalization). In other words, determine the Λ-values by making use of the experimental dynamical deformation parameter $\beta_\lambda = \sqrt{\frac{\hbar\omega_\lambda}{2C_\lambda}} \frac{1}{\sqrt{2\lambda+1}}$ and energies $\hbar\omega_\lambda$ (Broglia et al. (2016)), in conjunction with the expression of the RPA amplitudes X and Y and dispersion relation displayed in the caption to Fig. 1.2.3.

(see Sect. 2.5.4, Eq. (2.5.85)). Concerning the rules of NFT, they codify the way in which H_c (three-point vertices) and v (four-point vertices) are to be treated to all orders of perturbation theory. Also which processes (diagrams) are not allowed because they will imply overcounting of correlations already included in the basis states.

As will be shown in the following chapters, NFT allows, in an economic fashion, to sum to infinite order weakly convergent processes as well as particularly important ones, without at the same time being forced to do the same in connection with other processes that either are rapidly convergent or that lead to small contributions that can be neglected.[18]

Before proceeding, a simple estimate of the coupling strength κ is worked out, making use of a schematic separable interaction[19]

$$v = \kappa \hat{F} \hat{F}^\dagger \tag{1.2.11}$$

and of an expansion of the nuclear density $\rho(r, R)$ similar to (1.2.2), that is,

$$\delta\rho(\mathbf{r}) = -R_0 \frac{\partial\rho(r)}{\partial r} Y_{\lambda\mu}^*(\hat{r}). \tag{1.2.12}$$

With the help of the dynamical self-consistent relation (1.2.9), one obtains

$$\kappa = \int r^2 dr R_0 \frac{\partial\rho(r)}{\partial r} R_0 \frac{\partial U(r)}{\partial r}. \tag{1.2.13}$$

For attractive fields, both U and κ are negative.

Because of quantal zero-point fluctuations, a nucleon propagating in the nuclear medium moves through a cloud of (quasi) bosonic virtual excitations to which it couples, becoming dressed and acquiring an effective mass, charge, etc. (Fig. 1.2.4; see also Sects. 5.4 and 5.8).

Furthermore, the sharp transition expected to take place in the independent-particle motion (HF approximation), between occupied ($V_i^2 = 1$, Fig. 1.2.2 (f)) and empty states ($V_k^2 = 0$), becomes blurred due to the processes (b) and (c) (Fig. 1.2.4), which dress the nucleons. This is illustrated in Fig. 1.2.5, where a schematic representation of a detailed (NFT) calculation[20] of the occupation probability of neutron orbits in the closed shell system ^{208}Pb is displayed.[21] Similar results have been obtained for other nuclei, for example, the neutron open shell nucleus ^{120}Sn (Fig. 1.1.2). Within this context let us consider the example of the $1i_{11/2}$ neutron

---

[18] The reason for this flexibility is to be partially found in the fact that bare elementary modes of excitation used as basis states in NFT contain an important fraction of the many-body nuclear correlations, making the diagonalization of the nuclear Hamiltonian a low-dimensional problem (see also Sect. 2.1).

[19] See, e.g., Bohr and Mottelson (1975), Eqs. (6-37) and (6-74).

[20] See Bortignon et al. (1998), pp. 79, 80.

[21] See Sect. 5.8 as well as Mahaux et al. (1985), in particular Sect. 4.7 of this reference.

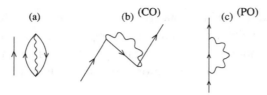

(a)   (b) (CO)   (c) (PO)

Figure 1.2.4 (a) A nucleon (single arrowed line pointing upward) moving in presence of the zero-point fluctuation of the nuclear ground state associated with a collective surface vibration. (b) Pauli principle leads to a dressing process of the nucleon (so-called correlation (CO) diagram) resulting from the exchange of the fermion line considered explicitly and that participating in the correlated particle-hole excitation of the ZPF. (c) Time ordering gives rise to the second possible lowest-order dressing process (so-called polarization (PO) diagram). The above are phenomena closely related to the Lamb shift of quantum electrodynamics (see, e.g., Weinberg (1996a), Sect. 14.3).

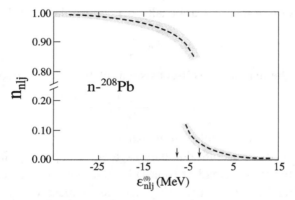

Figure 1.2.5 Schematic representation of the occupation probability (see Sect. 5.8) of ^{208}Pb neutron orbits: $1f_{7/2}$, $2p_{1/2}$, $1g_{7/2}$, $1h_{11/2}$, $1h_{9/2}$, $2f_{7/2}$, $2f_{5/2}$, $1i_{13/2}$, $3p_{3/2}$, $3p_{1/2}$ hole states with $\epsilon_i \leq \epsilon_F$, and of the $2g_{9/2}$, $1i_{11/2}$, $1j_{15/2}$, $3d_{5/2}$, $2g_{7/2}$, $42_{1/2}$, $2d_{3/2}$, $2h_{11/2}$, $2h_{9/2}$ particle states ($\epsilon_k > \epsilon_F$). The arrows show the location of $\epsilon_F^- = \epsilon_{3p_{1/2}}$ and $\epsilon_F^+ = \epsilon_{2g_{9/2}}$.

state, which, in the independent particle model of ^{208}Pb, would be unoccupied. Taking the particle vibration coupling mechanism into account, which we limit for simplicity to the consideration of $\beta = 0$, $p - h$ collective surface vibration, all those ZPF of the ^{208}Pb "vacuum" (see Fig. 1.2.4 (a)) that involves the $1i_{11/2}$ orbital in the structure of the collective mode would contribute to $n_{1i_{11/2}}$.

Vibrational modes can also become renormalized through the coupling to dressed nucleons, which, in intermediate virtual states, can exchange the vibrations that produce their clothing, with the second fermion (hole state). Such a

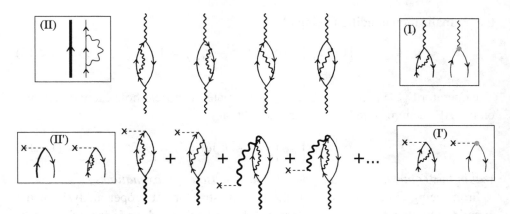

Figure 1.2.6 (top) Examples of renormalization processes dressing a (p-h) collective vibrational state. (bottom) Intervening the process with an external electromagnetic field ($E\lambda$: cross followed by dashed horizontal line; bold wavy lines, renormalized vibration of multipolarity λ), the $B(E\lambda)$ transition strength can be determined. In insets (I) and (I'), the hatched circle in the diagram to the right stands for the renormalized PVC strength resulting from the processes described by the corresponding diagrams to the left (vertex corrections). In (II') the boldface arrowed line (left diagram) represents the motion of a nucleon of effective mass m^* in a potential $(m/m^*)U(r)$, generated by the self-energy process shown to the right (see also (II) diagram to the right), $U(r)$ being the potential describing the motion of bare nucleons (drawn as a thin arrowed line) of mass m_k.

process leads to a renormalization of the PVC vertex[22] (Fig. 1.2.6). The exchange of vibrations between two particle states renormalizes the bare NN-interaction, in particular the $^1S_0$ component[23] (bare pairing interaction).[24]

The analytic procedures equivalent to the diagrammatic ones to obtain the HF and RPA solutions associated with the bare NN-interaction v are provided by the relations (1.1.16) and (1.1.15), respectively, replacing the corresponding Hamiltonians by $(T + v)$, where T is the kinetic energy operator. The phonon operator associated with surface vibrations is defined as

$$\Gamma_\alpha^\dagger = \sum_{ki} X_{ki}^\alpha \Gamma_{ki}^\dagger + Y_{ki}^\alpha \Gamma_{ki}, \qquad (1.2.14)$$

[22] Bertsch et al. (1983); Barranco et al. (2004), and refs. therein. It is to be noted that in the case in which the renormalized vibrational modes, i.e., the initial and final wavy lines in Fig. 1.2.6, have angular momentum and parity $\lambda^\pi = 0^+$, and one uses a model in which there is symmetry between the particle and the hole subspaces, the four diagrams sum to zero, because of particle (gauge) conservation, a fact connected with Furry's theorem of QED and generalized Ward's identity (Ward (1950)).

[23] See footnote 55, Ch. 3.

[24] See, e.g., Duguet (2013, 2004); Duguet and Lesinski (2008); Lesinski et al. (2009); Hebeler et al. (2009); Hergert and Roth (2009); Baroni et al. (2010); Duguet et al. (2010); Lesinski et al. (2011), and refs. therein.

the normalization condition being

$$\left[\Gamma_\alpha, \Gamma_\alpha^\dagger\right] = \sum_{ki} \left(X_{ki}^{\alpha 2} - Y_{ki}^{\alpha 2}\right) = 1. \tag{1.2.15}$$

The operator $\Gamma_{ki}^\dagger = a_k^\dagger a_i (\epsilon_k > \epsilon_F, \epsilon_i \le \epsilon_F)$ creates a particle-hole excitation acting on the HF vacuum state $|0\rangle_F$. It is assumed that

$$\left[\Gamma_{ki}, \Gamma_{k'i'}^\dagger\right] = \delta(k, k')\delta(i, i'). \tag{1.2.16}$$

Within this context, RPA is a harmonic, quasi-boson approximation.

From being antithetic views of the nuclear structure, a proper analysis of the experimental data testifies to the fact that the collective and the independent particle pictures of the nuclear structure require and support each other.[25] To obtain a quantitative description of nucleon motion and nuclear phonons (vibrations), one needs a proper description of the k- and ω-dependent "dielectric" function of the nuclear medium, in a similar way to which a proper description of the reaction processes used as probes of the nuclear structure requires the use of the optical potential (continuum "dielectric" function).

The NFT solutions that diagonalize (1.2.10) provide all the elements to calculate the structure properties of nuclei, and also of the optical potential needed to describe nucleon-nucleus as well as nucleus-nucleus elastic scattering and reaction processes.[26] Furthermore, the NFT solutions of (1.2.10) show that both single-particle (fermionic) and collective (quasi-bosonic) elementary modes of excitation emerge from the same properties of the NN-interaction. For example, a bunching of levels of the same or opposite parity just below and above the Fermi energy implies low-lying quadrupole (e.g., $^A_{50}\mathrm{Sn}_N$-isotopes) or octupole ($^{208}_{82}\mathrm{Pb}_{126}$) collective vibrational modes. And, as a bonus, one has the building blocks to construct the nuclear spectrum and bring quantitative simplicity into the experimental findings. Within this scenario, the NFT solutions of (1.2.10) also indicate the minimum set of experimental probes needed to have a "complete" picture of the nuclear structure associated with a given energy region. This is a consequence of the central role played by the quantal many-body renormalization process that interweaves the variety of elementary modes of excitation. Renormalization processes that act on par on the radial dependence of the wavefunctions (formfactors) and on the single-particle content (structure) of the orbitals involved in the reaction under discussion

[25] Bohr and Mottelson (1975).

[26] It is of note that in the present monograph the subject of the optical potential is not treated. There exists a vast literature concerning this subject. See, e.g., Feshbach (1958, 1962); Jackson (1970); Jeukenne et al. (1976); Sartor and Mahaux (1980); Pollarolo et al. (1983); Satchler (1983); Mahaux et al. (1985); Broglia and Winther (2004); Dickhoff and Van Neck (2005); Montanari et al. (2014); Dickhoff et al. (2017); Dickhoff and Charity (2019); Fernández-García et al. (2010a); Fernández-García et al. (2010b); Jenning (2011); Barbieri and Jennings (2005); Rotureau et al. (2017); Broglia et al. (1981), and refs. therein.

and, as a consequence, structure and reactions are to be treated on equal footing[27] (see, e.g., Sect. 5.2.2).

The development of experimental techniques and associated hardware has allowed for the identification of a rich variety of elementary modes of excitation aside from collective surface vibrations and independent particle motion: quadrupole and octupole rotational bands, giant resonances of varied multipolarity and isospin, as well as pairing vibrations and rotation, together with giant pairing vibrations of transfer quantum number $\beta = \pm 2$. These modes can be specifically excited in inelastic and Coulomb excitation processes and one- and two-particle transfer reactions.

1.3 Pairing Vibrations

This new elementary mode of excitation is introduced by making a parallel with quadrupole surface vibrations within the framework of RPA, namely,

$$\left[(H_{sp} + H_i), \Gamma_\alpha^\dagger\right] = \hbar\omega_\alpha \Gamma_\alpha^\dagger, \tag{1.3.1}$$

where for simplicity one uses, instead of v, a quadrupole-quadrupole ($i = QQ$) defined as[28]

$$H_{QQ} = -\kappa Q^\dagger Q, \tag{1.3.2}$$

with

$$Q^\dagger = \sum_{ki} \langle k|r^2 Y_{2\mu}|i\rangle a_k^\dagger a_i, \tag{1.3.3}$$

while H_{sp} and $\Gamma_\alpha^\dagger$ were defined in (1.2.14) supplemented by (1.2.15) and Eq. (1.2.16).

In connection with the pairing energy mentioned above (odd-even staggering), it is a consequence of correlation of pairs of like nucleons moving in time-reversed states $(v, \tilde{v})$, a similar phenomenon to that found in metals at low temperatures and giving rise to superconductivity.[29] The pairing interaction[30] can be written, within the approximation (1.3.2) used in the case of the quadrupole-quadrupole force, as

---

[27] Within this context, and referring to one-particle transfer reactions for concreteness, the prescription of using the ratio of the absolute experimental cross section and of the theoretical one calculated in the distorted wave Born approximation making use of Woods–Saxon single-particle wavefunctions as formfactors, to extract the single-particle content of the orbitals under study, may not be completely appropriate.

[28] Bohr and Mottelson (1975); see also Bayman (1961) and Bès (2016). It is of note that in this case, and at variance with (1.2.11)–(1.2.13), the fact that the force is attractive is explicitly expressed through the minus sign, assuming $\kappa > 0$. This is done for didactical purposes, to connect with the standard notation for the pairing interaction given in Eq. (1.3.4).

[29] Bardeen et al. (1957a,b). With regard to the extension of BCS to nuclei, see Bohr et al. (1958). See also Broglia and Zelevinsky (2013).

[30] See footnote 24 of this chapter.

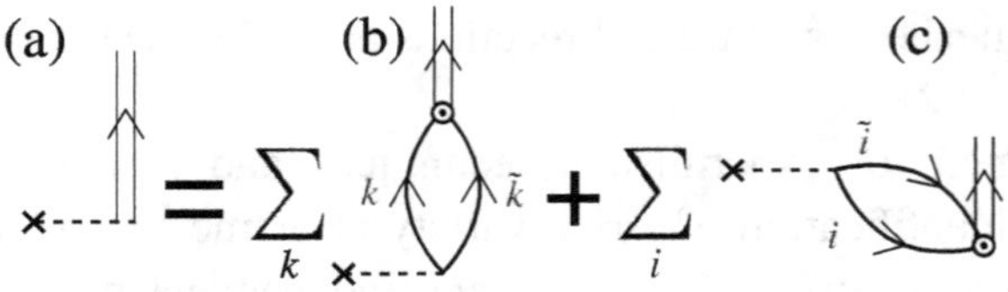

Figure 1.3.1 Graphical representation of the RPA dispersion relation describing the pair addition vibrational mode (see also Sect. 3.5.2), drawn as a double arrowed line. A cross followed by a dashed horizontal line stands for (see Eqs. (1.3.7) and (1.3.8)) (a) the collective operator $\Gamma_\alpha^\dagger$, (b) the operator $\Gamma_k^\dagger$ creating *a pair of nucleons moving in time-reversal states $(k, \tilde{k})$ above the Fermi energy* $(\epsilon_k > \epsilon_F)$, (c) the operator Γ_i annihilating *a pair of time-reversal nucleon holes* $(i, \tilde{i})$ associated with ground state correlations $(\epsilon_i \leq \epsilon_F)$.

$$H_P = -G P^\dagger P, \tag{1.3.4}$$

where

$$P^\dagger = \sum_{\nu > 0} P_\nu^\dagger \tag{1.3.5}$$

and

$$P_\nu^\dagger = a_\nu^\dagger a_{\tilde{\nu}}^\dagger, \tag{1.3.6}$$

the quantity G being the pairing coupling constant (attractive $G > 0$). Concerning the states $(\nu, \tilde{\nu})$, see the caption to Fig. 1.3.1. Consequently, in this case, the concept of independent particle field $\hat{Q}$ (see also (1.2.6) and (1.2.7)) associated with particle-hole (ph) excitations and carrying transfer quantum number $\beta = 0$ has to be generalized to include fields describing independent pair motion of different multipolarity and (normal) parity, in particular 0^+, in which case $\alpha \equiv (\beta = +2, J^\pi = 0^+)$. The associated RPA creation operator can be written as

$$\Gamma_a^\dagger(n) = \sum_k X_n^a(k)\Gamma_k^\dagger + \sum_i Y_n^a(i)\Gamma_i, \tag{1.3.7}$$

with

$$\Gamma_k^\dagger = a_k^\dagger a_{\tilde{k}}^\dagger \ (\epsilon_k > \epsilon_F), \quad \Gamma_i^\dagger = a_{\tilde{i}} a_i \ (\epsilon_i \leq \epsilon_F) \tag{1.3.8}$$

and

$$\sum_k \left(X_n^a(k)\right)^2 - \sum_i \left(Y_n^a(i)\right)^2 = 1, \tag{1.3.9}$$

for the pair addition $((pp), \beta = +2)$ mode, and a similar expression for the pair removal $((hh), \beta = -2)$ one.

In Fig. 1.3.1 the NFT graphical representation of the RPA equations for the pair addition mode is given. The state $\Gamma_a^\dagger(\beta = +2)|\tilde{0}\rangle$, where $|\tilde{0}\rangle$ is the correlated ground state of a closed shell nucleus, can be viewed as the nuclear parallel of a Cooper pair found at the basis of the microscopic theory of superconductivity.[31] While surface vibrations are associated with the normal ($\beta = 0$) nuclear density,

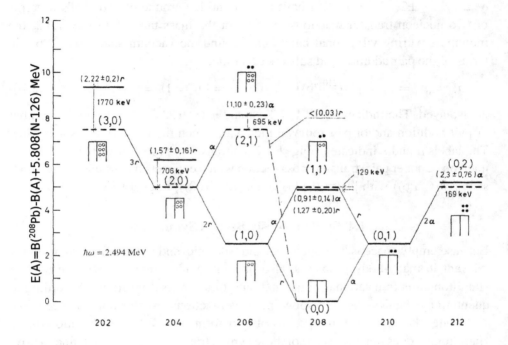

Figure 1.3.2 The many-phonon pairing vibrational spectrum around the "vacuum" $|gs(^{208}\text{Pb})\rangle$. The energies predicted by the pairing vibrational model (harmonic approximation) are displayed as dashed horizontal lines. The quantum numbers (n_r, n_a) indicate the number of pair addition and pair removal modes and are displayed for each level. A schematic representation of the many-particle many-hole structure of the state is also given. The transitions predicted by the model are indicated in units of r and a (see Eq. (1.3.10) and subsequent text). The corresponding experimental numbers are also given together with their errors. The dashed line between the states (0,0) and (2,1) indicates that the $^{208}\text{Pb}(p, t)^{206}\text{Pb}$ reaction to the three-phonon state in ^{206}Pb was carried out and an upper limit of $0.03r$ for the corresponding cross section determined (see Flynn et al. (1972a), Broglia et al. (1973); see also Lanford and McGrory (1973)). The quantities $B(A)$ (y-axis) stand for the binding energy of the corresponding isotope of Pb (see x-axis), N being the corresponding neutron number. The linear term in N has been introduced so as to make $\hbar\omega_a = \hbar\omega_r (= 2.494 \text{ MeV})$.

---

[31] The solution of the Cooper pair problem, being of TD-type, does not take into account ground state correlations. Within this context, see Ambegaokar (1969), p. 285.

pairing vibrations are connected with the so-called abnormal ($\beta = \pm 2$) nuclear density (density of Cooper pairs), both static and dynamic.

Similar to the quadrupole and octupole vibrational bands built out of n_α phonons of quantum numbers $\alpha \equiv (\beta = 0, \lambda^\pi = 2^+, 3^-)$ schematically shown in Fig. 1.1.3 and experimentally observed in inelastic and Coulomb excitation and associated γ-decay processes, pairing vibrational bands built of n_α phonons of quantum numbers $\alpha \equiv (\beta = \pm 2, \lambda^\pi = 0^+, 2^+)$ have been identified around closed shells in terms of two-nucleon transfer reactions throughout the mass table.[32] In Fig. 1.3.2, the monopole pairing vibrational band built around the vacuum state $|gs(^{208}\text{Pb})\rangle$ in terms of the pair addition and substraction modes

$$|a\rangle = |n_a = 1\rangle = |gs(^{210}\text{Pb})\rangle \quad \text{and} \quad |r\rangle = |n_r = 1\rangle = |gs(^{206}\text{Pb})\rangle \quad (1.3.10)$$

is displayed. The indices $n_a (= 1, 2, \ldots)$ and $n_r (= 1, 2, \ldots)$ indicate the number of pair addition and/or pair removal modes of which the $|n_r, n_a\rangle$ states are made. The labels a and r indicate in Fig. 1.3.2 the transition probability, that is, the two-nucleon (Cooper pair) transfer cross section which connects the ground ("vacuum") state $|gs(^{208}\text{Pb})\rangle$ with $|a\rangle (\equiv (0, 1))$ and $|r\rangle (\equiv (1, 0))$, respectively.

1.4 Spontaneously Broken Symmetry

Because empty space is homogeneous and isotropic, and to the degree of accuracy relevant in many-body problems, also invariant under time reversal, the nuclear Hamiltonian is translational, rotational, and time-reversal invariant. According to quantum mechanics, the corresponding wavefunctions transform in an irreducible way under the corresponding groups of transformation. When the solution of the Hamiltonian does not have some of these symmetries, for example, it defines a privileged direction in space violating rotational invariance, one is confronted with the phenomenon of spontaneously broken symmetry. Strictly speaking, this can take place only for idealized systems that are infinitely large. But when one sees similar phenomena in atomic nuclei, although not so clear or regular, one recognizes that this system is after all a finite quantum many-body system (FQMBS).[33]

1.4.1 Quadrupole Deformations in 3D Space

A nuclear embodiment of the spontaneous symmetry breaking phenomenon is provided by a quadrupole deformed mean field, a situation one is confronted with when the value of the lowest quadrupole frequency ω_2 of the RPA solution (1.1.15) tends to zero ($C_2 \to 0$, D_2 finite). It is a phenomenon resulting from the interplay

---

[32] See footnote 10 of the present chapter.
[33] Anderson (1972, 1984).

of the interaction v (H_{QQ} in (1.3.2)) and the nucleons outside the closed shell, leading to tidal-like polarization of the spherical core.

Coordinate and linear momentum $((x, p_x)$ single-particle motion) as well as Euler angles and angular momentum $(\omega, \mathbf{I})$ associated with rotation in three-dimensional (3D) space are conjugate variables. Similarly, the gauge angle and the number of particles $((\phi, N)$ rotation in gauge space) fulfill $[\phi, N] = i$. The operators $e^{-ip_x x}$, $e^{-i\omega \cdot \mathbf{I}}$, and $e^{-iN\phi}$ induce Galilean transformation in 1D space and rotations in 3D and in gauge (2D) space, respectively.

Making again use, for didactical purposes, of H_{QQ} instead of v, and denoting by $|N\rangle$ the Nilsson[34] state, that is, the mean field solution of the Hamiltonian $T+H_{QQ}$, one can write

$$\langle N|\hat{Q}|N\rangle = Q_0, \tag{1.4.1}$$

where, for simplicity, we assumed axial symmetry ($\lambda = 2, \mu = 0$). One is then confronted with the emergence of a static quadrupole deformation. Rewriting H_{QQ} in terms of $((\hat{Q}^\dagger - Q_0) + Q_0)$ and its Hermitian conjugate leads to

$$H = H_{sp} + H_{QQ} = H_{MF} + H_{fluct}, \tag{1.4.2}$$

where

$$H_{MF} = H_{sp} - \kappa Q_0(\hat{Q}^\dagger + \hat{Q}) + \kappa Q_0^2 \tag{1.4.3}$$

is the mean field and

$$H_{fluct} = -\kappa(\hat{Q}^\dagger - Q_0)(\hat{Q} - Q_0) \tag{1.4.4}$$

the residual interaction inducing fluctuations around Q_0.

Let us concentrate on the mean field Hamiltonian H_{MF}. It describes the motion of nucleons in a single-particle potential of radius $R_0(1+\beta_2 Y_{20}(\hat{r}))$, with β_2 proportional to the intrinsic quadrupole moment[35] Q_0 ($\beta_2 \approx Q_0/(ZR_0^2)$). The reflection invariance and axial symmetry of the Nilsson Hamiltonian implies that parity π and projection Ω of the total angular momentum along the symmetry axis are constants of motion for the one-particle Nilsson states. These states are twofold degenerate, since two orbits that differ only in the sign of Ω represent the same motion, apart from the clockwise and anticlockwise sense of revolution around the symmetry axis. One can thus write the Nilsson creation operators in terms of a linear combination of creation operators carrying good total angular momentum j,

$$\gamma_{a\Omega}^\dagger = \sum_j A_j^a a_{aj\Omega}^\dagger, \tag{1.4.5}$$

---

[34] Nilsson (1955).

[35] Mottelson and Nilsson (1959), where use has been made of $\beta_2 \approx 0.95\delta$; see Bohr and Mottelson (1975), p. 47 and Eq. (4-191). See also Nilsson (1955) and Ragnarsson and Nilsson (2005).

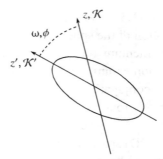

Figure 1.4.1 Schematic representation of a deformed state in 3D (gauge) space, where the laboratory ($\mathcal{K}$) and the intrinsic ($\mathcal{K}'$, body fixed) frames of reference make a relative angle ((ω, Euler angles); (ϕ, gauge angle)) (see Eqs. (1.4.7) and (1.4.50)).

where the label a stands for the quantum numbers which, aside from Ω, specify the orbital.

Expressing (1.4.5) in the intrinsic, body-fixed system of coordinates $\mathcal{K}'$ (Fig. 1.4.1), where the 3 (z') axis lies along the symmetry axis and the 1 and 2 (x', y') axes lie in a plane perpendicular to it, one can write

$$\gamma_{a\Omega}^{'\dagger} = \sum_{j} A_j^a \sum_{\Omega'} \mathcal{D}_{\Omega\Omega'}^2(\omega) a_{aj\Omega'}^\dagger. \tag{1.4.6}$$

The Nilsson state can then be expressed as

$$|N(\omega)\rangle_{\mathcal{K}'} = \prod_{a\Omega>0} \gamma_{a\Omega}^{'\dagger} \gamma_{a\widetilde{\Omega}}^{'\dagger} |0\rangle_F, \tag{1.4.7}$$

where ω represents the Euler angles, $|0\rangle_F$ is the particle vacuum, and $\gamma_{a\widetilde{\Omega}}^{'\dagger}|0\rangle_F$ is the state time reversed to $\gamma_{a\Omega}^{'\dagger}|0\rangle_F$. For well-deformed nuclei, a conventional description of the one-particle motion is based on the similarity of the nuclear potential to that of an anisotropic oscillator potential,

$$V = \frac{1}{2}M\left(\omega_3^2 x_3^2 + \omega_\perp^2(x_1^2 + x_2^2)\right) = \frac{1}{2}M\omega_0 r^2\left(1 - \frac{4}{3}\delta \times P_2(\cos\theta)\right), \tag{1.4.8}$$

with $\omega_3\omega_\perp^2 = \omega_0^3$, that is, a volume which is independent of the deformation $\delta \approx 0.95\beta_2$. The corresponding single-particle states have energy

$$\epsilon(n_3 n_\perp) = (n_3 + \tfrac{1}{2})\hbar\omega_3 + (n_\perp + 1)\hbar\omega_\perp, \tag{1.4.9}$$

where n_3 and $n_\perp = n_1 + n_2$ are the number of quanta along and perpendicular to the symmetry axis, $N = n_3 + n_\perp$ being the total oscillator quantum number. The degenerate states with the same value of $n_\perp$ can be specified by the component Λ of the orbital angular momentum along the symmetry axis,

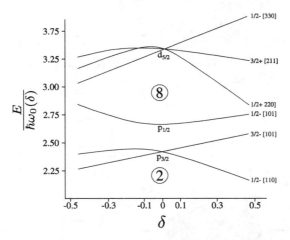

Figure 1.4.2 Nilsson single-particle levels in a quadrupole deformed potential in the regions $2 < Z < 20$ and $2 < N < 20$.

$$\Lambda = \pm n_\perp, \pm(n_\perp - 2), \ldots, \pm 1 \text{ or } 0. \tag{1.4.10}$$

The complete expression of the Nilsson potential includes, aside from the central term discussed above, a spin-orbit term ($\Sigma = \pm 1/2$, $\Omega = \Lambda + \Sigma$) and a term proportional to the orbital angular momentum quantity squared, so as to make the shape of the oscillator resemble more that of a Woods–Saxon potential. One can label the Nilsson levels in terms of the asymptotic quantum numbers $\Omega[Nn_3\Lambda]$. The resulting states provide an overall account of the experimental findings. An example of relevance for light nuclei (N and $Z < 20$) is given in[36] Fig. 1.4.2.

The Nilsson (intrinsic) state (1.4.7) does not have a definite angular momentum but is rather a superposition of such states,

$$|N(\omega)\rangle_{\mathcal{K}'} = \sum c_I |I\rangle. \tag{1.4.11}$$

Because there is no restoring force associated with different orientations of $|N(\omega)\rangle_{\mathcal{K}'}$, fluctuations in the Euler angles diverge in the right way so as to restore rotational invariance, leading to a rotational band whose members are

$$|IKM\rangle \sim \int d\omega \, \mathcal{D}^I_{MK}(\omega) \, |N(\omega)\rangle_{\mathcal{K}'}, \tag{1.4.12}$$

with energy

$$E_I = \frac{\hbar^2}{2\mathfrak{J}} I(I+1), \tag{1.4.13}$$

---

[36] Mottelson and Nilsson (1959).

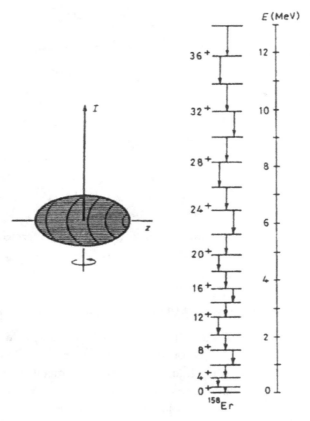

Figure 1.4.3 Ground state rotational band of ^{158}Er.

where $\mathfrak{J}$ is the moment of inertia. The quantum numbers I, M, K are the total angular momentum I and its third components M and K along the laboratory (z) and intrinsic (z') frames of reference, respectively. Rotational bands have been observed up to rather high angular momenta in terms of individual transitions, in particular, up to $I = 60\hbar$ in the case of[37] the nucleus ^{152}Dy. An example of quadrupole rotational bands[38] is given in Fig. 1.4.3.

1.4.2 Deformation in Gauge Space

Let us now return to the Hamiltonian $H = H_{sp} + H_P$ (Eqs. (1.1.13) and (1.3.4)) introduced in connection with pairing vibrations.

In the case in which $\hbar\omega_1(\beta = -2, \lambda^\pi = 0^+) = \hbar\omega_1(\beta = +2, \lambda^\pi = 0^+) = 0$, the system deforms, this time in gauge space. The mean field solution of the pairing

[37] Nolan and Twin (1988), and refs. therein.
[38] Burde et al. (1982).

Hamiltonian,[39] called $|BCS\rangle$, leads to the finite expectation value

$$\alpha_0 = \langle BCS|P^\dagger|BCS\rangle \tag{1.4.14}$$

of the pair creation operator $P^\dagger$, a quantity which can be viewed as the order parameter of the new gauge deformed phase of the system.

Similarly to (1.4.2), the pairing Hamiltonian can be written as

$$H = H_{MF} + H_{fluct}, \tag{1.4.15}$$

where

$$H_{MF} = H_{sp} - \Delta(P^\dagger + P) + \frac{\Delta^2}{G} \tag{1.4.16}$$

and

$$H_{fluct} = -G(P^\dagger - \alpha_0)(P - \alpha_0). \tag{1.4.17}$$

The quantity

$$\Delta = G\alpha_0 \tag{1.4.18}$$

is the pairing gap (Fig. 1.4.4), with 2Δ measuring the binding energy of Cooper pairs and the quantity α_0 being the number of them (see Eq. (1.4.42)).

The mean field pairing Hamiltonian

$$H_{MF} = \sum_{\nu>0}(\epsilon_\nu - \lambda)\,(a_\nu^\dagger a_\nu + a_{\tilde\nu}^\dagger a_{\tilde\nu}) - \Delta\sum_{\nu>0}(\epsilon_\nu - \lambda)\,(a_\nu^\dagger a_{\tilde\nu}^\dagger + a_{\tilde\nu}a_\nu) + \frac{\Delta^2}{G} \tag{1.4.19}$$

is a bilinear expression in the creation and annihilation operators, ν labeling the quantum numbers of the single-particle orbitals where nucleons are allowed to correlate (e.g., (nlj, m)), while $\tilde\nu$ denotes the time-reversal state which in this case is degenerate with ν and has quantum numbers $(nlj, -m)$. The condition $\nu > 0$ in Eq. (1.4.19) implies that one sums over $m > 0$. It is of note that

$$\hat N = \sum_{\nu>0}(a_\nu^\dagger a_\nu + a_{\tilde\nu}^\dagger a_{\tilde\nu}) \tag{1.4.20}$$

is the number operator and $\lambda\hat N$ in Eq. (1.4.19) acts as the Coriolis force in the body-fixed frame of reference in gauge space (Fig. 1.4.1).

One can diagonalize H_{MF} by a rotation in the $(a^\dagger, a)$ space. This can be accomplished through the Boguliubov–Valatin (quasiparticle) transformation[40]

$$\alpha_\nu^\dagger = U_\nu a_\nu^\dagger - V_\nu a_{\tilde\nu}, \tag{1.4.21}$$

[39] Bardeen et al. (1957a,b).
[40] Bogoljubov (1958a,b); Valatin (1958).

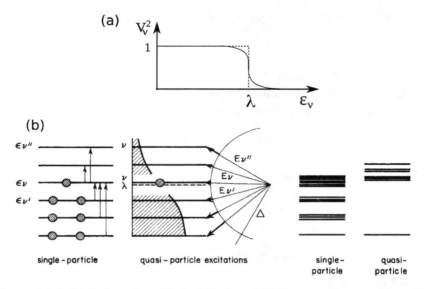

Figure 1.4.4 (a) Independent (dashed line) and BCS occupation numbers. (b) Ground state and excited states in the extreme single-particle model and in the pairing-correlated, superfluid model in the case of a system with an odd number of particles. In the first case, the energy of the ground state of the odd system differs from that of the even with one particle fewer by the energy difference $\epsilon_\nu - \epsilon_{\nu'}$ while in the second case by the energy $E_\nu = \sqrt{(\epsilon_\nu - \lambda)^2 + \Delta^2} \approx \Delta$, associated with the fact that the odd particle has no partner. Excited states can be obtained in the independent particle case, where it is assumed that levels are twofold degenerate (Kramers degeneracy) by promoting the odd particle to states above the level ϵ_ν, or by exciting one particle from the states below to the state ϵ_ν, or to one above it. To the left, only a selected number of these excitations are shown. In the superfluid case, excited states can be obtained by breaking of pairs in any orbit. The associated quasiparticle energy is drawn also here by an arrow of which the thin part indicates the contribution of the pairing gap and the thick part indicates the kinetic energy contribution, that is, the contribution arising from the single-particle motion. Note the very different densities of levels emerging from these two pictures, which are shown at the far right of the figure.

where the occupation numbers (U_ν, V_ν) fulfill $|U_\nu|^2 + |V_\nu|^2 = 1$. The BCS solution does not change the energies ϵ_ν of the single-particle levels (measured in (1.4.19) from the chemical potential or Fermi energy λ), nor the associated wavefunctions $\varphi_\nu(\mathbf{r})$, but the occupation probabilities of levels around the Fermi energy within an energy range 2Δ ($2\Delta/\epsilon_F \approx 3$ MeV/36 MeV≈ 0.06).[41] The quasiparticle operator $\alpha_\nu^\dagger$ creates a particle in the single-particle state ν with probability U_ν^2, while it creates a hole (annihilates a particle) with probability V_ν^2. To be able to create a particle, the state ν should be empty, while to annihilate a particle, it has to be filled, so U_ν^2 and V_ν^2 are the probabilities that the state ν is empty and is occupied, respectively. Within this context, the one-quasiparticle states

---

[41] See also the discussion starting before Eq. (1.4.53) and ending before Eq. (1.4.55).

$$|v\rangle = \alpha_v^\dagger |BCS\rangle \tag{1.4.22}$$

are orthonormal. In particular,

$$\langle v|v\rangle = 1 = \langle BCS|\alpha_v\alpha_v^\dagger|BCS\rangle = \langle BCS|\left\{\alpha_v, \alpha_v^\dagger\right\}|BCS\rangle = U_v^2 + V_v^2, \tag{1.4.23}$$

where the anticommutators

$$\left\{a_v, a_{v'}^\dagger\right\} = \delta(v, v') \tag{1.4.24}$$

and

$$\{a_v, a_{v'}\} = \left\{a_v^\dagger, a_{v'}^\dagger\right\} = 0 \tag{1.4.25}$$

have been used. Note that the $|BCS\rangle$ state is the quasiparticle vacuum

$$\alpha_v|BCS\rangle = 0, \tag{1.4.26}$$

in a similar way in which $|0\rangle_F$ is the particle vacuum.

Inverting the quasiparticle transformation (1.4.21) and its complex conjugate, that is, expressing $a_v^\dagger$ and a_v (and their time reversals (tr)) as functions of $\alpha_v^\dagger$ and α_v (and tr), one can rewrite (1.4.19) in terms of quasiparticles. Minimizing the energy $E_0 = \langle BCS|H|BCS\rangle$ with respect to[42] V_v,

$$\frac{\partial E_0}{\partial V_v} = 0, \tag{1.4.27}$$

and making use of the expression for the average number of particles

$$N_0 = \langle BCS|\hat{N}|BCS\rangle = 2\sum_{v>0} V_v^2, \tag{1.4.28}$$

of the number of Cooper pairs

$$\alpha_0 = \langle BCS|P^\dagger|BCS\rangle = \sum_{v>0} U_v V_v, \tag{1.4.29}$$

and of the pairing gap

$$\Delta = G\sum_{v>0} U_v V_v, \tag{1.4.30}$$

one obtains

$$H_{MF} = H_{11} + U, \tag{1.4.31}$$

where

$$H_{11} = \sum_v E_v \alpha_v^\dagger \alpha_v \tag{1.4.32}$$

---

[42] For details, see, e.g., Ragnarsson and Nilsson (2005). See also Nathan and Nilsson (1965) and Brink and Broglia (2005), App. G, Sect. G.3, and refs. therein.

and

$$U = 2 \sum_{\nu > 0} (\epsilon_\nu - \lambda) V_\nu^2 - \frac{\Delta^2}{G}. \qquad (1.4.33)$$

The quantity

$$E_\nu = \sqrt{(\epsilon_\nu - \lambda)^2 + \Delta^2} \qquad (1.4.34)$$

is the quasiparticle energy, while the occupation (emptiness) probability amplitudes
are

$$V_\nu = \frac{1}{\sqrt{2}} \left(1 - \frac{\epsilon_\nu - \lambda}{E_\nu} \right)^{1/2} \qquad (1.4.35)$$

$$U_\nu = \frac{1}{\sqrt{2}} \left(1 + \frac{\epsilon_\nu - \lambda}{E_\nu} \right)^{1/2}, \qquad (1.4.36)$$

respectively. From the relations (1.4.28), (1.4.30), and (1.4.34), one obtains

$$N_0 = 2 \sum_{\nu > 0} V_\nu^2 \qquad (1.4.37)$$

and

$$\frac{1}{G} = \sum_{\nu > 0} \frac{1}{2E_\nu}. \qquad (1.4.38)$$

These equations, known as the BCS number and gap equations, allow one to cal-
culate the parameters λ and Δ from the knowledge of N_0, G, and ϵ_ν, parameters
which completely determine E_ν, V_ν, and U_ν and thus the BCS mean field solution
(Fig. 1.4.4). The validity of the BCS description of superfluid open shell nuclei has
been extensively confirmed.[43]

The relation (1.4.26) implies, as a consequence of the Pauli principle, that

$$
\begin{aligned}
|BCS\rangle &= \frac{1}{\text{Norm}} \prod_{\nu > 0} \alpha_\nu \alpha_{\bar\nu} |0\rangle_F = \prod_{\nu > 0} (U_\nu + V_\nu P_\nu^\dagger) |0\rangle_F \\
&= \left(\prod_{\nu > 0} U_\nu \right) \sum_{N \text{ even}} \frac{\left(\sum_{\nu > 0} c_\nu P_\nu^\dagger \right)^{N/2}}{(N/2)!} |0\rangle_F \\
&= \left(\prod_{\nu > 0} U_\nu \right) \sum_{n = 0, 1, 2, \dots} \frac{\left(\sum_{\nu > 0} c_\nu P_\nu^\dagger \right)^n}{n!} |0\rangle_F \\
&= \left(\prod_{\nu > 0} U_\nu \right) \exp \left(\sum_{\nu > 0} c_\nu P_\nu^\dagger \right) |0\rangle_F, \qquad (1.4.39)
\end{aligned}
$$

[43] See Bohr and Mottelson (1975); Ring and Schuck (1980); Brink and Broglia (2005); Broglia and Zelevinsky (2013), and refs. therein.

where

$$c_\nu = V_\nu/U_\nu, \tag{1.4.40}$$

while n denotes the number of pairs of particles $(2n = N)$. *The last expression of (1.4.39) implies that $|BCS\rangle$ is a coherent state.*

In the above discussion of BCS we have treated in a rather cavalier fashion the fact that the amplitudes U_ν and V_ν are in fact complex quantities. A possible choice of phasing is[44]

$$U_\nu = U_\nu'; \quad V_\nu = V_\nu' e^{-2i\phi}, \tag{1.4.41}$$

and, as a consequence, the order parameter and the pairing gap can be written as

$$\alpha_0 = e^{-2i\phi}\alpha_0' \quad (\alpha_0' = \sum_{\nu>0} U_{\nu'} V_{\nu'}) \text{ and } \Delta = e^{-2i\phi}\Delta' \quad (\Delta' = G\alpha_0'), \tag{1.4.42}$$

U_ν' and V_ν' being real quantities, while ϕ is the gauge angle, conjugate variable to the number of particles operator. Then[45]

$$\hat{\phi} = i\partial/\partial \mathcal{N}, \quad \mathcal{N}, \tag{1.4.43}$$

and

$$\left[\hat{\phi}, \mathcal{N}\right] = i, \tag{1.4.44}$$

where $\mathcal{N} \equiv \hat{N}$ (Eq. (1.4.20)), the gauge transformations being induced by the gauge operator

$$\mathcal{G}(\phi) = e^{-i\mathcal{N}\phi}. \tag{1.4.45}$$

Let us replace the amplitudes (1.4.41) in (1.4.39). One then obtains

$$|BCS\rangle = \left(\prod_{\nu>0} U_\nu'\right) \sum_{N \text{ even}} e^{-iN\phi}|\Phi_N\rangle = \left(\prod_{\nu>0} U_\nu'\right) \sum_{N \text{ even}} e^{-iN\phi}|\Phi_N\rangle, \tag{1.4.46}$$

where

$$|\Phi_N\rangle = \frac{\left(\sum_{\nu>0} c_\nu' P_\nu^\dagger\right)^{N/2}}{(N/2)!}|0\rangle_F, \tag{1.4.47}$$

with $c_\nu' = V_\nu'/U_\nu'$. It is of note that

$$|\Phi_0\rangle = \sum_{\nu>0} c_\nu' P_\nu^\dagger|0\rangle_F \tag{1.4.48}$$

---

[44] Schrieffer (1973). The same results as those which will be derived are obtained with the alternative choice $U_\nu = U_\nu' e^{i\phi}$, $V_\nu = V_\nu' e^{-i\phi}$ (see, e.g., Potel et al. (2013b)).
[45] See, e.g., Brink and Broglia (2005), App. H, and refs. therein.

is the single Cooper pair state. The $|BCS\rangle$ state does not have a definite number of particles, but only in average, being a wavepacket in N. In fact, (1.4.46) defines a privileged direction in gauge space, being an eigenstate of $\hat{\phi}$:

$$\hat{\phi}|BCS\rangle = i\frac{\partial}{\partial N}\left(\prod_{\nu>0}U'_\nu\right)\sum_{N \text{ even}}e^{-iN\phi}|\Phi_N\rangle = \phi\,|BCS\rangle. \tag{1.4.49}$$

Expressing it differently, (1.4.46) can be viewed as an axially symmetric deformed system in gauge space, whose symmetry axis coincides with the z' component of the body-fixed frame of reference $\mathcal{K}'$, which makes an angle ϕ with the laboratory z-axis (Fig. 1.4.1).

With the help of Eq. (1.4.39) (first line), one can write

$$|BCS(\phi=0)\rangle_{\mathcal{K}'} = \prod_{\nu>0}\left(U'_\nu+V'_\nu P'^\dagger_\nu\right)|0\rangle_F = \prod_{\nu>0}\left(U_\nu+V_\nu P^\dagger_\nu\right)|0\rangle_F$$

$$= \prod_{\nu>0}\left(U'_\nu+e^{-2i\phi}V'_\nu P^\dagger_\nu\right)|0\rangle_F, \tag{1.4.50}$$

where use was made of the relations[46]

$$\mathcal{G}(\phi)a^\dagger_\nu\mathcal{G}^{-1}(\phi) = e^{-i\phi}a^\dagger_\nu = a'^\dagger_\nu \tag{1.4.51}$$

and

$$\mathcal{G}(\phi)P^\dagger_\nu\mathcal{G}^{-1}(\phi) = e^{-2i\phi}P^\dagger_\nu = P'^\dagger_\nu. \tag{1.4.52}$$

Spontaneously broken symmetry in nuclei is, as a rule, associated with the presence of rotational bands, as already found in the case of quadrupole deformed nuclei. *Consequently, one expects in nuclei with $\alpha_0 \neq 0$ rotational bands in which particle number plays the role of angular momentum, that is, pairing rotational bands (see Eq. (1.4.65)).*

Before proceeding with the discussion of the term H_{fluct}, let us return to the relation (1.4.26), that is, the fact that one can view the state $|BCS\rangle$ as the quasiparticle vacuum, implying that it can be written as

$$|BCS\rangle \sim \prod_\nu \alpha_\nu|0\rangle_F, \tag{1.4.53}$$

----

[46] It is to be noted that $\mathcal{G}$ induces a counterclockwise rotation, $\mathcal{G}(\chi)\hat{\phi}\,\mathcal{G}^{-1}(\chi) = \hat{\phi} - \chi$,

$$[N,\phi] = -i, \quad \mathcal{G}(\chi) = e^{-iN\chi} \quad \chi : \text{c-number}, \quad [e^{-iN\chi},\phi] = e^{-iN\chi}\phi - \phi e^{-iN\chi},$$
$$[e^{-iN\chi},\phi]e^{iN\chi} = e^{-iN\chi}\phi e^{iN\chi} - \phi, \quad e^{-iN\chi} = 1-iN\chi \quad \chi^2 \ll \chi,$$
$$[1-iN\chi,\phi] = -i[\chi,\phi] = -i\,(N[\chi,\phi]+[N,\phi]\chi) = -i(0-i\chi) = -\chi,$$
$$[e^{-iN\chi},\phi] = -\chi e^{-iN\chi},$$
$$[e^{-iN\chi},\phi]e^{iN\chi} = -\chi e^{-iN\chi}e^{iN\chi} = -\chi, \quad -\chi = \mathcal{G}(\chi)\phi\mathcal{G}^{-1}(\chi) - \phi,$$
$$\mathcal{G}(\chi)\phi\mathcal{G}^{-1}(\chi) = \phi - \chi.$$

as well as

$$|BCS\rangle \sim \prod_{\nu>0} \alpha_\nu \alpha_{\tilde{\nu}} |0\rangle_F. \qquad (1.4.54)$$

The state $|\nu'\rangle \sim \prod_{\nu\neq\nu'} \alpha_\nu |0\rangle_F$ corresponds to a one-quasiparticle state, while $|\nu'\tilde{\nu}'\rangle \sim \prod_{\nu>0,\nu\neq\nu'} \alpha_\nu \alpha_{\tilde{\nu}} |0\rangle_F$ is proportional to a two-quasiparticle state. Both states display an energy (pairing) gap with respect to the ground state, in the first case of the order of Δ, in the second of 2Δ. This is a property closely connected with the fact that in BCS condensation as described by $|BCS\rangle$, a macroscopic number of Cooper pairs are phase coherent, each being made out of pairs of fermions entangled in time-reversal states (see Eq. (3.4.2)), implying an energy gap for both pair dissociation and pair breaking. This is a property referred to as off-diagonal long-range order (ODLRO)[47] (see Sect. 4.9.3).

1.4.3 The Normal and the Superconducting State

Within this scenario it is useful in discussing BCS condensation in nuclei to go back to its origin[48], namely, low-temperature superconductivity. It is found that for superconductors based on metals displaying a strong electron-phonon coupling[49], the behavior of the BCS occupation probability V_k^2 resembles very much that of the same quantity, but for the metal at temperatures close to but above the critical temperature, that is, the metal in the normal state – a consequence of processes like those shown in graphs (b) and (c) of Fig. 1.2.4, where, in that case, the fermions are electrons (or electron holes) and the bosons are lattice phonons.[50] Such similitude is also observed in the nuclear case (Figs. 1.4.4 and 1.2.5).

Concerning superconductors, it is found that the changes observed in the behavior of the systems upon cooling below the critical temperature cannot be described at profit just in terms of changes in the occupation probability of a one-electron momentum eigenstate (from the sharp line to the smooth curve around λ ($\approx \epsilon_F$); Fig. 1.4.4 (a)), in keeping with the fact that in the process, no gap opens up in $k(\nu)$ space. In fact, the partial occupation V_ν^2 of the normal state carrying random (gauge) phases is, by lowering the temperature through T_c, progressively replaced by a single quantum state. In it, about the same set of many-body states, in which fermions are dressed and interact pairwise by exchanging bosons (aside from through the bare interaction: screened Coulomb potential in the case of superconductors, $NN^{-1}S_0$ potential in the case of superfluid nuclei), become coherently

---

[47] Penrose (1951); Penrose and Onsager (1956); Yang (1962); see also Anderson (1996).

[48] Bohr et al. (1958).

[49] These systems display, in the normal state, a consistent electric resistance. In other words, bad conductors are, as a rule, good superconductors, for example, Pb. On the other hand, Au, an excellent conductor, does not became superconducting at any measured temperature.

[50] See, e.g., Tinkham (1980), p. 28.

superposed with a fixed phase relation leading to ODLRO. It is then not so much, or better, not only the superconducting (superfluid) state that is special but also the so-called normal state.

In what follows we discuss the structure of H_{fluct} and single out the term responsible for restoring gauge invariance to the mean field solution, giving rise to pairing rotational bands for discrete values of the angular momentum in gauge space, namely, N, differing in two units from each other.

In terms of quasiparticles, H_{fluct} can be expressed as

$$H_{fluct} = H'_p + H''_p + C, \tag{1.4.55}$$

where

$$H'_p = -\frac{G}{4} \left(\sum_{\nu>0} (U_\nu^2 - V_\nu^2) \left(\Gamma_\nu^\dagger + \Gamma_\nu \right) \right)^2 \tag{1.4.56}$$

and

$$H''_p = \frac{G}{4} \left(\sum_\nu \left(\Gamma_\nu^\dagger - \Gamma_\nu \right) \right)^2, \tag{1.4.57}$$

with

$$\Gamma_\nu^\dagger = \alpha_\nu^\dagger \alpha_{\bar{\nu}}^\dagger. \tag{1.4.58}$$

The term C stands for constant terms, as well as for terms proportional to the number of quasiparticles, which consequently vanish when acting on $|BCS\rangle$. The term H'_p gives rise to two-quasiparticle pairing vibrations with energies $\gtrsim 2\Delta$. It can be shown that it is the term H''_p which restores gauge invariance.[51] In fact,

$$\left[H_{MF} + H''_p, \tilde{N} \right] = 0. \tag{1.4.59}$$

We now diagonalize $H_{MF} + H''_p$ in the quasiparticle RPA (QRPA),

$$\left[H_{MF} + H''_p, \Gamma_n^\dagger \right] = \hbar\omega_n \Gamma_n^\dagger, \quad \left[\Gamma_n, \Gamma_{n'}^\dagger \right] = \delta(n, n'), \tag{1.4.60}$$

where

$$\Gamma_n^\dagger = \sum_\nu \left(a_{n\nu} \Gamma_\nu^\dagger + b_{n\nu} \Gamma_\nu \right), \quad \Gamma_\nu^\dagger = \alpha_\nu^\dagger \alpha_{\bar{\nu}}^\dagger, \tag{1.4.61}$$

is the creation operator of the nth vibrational mode. The lowest energy root $n = 1$ can be written as

$$|1''\rangle = \Gamma_1^\dagger |0''\rangle = \frac{\Lambda_1''}{2\Delta} (\tilde{N} - N_0) |0''\rangle, \tag{1.4.62}$$

---

[51] Högaasen-Feldman (1961); Bès and Broglia (1966). For details, see Brink and Broglia (2005), App. I (Sect. I.4) and App. J.

where $\tilde{N}$ is the particle number operator written in terms of $\Gamma_\nu^\dagger$ and Γ_ν, and Λ_1'' is the strength of the quasiparticle-mode coupling.

It can be shown that the dimensionless prefactor $(\Lambda_1''/2\Delta)$ is the zero-point amplitude of the mode[52], that is,

$$\sqrt{\frac{\hbar\omega_1''}{2C_1''}} = \sqrt{\frac{\hbar^2}{2D_1''\hbar\omega_1''}}. \tag{1.4.63}$$

Because the lowest frequency is $\omega_1'' = 0$, the associated ZPF diverge ($\Lambda_1'' \sim (\hbar\omega_1'')^{-1/2}$). It can be seen that this is because $C_1'' \to 0$, while D_1'' remains finite. In fact,

$$\frac{D_1''}{\hbar^2} = \frac{2\Delta^2}{\Lambda_1''^2\hbar\omega_1''} = 4\sum_{\nu>0}\frac{U_\nu^2 V_\nu^2}{E_\nu}. \tag{1.4.64}$$

In keeping with the fact that a finite rotation in gauge space can be generated by a series of infinitesimal operations of type $\mathcal{G}(\delta\phi) = e^{i(\hat{N}-N_0)\delta\phi}$, the one phonon state $|1''\rangle = \Gamma_1^\dagger|0''\rangle$ results from rotations of the deformed (correlated) intrinsic state of divergent amplitude, that is, fluctuations of ϕ over the whole $0 - 2\pi$ range. By proper inclusion of these fluctuations, one can restore gauge invariance violated by $|BCS\rangle_{\mathcal{K}'}$. The resulting states[53]

$$|N\rangle \sim \int_0^{2\pi} d\phi\, e^{iN\phi}|BCS(\phi)\rangle_{\mathcal{K}'} \sim \left(\sum_{\nu>0} c_\nu' P_\nu^\dagger\right)^{N/2}|0\rangle_F \tag{1.4.65}$$

have fixed particle number and constitute the members of a pairing rotational band. Making use of a simplified model[54] (single j-shell), it can be shown that the energies of these states can be written as

$$E_N = \lambda(N - N_0) + \frac{G}{4}(N - N_0)^2, \tag{1.4.66}$$

where

$$\frac{G}{4} = \frac{\hbar^2}{2D_1''}. \tag{1.4.67}$$

An example of pairing rotational bands is provided by the ground states of the single open closed shell superfluid $^A_{50}\mathrm{Sn}_N$-isotopes (Fig. 1.4.5). The neutron number $N_0 = 68$ has been found to minimize $E_0 = \langle BCS|H|BCS\rangle$ (see Eq. (1.4.27)) and has thus been used in the solution of the BCS number equation (1.4.37).

---

[52] Amplitude, which in the case of surface vibrations was defined in Eq. (1.1.11). See Brink and Broglia (2005), App. I.

[53] Where use has been made of Eqs. (1.4.50), (1.4.46), and (4.7.27).

[54] See Brink and Broglia (2005), App. H, and refs. therein.

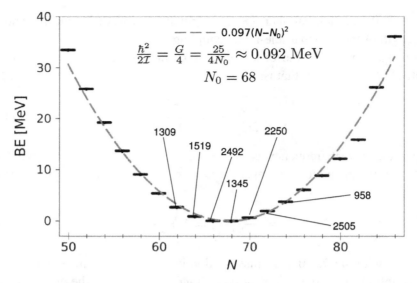

Figure 1.4.5 Pairing rotational band associated with the ground states of the Sn-isotopes. The horizontal lines represent the energies calculated according to the expression $BE = B(^{50+N}\text{Sn}_N) - 8.124N + 46.33$, subtracting the contribution of the single nucleon addition to the nuclear binding energy obtained by a linear fitting of the binding energies of the whole Sn chain. The estimate of $\hbar^2/2\mathcal{I}$ was obtained using the single j-shell model Eq. (1.4.67) (see also Brink and Broglia (2005), Appendix H and refs. therein). The numbers indicated with thin lines are the absolute value of the experimental $^A_Z\text{Sn}_N(p,t)^{A-2}_Z\text{Sn}_{N-2}$ (gs), cross sections integrated over angles (in units of μb) (Guazzoni et al. (1999, 2004, 2006, 2008, 2011, 2012)). After Potel et al. (2013b). For the color version of this figure, refer to cambridge.org/nuclearcooperpair.

Making use of the BCS pair transfer amplitudes,

$$B_\nu \sim \langle BCS|P_\nu^\dagger|BCS \rangle = U_\nu V_\nu, \tag{1.4.68}$$

in combination with a computer code (see App. 7.A) and of global optical parameters, one can account for the absolute value of the Cooper pair transfer differential cross section, within experimental error (see, e.g., Figs. 3.1.7 and 7.2.1).

Projecting out the ground state of the different Sn-isotopes from the intrinsic BCS state describing ^{118}Sn, one obtains a quantitative description of observations carried out with the help of the specific probe of pairing correlations (Cooper pair transfer), a result which testifies to the fact that pairing rotational bands can be considered elementary modes of nuclear excitation, emergent properties of restoration of spontaneous symmetry breaking of gauge invariance.[55]

---

[55] A finding related to the fact that $|BCS\rangle$ is a coherent state.

The fact that this insight follows from the use of QRPA[56] in the calculation of the ZPF of the collective solutions of the pairing Hamiltonian underscores the importance of conserving approximations[57] to describe quantum many-body problems, in particular, finite-size quantum many-body problems (FSQMB), of which the nucleus represents a paradigmatic example.

1.5 Giant Dipole Resonance

Aside from low-lying collective states, that is, rotations and low-energy vibrations, nuclei also display high-lying collective modes known as giant resonances.

If one shines a beam of photons on a nucleus, it is observed that the system absorbs energy resonantly essentially at a single frequency, of the order of[58] $\nu = 5 \times 10^{21}$ Hz, corresponding to an energy of $h\nu \approx 20$ MeV.

Although the wavelength of a 20 MeV γ-ray is smaller than that of other forms of electromagnetic radiation, such as visible light, it is still large ($\lambda \approx 63$ fm) compared to the dimensions of, for example, ^{40}Ca ($R_0 \approx 4.1$ fm). As a result, the photon electric field is nearly uniform across the nucleus at any time. This field exerts a force on the positively charged protons. Consequently, it can set the center of mass of the nucleus into an antenna like, dipole oscillation (Thompson scattering). Another possibility is that it leads to an internal excitation of the system. In this case, because the center of mass of the system has to remain at rest, the neutrons oscillate against the protons. The restoring force of the vibration, known as the giant dipole resonance (GDR), is provided by the attractive force between protons and neutrons.

The connotation of giant is in keeping with the fact that it essentially carries the full photo-absorption cross section (energy weighted sum rule; see below) and resonance because it displays a Lorentzian-like shape with a full width at half maximum of a few MeV ($\lesssim 5$ MeV), considerably smaller than the energy centroid[59] $\hbar\omega_{GDR} \approx 80/A^{1/3}$ MeV. Microscopically, the GDR can be viewed as a correlated

---

[56] Using the Tamm–Dancoff approximation, i.e., setting $Y \equiv 0$ (and thus $\sum X^2 = 1$) in the QRPA solution, does not lead to particle number conservation, in keeping with the fact that the amplitudes Y are closely connected to ZPF.

[57] That is, approximations respecting sum rules; see Eq. (1.4.59).

[58] Making use of $h = 4.1357 \times 10^{-15}$ eV×s, one obtains for $h\nu = 1$ eV the frequency $\nu = 2.42 \times 10^{14}$ Hz and thus $\nu = 4.8 \times 10^{21}$ Hz for $h\nu = 20$ MeV. The wavelength of a photon of energy E is $\lambda = hc/E \approx 2\pi \times 200$ MeV fm/E, which for $E = 20$ MeV leads to $\lambda \approx 63$ fm.

[59] Within this context we refer to the discussion concerning the renormalization of collective modes carried out in the text before Eq. (1.2.14), in particular regarding the cancellation between self-energy and vertex corrections (see Fig. 1.2.6 and footnote 22). This is a basic result of NFT – as discussed in later chapters (see in particular Sect. 5.5.1) – being connected with sum rule arguments in general and particle conservation in particular. This argument has been extended to the case of finite temperature as well as to include relativistic effects (Ward (1950); Nambu (1960); Bortignon and Broglia (1981); Bertsch et al. (1983); Bortignon et al. (1998); Litvinova and Wibowo (2018); Wibowo and Litvinova (2019); Lalazissis and Ring (2019), and refs. therein).

particle-hole excitation, that is, a state made out of a coherent linear combination of proton and neutron particle-hole excitations with essentially $\Delta N = 1$ character, as well as small $\Delta N = 3, 5, \ldots$ components (Fig. 1.2.3). Because the difference in energy between major shells is $\hbar\omega_0 \approx 41 A^{1/3}$ MeV, one expects that about half of the contribution to the energy of the GDR arises from the neutron–proton inter-action, more precisely, from the so-called symmetry potential V_1 (see Eq. (1.1.4)), which measures the energy price the system has to pay to separate protons from neutrons. Theoretical estimates lead to[60]

$$(\hbar\omega_{GDR})^2 = (\hbar\omega_0)^2 + \frac{3\hbar^2 V_1}{m\langle r^2\rangle} = \frac{1}{A^{2/3}}\left[(41)^2 + (60)^2\right] \text{ MeV}^2, \qquad (1.5.1)$$

resulting in

$$\hbar\omega_{GDR} \approx \frac{73}{A^{1/3}} \approx \frac{87}{R} \text{ MeV}, \qquad (1.5.2)$$

where $R = 1.2 A^{1/3}$ is the numerical value of the nuclear radius when measured in fm. The above quantity is to be compared with the empirical value $\hbar\omega_{GDR} \approx (80/A^{1/3})$ MeV $\approx (95/R)$ MeV.

It is of note that the elastic vibrational frequency of a spherical solid made out of particles of mass m can be written as $\omega_{el}^2 \sim \mu/(m\rho R^2) \sim v_t^2/R^2$, where R is the radius, ρ the density, and v_t the transverse sound velocity proportional to the Lamé shear modulus of elasticity μ. In other words, giant resonances in general, and the GDR in particular, are embodiments of the elastic response of the nucleus to an impulsive external field, like that provided by a photon. The nuclear rigidity to sudden solicitations is provided by the shell structure, quantitatively measured by the energy separation between major shells.

1.6 Giant Pairing Vibrations

Owing to spatial quantization, in particular to the existence of major shells of pair degeneracy $\Omega (\equiv (2j + 1)/2)$ separated by an energy $\hbar\omega_0 \approx 41/A^{1/3}$ MeV (see Fig. 1.1.2), the Cooper pair model can be extended to encompass pair addition and pair substraction modes across major shells.[61] Assuming that the single-particle energies ϵ_k and ϵ_i appearing in Fig. 1.3.1 are both at an energy $\hbar\omega_0$ away from the Fermi energy, one obtains the dispersion relation

$$-\frac{1}{G} = \frac{\Omega}{W - 2\hbar\omega_0} - \frac{\Omega}{W + 2\hbar\omega_0}, \qquad (1.6.1)$$

---

[60] See, e.g., Bohr and Mottelson (1975); see also Bertsch and Broglia (2005), Ch. 5, and refs. therein.
[61] Broglia and Bes (1977).

leading to

$$(2\hbar\omega_0)^2 - W^2 = 4\hbar\omega_0 G\Omega \qquad (1.6.2)$$

and implying a high-lying pair addition mode of energy

$$W = 2\hbar\omega_0 \left(1 - \frac{G\Omega}{\hbar\omega_0}\right)^{1/2}. \qquad (1.6.3)$$

The forward (backward) going RPA amplitudes are, in the present case,

$$X = \frac{\Lambda_0\Omega^{1/2}}{2\hbar\omega_0 - W} \quad \text{and} \quad Y = \frac{\Lambda_0\Omega^{1/2}}{2\hbar\omega_0 + W}, \qquad (1.6.4)$$

normalized according to the relation

$$1 = X^2 - Y^2 = \Lambda_0^2\Omega\frac{8\hbar\omega_0 W}{\left((2\hbar\omega_0)^2 - W^2\right)^2}, \qquad (1.6.5)$$

where Λ_0 stands for the particle–pair vibration coupling vertex. Making use of (1.6.2), one obtains

$$\left(\frac{\Lambda_0}{G}\right)^2 = \Omega \left(1 - \frac{G\Omega}{\hbar\omega_0}\right)^{-1/2}, \qquad (1.6.6)$$

a quantity corresponding, within the framework of the simplified model used, to the two-nucleon transfer cross section. Summing up, the monopole giant pairing vibration has an energy close to $2\hbar\omega_0$ and is expected to be populated in two-particle transfer processes with a cross section of the order of that associated with the (ground state) pair addition mode, being this one of the order of Ω. Simple estimates of (1.6.3) and (1.6.6) can be obtained making use of $\Omega \approx \frac{2}{3}A^{2/3}$ and $G \approx 17/A$ MeV, namely,

$$W = 0.85 \times 2\hbar\omega_0, \quad \left(\frac{\Lambda_0}{G}\right)^2 \approx 1.2\Omega. \qquad (1.6.7)$$

Experimental evidence of GPV in light nuclei has been reported.[62]

1.7 Sum Rules

There are important operator identities which restrict the possible matrix elements in a physical system. Let us calculate the double commutator of the Hamiltonian describing the system and a single-particle operator F. That is,

$$\left[\hat{F}, \left[H, \hat{F}\right]\right] = \left(2\hat{F}H\hat{F} - \hat{F}^2H - H\hat{F}^2\right). \qquad (1.7.1)$$

[62] Cappuzzello et al. (2021); Cappuzzello et al. (2015); Cavallaro et al. (2019); Bortignon and Broglia (2016). See also Laskin et al. (2016); Id Betan et al. (2002); Dussel et al. (2009); Mouginot et al. (2011); Khan et al. (2004); Avez et al. (2008); Khan et al. (2009); Dasso et al. (2015); Fortunato et al. (2002); Herzog et al. (1986).

We assume that $\hat{F} = \sum_k F(\mathbf{r}_k)$ and $H = T + v(\mathbf{r}, \mathbf{r}')$, where $v(\mathbf{r}, \mathbf{r}') = - \kappa_1$ $\hat{F}(\mathbf{r}) \hat{F}(\mathbf{r}')$. Thus

$$\left[\hat{F}, \left[H, \hat{F} \right] \right] = \sum_k \frac{\hbar^2}{m} \left(\nabla_k F(\mathbf{r}_k) \right)^2. \tag{1.7.2}$$

Taking the average value on the correlated ground state,

$$S(F) = \sum_\alpha |\langle \alpha | F | \tilde{0} \rangle|^2 (E_\alpha - E_0) = \frac{\hbar^2}{2m} \int d\mathbf{r} \, |\nabla F|^2 \rho(\mathbf{r}), \tag{1.7.3}$$

where $H|\alpha\rangle = E_\alpha |\alpha\rangle$, $H|\tilde{0}\rangle = E_0 |\tilde{0}\rangle$, and the sum $\sum_\alpha$ is over a complete set of eigenstates. The above result describes the reaction of a system at equilibrium to which one applies an impulsive field, giving to the particles a momentum ∇F. On the average, the particles started at rest, so their mean energy after the sudden impulse is $\hbar^2 |\nabla F|^2 / 2m$, a result which does not depend on the interaction acting among the nucleons, the energy being absorbed from the (instantaneous) external field before the system is disturbed from equilibrium. The relation (1.7.3) is known as the energy-weighted sum rule (EWSR) associated with the operator $\hat{F}$.

An important application of (1.7.3) implies a situation where F has a constant gradient. Inserting $\mathbf{F} = z \times \mathbf{z}$ in (1.7.3), the integral simplifies because $\nabla F = 1$, and it leads to a quantity proportional to the number of particles,

$$\sum_\alpha |\langle \alpha | F | \tilde{0} \rangle|^2 (E_\alpha - E_0) = \frac{\hbar^2 N}{2m}. \tag{1.7.4}$$

The electric field of a photon is of this form in the dipole approximation, which is valid when the size of the system is small compared to the wavelength of the photon. For the dipole operator referred to the nuclear center of mass, one obtains

$$F(\mathbf{r}) = e \sum_k \left[\frac{N - Z}{A} - t_z(k) \right] r_k Y_{1\mu}(\hat{r}_k), \tag{1.7.5}$$

with $t_z = -1/2$ for protons and $+1/2$ for neutrons, t_z being the z-component of the isospin operator. Using this expression in connection with (1.7.3), one obtains

$$\sum_{\alpha'} |\langle \alpha' | F | \tilde{0} \rangle|^2 (E_{\alpha'} - E_0) = \frac{9}{4\pi} \frac{\hbar^2 e^2}{2m} \frac{NZ}{A}. \tag{1.7.6}$$

The above relation is known as the Thomas–Reiche–Kuhn (TRK) sum rule[63] and is equal to the maximum energy a system can absorb from the dipole field. The RPA solution respects the EWSR.

---

[63] See Bohr and Mottelson (1975) and Bertsch and Broglia (2005); see also Bortignon et al. (1998), Sect. 10.1.

1.8 Ground State Correlations

The zero-point fluctuations associated with collective nuclear vibrations affect, among other things, the nuclear mean field properties.[64] Within the harmonic approximation discussed in Sections 1.1.1 and 1.1.2 in connection with ($\beta = 0$) surface vibrational modes, the ground state Ψ_0 of the Hamiltonian (1.1.9) has an energy $\frac{1}{2}\hbar \left(\frac{C_\lambda}{D_\lambda}\right)^{1/2} = \hbar\omega_\lambda/2$ for each degree of freedom (i.e., $(2\lambda + 1)$ in the case of a vibration of multipolarity λ). The associated coordinate-momentum indeterminacy relation assumes its minimum value[65]

$$\Delta\alpha_{\lambda\mu}(n)\Delta\pi_{\lambda\mu}(n) = \frac{\hbar}{2}, \qquad (1.8.1)$$

as

$$\Delta\alpha_{\lambda\mu}(n) = \sqrt{\frac{\hbar\omega_\lambda(n)}{2C_\lambda(n)}}, \qquad (1.8.2)$$

and $\Delta\pi_{\lambda\mu} = \sqrt{\hbar D\omega_\lambda/2}$, in keeping with the fact that $|\Psi_0|^2$ is, mathematically, a Poisson distribution. The above expression implies that the mean square radius will be modified from its mean field value $\sqrt{\frac{3}{5}}R_0$ and thus also the nuclear density distribution. The value of $\hbar\omega_\lambda(n)/2C_\lambda(n)$ can be determined by working out the collective mode $|n_\lambda(n) = 1\rangle = \Gamma^\dagger_{\lambda\mu}(n)|\tilde{0}\rangle$ in RPA, the zero-point fluctuations being closely connected with the Y-amplitudes of the mode.

Let us calculate the effect of the zero-point fluctuations on the nuclear density distribution. The corresponding operator can be written as

$$\hat{\rho}(\mathbf{r}) = a^\dagger(\mathbf{r})a(\mathbf{r}), \qquad (1.8.3)$$

---

[64] Gogny (1979); Esbensen and Bertsch (1983); Reinhard and Drechsel (1979); Khodel et al. (1982); Barranco and Broglia (1987, 1985). See also Brown and Jacob (1963); Anderson and Thouless (1962) (it is likely a coincidence in connection with this inaugural issue of *Phys. Lett.* that short of a hundred pages after, one finds the paper of B. D. Josephson, Possible new effects in superconducting tunneling, *Phys. Lett.* **1**, 251 (1962)).

[65] The same result is found for Ψ_n describing a state with n-quanta, being at the basis that solutions with $n \gg 1$ behave as "quasiclassical" or "coherent" states of the harmonic oscillator (Glauber (2007)), in keeping with the fact that the contribution of the zero-point energy is negligible in such a case ($(n + 1/2)\hbar\omega \approx n\hbar\omega$) and that the many-quanta wavepacket always attains the lower limit of (1.8.1) (Basdevant and Dalibard (2005), pp. 153, 465) (discussions with Pier Francesco Bortignon in March 2018 concerning coherent states are gratefully acknowledged). Schrödinger used this result in a paper (Schrödinger (1926b)) to suggest that waves (material waves) described by his wave function were the only reality, particles being only derivative things. In support of his view he considered a superposition of linear harmonic oscillator wavefunctions and showed that the wave group holds permanently together in the course of time. And he adds that the same will be true for the electron as it moves in high orbits of the hydrogen atom, hoping that wave mechanics would turn out to be a branch of classical physics (Pais (2000)). It was Born who first provided the correct interpretation of Scrhödinger's wavefunction (modulus square) in his paper "Quantum mechanical collision phenomena" (Born (1926)). In it is stated that the result of solving with wave mechanics the process of elastic scattering of a beam of particles by a static potential is not what the state after the collision is, but how probable is a given effect of the collision.

where $a^\dagger(\mathbf{r})$ is the creation operator of a nucleon at point $\mathbf{r}$. It can be expressed in terms of the phase space creation operators $a_\nu^\dagger (\nu \equiv n, l, j, m)$ as

$$a^\dagger(\mathbf{r}) = \sum_\nu \varphi_\nu^*(\mathbf{r}) a_\nu^\dagger, \qquad (1.8.4)$$

where $\varphi_\nu(\mathbf{r})$ are the single-particle wavefunctions. Thus

$$\hat{\rho}(\mathbf{r}) = \sum_{\nu\nu'} \varphi_\nu^*(\mathbf{r}) \varphi_{\nu'}(\mathbf{r}) a_\nu^\dagger a_{\nu'}. \qquad (1.8.5)$$

The matrix element in the HF ground state is (Fig. 1.2.2 (h))

$$\rho_0(r) =_F \langle 0|\hat{\rho}(\mathbf{r})|0 \rangle_F = \sum_{i,(\epsilon_i \leq \epsilon_F)} |\varphi_i(\mathbf{r})|^2. \qquad (1.8.6)$$

To lowest order of perturbation theory in the particle–vibration coupling vertex, the NFT diagrams associated with the change of ρ_0 due to ZPF are shown in Fig. 1.8.1. Graphs (a), (b), (c), and (d) describe the changes in the density operator and in the single-particle potential, respectively.[66] This can be seen from the insets (I) and (II). The dashed horizontal line starting with a cross and ending at a hatched circle represents the renormalized density operator. This phenomenon is similar to that encountered in connection with vertex renormalization in Fig. 1.2.6, that is, the renormalization of the particle–vibration coupling (insets (I) and (I')). Concerning potential renormalization, the boldface arrowed line shown in inset (II) of Fig. 1.8.1 represents the motion of a renormalized nucleon due to the self-energy process induced by the coupling to vibrational modes, a phenomenon which can be described at profit through an effective mass, the so-called ω-mass m_ω, in which case particle motion is described by the Hamiltonian[67] $(\hbar^2/2m_\omega) \nabla^2 + \left(\frac{m}{m_\omega}\right) U(r)$. The ω-mass can be written as $m_\omega = (1 + \lambda)m$, where m is the nucleon mass and λ is the so-called mass enhancement factor $\lambda = N(0)\Lambda$, $N(0)$ being the density of levels at the Fermi energy and Λ the PVC strength, typical values being $\lambda = 0.4$.

The fact that in calculating $\delta\rho$, that is, the correction to the nuclear density distribution (renormalization of the density operator), one finds to the same order of perturbation a correction to the potential is in keeping with the self-consistency existing between the two quantities (Eq. (1.2.8); see also Eq. (1.2.9)). *Now, what changes is not only the single-particle energy but also the single-particle content measured by $Z_\omega = m/m_\omega$, as well as the radial dependence of the wavefunctions of the states.* It is of note that the effective mass approximation, although being quite useful, cannot take care of the energy dependence of the renormalization

---

[66] Barranco (1985); Barranco and Broglia (1985).

[67] The ratio (m/m_ω) gives a measure of the single-particle energy content (Mahaux et al. (1985), Eq. (3.5.18)), while m_ω is proportional to the (energy) slope of the (single-particle) self-energy dispersion relation (Sect. 5.4.1); see also Brink and Broglia (2005), and refs. therein.

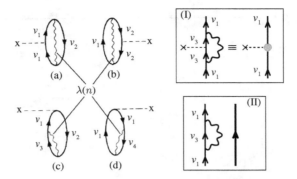

Figure 1.8.1 Lowest-order corrections in the particle–vibration coupling vertex of the nuclear density due to the presence of zero-point fluctuations associated with density vibrations. An arrowed line pointing upward denotes a particle, while one pointing downward denotes a hole. A wavy line labeled $\lambda(n)$ represents the nth surface vibrational phonon of multipolarity λ, in order of excitation energy ($n = 1, 2 \dots$). The density operator is described through a dotted horizontal line starting with a cross. Graphs (a) and (b) are typical examples of contributions to $\delta\rho$ (see inset (**I**); the dashed horizontal line starting with a cross and ending at a hatched circle in the diagram to the right represents the renormalized density operator, resulting from the processes displayed to the left); (c) and (d) are examples of potential contributions (see inset (**II**); the boldface arrowed line represents the renormalized single-particle state due to the coupling to the vibrations leading to the self energy process shown to the left).

process, which leads, in the case of single-particle motion, to renormalized energies, spectroscopic amplitudes, and radial wavefunctions, closely connected with the formfactors associated with one-particle transfer reactions.

The analytic expressions associated with diagrams (a) and (c) of Fig. 1.8.1 are

$$\delta\rho(r)_{(a)} = \frac{(2\lambda + 1)}{4\pi} \sum_{\nu_1 \nu_2} [Y_n(\nu_1 \nu_2; \lambda)]^2 \, R_{\nu_1}(r) R_{\nu_2}(r) \tag{1.8.7}$$

and

$$\delta\rho(r)_{(c)} = (2\lambda + 1)\Lambda_n(\lambda) \sum_{\nu_1 \nu_2 \nu_3} \frac{M(\nu_1, \nu_3; \lambda)}{\epsilon_{\nu_1} - \epsilon_{\nu_2}} (2j_1 + 1)^{-1/2}$$
$$\times \, Y_n(\nu_3 \nu_2; \lambda) \times R_{\nu_1}(r) R_{\nu_2}(r), \tag{1.8.8}$$

where M is the matrix element of $-\frac{R_0}{\kappa}\frac{\partial U}{\partial r} Y_{\lambda\mu}(\hat{r})$, $n = 1, 2 \dots$ labeling the first, second, and so on vibrational modes as a function of increasing energy and Λ_n is the strength of the particle–vibration coupling associated with the n-mode of multipolarity λ. The functions $R(r)$ are the radial wavefunctions associated with the states ν_1 and ν_2. While $\delta\rho_{(a)}$ can be written in terms of the RPA Y-amplitudes, which are directly associated with the zero-point fluctuations of harmonic motion, $\delta\rho_{(c)}$ contains a scattering vertex not found in RPA – that is, going beyond the

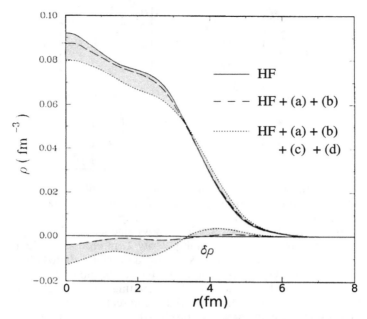

Figure 1.8.2 Modification in the charge density of ^{40}Ca induced by the zero-point fluctuations associated with vibrations of the surface modes. The results labeled HF, HF+(a)+(b), and HF+(a)+(b)+(c)+(d) are the Hartree–Fock density, and that resulting from adding to it the corrections $\delta\rho$ associated with the processes (a)+(b) and (a)+(b)+(c)+(d) displayed in Fig. 1.8.1. In the lower part of the figure the corresponding quantities $\delta\rho$ are displayed. After Barranco and Broglia (1987).

harmonic approximation – and essential to describing renormalization processes of the different degrees of freedom, namely, single-particle (energy, single-particle content and radial dependence of the wavefunction) and collective motion, as well as medium polarization interactions.

In Fig. 1.8.2, results of calculations of $\delta\rho$ carried out for the closed shell nucleus ^{40}Ca are shown. The vibrations were determined by diagonalizing multipole-multipole separable interactions in the RPA. All the roots of multipolarity and parity $\lambda^\pi = 2^+, 3^-, 4^+$, and 5^- which exhaust the EWSR were included in the calculations. Both isoscalar and isovector degrees of freedom were considered, as well as low-lying and giant resonances.

From the point of view of single-particle motion, the vibrations associated with low-lying modes display rather low frequency ($\hbar\omega_\lambda/\epsilon_F \approx 0.1$) and lead to an ensemble of essentially adiabatic deformed shapes. Nucleons can thus reach to distances from the nuclear center which are considerably larger than the radius R of the static spherical potential. Because the frequency of the giant resonances are of similar magnitude to those corresponding to the single-particle motion, the associated surface deformations average out. Thus low-lying vibrational modes account for most of the contributions to the changes in the density distribution.[68]

Making use of $(\delta\rho)_{low-lying}$, the change in the mean square radius was calculated, leading[69] to $\delta\langle r^2\rangle = 0.494$ fm². It implies a relative correction $\delta\langle r^2\rangle/((3/5)R_0^2) \approx 5\%$, where $R_0 = r_0 A^{1/3} = 4.1$ fm is the value of ^{40}Ca radius obtained from systematics ($r_0 = 1.2$ fm), $(3/5)R_0^2 \approx 10.11$fm² being the associated value of the mean square radius.

Similar calculations to the ones discussed above, but in this case for the nucleus ^{208}Pb and taking into account only the contributions of the low-lying octupole vibration[70], indicate that nucleons are to be found a reasonable part of the time in higher shells than those assigned to them by the shell model, the average number of "excited" particles being ≈ 2.4. Correspondingly, hole states will be left behind. Because of their presence in the closed shell nucleus, one can transfer a nucleon to states below the Fermi energy in, for example, (d, p) or $(^3\mathrm{He},d)$ one-neutron or one-proton stripping reactions, respectively, leaving the final nucleus with one nucleon above closed shell coupled to a vibration.

Systematic studies of such multiplets have been carried out throughout the mass table, in particular around the closed shell nucleus $^{208}_{82}$Pb$_{126}$ (Fig. 1.8.3). Within this context it is not only quite natural but also necessary to deal with structure and reactions on equal footing. This is one of the main goals of the present monograph.

1.9 Interactions

A number of subjects are not touched upon in the present monograph, in particular, the role temperature plays in the structure and decay of nuclei and that of the bare, NN-potential and/or effective forces. In what follows we briefly elaborate on this last point, referring to Bortignon et al. (1998) and references therein concerning the first one.

The Coulomb interaction resulting from the exchange of photons between charged particles defines the domain area of quantum electrodynamics (QED). Being the electron and the photon fields in interaction, what we call an electron is only partially to be associated with the electron field alone. It is also partially to be associated with the photon field which dresses the electron field. And conversely,

---

[68] Another example of the recurrent central role played by low-frequency modes in determining the properties and behavior of systems at all levels of organization, from the atomic nucleus to the Casimir effect in QED (see Fig. 7.E.3), to superconductivity in metals (Sect. 4.8.2) as well as to the folding of proteins (Micheletti et al. (2004)) in connection with metabolism, and of brain activity ($\nu < 0.1$ Hz) (Mitra et al. (2018); Vyazovskiy and Harris (2013)) in different phases of sleep.

[69] $\delta\langle r^2\rangle = (0.371(2^+ + 3^-) + 0.123(4^+ + 5^-))$fm²=0.494fm²; see the fourth column (Micr. (a)+(b)+(c)+(d)) of Table 1 of Barranco and Broglia (1987).

[70] Brown and Jacob (1963). See also Anderson and Thouless (1962).

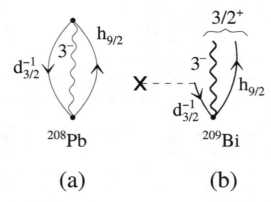

(a)   (b)

Figure 1.8.3 (a) Example of zero-point fluctuation of the ground state of the double-magic nucleus $^{208}_{82}\text{Pb}_{126}$ associated with the low-lying octupole vibration of this system, observed at an energy of 2.615 MeV and displaying a collective electromagnetic decay to the ground state of 32 B_{sp}. The **proton** particle–hole component $(h_{9/2}, d^{-1}_{3/2})_{3^-}$ displayed carries a large amplitude in the octupole vibration wavefunction. (b) Diagram representing the transfer of one proton to ^{208}Pb, which fills the $d^{-1}_{3/2}$ hole state leading to a $3/2^+$ in $^{209}_{83}\text{Bi}_{126}$, member of the septuplet of states $|(3^- \otimes h_{9/2}J^{\pi})\rangle$ with $J^{\pi} = 3/2^+, 5/2^+, \ldots, 15/2^+$. The horizontal dashed line starting with a cross stands for the stripping process $(^3\text{He}, d)$; see also Sect. 2.5.4, Fig. 2.5.11(f).

what is called the photon field can materialize itself in space in terms of an electron and a positron.[71]

The Coulomb interaction is the best known of all physical interactions, and QED constitutes the paradigm of theories which can be considered correct. In natural units in which the magnetic moment of the electron emerging from the Dirac equation has the value of 1, the results of QED corrections agree – once divergent contributions have been renormalized by properly adjusting the value of the bare electron mass and charge – within experimental error, namely, down to the eleventh decimal figure, with observation, that is, $1.00115965221 \pm 0.00000000003$ (exp.); $1.00115965246 \pm 0.00000000020$ (QED).[72]

The energy difference between the $2S_{1/2}$ and the $2P_{1/2}$ states of the hydrogen atom, which, according to Dirac's theory, should be degenerate, emerges naturally, according to QED, from the dressing of the hydrogen's electron by the photon associated with the zero-point fluctuation of the electromagnetic vacuum. The measured value of the Lamb shift[73] is 1057.845 (9) MHz, an experimental value[74]

[71] Schwinger (2001).
[72] Kinoshita (1990).
[73] Lamb (1964).
[74] Lundeen and Pipkin (1981, 1986); Pipkin (1990).

whose accuracy is limited by the ≈ 100 MHz natural line width of the $2P_{1/2}$ state. The best QED value, limited by uncertainties in the radius of the proton, is 1057.865 MHz.[75]

The above two numbers (electron magnetic moment and Lamb shift) are results of *ab initio* calculations within the framework of a field theory (QED) to the extent that quantum mechanics in general[76] and the Dirac equation[77] in particular can be considered, on grounds of their universal validity, to be so, including QED with its bare mass and charge parameters used to cure infinities (renormalization).[78]

Nuclear field theory (NFT) was developed following the (graphical) approach adopted by Feynman in his formulation of QED, making use of the correspondence electron→nucleon, photons→collective vibrations, fine structure constant→particle–vibration coupling. In Feynman's QED, *nothing is really free*[79] *(bare)*, a fact that transfers to NFT. This, together with the very large relative value of the induced (retarded) nuclear interaction – as large or larger than that of the bare NN-interaction in, for example, the case of the $^1S_0$ (pairing channel)[80] – arising from the exchange of collective vibrations between nucleons, makes the role which specific (bare) nuclear forces (four-point vertices v) may play in the results of NFT calculations somewhat blurred.

The program one develops in the present monograph concerning nuclear structure is that of renormalized nuclear field theory NFT_{ren}. Starting from the mean field at the basis of the nonobservable *bare single-particle energies*, one employs a generic shape (Woods–Saxon), adjusting the depth, radius, and diffusivity of the central potential and the depth of the spin-orbit one (making use of an r-dependent effective k-mass (exchange potential)), so that the dressed single-particle states best reproduce the experimental findings,[81] a procedure which parallels that associated with the bare mass parameter entering renormalized[82] QED. In connection with the coupling between nucleons and surface vibrations, namely, the strength Λ_α, one adjusts κ so that the dressed modes (see Fig. 1.2.6 as well as inset of Fig. 1.2.3)

---

[75] Sapirstein and Yennie (1990); Grotch (1994).

[76] Heisenberg (1925, 1927); Born and Jordan (1925); Born et al. (1926); Dirac (1925); Schrödinger (1926b); Born (1926); Pauli (1925).

[77] Dirac (1928a,b).

[78] Feynman (1949); Schwinger (1948); Tomonaga (1946); see also Dyson (1949); Schwinger (1958); Schweber (1994).

[79] Feynman (1975).

[80] E.g., the contribution to the pairing gap of ^{120}Sn ($\Delta \approx 1.45$ MeV) arising from the bare $^1S_0$ (pairing) interaction v_{14} (Argonne NN-potential) and from the exchange of collective vibrations is about equal (Idini et al. (2015)). In the case of ^{11}Li, essentially all of the pairing interaction which binds the two halo neutrons to the core ^{9}Li arises from the exchange between them of the soft $E1$-mode (pygmy dipole resonance of ^{11}Li) (Barranco et al. (2001); Potel et al. (2010); Broglia et al. (2019a)).

[81] See, e.g., Barranco et al. (2017) and Barranco et al. (2020).

[82] See Sect. 5.3.3.

best reproduce the observed properties of the collective modes, a procedure which parallels the tuning of the bare charge in renormalized QED.

The Mayer and Jensen sequence of levels around the $N = 8$ magic nuclei is $1p_{1/2}, 1d_{5/2}, 2s_{1/2}$ (Fig. 1.1.2), while experimentally, $^{11}_{4}\text{Be}_{7}$ displays the sequence[83] $1/2^{+}, 1/2^{-}$ (bound), $5/2^{+}$ (resonance) – a consequence of the dressing of the bare $1p_{1/2}, 1d_{5/2}$, and $2s_{1/2}$ states by the quadrupole vibration[84] of ^{10}Be ($\beta_{2} \approx 0.9$).[85] This extremely large value of the dynamical deformation parameter implies a Lamb-shift-like mechanism which shifts the bare $1p_{1/2}$ and $2s_{1/2}$ orbitals by more than 3 MeV with respect to each other, the resulting dressed levels reproducing the experimental ones ($\epsilon_{\widetilde{1/2}^{+}} = -0.5$ MeV, $\epsilon_{\widetilde{1/2}^{-}} = -0.18$ MeV, i.e.,

$$\Delta\epsilon = \left(\epsilon_{\widetilde{1/2}^{-}} - \epsilon_{\widetilde{1/2}^{+}}\right) = 0.32 \text{ MeV; see Fig. 5.2.4}).$$

The isotopic shift of the charge radius $\langle r^{2}\rangle^{1/2}_{^{10}\text{Be}}$ arising from the addition of a neutron to ^{10}Be receives contributions from the $|s_{1/2}\rangle$ and $|(d_{5/2} \otimes 2^{+})_{1/2^{+}}\rangle$ components of the $|\widetilde{1/2^{+}}\rangle$ ground state of ^{11}Be. It leads to the prediction $\langle r^{2}\rangle^{1/2}_{^{11}\text{Be}} = 2.48$ fm, to be compared with the experimental value[86] of 2.44 ± 0.06 fm. Summing up, the values $\epsilon_{\widetilde{1/2}^{+}}$, $\epsilon_{\widetilde{1/2}^{-}}$, and $\langle r^{2}\rangle^{1/2}_{^{11}\text{Be}}$ can be used to assess the accuracy of renormalized NFT.

It would be useful to have a bare NN-potential which, employed in nuclear structure calculations, leads to bare static and time-dependent mean fields whose solutions (single particles and collective modes), interweaved according to the rules of a field theory, for example, NFT (Sect. 2.5.2), accounted for the value of the observables resulting from a "complete" set of probes (see, e.g., Figs. 2.8.1 and 5.2.6). Before being able to do so, the aim at shedding light onto the physics at the basis of new experimental results and that of providing guidance in the quest to achieve a unified picture of nuclear structure and reactions can be carried out in terms of empirical renormalized NFT.

In keeping with the fact that the domain area of QED is all of chemistry and much of biology[87], the important *renormalization effects* undergone by *bare forces* when used to describe even one of the simplest (real) many-body systems – that is, the hydrogen atom in the case of the Coulomb interaction – is paradigmatic.

Because of the importance this issue has in connection with the unified nuclear field theory of structure and reactions we discuss in the present monograph, we also

---

[83] See footnote 23 of Ch. 5.

[84] Barranco et al. (2017).

[85] When the objective of the calculation is not the study of a collective mode in itself, as it was in the case of, e.g., Barranco et al. (2004), but its renormalizing effects on single-particle motion, one adjusts κ (strength of the corresponding separable interaction) so that the RPA (QRPA) result reproduces the observed properties of the collective mode. In such a case, one refers to empirical renormalization (Barranco et al. (2017, 2020); Broglia et al. (2016)).

[86] Nörtershäuser et al. (2009).

[87] Feynman (2006).

attempt to shed light on bare force-renormalization process phenomena through a number of interdisciplinary examples. An important one is provided by dispersive (retarded) forces like the van der Waals interaction (App. 3.B), and also by the Casimir effect and the hydrophobic force (App. 7.E), let alone by the renormalization the Coulomb force undergoes in metals (Sect. 4.8). This is a dressing which in this case changes not only the value but also the sign of the interaction, resulting eventually in the phenomenon of superconductivity.

One can mention that at the basis of the mechanism which inverts, in the odd $N = 7$ isotones ${}^{10}_{3}$Li and ${}^{11}_{4}$Be, the standard Mayer–Jensen sequence of levels $1p_{1/2}$ and $2s_{1/2}$, one finds what likely is the largest Lamb-shift-like phenomenon displayed by a physical system.[88]

1.10 Hindsight

In order that a nucleon, moving in a level close to the Fermi energy, can display a mean free path larger than nuclear dimensions, all other nucleons must move in a rather ordered, correlated fashion. Within this context, to posit that single-particle motion is the most collective of all nuclear motions[89] seems natural. Associated with mean field and single-particle motion, one finds the typical bunching of the corresponding levels closely connected with the major shells lying above and below the Fermi energy, a fact which determines the parity (and angular momentum) of the lowest excitations associated with the promotion of a nucleon across the Fermi surface. By correlating these (p-h) excitations through the same components of the NN-interaction leading to the single-particle bunching, one obtains low-lying collective multipole vibrations and the associated formfactors and coupling strengths to single particles. Similar arguments result in the presence of multipole pairing vibrations in the low-energy spectrum and of the corresponding particle–pair vibration coupling vertices.

It is then natural to consider single-particle motion and collective states on equal footing and as basis states of a physical description of the atomic nucleus in which to diagonalize both three-point (PVC) and four-point (v) vertices. Such an approach provides also an indication concerning the minimum set of experiments needed to obtain a "complete" picture (test) of the low-energy properties of the atomic nucleus. They constitute the specific probes of each of the basis states (elementary modes of excitation), namely, inelastic scattering and Coulomb excitation (particle-hole collective vibrations and rotations), one-particle transfer processes (independent-particle motion), and two-particle transfer reactions (pairing vibrations and rotations).

---

[88] Barranco et al. (2017).
[89] Mottelson (1962).

Summing up, because of the interweaving existing between the variety of elementary modes of excitation, experimental probes associated with fields which carry transfer quantum number $\beta = 0, \pm 1$ and ± 2, and different multipolarities, spins and isospins are needed to characterize the structure of nuclei. This is what we attempt to explain and formulate in the following chapters, giving special emphasis to transfer processes, in which case the relative motion of the reacting nuclei and the intrinsic motion of the nucleons in target and projectile cannot be separated, and one is forced to treat structure and reactions in a unified way.

2

Structure and Reactions

To introduce the quanta of nuclear physics, let us consider, as an example the spectrum of elementary excitations based on the ground state of ^{208}Pb. This "vacuum" state has specially simple properties (has a minimum of degeneracy) on account of the closed-shell configuration of 82 protons and 126 neutrons.

Aage Bohr

In what follows, the connection between the concept of elementary modes of excitation and associated specific probes is discussed in terms of selected experiments, a connection which finds its ultimate test in terms of predicted and measured absolute transition probabilities and differential cross sections.

2.1 Elementary Modes of Excitation and Specific Probes

Subject to external probes which couple weakly to the nucleus, that is, in such a way that the system can be expressed in terms of the properties of the excitation in the absence of probes[1], the nucleus reacts in terms of single-particle (-hole) motion (one-particle transfer), vibrations (surface, spin, etc.), and rotations (Coulomb excitation and inelastic scattering) and pairing vibrations and rotations (two-nucleon transfer reactions), that is, in terms of elementary modes of excitation.

Collective vibrations in nuclei can be characterized by a variety of quantum numbers, in particular angular momentum (J), parity (π), and transfer quantum number (β). Let us consider the doubly magic nucleus $^{208}_{82}$Pb$_{126}$ to illustrate some aspects of the transfer quantum number β. The ground state of this nucleus ^{208}Pb, namely, $|\mathrm{gs}(^{208}\mathrm{Pb})\rangle$, has angular momentum zero and positive parity ($J^{\pi} = 0^{+}$). It can be viewed as the vacuum state of the variety of elementary modes of excitation at the basis of the nuclear field theory (NFT) description of this system. The

---

[1] See, e.g., Pines and Nozières (1966); Bohr and Mottelson (1975), and refs. therein. Within the context of linear response, see also Sect. 7.4, text after Eq. (7.4.1).

lowest-lying collective vibration ($E_x \approx 2.6$ MeV) is an octupole surface vibration. Microscopically, it can be viewed as a correlated particle-hole (p, h) excitation. Consequently, this state is characterized by the quantum numbers $J^\pi = 3^-, \beta = 0$. A neutron moving around ^{208}Pb in levels above the Fermi energy can have quantum numbers $j^\pi = 9/2^+, 11/2^+, 5/2^+, \ldots$ ($2g_{9/2}, 1i_{11/2}, 3d_{5/2}, \ldots$), all of them having $\beta = 1$.

In connection with quantum electrodynamics (Feynman graphical formulation[2]), these two modes (octupole and single-particle modes) parallel the photon and the electron. The nucleus ^{208}Pb, aside from having a rich variety of $\beta = 0$ collective (p, h) correlated modes ("photons") like the $J^\pi = 2^+$ (4.1 MeV), 5^- (3.2 MeV), and the octupole vibration, also displays collective $\beta = \pm 2$ modes. These pairing vibrations can be viewed as correlated two-particle ($\beta = +2$) or two-hole ($\beta = -2$) vibrations. For example, pair addition modes ($\beta = +2$) with quantum numbers $J^\pi = 0^+, 2^+, 4^+, \ldots$ correspond to the ground state and to the lowest $2^+, 4^+, \ldots$ states of ^{210}Pb. In particular, the $|J^\pi = 0_1^+, \beta = 2\rangle \equiv |\text{gs}(^{210}\text{Pb})\rangle$ can be viewed as a nuclear Cooper pair[3]: a weakly correlated pair of fermions moving in time-reversal states lying close to the Fermi energy. Cooper pairs are the building blocks of the microscopic theory of superconductivity developed by Bardeen, Cooper, and Schrieffer (BCS)[4] and of its extension to the nuclear case.[5]

In[6] Figs. 2.1.1–2.1.3, examples of specific reactions which have identified different elementary modes of excitation of ^{208}Pb mentioned above are given. In particular, a cartoon representation of elastic, inelastic, one- and two-particle direct transfer reactions induced by alpha, deuteron, and triton projectiles impinging on ^{208}Pb are shown. In all cases a standard setup is used, in which a light projectile is aimed at a fixed target (thin foil made out of ^{208}Pb). The outgoing particles carrying the corresponding physical information, that is, momentum, angular momentum, energy, and so on, transferred to or from the target, are deflected by the electromagnetic fields of a spectrometer and eventually recorded at a given angle by particle detectors. These events provide structural information as shown in the two-dimensional strength function displayed below the cartoon laboratory setup.

The associated strength functions, recorded at a wide range of angles, provide the absolute differential cross sections associated with each of the nuclear states populated in the process. They are typically measured in[7] milibarns per steradian

---

[2] "The practical usefulness of the Feynman rules and diagrams made them one of the most essential elements of the scientific training of every theoretical physicist" (Mehra (1996)).

[3] See footnote 31 in Ch. 1.

[4] Bardeen et al. (1957a,b).

[5] Bohr et al. (1958).

[6] Mottelson (1976b).

[7] A barn is defined as 1 b = 10^{-28} m^2 = 100 fm^2. In three dimensions, the solid angle Ω is related to the area of the spherical surface A it spans ($\Omega = A/R^2$ sr), in a similar way in which in two dimensions, an angle θ is related to the length L of the circular arc it spans. $\theta = L/R$ rad.

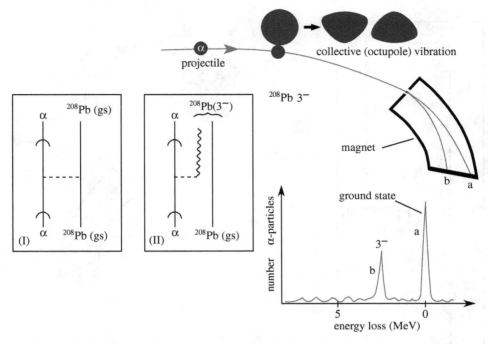

Figure 2.1.1 Schematic representation of: an ***elastic*** process labeled a in both the spectrograph and in the strength function plotted below it (population of the ground state), and ***inelastic*** b (population of the lowest octupole vibration lying at 2.62 MeV) processes associated with the reaction $^{208}\mathrm{Pb}(\alpha, \alpha')^{208}\mathrm{Pb}^*$. In the inset (I) a schematic, nuclear field theory diagram of reaction plus structure (NFT$_{(s+r)}$), describing the elastic process (potential scattering, dashed horizontal line), is displayed. The α-projectile moving in the continuum is represented by an arrowed (curved) line. From the measurement of the elastic differential cross section, one can deduce the partial wave phase shifts (Sect. 4.2). In the inset (II) a schematic NFT$_{(s+r)}$ diagram describing the inelastic excitation (see Fig. 1.1.1 (inset (I))) of the low-lying octupole vibration (wavy line) of $^{208}\mathrm{Pb}$ by the action of the transient field created by the α-particle on the target (horizontal dashed line) is given. Outgoing α particles are deflected in a magnetic spectrograph and recorded in a detector. The corresponding excitation function is given in the lowest part of the figure. For the color version of this figure, refer to cambridge.org/nuclearcooperpair.

(mb/sr). To translate these quantities into nuclear structure information, a model of structure and of reactions is needed to calculate the absolute cross sections, to be compared with the data. The risk of using relative cross section is in overlooking limitations in the description of the reaction mechanism or in that of the structure description of the states involved in the reaction under study – or in both.

In this connection, it is of note that either one sets equal weight in correctly calculating the static and dynamic properties of the single particles and of the collective modes, respectively, and on their interweaving, leading to dressed (renormalized), physical modes, than in working out the reaction mechanism, or the

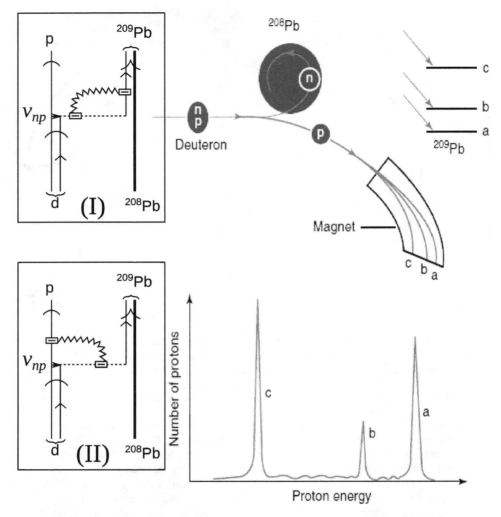

Figure 2.1.2 Schematic representation of the one-nucleon transfer reaction
^{208}Pb$(d, p)^{209}$Pb populating valence single-particle states of ^{209}Pb. In the insets
a schematic NFT$_{(s+r)}$ diagram describing the process is shown. Curved arrowed
lines describe the projectile d (deuteron) and outgoing particle p (proton) moving
in the continuum. The short horizontal arrowed line labeled v_{np} represents the
proton–neutron interaction inducing the transfer process (dashed horizontal line)
while the open dashed rectangle indicates the particle–recoil coupling (PRC) vertex. That is, the coupling of the relative motion to the recoil process described in
terms of a jagged line (App. 2.B). This information is carried out in the center of
mass system by the outgoing particle in the final channel. Within this context the
jagged line is involved in a virtual process (insets (I) and (II)). The energy and
momentum of the outgoing proton reflects the recoil, the Q-value of the reaction
and the angular momentum and excitation energy of the final state as analyzed in
the magnet and recorded in the particle detector (a,b,c). For the color version of
this figure, refer to cambridge.org/nuclearcooperpair.

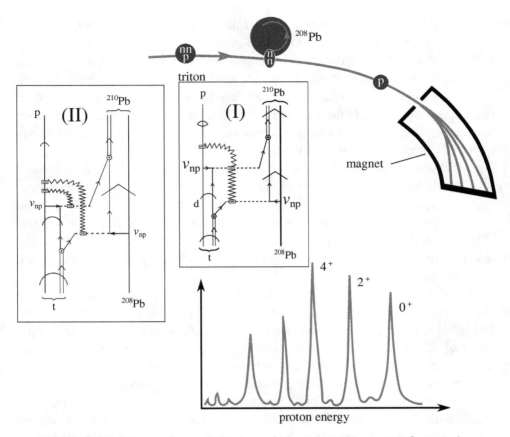

Figure 2.1.3 Schematic representation of the two-nucleon transfer reaction ^{208}Pb$(t, p)^{210}$Pb populating the ground state 0^+, and lowest lying excited states 2^+ and 4^+, that is, monopole, quadrupole, and hexadecapole pair addition modes (multipole pairing vibrations) of ^{208}Pb (App. 7.D; see also Brink and Broglia (2005) Sect. 5.3.1 p. 108). In the inset (I) a NFT$_{(s+r)}$ diagram associated with the (successive) transfer process is displayed. The jagged line brings information to the outgoing particle in the final channel, of the change in scaling in the asymptotic outgoing waves with respect to the incoming ones, concerning the different mass partitions (recoil) of summed value $2m$ (App. 2.B; concerning the apparent non-linearity that is the direct coupling of two recoil modes drawn for simplicity, it can be avoided drawing the process as shown in inset (II); see also App. 2.A). For the color version of this figure, refer to cambridge.org/nuclearcooperpair.

confrontation between theoretical predictions and experimental observation may not be fruitful.[8]

---

[8] *Structure and Reactions.* Within this context one can ask, how does one understand which the correct elements are to describe a reaction process, if one does not know in detail the structure of the initial and final states? In a nutshell, how can one understand reaction without knowing structure (eyes without object)? Vice versa, how can one understand what the elements needed for a correct description of the structure of nuclear states are, if one does not know how to observe them (specific probe) or how to bring that information to the detector? In other words, how can one understand structure without knowing reaction (object without eyes)? The answer to both questions is, likely, that one could find it not particularly simple.

Echoing Heisenberg's requirement[9] that no concept enters the quantal description of a physical system which cannot, in principle, be observable, and Landau's finding that weakly excited states of a quantal many-body system may be regarded as a gas of weakly interacting elementary modes of excitation[10], Bohr, Mottelson,[11] and coworkers developed a unified description of the nuclear structure, in particular, a nuclear field theory (NFT),[12] in terms of Feynman diagrams describing the behavior of quasiparticles, vibrations, and rotations, and their couplings both in 3D[13] and in gauge[14] and other "abstract" spaces, which had close connections with direct nuclear reactions.[15] Within this context Figs. 2.1.1–2.1.3 (insets) display schematic representations of *unified NFT diagrams of structure and reactions (NFT$_{(s+r)}$)*[16], which microscopically describe the variety of structure and reaction processes in terms of Feynman diagrams in a basis of elementary modes of excitation, that is, in the present case in which the target is a closed shell system, particle-hole (inelastic scattering), one-particle (single-particle stripping), and two-particle (Cooper pair transfer) modes.

In inset (I) of Fig. 2.1.1 a diagram describing elastic scattering is shown, while in inset (II) a NFT$_{(s+r)}$ diagram describing the inelastic excitation of the low-lying octupole vibration of ^{208}Pb is displayed. A curved arrow on a line indicates propagation of the projectile in the continuum (asymptotic waves). The horizontal dashed line represents the action of the mean field or of the bare interaction.

In the insets of Fig. 2.1.2 NFT$_{(s+r)}$ diagrams describing the process ^{208}Pb$(d, p)^{209}$Pb are schematically shown. A standard pointed arrowed line indicates the neutron moving with the proton in the deuteron ((curved) arrowed double line) or around the ^{208}Pb core (boldface line). The jagged curve represents the recoil mode coupling the intrinsic and the relative motion, thus accounting for the different mass partition and the associated change in scaling between entrance and exit channel distorted waves, the corresponding momentum mismatch being taken care of by a Galilean transformation (recoil effect). The jagged curve can transfer the information of momentum mismatch to either the residual nucleus (inset (I)) or to the outgoing particle, that is, the proton (inset (II)). Horizontal short arrowed lines

[9] Heisenberg (1949).
[10] Landau (1941).
[11] Bohr (1964, 1976); Bohr and Mottelson (1969); Mottelson (1976a); Bohr and Mottelson (1975); Bohr et al. (1958), and refs. therein.
[12] Bès et al. (1974); Broglia et al. (1976); Bohr and Mottelson (1975); Bès and Kurchan (1990); Mottelson (1976a), and refs. therein. See also Broglia (2020).
[13] Nilsson (1955); Bohr and Mottelson (1975), and refs. therein.
[14] Bohr et al. (1958); Belyaev (1959); Högaasen-Feldman (1961); Bès and Broglia (1966); Bjerregaard et al. (1966); Broglia and Riedel (1967); Bohr and Mottelson (1975).
[15] Alder et al. (1956); Alder and Winther (1975); Broglia and Winther (2004); concerning the general development of direct nuclear reactions, see Austern (1970); Jackson (1970); Satchler (1980); Satchler (1983); Brink (1985); Glendenning (2004); Thompson and Nunes (2009), and refs. therein.
[16] Broglia (1975); Broglia and Winther (2004); Potel et al. (2013a); Broglia et al. (2016).

stand for the proton–neutron (nucleon–nucleon) bare interaction inducing transfer. It is of note that choosing in an appropriate way the (post or prior) representation to describe the reaction process (energy conservation), one can evidence the single-particle mean field or the proton–neutron interaction as inducing the transfer process.

In Fig. 2.1.3 a cartoon representation of the ^{208}Pb$(t, p)^{210}$Pb process is given, while in the inset the corresponding NFT diagram is displayed. The dineutron moving in the triton and around the ^{208}Pb core (pair addition mode) is represented by a double arrowed line. Each individual transferred neutron is indicated with a single arrowed line. The curved arrows on the triton and on the proton indicate motion in the continuum with incoming and outgoing asymptotic waves, respectively. The pointed arrow encompassing the pair addition mode and the core ^{208}Pb indicate intrinsic (structure) motion. In selecting this NFT diagram the assumption was made, following the results of detailed calculations, that the main contribution to the process arises from the successive transfer of the nucleons. Two jagged curves are shown, one connecting the first and the second transfer processes at the particle–recoil coupling vertex (PRCV, dashed open square), in keeping with the fact that the channel ^{209}Pb+d has no asymptotic waves. The second jagged curve emerges from the second PRCV (inset (I)) or from a new one (inset (II)) and carries the mismatch information to the outgoing proton (see App. 2.B, in particular the paragraph after Eq. (2.B.36)).[17]

It is of note that the successive transfer of neutrons described in insets (I) and (II) can be related to the successive tunneling of pairs of electrons involved in the Josephson effect[18]; see Sect. 4.6, in particular the discussion following Eq. (4.6.1).

As discussed in the following chapters, Cooper pairs are extended objects, the fermionic partners being correlated over distances much larger than nuclear dimensions (correlation length $\xi \approx 14$ fm $\gg R_0 \approx 6$ fm ($A = 120$)). Because the single particle potential acts on these pairs as a rather strong external field, this correlation length feature is not obvious in structure calculations, becoming apparent by connecting structure with Cooper pair transfer processes.

Elementary modes of excitation, that is, single-particle and collective motion, are the way nuclei react to external probes and thus are closely connected with the physical observables, namely, absolute transition probabilities and absolute differential cross sections. Within this context, bare elementary modes of excitation already contain an important fraction of nuclear many-body correlations, thus making the diagonalization of the nuclear Hamiltonian leading to dressed elementary

---

[17] Concerning the apparent nonlinearity of the recoil process displayed in inset (I) (two jagged lines associated with a single PRCV), we refer to Fig. 2.A.1 and associated discussion.

[18] Josephson (1962); Anderson (1964b). See also Brink and Broglia (2005), App. L and refs. therein.

modes of excitation a low-dimensional problem. For example, in terms of $10^2 \times 10^2$ matrices, each matrix element contains already much physical insight into nuclear structure at large.[19]

In keeping with the fact that all the nuclear degrees of freedom are exhausted by those of the nucleons, and that the different reactions, that is, elastic, Coulomb, and inelastic excitations, as well as one- and two-particle transfer reactions, project particular but somewhat overlapping components of the total wavefunction, the nuclear elementary modes of excitation give rise to an overcomplete, nonorthogonal, Pauli principle violating basis, concerning both structure and reactions. Nuclear field theory[20] provides the conserving sum rules protocol to diagonalize in this basis the three- and four-point vertices of the nuclear Hamiltonian (Eq. (1.2.10)) to any order of perturbation theory, also infinite if so required for specific processes (see Sect. 2.5). The dressed physical elementary modes resulting from the interweaving of the bare modes are orthogonal to each other and fulfill the Pauli principle, providing a microscopic solution to the many-body nuclear problem.

$NFT_{(s+r)}$ diagrams[21] (see, e.g., Figs. 2.7.2, 2.7.3 (a), (b), and 2.7.5) embody predictions in terms of absolute differential cross sections and transition probabilities, quantities which can be directly compared with the observables obtained by studying the nuclear system with the variety of the ever more precise arsenal of experimental probes (see, e.g., Fig. 2.8.1).

At this point a proviso or two are in place. The original elementary modes of nuclear excitation melt together, due to their interweaving, into effective fields.[22] Each of them displays properties which reflect that of the others with which it couples, their individuality resulting from the actual relative importance of each one of them. What one calls a physically (dressed) particle is only partially to be associated with that particle field alone. It is also partially to be associated with the vibrational fields (surface, density, spin[23], pairing[24], etc. vibrational modes), because they are in interaction through the bare NN-potential as well as the particle–vibration coupling vertices.[25] And conversely, what one calls a nuclear vibration can couple to particle-hole (in the case of a surface vibration), two-particle (in the case of a pair addition), or two-hole (in the case of a pair removal) configurations and materialize in specific fermionic states.

---

[19] Illuminating discussion with B. A. Brown during the fifteenth International Conference on Nuclear Reactions Mechanisms (Varenna, June 2018) regarding this, and related issues, is gratefully acknowledged (RAB).

[20] Bès et al. (1974); Bès et al. (1976a,b,c), Bès and Broglia (1975); Broglia et al. (1976); Bès et al. (1975a); Bohr and Mottelson (1975); Mottelson (1976a); Bès and Broglia (1977); Bortignon et al. (1977); Bortignon et al. (1978); Bès (1978); Broglia and Winther (2004); Reinhardt (1975); Reinhardt (1978a,b); Reinhardt (1980); see also Broglia (2020).

[21] Broglia (1975); Broglia and Winther (2004); Broglia et al. (2016).

[22] Within this context see Dickhoff and Van Neck (2005), and refs. therein.

[23] Bertsch and Broglia (2005), Chs. 6–8.

[24] Brink and Broglia (2005), Ch. 5.

[25] Bohr and Mottelson (1975).

Thus nucleons (fermions) couple to vibrational modes (quasi-bosons) and, eventually, can reabsorb them, returning to the original state. The outcome of such processes, namely, the dressed physical elementary modes of excitation, is closely connected with the renormalization program of quantum electrodynamics (QED)[26] implemented in NFT in terms of Feynman diagrams. Renormalized NFT, that is, (NFT)$_{ren}$, implies that the intermediate, virtual states clothing the elementary modes of excitation should be fully dressed.[27] They can thus, in principle, be taken from experiment.

The specific experimental probes of the bare elementary modes of nuclear excitation reveal only one aspect, in most cases likely the most important one of the physical (dressed) elementary modes. Renormalized NFT reflects the physical unity of low-energy nuclear research requiring the melting not only of elementary modes of excitation but also of structure and reaction theory, let alone of the different experimental techniques developed to study the atomic nucleus, in other words, the need for a "complete" set of experimental probes to reveal the multifaceted properties of dressed elementary modes of excitation.

As seen from the contents of the present monograph, the accent is set at relating theoretical predictions with experimental findings, through the unification of structure and reactions, in particular, the unification of pairing and two-nucleon transfer processes, where the two subjects are blended together. Once the NFT rules to work out the variety of elements (spectroscopic and, with the help of them, reaction amplitudes) have been laid out and/or the pertinent literature referred to, concrete embodiments are provided and eventual absolute cross sections and transition probabilities calculated and confronted with the experimental data.

As already stated in the previous chapter, an essential test theory to pass is connected with some operator identities which are known as sum rules.[28] An overcomplete (nonorthogonal) set of basis states, like that provided by elementary modes of excitation, will violate sum rules. For example, a hole and a pair addition is nonorthogonal to a particle and a surface ($p - h$) vibration (see Sect. 2.5.4). As we shall see, taking into account the couplings between these modes following the rules of NFT, such violations are eliminated and sum rules are fulfilled (Sect. 2.5).

A similar situation is found in the case of, for example, two-nucleon transfer, because the single-particle states in projectile and target are, as a rule, nonorthogonal. Taking into account the corresponding overlaps, one can remove the overcounting (see Sects. 4.1, 6.2.9, and 6.5).

---

[26] Feynman (1975); Schwinger (2001).

[27] Barranco et al. (2017); see also Broglia et al. (2016) and Chs. 6 and 7.

[28] See, e.g., Bohr and Mottelson (1975); Ring and Schuck (1980); Bertsch and Broglia (2005), Ch. 4; Bortignon et al. (1998), Chs. 3 and 8 and refs. therein.

2.2 Sum Rules Revisited

A quantitative measure of the overcompleteness of the elementary modes of excitation basis is provided by the use of exact and of empirical[29] sum rules that the observables (cross sections) associated with the variety of probes to which the nucleus is subject have to fulfill. Examples of the first type are given by the Thomas–Reiche–Kuhn (TRK) sum rule[30,31] and by a sum rule which relates one- with two-particle transfer processes.[32] Within the framework of simple but still physically relevant models, connected with inelastic scattering processes, we refer to Sect. 2.5.2 in general and Eqs. (2.5.40)–(2.5.44) in particular. With regard to one-particle transfer sum rules, we refer to Sect. 2.5.3 and Eqs. (2.5.75) and (2.5.78).

Concerning examples of the second type (empirical, physical sum rules), we refer to sum rules associated with two-nucleon transfer reactions (TNTR)[33] discussed in Sects. 2.2.1 and[34] 7.2.2.

Physically, they provide information concerning (1) the maximum amount of energy which the quantal system can absorb from a beam of photons (γ–rays) shone on it; (2) the total two-nucleon transfer cross section exhausted by the final ($A \pm 2$) states populated in the transfer process.

In other words, these sum rules provide a quantitative measure of the single-particle subspace the quantal system under study, in particular, the nucleus, uses to correlate particle-hole excitations and thus induce the antenna-like motion of protons against neutrons, and also to correlate pairs of nucleons moving in time-reversal states around the Fermi energy, leading to a static or dynamic order parameter in gauge space (Fig. 4.4.2), and resulting in a sigmoidal distribution of the associated pair (ν, $\tilde{\nu}$) level occupancy around ϵ_F (Fig. 1.4.4).

As shown in Sect. 1.7, the TRK sum rule can, in the nuclear case, be written as

$$S(E1) = \sum_{\alpha} \left| \langle \alpha | F | \tilde{0} \rangle \right|^2 (E_\alpha - E_0) = \frac{9}{4\pi} \frac{\hbar^2 e^2}{2m} \frac{NZ}{A}, \qquad (2.2.1)$$

where $|\alpha\rangle$ labels a complete set of excited states populated by acting with the dipole operator F (Eq. (1.7.5)) on the correlated vacuum state $|\tilde{0}\rangle$. Each (quasi)

---

[29] In this connection see paragraph after Eq. (2.B.44).

[30] Bohr and Mottelson (1975), Sect. 6-4.

[31] Bertsch and Broglia (2005), Ch. 3, in particular Sect. 3.3.

[32] Bayman and Clement (1972); Lanford (1977).

[33] Broglia et al. (1972).

[34] Within this context, the absolute two-nucleon transfer cross section populating the ground state of a superfluid nucleus is proportional to the number of Cooper pairs contributing to the nuclear condensate (modulus squared). This quantity is rather stable along a pairing rotational band (see however Fig. 3.1.3 and associated discussion in Sect. 3.1.1), in keeping with the fact that the "intrinsic" $|BCS\rangle$-state of the deformed system in gauge space, is a coherent state and essentially the same for all members of the band. This fact is at the basis of a newly found physical sum rule (Potel et al. (2017)).

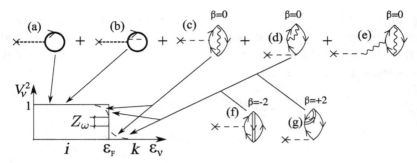

Figure 2.2.1 Schematic representation of the Fermi distribution. The sharp (continuous line) step function schematically represents the Hartree–Fock occupation numbers. The associated nuclear density is measured with the help of an external field (cross attached to a dashed line) through processes of type (a) (Hartree: H) and (b) (Fock: F) is expected to display a diffusivity of the order of the strong force range. Zero-point fluctuations (ZPF) associated with collective particle-hole vibrations, i.e., processes with transfer quantum number $\beta = 0$ and shown in (c), (d), and (e), and with pairing vibrations, i.e., pair removal (graph (f)) and pair addition modes (graph (g)), smooth out the occupation numbers around the Fermi energy (dashed curve; the discontinuity gives a measure of the single-particle content Z_ω) and lead to a nuclear density of larger (dynamical) diffusivity and radius than that associated with HF (see Sect. 1.8). One- and two-particle strengths which in the (mean field) approximation are found in a single A-mass system are, as a result of ZPF ($\beta = 0, \pm1, \pm2$), distributed over a number of nuclei ($A, A \pm 1, A \pm 2$) (see also Sect. 5.8, Fig. 5.8.1).

bosonic elementary mode of excitation provides, within the harmonic approximation, a specific contribution to the total zero-point fluctuations of the ground state, that is,

$$\langle\tilde{0}|F^2|\tilde{0}\rangle = \sum_\alpha \frac{\hbar\omega}{2C_\alpha} = \sum_\alpha \frac{\hbar^2}{2D_\alpha}\frac{1}{\hbar\omega_\alpha}. \tag{2.2.2}$$

As mentioned in the previous chapter, ZPF perturb the static nucleon Fermi sea, that is, the set of occupied levels of the mean field potential (see Fig. 2.2.1), inducing virtual particle-hole excitations (k, i, i.e., $\epsilon_i \leq \epsilon_F$ and $\epsilon_k > \epsilon_F$). Because $|\langle\alpha|F|\tilde{0}\rangle|^2$ measures the probability with which the state $|\alpha\rangle$ is populated, the α-sum in (2.2.1) gives a measure of the maximum energy that the nucleus can absorb from the γ-beam. It is customary to measure $|\langle\alpha|F|\tilde{0}\rangle|^2$ in single-particle (sp) units (Weisskopf (W) units)

$$B_{sp}(E1; j_1 \to j_2) = \frac{3}{4\pi}e_{E1}^2\langle j_1 \tfrac{1}{2} 1 0|j_2 \tfrac{1}{2}\rangle^2 \times \langle j_2|r|j_1\rangle^2,$$

$$\approx \frac{0.81}{4\pi}A^{2/3}e_{E1}^2 \text{ fm}^2 = B_W(E1), \tag{2.2.3}$$

where $(e)_{E1} = (1 - Z/A)e = (N/A)e$ for protons and $(e)_{E1} = -(Z/A)e$ for neutrons, in keeping with the fact that the motion of a nucleon is associated with a recoil of the rest of the nucleus and that the center of mass of the system must remain, in an intrinsic excitation[35], at rest. It is of note that

$$S(E1) = \sum_n |\langle n|F|\tilde{0}\rangle|^2 (E_n - E_0)$$

$$= \sum_{k,i} |\langle k, i|F|\text{gs(MF)}\rangle|^2 (\epsilon_k - \epsilon_i), \qquad (2.2.4)$$

provided $|\tilde{0}\rangle$ contains the ground state correlations mentioned in connection with Eq. (2.2.2) and $|\text{gs(MF)}\rangle$ those associated with HF (see Fig. 1.2.2), in other words, provided

$$|\text{HF}\rangle = |\text{gs(MF)}\rangle = \prod_{i \in occup} a_i^\dagger |0\rangle, \qquad (2.2.5)$$

where $|0\rangle$ is the particle vacuum state $(a_j|0\rangle = 0)$, and $\Gamma_\alpha|\tilde{0}\rangle = 0$, $\Gamma_\alpha^\dagger$ being the creation operator of a dipole (RPA) correlated particle-hole-like mode $(\Gamma_\alpha^\dagger = \sum_{ki} X_{ki}^\alpha a_k^\dagger a_i + Y_{ki}^\alpha (a_k^\dagger a_i)^\dagger)$.[36]

Relation (2.2.4) is a consequence of the fact that $S(E1)$ is proportional to the average value of the double commutator $[[H, F], F]$ in the ground state of the system ($|\tilde{0}\rangle$ or $|HF\rangle$). Because F is a function of only the nucleon coordinates, and assuming $v(|\mathbf{r} - \mathbf{r}'|)$ to be velocity independent, the only contribution to the double commutator arises from the (universal) kinetic energy term of the Hamiltonian. Thus the value (2.2.1) is model independent. In other words, this value does not depend on the correlations acting among the nucleons but on the number of them participating in the motion and on their mass (inertia) as testified by the fact that $\sum_\alpha \hbar\omega_\alpha \left(\frac{\hbar\omega_\alpha}{2C_\alpha}\right) = \sum_\alpha \left(\frac{\hbar^2}{2D_\alpha}\right)$. *It is then not surprising that the TRK sum rule was used in the early stages of quantum mechanics to determine the number of electrons in atoms.*

2.2.1 Empirical Two-Nucleon Transfer Sum Rules

Let us now consider two-nucleon transfer processes. The absolute cross sections associated with the population of final states can be set essentially on equal footing with respect to each other concerning Q-value and recoil effects, with the help of empirically determined global functions[37] (see also Eqs. (2.B.38)–(2.B.40)

---

[35] Bohr and Mottelson (1969).

[36] See footnote 14 in Ch. 1. See also Bertsch and Broglia (2005), Ch. 4, and Brink and Broglia (2005), Ch. 8, Sect. 8.3; Bohr and Mottelson (1975), Sect. 6-5 h. As can be seen from Fig 1.2.3 (inset), in the RPA approximation no scattering vertices (Fig. 1.2.1) are present. Consequently, the coupling between one- and two-phonon states is, within this (harmonic) approximation, not possible.

[37] See Broglia et al. (1972).

and subsequent discussion). In this way, the theoretical absolute cross section associated with, for example, the $A(t, p)A + 2$ population (we assume N to be even) of the nth final state of spin J can be written as

$$\sigma^{(n)}(J, Q_0) = \left| \sum_{j_1 \geq j_2} B^{(n)}(j_1 j_2; J) S(j_1 j_2; J, Q_0) \right|^2, \qquad (2.2.6)$$

where

$$S^2(j_1 j_2; J, Q_0) = \sigma(j_1, j_2; J, Q_0), \qquad (2.2.7)$$

the quantity $\sigma(j_1, j_2; J, Q_0)$ being the absolute two-nucleon transfer cross section associated with the pure two-particle configuration $(j_1, j_2)_J$, while

$$B(j_1 j_2; J) = \left\langle \Phi_{J_f}(\xi_{A+2}) \left| \left[\Phi_{J_i}(\xi_A) \frac{\left[a_{j_1}^{\dagger} a_{j_2}^{\dagger} \right]_J}{\left[1 + \delta(j_1, j_2) \right]^{1/2}} \right]_{J_f} \right\rangle \qquad (2.2.8)$$

is the two-nucleon spectroscopic amplitude, $\Phi_{J_i}(\xi_A)$ being the wavefunction describing the ground state of the initial nucleus, $\Phi_{J_f}(\xi_{A+2})$ that of the final state, ξ labeling the relative radial and the spin intrinsic coordinates. Assuming A to be a closed shell system, and $J = 0$, one can write

$$|0_n^+\rangle = \sum_j c^{(n)}(j, j; J = 0)|j, j; J = 0\rangle, \qquad (2.2.9)$$

where $n = 1, 2, 3, \ldots$ labels the final nucleus states of spin and parity $J^{\pi} = 0^+$ in increasing energy order and $B^{(n)} = c^{(n)}$. Making use of the completeness relation of the coefficients $c^{(n)}$, one obtains

$$\sum_n \sigma^{(n)}(J = 0, Q_0) = \sum_j \sigma(j, j; J = 0, Q_0). \qquad (2.2.10)$$

The above equation parallels (2.2.4), aside from the fact that the Q-value effect is, in connection with (2.2.4), analytically dealt with, while $\sigma(\{Q\})$ is a functional of Q. The complete separation of the relative and intrinsic motion coordinates taking place in, for example, (2.2.4) is in keeping with the fact that in elastic and inelastic processes, the mass partition is equal in both entrance and exit channels. Thus the intrinsic (structure) and the relative motion (reaction) coordinates can be treated separately. This is not the case for transfer processes, both intrinsic and reaction coordinates being interweaved through the recoil process (particle-recoil mode coupling (dashed open square), jagged curves (recoil mode); see, e.g., Fig. 2.1.3 as well as Figs. 2.7.2 and 2.7.3; see also Sect. 2.B.1 of App. 2.B).

A parallel with the discussion carried out in connection with (2.2.3) regarding the TRK sum rule can be drawn defining two-particle transfer units as

$$\sigma_{2pu}^{max}(J, Q_0) = \max\left[\sigma(j_1, j_2; J, Q_0)\right], \tag{2.2.11}$$

where max[ ] indicates that the largest two-particle absolute cross section in the single-particle subspace (*hot orbital*) is to be considered. In this way, one can write the relation (2.2.10) in dimensionless units. Another quite useful, this time exact, two-particle transfer sum rule has been introduced in the literature,[38] which relates two-nucleon stripping and pickup reactions cross sections with single-particle transfer processes.[39]

The above arguments carried out for nuclei around closed shells can, equally well, be applied to the case of open shell nuclei, making use of the corresponding two-nucleon spectroscopic amplitudes[40], in particular, in the case of phase-correlated, independent pair motion, that is, the BCS mean field solution of the pair problem (pairing Hamiltonian). The summed spectroscopic pair transfer amplitudes, each term weighted with $(j + 1/2)^{1/2}$, is given by

$$\alpha_0' = \langle BCS(N + 2)|P'^\dagger|BCS(N)\rangle$$
$$= \sum_j \frac{2j + 1}{2} U_j'(N) V_j'(N + 2), \tag{2.2.12}$$

where[41]

$$P'^\dagger = \sum_j \sum_{m>0} a_{jm}'^\dagger a_{\widetilde{jm}}'^\dagger = \sum_j \sqrt{\frac{2j + 1}{2}} T'^\dagger(j^2(0)) \tag{2.2.13}$$

creates two nucleons in time-reversal states,

$$T'^\dagger(j^2(0)) = \frac{\left[a_j'^\dagger a_j'^\dagger\right]_0}{\sqrt{2}}, \tag{2.2.14}$$

being the $J = 0$ two-nucleon transfer operator. The associated expectation value in the $|BCS\rangle$ ground state is the two-nucleon transfer spectroscopic amplitude

$$B(j^2(0), N \rightarrow N + 2) = \langle BCS(N + 2)|T'^\dagger(j^2(0))|BCS(N)\rangle$$
$$= \sqrt{\frac{2j + 1}{2}} U_j'(N) V_j'(N + 2), \tag{2.2.15}$$

---

[38] Bayman and Clement (1972). In this connection, and within the context of a schematic model, see Eq. (2.5.75) and subsequent discussion.

[39] Lanford (1977).

[40] See Yoshida (1962); see also App. 2 of Broglia et al. (1973).

[41] The primed quantities are the particle creation operators and BCS occupation amplitudes referred to the intrinsic system of reference in gauge space (see Fig. 1.4.1 and Sect. 4.7). See also Potel et al. (2013b).

implying that to create a pair of particles in a quantal state j, it has to be empty in the initial system (i.e., N, probability $U_j'^2(N)$) and has to result occupied in the final nucleus (i.e., $N + 2$, probability $V_j'^2(N + 2)$).

The ZPF associated with pairing vibrations smooth out the sharp HF discontinuity at the Fermi surface (Fig. 2.2.1), the number of pairs in each level participating in this smoothing is $(2j + 1)/2$, their occupancy being measured by the simultaneous, and apparently contradictory, property of being a particle (amplitude V_j') and a hole (amplitude U_j'). In other words, α_0' measures the number of pairs of nucleons (Cooper pairs) participating in the smoothing out of the Fermi surface[42], being the spectroscopic amplitude associated with the population of pairing rotational bands in two-nucleon transfer processes. It is expected that α_0' depends weakly on N and is about conserved along a pairing rotational band. Because $d\sigma(gs(N) \rightarrow gs(N + 2))/d\Omega \approx |\alpha_0'|^2$, conservation is also expected for these absolute cross sections. But in this case, it is a conservation of a physical character and not a strict mathematical relation (see Sect. 7.2.1).

If one finds that at the angle where the $L = 0$, two-nucleon transfer differential cross sections have the first maximum, as a rule close to $0°$ in (t, p) or (p, t) reactions, the strength function is dominated by a single peak, that is, that associated with the ground state transition, and this is so for a number of isotopes differing in mass number by two units, so it can be concluded that one is in the presence of a pairing rotational band (see Figs. 3.1.3 and 3.1.4). This is one of the reasons why (2.2.12) can be viewed as the order parameter of the nuclear superfluid phase and, in keeping with (2.2.13) and (2.2.14), that two-nucleon transfer reactions can be regarded as the specific tool to probe pairing in nuclei.

Because in a finite many-body system like the nucleus, quantal fluctuations in general and those of particle number in particular (pairing vibrations) are much larger than in bulk systems, a dynamic parallel to (the static deformation value) α_0, namely, α_{dyn}, can be defined at profit (see Fig. 3.5.7).[43]

2.3 Fluctuations and Damping

In this section, we discuss some concepts and applications of the particle–vibration coupling mechanism[44] found at the basis of the development of NFT. The Hamiltonian describing a system of independent particles and of collective surface vibrations can be written as[45]

---

[42] Schrieffer (1964); Potel et al. (2017), and refs. therein.

[43] See also Potel et al. (2017).

[44] See footnote 16 in Ch. 1.

[45] That is, essentially (1.2.10) without v, four-point vertex shown in Sect. 2.5 to be necessary for working out the (sum rule) conserving nuclear field theory rules. Within this framework, one can consider the present section as a warm-up to the introduction of NFT, aiming only at providing evidence of the importance of particle vibration coupling PVC renormalization processes.

$$H = H_{MF} + H_{coll} + H_c, \tag{2.3.1}$$

where

$$H_{MF} = T + U \tag{2.3.2}$$

is the mean field Hamiltonian, sum of the single-particle kinetic energy and of the self-consistent potential $U = f(\rho)$, functional of the density. That is,

$$U = U_H + U_x, \tag{2.3.3}$$

where the Hartree potential U_H was defined in Eq. (1.2.8) ($\rho(r) = \sum_{\epsilon_i \le \epsilon_F} |\varphi_i(\mathbf{r})|^2$) and

$$U_x = - \sum_{i(\epsilon_i \le \epsilon_F)} \varphi_i^*(\mathbf{r}')v(|\mathbf{r} - \mathbf{r}'|)\varphi_i(\mathbf{r}) \tag{2.3.4}$$

is the exchange (Fock) potential.

To second order in the particle–vibration coupling strength, one finds[46] that the nucleons move in an ω-dependent, nonlocal dielectric medium, that is,

$$\left(-\frac{\hbar^2}{2m}\nabla_r^2 + U_H(r)\right)\varphi_j(\mathbf{r}) + \int d\mathbf{r}' U_x(\mathbf{r}, \mathbf{r}')\varphi_j(\mathbf{r}'),$$
$$+ (\Delta E_j + i W_j)\varphi_j(\mathbf{r}),$$

which can approximately be written as

$$\left(-\frac{\hbar^2}{2m_k}\nabla_r^2 + U_H''(r) + \Delta E_j'' + i W_j''\right)\varphi_j(\mathbf{r}) = \varepsilon_j \varphi_j(\mathbf{r}),$$
$$\left(U_H'' = \frac{m}{m_k}U \text{ and similarly for } \Delta E'' \text{ and} W''\right), \tag{2.3.5}$$

where $m_k = m\left(1 + \frac{m}{\hbar^2 k}\partial U_x/\partial k\right)^{-1} \approx 0.7m$ is the k-mass[47], while

$$\Delta E_v(\omega) = \lim_{\Delta \to 0} \sum_{\alpha'} \frac{V_{v,\alpha'}^2(\omega - E_{\alpha'})}{(\omega - E_{\alpha'})^2 + (\frac{\Delta}{2})^2} \tag{2.3.6}$$

and

$$W_v(\omega) = \lim_{\Delta \to 0} \Delta \sum_{\alpha'} \frac{V_{v,\alpha'}^2}{(\omega - E_{\alpha'})^2 + (\frac{\Delta}{2})^2} \tag{2.3.7}$$

---

[46] See, e.g., Brink and Broglia (2005); Mahaux et al. (1985), and refs. therein. See also Bernard and Giai (1981).
[47] This is in keeping with the fact that the nonlocal component of the mean field can be parameterized at profit as $0.4E$, where $E = |(\hbar^2 k^2/2m) - \epsilon_F|$ (Perey and Buck (1962); see also Sect. 2.7.1).

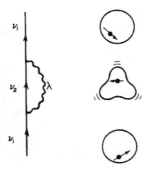

Figure 2.3.1 Self-energy (polarization, PO) graph renormalizing a single-particle (Fig. 1.2.4 (c)). Time ordering leads to the corresponding correlation diagram (CO) (see, e.g., Fig. 1.2.4 (b); within this context see also Brink and Broglia (2005), Fig. 9.2).

are the real and imaginary contributions to the self-energy calculated in second order of perturbation theory, ω standing for a generic value of the single-particle energy[48] (see Fig. 2.3.1; note that in this case, $E_{\alpha'} = \epsilon_{\nu_2} + \hbar\omega_\alpha$, where $\alpha \equiv \lambda$).

For many purposes, ΔE can be treated in terms of an effective mass

$$m_\omega = m(1 + \lambda), \tag{2.3.8}$$

where

$$\lambda = -\frac{\partial \Delta E}{\partial \omega}\bigg|_{\epsilon_F} \tag{2.3.9}$$

is the *mass enhancement factor*, while

$$Z_\omega = m/m_\omega$$

measures the single–particle content (discontinuity) at the Fermi energy.

---

[48] Given a Hamiltonian H_c, the contribution to the energy in second-order perturbation theory can be written as

$$\Sigma_\nu(\omega) = \sum_{\alpha'} \frac{V^2_{\nu,\alpha'}}{\omega - E_{\alpha'}},$$

where $|\alpha'\rangle \equiv |n_\alpha = 1, \nu'\rangle$ are the intermediate states which can couple to the initial single-particle state ν. Note that the expression above is not well defined, in that the energy denominator may vanish. Now, we are not confronted with an accidental degeneracy but with a case where there can be many intermediate states with $E_{\alpha'} \approx \omega$, in other words, where the particle can decay into more complicated (doorway-) states $|\alpha'\rangle$ (Feshbach (1958)), starting in the single-particle level ν of energy ω, without changing its energy (real, on-the-energy shell process). This is a typical dissipative process and has to be solved by direct diagonalization (see Fig. 2.3.4). Another way around is to extend the function $\sum_\nu(\omega)$ into the complex plane ($E_{\alpha'} \to E_{\alpha'} + \frac{i\Delta}{2}$), thus *regularizing the divergence* through a coarse grain approximation, determining the finite contributions, and then taking the limit for $\Delta \to 0$ (Eqs. (2.3.6) and (2.3.7)). The resulting complex potential (*optical potential* from the *complex dielectric function* of optics) parameterizes in simple terms the shift of the centroid of the single-particle state and its finite lifetime.

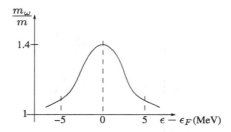

Figure 2.3.2 Schematic representation of the nucleon ω-mass as a function of the single-particle energy measured with respect to the Fermi energy.

Consequently, Eq.(2.3.5) can be rewritten as

$$\left(-\frac{\hbar^2}{2m^*}\nabla_r^2 + U_H' + iW'(\omega)\right)\varphi_j(\mathbf{r}) = \varepsilon_j\varphi_j(\mathbf{r}), \qquad (2.3.10)$$

with

$$m^* = \frac{m_k m_\omega}{m}, \qquad (2.3.11)$$

where $U_H' = (m/m^*)U$ and similarly for W'. Because $\lambda \approx 0.4$ (i.e., the dressed single-particle m_ω is heavier than the bare nucleon, as it has to carry a phonon along or, better, move through a cloud of virtual phonons) and $m_k = 0.7m$, $m^*/m \approx 1$ and $Z_\omega = m/m_\omega \approx 0.7$. Furthermore, due to the fact that $\hbar\omega_\alpha \approx 2\text{–}2.5 MeV$, *the range* of single-particle energies $E = \varepsilon - \varepsilon_F$ over which the particle–vibration coupling process displayed in Fig. 2.3.1 is effective is $\approx \pm 2\hbar\omega_\alpha \approx 4\text{–}5$ MeV around the Fermi energy (see Figs. 2.3.2 and 2.3.3).

It is of note that ΔE_j indicates the shift of the centroid of the "dressed" single-particle state due to the coupling to the doorway states, while $\Gamma = 2W$ measures the energy range over which the single-particle state spreads due to this coupling (see Fig. 2.3.4). While a large number of states contribute to ΔE ("off-the-energy-shell process," i.e., intermediate, virtual processes in which energy is not conserved), only "on-the-energy-shell processes," that is, processes which conserve energy, contribute to Γ. In fact,

$$\lim_{\Delta \to 0} \frac{\Delta}{(\omega - E_{\alpha'})^2 + \left(\frac{\Delta}{2}\right)^2} = 2\pi\delta(\omega - E_{\alpha'})$$

and

$$\Gamma(\omega) \approx 2\pi\bar{V}^2 N(\omega), \qquad (2.3.12)$$

where $\bar{V}^2$ is the average value of $V_{\nu,\alpha'}^2$, while

$$N(\omega) = \sum_{\alpha'}\delta(\omega - E_{\alpha'}), \qquad (2.3.13)$$

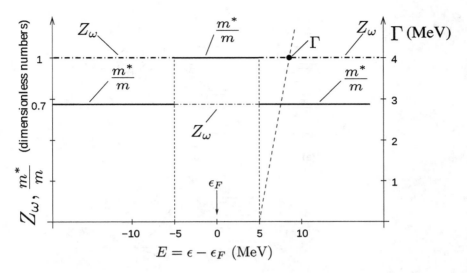

Figure 2.3.3 Schematic representation of the behavior of m^*/m, $Z_\omega = (m_\omega/m)^{-1}$ and Γ (Eq. (2.3.20)) as a function of $E = \varepsilon - \varepsilon_F$. Note the scales for Z_ω and m^*/m (left) and Γ (right).

is the density of energy-conserving states α'. Eq. (2.3.12) is known as the *Fermi golden rule*.

Assuming the distribution of single-particle levels is symmetric with respect to the Fermi energy,

$$\Delta E_\nu(\omega) = \lim_{\Delta \to 0} \sum_{\alpha'} \frac{V_{\nu,\alpha'}^2(\omega - E_{\alpha'})}{(\omega - E_{\alpha'})^2 + \left(\frac{\Delta}{2}\right)^2} = 0, \qquad (2.3.14)$$

as there are equally many states pushing the state down than up in energy.

In the above discussion, the imaginary potential was introduced as an approximation to the breaking of a stationary state into many, more complicated stationary states through the coupling to doorway states (Fig. 2.3.4(b)). This is the correct picture to describe the coupling of a nucleon moving in a single-particle state with more complicated configurations[49,50]. However, such a description can become quite involved. On the other hand, to account for the change in the centroid energy

---

[49] See, however, Caldeira and Leggett (1981, 1983), and refs. therein.

[50] To be noted that if we spread the strength of a stationary quantal state in a number of doorway stationary states over an energy range Γ (of the order of few MeV in the case of the GDR, see Sects. 1.5 and 5.5.1), and set all components in phase at $t = 0$, they will essentially be out of phase at time $t = \tau = \hbar/\Gamma$. In other words, each component will behave independently of others, and the original correlated state, created at $t = 0$ with probability 1, essentially ceases to exist at $t = \tau$. This does not imply that each of the incoherent members of the original coherent state cannot, e.g., γ-decay at a much later stage ($\Gamma_{\gamma_0}/\Gamma \lesssim 10^{-2}$); see Bortignon et al. (1998). RAB acknowledges discussions with B. Herskind on this issue.

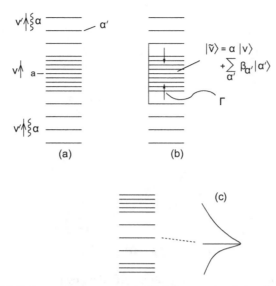

Figure 2.3.4 (a) Schematic representation of the result of the diagonalization of H_c in a basis consisting of the single-particle states $|\nu\rangle$ and the $|\alpha'\rangle = |\nu', \alpha\rangle$ door-way states. (b) Range of doorway mixing leading to the dressed single-particle state $|\tilde{\nu}\rangle$. In (c) we show a situation where there are more states $|\alpha'\rangle$ above $|a\rangle$ than below.

and the spreading width of a single-particle state in terms of an *optical potential* $(\Delta E_\nu + i W_\nu)$ is very economic. In this case, Γ measures the range of energy over which the "pure" single-particle state $|a\rangle$ spreads due to the coupling to the more complicated doorway states $|\alpha'\rangle$. In other words, a stationary state

$$\varphi_\nu(\mathbf{r}, t) = e^{i\omega t} \varphi_\nu(\mathbf{r}), \tag{2.3.15}$$

has a probability density

$$\int d\mathbf{r} |\varphi_\nu(\mathbf{r}, t)|^2 = \int d\mathbf{r} |\varphi_\nu(\mathbf{r})|^2 = 1, \tag{2.3.16}$$

which does not depend on time. That is, if at $t = 0$, the probability that the particle is in the state ν is 1, it will also have this probability at $t = \infty$, implying an infinite lifetime. If, however,

$$\hbar\omega = \varepsilon_\nu^{(0)} + \Delta E_\nu(\omega) + i\frac{\Gamma_\nu}{2}(\omega), = \varepsilon_\nu + i\frac{\Gamma_\nu}{2}(\omega), \quad (\varepsilon_\nu = \varepsilon_\nu^{(0)} + \Delta E_\nu; \Gamma_\nu = 2W_\nu),$$

$$\varphi_\nu(\mathbf{r}, t) = e^{i\frac{\varepsilon_\nu t}{\hbar}} e^{-\frac{\Gamma_\nu t}{2\hbar}} \varphi_\nu(\mathbf{r}),$$

$$\int d\mathbf{r} |\varphi_\nu(\mathbf{r}, t)|^2 = e^{-\frac{\Gamma_\nu t}{\hbar}}, \tag{2.3.17}$$

implying a lifetime for the single-particle state given by

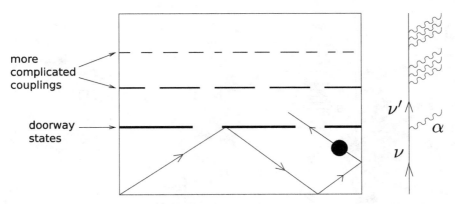

Figure 2.3.5 Schematic representation of the different levels of couplings leading to the damping of a single-particle state. It is essentially the first doorway coupling which controls the probability the ball (black dot), reflecting elastically on the walls of the box has to return to the first compartment (initial state) once it left it. To the right, the coupling of the nucleon, moving originally in the state $|\nu\rangle$ (arrowed line), with an increasing number of phonons $|\alpha\rangle$ (wavy lines).

$$\tau = \hbar / \Gamma_\nu. \tag{2.3.18}$$

One may ask, how is it possible that the coupling to complicated (but still simple) states like $|\alpha'\rangle = |n_\alpha = 1, \nu'\rangle$ made out of a nucleon in the state ν' and a one phonon state of quantum numbers α can explain the damping of a single-particle state lying 8–10 MeV from the Fermi energy ε_F, where the density of levels is expected to be conspicuous? This is because the Hamiltonian (2.3.1) contains the basic physics needed to describe the dressed single-particle motion as far as collective surface modes are concerned, modes which, as a rule, play the central role in the dressing of the single-particle motion in nuclei.[51] Couplings to more complicated states go through a hierarchy of processes. In other words, the variety of couplings should first go through the coupling to states of type $|\alpha'\rangle$, which act as proper doorway states (see Fig. 2.3.5).[52] Summing up, in the nuclear case, the *doorway coupling provides the basic mechanism to break the single-particle strength, while, as a rule, higher-order couplings essentially fill in valleys* (see Fig. 2.3.6).

In the case of the bound $2s_{1/2}$ proton orbital[53] of ^{40}Ca ($\varepsilon - \varepsilon_F = -8$ MeV), simple estimates[54] lead to $\bar{V}^2 \approx 0.3 MeV$ for the coupling to an $L = 2$ phonon, and $N(\epsilon_F) \approx 2$ MeV^{-1}. Consequently, making use of Eq. (2.3.12), one obtains

[51] Something which was learned in the study of the nuclear spectrum in nuclei along the stability valley, and which has been found to remain correct also in connection with exotic nuclei, in particular, light halo exotic nuclei regarding the phenomenon of parity inversion.

[52] Feshbach (1958).

[53] See footnote 118 of this Chapter.

[54] Mahaux et al. (1985); Mougey et al. (1976).

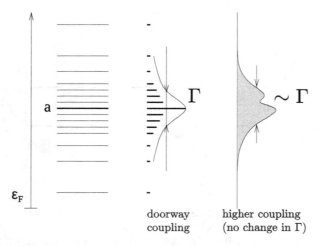

doorway   higher coupling
coupling   (no change in Γ)

Figure 2.3.6 Schematic representation of the breaking of a single-particle state $|a\rangle$ (heavy black horizontal line) through the coupling to doorway states ($|\alpha'\rangle = |n_\alpha = 1, \nu'\rangle$; thin horizontal lines) and eventually to increasingly more complicated (many-particle)–(many-hole) configurations.

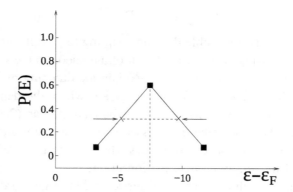

Figure 2.3.7 Schematic representation of the experimental strength function (solid squares) associated with the bound $2s$ proton state of ^{40}Ca. Also indicated is the full width at half maximum (FWHM).

$$\Gamma \approx 3.8 \text{ MeV}, \qquad\qquad (2.3.19)$$

in overall agreement with the experimental findings (see Fig. 2.3.7).

The estimate given in Eq. (2.3.19) is a particular example of the general (empirical) result (see Fig. 2.3.3)[55]

---

[55] Bertsch et al. (1983).

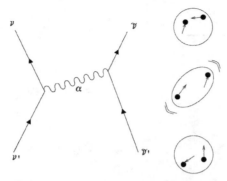

Figure 2.3.8 Schematic representation of the exchange of a (quadrupole) surface phonon between nucleons.

$$\Gamma_{sp}(E) = \begin{cases} 0.5|E| & |E| > 5\ \text{MeV} \\ 0 & E \leq 5\ \text{MeV}, \end{cases} \tag{2.3.20}$$

where

$$E = |\varepsilon - \varepsilon_F|. \tag{2.3.21}$$

2.3.1 Induced Interaction

A nucleon close to the Fermi energy, which, by bouncing inelastically off the nuclear surface, excites a collective mode, moving in the process to another, or remaining in the same state (Fig. 1.2.1), has no other choice, within this particle vibration scenario, than to continue in such a state or to reabsorb the vibration at a later instant of time (self-energy; Fig. 2.3.1). In the presence of another nucleon, the collective vibration excited by one nucleon may be absorbed by the second one (Fig. 2.3.8), the exchange of a vibration leading to an (induced) interaction.

Simple estimates of this induced interaction lead, in the case of ^{210}Pb, that is, two neutrons above the $N = 126$ closed shell, to correlation energies for pairs of particles moving around it and coupled to angular momentum $J^{\pi} = 0^{+}$ of ≈ -1.5 MeV (see Fig. 2.3.9), when α is summed over the different multipolarities ($\lambda^{\pi} = 2^{+}, 3^{-}, 5^{-}$; label α in Fig. 2.3.8), a value not far from the experimental correlation energy of $|^{210}\text{Pb}(gs)\rangle$ (see Fig. 3.5.1, Eq. (3.5.2), and text following it, as well as footnote 50 of Ch. 3; see also text after Eq. (3.5.5)).

It is about half the above value if one takes into account the fact that the single-particle content for each of the interacting particles is about $Z_{\omega} \approx 0.7$. From this result one can conclude that the pairing interaction induced by the process depicted

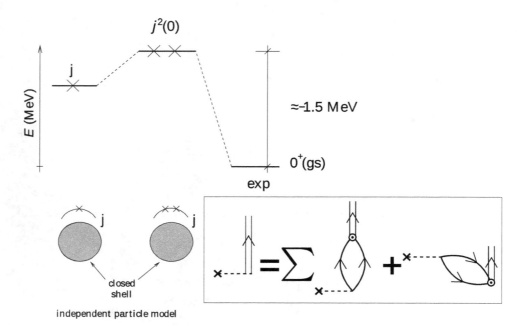

Figure 2.3.9 Schematic representation of the predictions of the independent particle model for one and two particles outside a closed shell, in comparison with the experimental findings (e.g., for the case of ^{210}Pb, where $j = g_{9/2}$). In the inset, the graphical RPA dispersion relation of the monopole pair addition mode of ^{208}Pb, i.e., $|^{210}\text{Pb(gs)}\rangle$ is shown (see Sect. 3.5.2, Eqs. (3.5.8)–(3.5.10)).

in Fig.2.3.8 renormalizes in an important way the bare, $NN(^1S_0)$–short range pairing interaction. This issue is taken up in more detail later on, in particular in connection with exotic halo nuclei (see, e.g., Fig. 2.7.1 and Sect. 3.6). Concerning nuclei lying along the stability valley, see the next section.

2.4 Well-Funneled Nuclear Structure Landscape

Collective vibrations are composite modes. Being made out of fermions (nucleons), they contain in their microscopic structure a consistent fraction of the nuclear correlations. One is then confronted to deal, in working out their interweaving through three- (PVC) and four- (v) point vertices, with an overcomplete set of states and, as a consequence, the need to correct for eventual Pauli principle violations and nonorthogonality contributions.

The rules to do so have been cast into a graphical field theory, namely, the nuclear field theory (NFT). In it, the free fields are to be calculated in the HF (HFB) approximation (particle (quasiparticle)) and in the RPA (QRPA) (vibrations). These elementary modes of excitation interact through the particle–vibration coupling vertices, while particles can also interact through four-point vertices (NN-bare interaction).[56]

A similar situation is found in the case of transfer processes in general and of two-nucleon transfer in particular. One can work out the associated transfer amplitude by orthogonalizing, making use of second-order perturbation theory, the single-particle wavefunctions of target and projectile, in other words, correct the corresponding simultaneous and successive transfer amplitude expressions for nonorthogonality to the order of perturbation needed. This can be done within the semiclassical approximation[57] or the DWBA.[58]

The NFT rules for evaluating the effect of the couplings between fermions and (quasi) bosons involve a number of restrictions concerning initial and intermediate states as compared with the usual rules of perturbation theory. This is in keeping with the fact that the collective modes contain, from the start, the correlations arising from forward and backward going particle–hole ($\beta = 0$) as well as particle–particle ($\beta = +2$) and hole–hole ($\beta = -2$) bubbles, where β is the transfer quantum number.

The general validity of NFT rules has been demonstrated by proving the equivalence existing, to each order of perturbation theory, between the many-body finite nuclear system propagator calculated in terms of Feynman diagrams involving only the fermionic degrees of freedom that is, explicitly respecting the Pauli principle in a complete and not overcomplete basis, also known as the Feynman–Goldstone propagator, and the propagator constructed in terms of Feynman diagrams involving fermion and vibrational degrees of freedom (NFT Feynman diagrams) in the case of a general two-body interaction and an arbitrary distribution of single-particle levels.[59]

Concerning the actual embodiment of NFT, one can recognize the practical difficulties of respecting the corresponding rules. This is because, at present, there is not a bare, well-behaved NN-force with which it is possible to generate a mean field, to determine the single-particle states (sp) and, by introducing a periodic time-dependence, calculate the variety of collective modes which, interweaved with (sp), lead to renormalized states, providing an overall account of the data.

On the other hand, implementation of the NFT rules (renormalization)[60] has been carried out making use of the Skyrme force[61] SLy4 to determine the

---

[56] In connection with the reaction processes one finds again four-point (e.g., the proton–neutron interaction in, e.g., the (p, d) reactions) and three-point vertices (e.g., particle–recoil coupling vertices and mean field–nucleon interaction).

[57] See Sects. 6.1 and 6.5 and Sect. 2.B.1 of App. 2.B.

[58] Sects. 4.2 and 6.1; see also the discussion connected with the tunneling Hamiltonian (4.6.1), in connection with weak coupled metallic superconductors and the Josephson effect.

[59] Bès and Broglia (1975) and Bès et al. (1976a); see also Baranger (1969) and the lecture notes of McFarlane (1969).

[60] See, e.g., Broglia et al. (2016), and refs. therein.

[61] Chabanat et al. (1997).

mean field and spin vibrational channels of separable multipole–multipole forces with self-consistent coupling constants to calculate the variety of density vibrational channels and of the Argonne v_{14} potential to describe the bare pairing interaction.

The resulting structure predictions can, as a rule, together with the corresponding reaction computational codes, in particular COOPER and ONE (see App. 7.A), provide an overall account of "complete" sets of experimental data, obtained with the help of Coulomb, inelastic, γ-decay, and one- and two-nucleon transfer data, which extensively maps out the nuclear structure and reaction landscape[62], as discussed below.

Similar but more accurate results are obtained making use of renormalized NFT, which parallels renormalized QED (see, e.g., Sects. 5.2.2–5.2.3; see also Sect. 5.3.3 and Table 5.3.1), namely, (1) adjusting the bare potential parameters to be used together with a k-mass (similar to the renormalization of the bare mass in QED); (2) making use of empirical values of the deformation parameters (PVC vertices) and phonon energies (parallel to renormalization of the bare charge).[63]

2.4.1 The Island of Open-Shell Superfluid Nuclei $^{118,119,120,121,122}Sn$

In what follows we confront theoretical predictions with experimental observations, concerning one- and two-particle transfer reactions together with an essentially "complete" set of other observables for the open-shell nucleus ^{120}Sn, lying along the stability valley. From this example one can conclude that (1) it is possible to predict, with few free parameters, most of them strongly constrained by empirical input, the experimental findings within a 10 percent level of accuracy; (2) the nuclear landscape, as it emerges from NFT based on elementary modes of excitation and of their interweaving, is well funneled (Fig. 2.4.1; see also Table 2.4.1).[64] Its minimum essentially coincides with the global minimum resulting from the empirical renormalization choice of basic quantities (m_k, strength bare pairing, properties of few low-lying collective ($p - h$) modes). An important proviso concerning the above parlance is that one considers a group of homogeneous nuclei as, for example, open-shell spherical superfluid nuclei (like the Sn isotopes)[65] or nuclei around closed shells (like ^{208,209,210}Pb, ^{209}Bi, ^{210}Po, etc., or ^{10,11,12}Be, ^{9,10,11}Li, etc.).[66]

---

[62] Idini et al. (2015, 2014); Potel et al. (2013a).

[63] Barranco et al. (2017, 2020). See also Sects. 5.2.2–5.2.3.

[64] For details, see Idini et al. (2015).

[65] Within this context we refer to the conclusions of Idini et al. (2015).

[66] In connection with light exotic halo nuclei we refer to Sect. 2.7 of the present chapter and to Sects. 5.2.3 and 7.1.

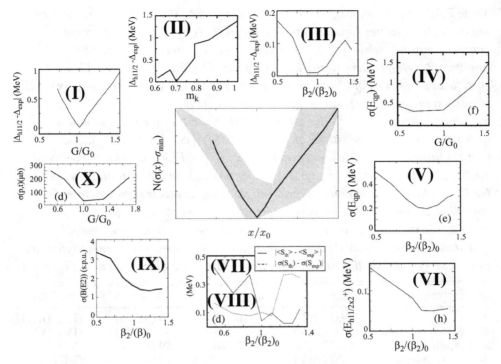

Figure 2.4.1 Root-mean-square deviations $\sigma(x)$ between theoretical predictions and experimental values of the different structural properties which characterize the open-shell nucleus ^{120}Sn (see text as well as Table 2.4.1 for the characterization of (I)–(X); see also Fig. 2.8.1); (central figure) the bold black line and gray area provide a schematic representation of $\langle\sigma\rangle$ and associated fluctuations, respectively.

Let us now comment on Fig. 2.4.1. In it, the root mean square deviations $\sigma(x)$ (see also Table 2.4.1) between theoretical predictions and experimental values of the different structural properties which characterize ^{120}Sn are displayed. The calculations involve the island of open-shell superfluid nuclei ^{118,119,120,121,122}Sn. The root mean square deviations are displayed as a function of the pairing coupling constant G (referred to the employed value $G_0 = 0.22$ MeV), the k-mass m_k $((m_k)_0 = 0.7m)$, and the dynamical quadrupole deformation parameter β_2 $((\beta_2)_0 = 0.13)$, that is, in general, as a function of x (x_0), measured with respect to the minimum value $\sigma_{min} = \sigma(x_{min})$, displayed in the interval $0.5 \leq x/x_0 \leq 1.5$ and normalized according to $0 \leq N(\sigma(x) - \sigma_{min}) \leq 1$.

The curves represent the deviation from experiment of the pairing gap associated with the $h_{11/2}$ orbital $(\Delta_{h_{11/2}}(G/G_0)$ (I); $\Delta_{h_{11/2}}(m_k/(m_k)_0)$ (II); $\Delta_{h_{11/2}}(\beta_2/(\beta_2)_0)$ (III); the deviation of the quasiparticle spectrum $(E_{qp}(G/G_0))$ (IV); $E_{qp}(\beta_2/(\beta_2)_0)$ (V); the deviation of the $h_{11/2} \otimes 2^+$ multiplet splitting

Table 2.4.1 *Root-mean-square deviation σ between the experimental data and the theoretical values expressed in keV for the pairing gap, quasiparticle energies, multiplet ($h_{11/2} \otimes 2^+$) splitting, centroid, and width of the $5/2^+$ low-lying single-particle strength distribution (Fig. 5.2.3) associated with the structural properties of ^{120}Sn. In single-particle units B_{sp} for the γ-decay (B(E2) transition probabilities) and in mb for $\sigma_{2n}(p, t)$ (Fig. 7.2.1). In brackets, the ratio $\sigma_{rel} = \sigma/L$ between σ and the experimental range L of the corresponding quantities, is given. The values of L are: 1.4 MeV (Δ), 1 MeV (E_{qp}), 700 keV (mult. splitting), 1 MeV ($d_{5/2}$ centroid), 809 keV (= 1730–921) keV; $d_{5/2}$ width), 10 B_{sp} (B(E2)), 2250 μb ($\sigma_{2n}(p, t)$)). Columns 2, 3, and 4 contain the results of NFT calculations making use of bare single-particle levels from Hartree–Fock with Sly4. Same, but for a 600keV shift toward the Fermi energy of the $\epsilon_{d5/2}$ orbital, and optimal values of ϵ_j for all valence levels so that the dressed quasiparticle states provide the best fit to that data. In the last column reference to the corresponding diagrams shown in Fig. 2.4.1 are given. See also Fig. 2.8.1.*

Observables	SLy4	$d_{5/2}$ shifted	Opt. levels	Fig. 2.4.1
Δ	10 (0.7%)	10 (0.7 %)	50 (3.5 %)	**(I), (II), (III)**
E_{qp}	190 (19%)	160 (16%)	45 (4.5 %)	**(IV), (V)**
Mult. splitt.	50 (7%)	70 (10%)	59 (8.4 %)	**(VI)**
$d_{5/2}$ strength (centr.)	200 (20%)	40 (4%)	40 (4%)	**(VII)**
$d_{5/2}$ strength (width)	160 (20%)	75 (9.3%)	8 (1%)	**(VIII)**
B(E2)	1.4 (14%)	1.34 (13%)	1.43 (14%)	**(IX)**
$\sigma_{2n}(p, t)$	40 (2%)	40 (2%)	40 (2%)	**(X)**

$E_{h_{11/2}\otimes 2^+}(\beta_2/(\beta_2)_0)$ **(VI)**; the deviation of the centroid position of the $d_{5/2}$ strength function $S_{d_{5/2}}(\beta_2/(\beta_2)_0)$ **(VII)**; the deviation of the width of the $d_{5/2}$ strength function $S_{d_{5/2}}(\beta_2/(\beta_2)_0)$ **(VIII)**; the deviation of the quadrupole transition strength $B(E2)(\beta_2/(\beta_2)_0)$ **(IX)**; the deviation of the two-neutron transfer cross section $\sigma_{2n}(p, t)$ (G/G_0) **(X)**. For an overview, see Fig. 2.8.1. The basic feature displayed by Fig. 2.4.1 is the fact that, in spite of the fluctuations of the results typical of finite many-body systems, they clearly define a funnel in which all minima fall within a narrow window of x/x_0 values (1 ± 0.2).

This can be considered an emergent property of a description of structure and reactions carried out in a basis of elementary modes of excitation interacting, according to the NFT rules, through the NN-interaction as well as through PVC and PRC vertices.

The concept of a well-funneled energy landscape is easy to understand in the case in which the number of particles $N \to \infty$ (thermodynamic limit). For example, a swing has a very simple and well-funneled potential energy landscape. A similar concept which may still retain the (semi) classical viewpoint, but now referred to the free energy of large molecules, has been used in an attempt to describe protein folding.[67] One has hypothesized that the results of all-atom, explicit solvent classical molecular dynamic simulations can be interpreted in terms of a somewhat rugged but still well funneled free energy landscape.

When we see such behavior in the nuclear case, even not so well defined, and somehow imperfect, we recognize that the nucleus is, after all, not macroscopic.[68] Concepts strictly valid for $N \to \infty$ are strongly renormalized by quantal finite size effects, in particular zero-point fluctuations (ZPF; Fig. 2.4.2).[69]

On the other hand, it is likely that the "imperfect" nature of the nuclear structure landscape funnel, an example of which is shown in Fig. 2.4.1, embodies more accurately the physics of quantal many-body systems than a smooth, more pedagogical construct essentially based on potential energy, even with the entropic contribution (free energy). This is in keeping with the fact that the interaction terms (potential energy) of the Hamiltonian contain the last vestiges of Newton's conception of force or causation, being thus too much anchored to classical mechanics.[70]

Within this context we refer to Fig. 2.4.2 and to the fact that the ground state (nuclear vacuum) contains the physics of the system in terms of virtual processes. This is in keeping with the fact that acting on the system with the variety of probes available in the laboratory, one obtains as on-shell final states, whose properties can eventually be observed with the help of the appropriate detectors, the variety of dressed (renormalized) elementary modes of excitation of the nucleus under study. In particular, diagrams (b), (c) of Fig 2.4.2 describe inelastic and two-nucleon transfer processes, while diagrams (d)–(f) portray one-nucleon transfer reactions.

In all orders in the particle–vibration coupling vertex ((PVC), solid dot), starting from second order (graph (f)), *NFT diagrams take care of the Pauli principle acting between the quasiparticles considered explicitly (continuous curves) and*

---

[67] Wolynes (2016); Wolynes et al. (2012); see also the cover figure of the issue of *PNAS* in which this reference was published.

[68] Anderson (1972).

[69] This nuclear result, in turn, may be used at profit to shed light into the physics which is at the basis of protein folding. These molecules are after all quantal systems, and the associated quantal fluctuations are likely to play an important role in the associated fleeting transition states. The fact that average values and shapes of the transit-time distributions agree well with the simplest one-dimensional theory may only reflect the large uncertainties of the tunneling probabilities, small changes in the barrier's parameters compensating for the lack of quantal effects, let alone the fact that being most high-dimensional models, as well as real processes, "sloppy," their behavior depend on very few parameters (collective variables (CV); see Buchanan (2015), Transtrum et al. (2015), and references therein). It is unlikely that one or few of them are not related to quantal fluctuations. This is particularly true if one considers a NMR chemical shift biased molecular dynamic simulation, in keeping with the fact that NMR, based on the precession of nuclear spins is by its essence, a quantal phenomenon. Within this context one is reminded of α- and exotic-decay (see, e.g., Ch. 7 of Brink and Broglia (2005), and refs. therein). While setting the formation probability of an alpha particle equal to 1 can be compensated by modest changes in the associated tunneling probability (barrier), this is not possible in the case of, e.g., exotic decay.

[70] Born (1948), pp. 95, 103; Pais (1986), p. 258.

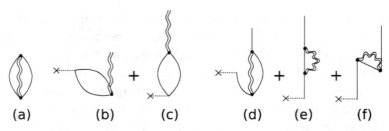

| (a) | (b) | (c) | (d) | (e) | (f) |

Figure 2.4.2 Schematic representation of the NFT$_{(s+r)}$ diagrams at the basis of the characterization of a superfluid nucleus like ^{120}Sn. (a) Nuclear structure (NFT$_{(s)}$). Zero-point fluctuations (ZPF) characterizing the nucleus ground state. Continuous lines describe quasiparticle (*qp*) states; double wavy curves stand for correlated two quasiparticle (*2qp*) vibrational modes. Because $\alpha_\nu^\dagger = U_\nu a_\nu^\dagger - V_\nu a_{\bar\nu}$, these modes encompass both particle-hole (*ph*) like vibrations, e.g., surface quadrupole vibrations, and correlated (*pp*) and (*hh*) monopole and multipole pairing vibrations. Intervening with an external field (cross followed by dashed line), one can excite (b) and (c) multipole (*ph*-like; inelastic scattering) and pairing (*2p*-like, *2h*-like) vibrations (two-particle transfer). Graphs (d)–(f) describe possible one-quasiparticle transfer processes.

those participating in the modes (double wavy lines), as well as between modes. As a consequence, self-energy processes based on pure or little collective two quasiparticle excitations are correctly screened out.

2.5 Nuclear Field Theory for Pedestrians

The diagrams were intended to represent physical processes and the mathematical expressions to describe them.... I would see electrons going along, being scattered at one point ...emitting a photon and the photon goes over there.... I thought that if they really turn out to be useful it would be fun to see them in the pages of *Physical Review*.

R. P. Feynman

Nuclear field theory (NFT) was tailored after Feynman's graphical version of quantum electrodynamics (QED).[71] It is then natural that in discussing NFT, analogies with QED are recurrent. Arguably, as a consequence of special relativity, which put an end to the concept of ether, the field-free and matter-free vacuum was rightly considered as *bona fide* empty space. The advent of quantum mechanics changed this situation, the vacuum becoming populated.

In quantum mechanics, an oscillator, for example, cannot be at rest. The oscillatory nature of the radiation field requires zero-point fluctuations (ZPF) of the electromagnetic fields in the vacuum state. The occupation of the negative

[71] Bès and Broglia (1977), with kind permission of Società Italiana di Fisica. Copyright 1977 by the Italian Physical Society.

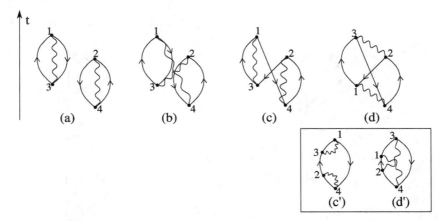

Figure 2.5.1 Oyster diagrams describing the correlation of the nuclear ground state associated with the ZPF of collective particle-hole-like excitations. In (a) we show two such diagrams. In (b) and (c) we display a symmetrized (boson exchange), and antisymmetrized (fermion exchange) correction to (a), while (d) contains a simultaneous boson and fermion exchange. In all the diagrams shown, only ground state correlation vertices are present. They are connected with the Y_{ki}^{α}-components ($\epsilon_k > \epsilon_F$, $\epsilon_i \leq \epsilon_F$) of the RPA (QRPA) wavefunction describing the collective mode (wavy line). While this is so for any time ordering, i.e., the sequence with which the particle–vibration coupling vertex (black dots) appears in the case of the processes shown in (a) and (b), this is not the case in connection with processes shown in (c) and (d), as can be seen from the corresponding diagrams (c') and (d') shown in the inset, the reason for which one has numbered the four PVC vertices. Because of the Pauli principle between particles (holes) explicitly present and those involved in the collective modes, the harmonic approximation has to be corrected. This is diagrammatically reflected by the emergence of scattering vertices.

kinetic energy electron states and the subsequent calculation of the cross section for pair creation by photons contributed another step in the understanding of the QED vacuum, let alone the Lamb shift.[72]

When the fields are expressed in terms of creation and annihilation operators, the interaction between fermion and boson fields is proportional to the product of two fermion creation or destruction operator $a^{\dagger}$ or a, and of one boson operator $\Gamma^{\dagger}$ or Γ, for example, $a_{\nu'}^{\dagger} a_{\nu} \Gamma_{\alpha}^{\dagger}$, (see Fig. 2.5.1), *that is, bilinear in the fermion fields and linear in the (quasi-) boson fields.*

A detailed graphical NFT treatment of the vacuum ZPF has an important consequence concerning the probing of nuclear structure with reactions. By intervening these ZPF with an external field, one will excite the modes whose properties can be compared with experimental findings. In other words, if one is in doubt of which are the properly dressed elementary, physical modes of excitation, one should find

---

[72] In high-energy collisions and accelerator laboratories, some of the original beam energy can be consumed by ripping electron–positron pairs out of the vacuum (Bruce et al. (2007)).

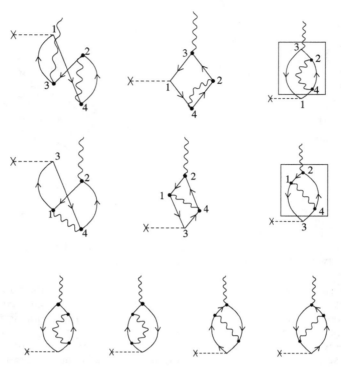

Figure 2.5.2 Some of the possible outcomes resulting from acting with an external single-particle field, i.e., that associated with inelastic processes (represented by a horizontal dashed line starting with a ×) on the ZPF of a nucleus ground state associated with particle–hole correlated vibrations. Within this context one returns to the question of renormalization mentioned in the text (see also Idini et al. (2015); Broglia et al. (2016); Barranco et al. (2017); see also Sect. 5.3). The diagrams of the first row result by intervening the virtual process shown in Fig. 2.5.1 (c) and eventual time orderings. Similar for those of the second row but in connection with diagram (d) of Fig. 2.5.1. The boxed processes correspond to particle self-energy (first row) and vertex correction (second row). Reversing the sense in which the fermions (arrowed lines) circle the loop from anticlockwise to clockwise, one obtains two new graphs. The set of processes obtained in this way are shown in the third and last row, and constitute a sum rule conserving set of diagrams (see first row Fig. 1.2.6).

out how to specifically excite the mode in question, by acting with an external field on the ZPF of the vacuum (Hawking-like radiation[73]; see also Sect. 5.3, in particular Fig. 5.3.2), that is, carry out a NFT-like *gedankenexperiment*,[74] as in Fig. 2.5.2, for *p-h*-vibrations and in Fig. 2.5.3 regarding pairing vibrations. Because the corresponding processes deal with physical states, they translate with ease into

[73] Barranco et al. (2019b).
[74] H. C. Øersted, circa 1812.

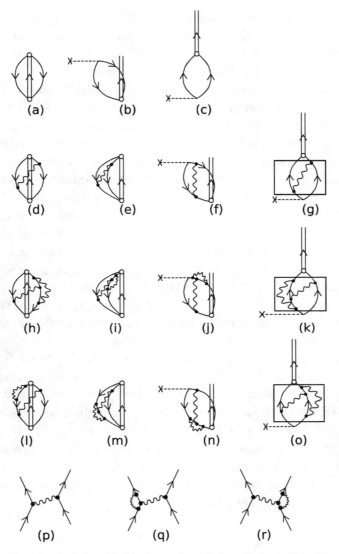

Figure 2.5.3 ZPF associated with the pair addition mode, taking into account the interweaving of nucleons with density modes. The processes boxed in (g), (k), and (o) are associated with the induced pairing interaction (medium polarization effects; (p), (q), (r)) resulting from the exchange of density modes between nucleons moving in time-reversal states, including also vertex corrections. The two-nucleon stripping and pickup external field is labeled by a dashed horizontal line which starts with an ×. The possibility of using pairing vibrational modes as intermediate bosons contributing to the induced pairing interaction, not only in $^1S_0$ channels but also in other channels (multipole pairing modes), is discussed in App. 7.D, in particular, in connection with the possible presence of "vortex-like" pair addition modes, in exotic, halo nuclei, with $J^\pi = 1^-$ and $\beta = +2$ quantum numbers, connected also with the pygmy dipole mode in ^{11}Li (see, e.g., Broglia et al. (2019a), and refs. therein).

a laboratory setup. In keeping with the fact that the vacuum contains all the information (physical degrees of freedom) of the quantal system under study, forcing virtual processes associated with vacuum ZPF to become real, one gets, in each instance, the real, dressed, physical mode of excitation.

The fact that one can treat fully quantum mechanically and on equal footing both structure and reaction processes is apparent from these figures and considerations.[75] Thus unification of structure and reactions, and dressing of energies, vertices (interactions), and formfactors (single-particle radial wavefunctions and associated transition densities), results in a single vacuum correlated state (see, e.g., Figs. 2.5.1 and 2.6.2), vacuum states which through their fluctuations reflect single-particle, normal, and abnormal (pairing) density vibrations and their interweaving. It also indicates the set of specific probes which make these virtual states collapse into on-the-energy shell states, providing the corresponding physical information to the outgoing particles, also photons, which eventually interact with the corresponding detectors – structure and reaction processes free of nonorthogonality, overcompleteness, and Pauli principle violating contributions.

Summing up, the last line of Fig. 2.5.2 displays, together with the corresponding time orderings, the lowest-order self-energy and vertex correction renormalizing (particle-hole) modes. It thus gives rise to the physical collective vibrations whose properties can be directly compared with the experimental findings. In other words, the processes shown in the last line of Fig. 2.5.2 imply that the elementary modes participating in the virtual states have to display, with an exception made for energy (off-shell modes), the same properties of the physical, dressed (renormalized), on-shell modes whose properties can be directly compared with experiment[76] (renormalized NFT). This is because one can, through an experiment, force such virtual states to become real (on-shell) on short call.

Let us now provide an introduction to NFT for pedestrians and see how the above considerations become implemented.[77]

2.5.1 The Concept of Elementary Modes of Excitation

The Hamiltonian of a many-body system of noninteracting particles, bosons, or fermions can be written as

$$H = \sum_i H_i, \tag{2.5.1}$$

---

[75] Embodiments being found in, e.g., Figs. 2.7.2, 2.7.3, and 5.3.2.

[76] Examples of these processes in the case of giant resonances are found in Bortignon and Broglia (1981); Bertsch et al. (1983), and refs. therein. For low-lying states, see Barranco et al. (2004).

[77] For further details, we refer the reader to Bortignon et al. (1977), and refs. therein.

where the summation is over all the particles of the system and where each H_i depends only on the variables of the ith particle. The single-particle Schrödinger equation is

$$H_i \psi_k(\mathbf{r}_i) = \epsilon_k \psi_k(\mathbf{r}_i), \tag{2.5.2}$$

where ϵ_k is the single-particle energy eigenvalue and

$$\psi_k(\mathbf{r}_i) \equiv \langle \mathbf{r}_i | a_k^\dagger | 0 \rangle \tag{2.5.3}$$

is the corresponding wave function. The operator $a_k^\dagger$ creates a particle in the state k when acting in the vacuum state $|0\rangle$. The energy levels of the system are given by the equation

$$E_n = \sum_k n_k \epsilon_k, \tag{2.5.4}$$

the corresponding eigenstates being

$$|n\rangle = \prod_k \frac{(a_k^\dagger)^{n_k}}{\sqrt{n_k!}} |0\rangle, \tag{2.5.5}$$

where $n_k = 0$ or 1 in the case of fermions and $n_k = 0, 1, 2, \ldots$ in the case of bosons.

We now consider a system of interacting particles. The Hamiltonian will in this case be

$$H = \sum_i H_i + \frac{1}{2} \sum_{i,j} H_{ij}, \tag{2.5.6}$$

where i, j label the coordinates of the ith and jth particle, respectively.

In some cases it is possible to recast the two-body Hamiltonian in the form

$$H = \sum_\tau H_\tau', \tag{2.5.7}$$

with the associated Schrödinger equation

$$H_\tau' \psi_\tau(\zeta) = \epsilon_\tau \psi_\tau(\zeta), \tag{2.5.8}$$

ζ representing a general variable (e.g., the single-particle coordinate, the gap parameter, the shape of the nucleus). The wave function $\psi_\tau(\zeta)$ is the ζ-coordinate representation of the eigenstate $\Gamma_\tau^\dagger | \tilde{0} \rangle$. The operator $\Gamma_\tau^\dagger$ creates an excitation with the quantum number τ when acting in the state $|\tilde{0}\rangle$, the correlated vacuum of all the excitations τ.

The energy of the levels of the system, or at any rate of the most important ones to determine the physical response of it to external probes, can be expressed in the form

$$E_m = \sum_\tau n_\tau \epsilon_\tau. \tag{2.5.9}$$

The corresponding eigenstate can be written in the same way as before, that is,

$$|n\rangle = \prod_\tau \frac{(\Gamma_\tau^\dagger)^{n_\tau}}{\sqrt{n_\tau!}} |\tilde{0}\rangle. \tag{2.5.10}$$

Additivity features similar to (2.5.9) hold for other physical quantities, that is,

$$\langle n|\mathcal{O}|m\rangle = \sum_\tau A_\tau \sqrt{n_\tau} \delta(n_\tau, m_\tau + 1), \tag{2.5.11}$$

where

$$O = \sum_\tau A_\tau \Gamma_\tau^\dagger \tag{2.5.12}$$

is the operator which specifically excites the eigenstates described by $\psi_\tau(\zeta)$. Because the excitation energies E_m and observables $|\langle m'|\mathcal{O}|m\rangle|^2$ (e.g., absolute two-particle transfer cross section, electromagnetic-transition probabilities) are linear combinations of ϵ_τ and A_τ, respectively, the eigenstates with energy ϵ_τ and associated observable A_τ are called the *elementary excitations of the system*.

The elementary modes of excitation of a many-body system represent a generalization of the idea of normal modes of vibration. They provide the building blocks of the excitation spectra, giving insight into the deep nature of the system one is studying, aside from allowing for an economic description of complicated spectra in terms of a gas of, as a rule, weakly interacting bosons and fermions. In the nuclear case they correspond to dressed particles and empirically renormalized vibrations.

Two ideas lie behind the concept of elementary modes of excitation.[78] First, one does not need to be able to calculate the total binding energy of a nucleus to accurately describe the low-energy excitation spectrum, in much the same way one can calculate the normal modes of a metal rod not knowing how to calculate its total cohesive energy. The second idea is that low-lying states ($\hbar\omega \ll \epsilon_F \ll BE$, i.e., binding energy) are of a particularly simple character, and amenable to a simple treatment, their interweaving being carried out at profit, in many cases, in perturbation theory.[79] Within this context it is necessary to have a microscopic description of the ground state of the system which ensures that it acts as the vacuum state

---

[78] This concept was introduced by Landau (1941) to describe the spectrum of He II. It was utilized by Bohr and Mottelson (1975) to obtain a unified description of the nuclear spectrum. See also ter Haar (1965, 1969).

$|\tilde{0}\rangle$ of the elementary modes of excitation. In other words, $a_\nu|\tilde{0}\rangle = 0$, $\Gamma_\alpha|\tilde{0}\rangle = 0$, where $a_\nu^+|\tilde{0}\rangle = |\nu\rangle$, and $\Gamma_\alpha^+|\tilde{0}\rangle = |\alpha\rangle$ represent a single-particle and a one-phonon state. This implies, in keeping with the indeterminacy relations[80] $\Delta x \Delta p \geq \hbar/2$, $\Delta I \Delta \Omega \geq 1$, $\Delta N \Delta \phi \geq 1$, and so on, that $|\tilde{0}\rangle = |0\rangle_F|0\rangle_B$ displays the quantal zero-point fluctuations (ZPF) of the many-body system under study.

Within the framework of nuclear field theory (NFT) used below, in which single-particle (fermionic, F) and vibrational (quasi-bosonic, B) elementary modes of excitation are to be calculated making use of HFB and QRPA, respectively[81], $|\tilde{0}\rangle$ must display the associated ZPF (cf. App. 2.C). In particular for (harmonic) vibrational modes, the indeterminacy relation achieves its lowest possible value $\Delta x \Delta p = \hbar/2$, the associated zero-point energy amounting to $\hbar\omega/2$ for each degree of freedom, for example, $5\hbar\omega/2$ for quadrupole vibrations, $\hbar\omega$ being the energy of the collective vibrational mode under consideration.

An illustrative example of the above arguments is provided by the low-lying quadrupole vibrational state of ^{120}Sn. Diagonalizing the effective SLy4 Skyrme interaction[82] in QRPA leads to a value of $B(E2)$ (890 e^2 fm^2), which is about a factor of 2 smaller than experimentally observed (2030 e^2 fm^2). Taking into account renormalization effects in NFT, namely, in a conserving approximation (self-energy and vertex corrections; see Figs. 1.2.6 (first line) and 2.5.2 (last line)), one obtains[83] a value of 2150 e^2 fm^2.

If the collective phonons are not the main object of study but are to be used to dress the single-particle states and/or give rise to the induced pairing interaction, one can make use of phonons which account for the experimental findings (empirical renormalization[84]).[85]

---

[79] More precisely, and in keeping with the fact that (quasi) boson degrees of freedom have to decay through linear particle–vibration coupling vertices of strength Λ into their fermionic components to interact with another vibrational mode, the interweaving between the variety of many-body components dressing a single-particle state or a collective vibration will be described at profit in terms of an arrowed matrix. Assuming perturbation theory to be valid, such a matrix can be transformed, neglecting contributions of the order of Λ^3 or higher, into a codiagonal matrix, namely, a matrix whose nonzero elements are $(i, i-1)$ and $(i, i+1)$, aside from the diagonal ones (i, i).

[80] The quantities I and N are the angular momentum and particle number, conjugate variables to the Euler and gauge angles Ω, ϕ, respectively.

[81] That is, Hartree–Fock (HF) and random phase (RPA) approximations extended to open-shell superfluid nuclei, namely, Hartree–Fock–Bogoliubov and quasi random phase approximation.

[82] Ring and Schuck (1980).

[83] Barranco et al. (2004).

[84] Idini et al. (2015); Broglia et al. (2016); Barranco et al. (2017).

[85] As already mentioned, with the help of experimental probes which couple weakly to the nucleus, i.e., in such a way that the system can be expressed in terms of the properties of the excitation in the absence of probes (see however Sect. 7.4), it has been possible to identify, among others, the following elementary modes of excitation in systems around closed shells:

1. single-particles and -holes,
2. shape vibrations,
3. spin and isospin vibrations and charge exchange modes,

2.5.2 NFT Rules

A field theory can be formulated in which the nuclear elementary modes of excitation play the role of the free fields and in which their mutual interweaving takes place through four-point (v) as well as three-point, particle–vibration coupling vertices.[86] This theory provides a graphical perturbative approach to obtain a solution of the many-body nuclear structure problem in the product basis $\psi_\tau(\zeta)\psi_\eta(\Delta)\ldots\psi_\gamma(\Gamma)$.

In what follows we state and apply the nuclear field theory rules to calculate the interactions between the nuclear free fields and the reaction processes populating the resulting (renormalized) physical states, making use of a simple model.

Schematic Model

The model considered consists of two single-particle levels, each with degeneracy[87] Ω and with a schematic monopole particle-hole interaction coupling the particles in the two levels.

The total Hamiltonian is equal to

$$H = H_{sp} + H_{TB}, \tag{2.5.13}$$

where

$$H_{sp} = \frac{\epsilon}{2}N_0, \qquad N_0 = \sum_{\sigma=\pm 1,m} \sigma a^\dagger_{m,\sigma} a_{m,\sigma}, \tag{2.5.14}$$

and

$$H_{TB} = -\frac{V}{2}\left(A^\dagger A + A A^\dagger\right), \qquad A^\dagger = \sum_m a^\dagger_{m,1} a_{m,-1}. \tag{2.5.15}$$

The index σ ($=\pm 1$) labels the two levels, while m labels the degenerate states within each level. The strength of the monopole coupling is denoted V and the energy difference between the two levels ϵ. The matrix element of (2.5.15) is given by

$$\langle m, 1; m', -1 | H_{TB} | m'', 1; m''', -1 \rangle = -V \delta(m, m') \delta(m'', m'''). \tag{2.5.16}$$

4. pairing vibrations.

Away from closed shells, one has to add to the above modes

5. rotations and $2qp$-like vibrations in 3D space (e.g., quadrupole rotations and β- and γ-vibrations)
6. rotations and $2qp$-like vibrations in gauge space (pairing rotations and pairing vibrations).

Different probes have been utilized in the process of the identification of the different modes. In particular two-neutron transfer reactions induced by tritons and protons have played a central role in unraveling the basic features of the pairing modes.

[86] Bès et al. (1974); Broglia et al. (1976); Bohr and Mottelson (1975); Mottelson (1976a).

[87] It is of note the difference of a factor of 2 in the degeneracy of each level as compared to Sect. 2 of Bortignon et al. (1977), in which case it is 2Ω. This is in keeping with the fact that, as a rule, $\Omega = (2j + 1)/2$. See also Eq. (2.5.85) and related discussion.

Field-Theoretical Solutions

The (quasi-) boson fields are defined through the random-phase approximation, in terms of particle-hole excitations. The basis utilized to describe the nuclear systems is a product of the different free fields. The closed-shell system of the model under consideration corresponds to the lowest ($\sigma = -1$) level filled with Ω particles, while the upper ($\sigma = 1$) level remains empty. The basis particle and hole states are obtained by adding or removing a single particle to/from this closed-shell configuration. The corresponding wave functions and energies, which should include the Hartree–Fock corrections (see Fig. 1.2.2 (b), (c)) generated by the residual interaction[88], are

$$\begin{cases} |m, 1\rangle = a^\dagger_{m,1}|0\rangle, & E(m, 1) = \frac{1}{2}(\epsilon + V) \\ |m, -1\rangle = a_{m,-1}|0\rangle, & E(m, -1) = \frac{1}{2}(\epsilon + V). \end{cases} \tag{2.5.17}$$

Thus the unperturbed energy for producing a particle-hole excitation with respect to the ground state is

$$\epsilon' = E(m, 1) + E(m, -1) = \epsilon + V. \tag{2.5.18}$$

The term V in (2.5.18) is the Hartree–Fock contribution to the particle-hole excitation.

If we define the creation operator of the normal modes as

$$\Gamma^\dagger_\nu = \sum_m \lambda^\nu_m a^\dagger_{m,1} a_{m,-1}, \tag{2.5.19}$$

the linearization equation (see Eq. (1.1.15)),

$$[H, \Gamma^\dagger_\nu] = \omega_\nu \Gamma^\dagger_\nu, \tag{2.5.20}$$

yields

$$\begin{cases} \omega_1 = \epsilon' - V\Omega \\ \omega_\nu = \epsilon' \quad (\nu = 2, 3, \ldots, \Omega). \end{cases} \tag{2.5.21}$$

Utilizing (2.5.20) and the normalization condition

$$[\Gamma_\nu, \Gamma^\dagger_{\nu'}] = \delta(\nu, \nu'), \tag{2.5.22}$$

we obtain for the amplitudes associated with the lowest mode

$$\lambda^1_m = \frac{1}{\sqrt{\Omega}}. \tag{2.5.23}$$

---

[88] The Hartree–Fock energy associated with the Hamiltonian (2.5.13) can be obtained from the linearization relation $[H, a^\dagger_{\sigma,m}] = E(m, \sigma)a^\dagger_{\sigma,m}$ acting on the Hartree–Fock vacuum (see Eq. (1.1.16)), which in this case coincides with the closed shell, that is, the vacuum defined by $a^\dagger_{m,-1}|0\rangle = 0$, $a_{m,1}|0\rangle = 0$.

Figure 2.5.4 Graphical representation of the amplitude of the collective phonon (wavy line) on a given particle-hole excitation $((m, 1), (m, -1))$. This amplitude can be written in terms of the interaction vertex, denoted Λ_i, and the energy denominator $\omega_i - \epsilon'$. The particles (holes) are depicted by upward (downward) going arrowed lines. Time is assumed to run upward.

One can also write this amplitude as the ratio between a coupling matrix element and an energy denominator, that is,

$$\lambda_m^1 = \frac{\Lambda_1}{\omega_1 - \epsilon'}. \tag{2.5.24}$$

From (2.5.21), (2.5.23), and (2.5.24) we obtain

$$\Lambda_1 = -V \sqrt{\Omega}, \tag{2.5.25}$$

which is the strength with which a particle-hole excitation $(m, 1; m, -1)$ couples to the collective phonon (see Fig. 2.5.4). This can also be seen by calculating the matrix element of the interaction Hamiltonian (2.5.15) between the normal modes and the single particle-hole state

$$\Lambda_\nu = \langle n_\nu = 1 | H_{TB} | m, 1; m', -1 \rangle = -V \sqrt{\Omega}\, \delta(m, m')\, \delta(\nu, 1). \tag{2.5.26}$$

Note that the particle–vibration coupling strengths associated with the other normal modes lying at an energy ϵ' (see (2.5.21)) are equal to zero. The exact solution of (2.5.13) is reproduced by utilizing as the basic degrees of freedom both the vibrations (see (2.5.21)) and the particles (see (2.5.17)) coupled through the interaction (2.5.16) (four-point vertex) and (2.5.26) (particle–vibration coupling). A significant part of the original interaction has already been included in generating the collective mode (2.5.21). This implies that the rules for evaluating the effect of the couplings (2.5.16) and (2.5.26) between fermions and bosons involve a number of restrictions as compared with the usual rules of perturbation theory that are to be utilized in evaluating the effect of the original interaction (2.5.15) acting in a fermion space. They read as follows:

I. In initial and final states, proper diagrams involve collective modes and particle modes, but not any particle configuration that can be replaced by a combination of collective modes. This restriction permits an initial state comprising the configuration $(n_\nu = 1; m)$ but excludes $(m', 1; m', -1; m, 1)$.

II. The couplings (2.5.16) and (2.5.26) are allowed to act in all orders to generate the different diagrams of perturbation theory; the restriction I does not apply to internal lines of these diagrams.

III. The internal lines of diagrams are, however, restricted by the exclusion of diagrams in which a particle-hole pair is created and subsequently annihilated without having participated in subsequent interactions.

IV. The energies of the uncoupled particle and phonon fields are to be calculated by utilizing the Hartree–Fock approximation (see Eq. (2.5.17)) and the RPA (see Eq. (2.5.21)), respectively. The contributions of all allowed diagrams are evaluated by the usual rules of perturbation theory.

We note that the external fields acting on the system are allowed to create any state which may generate the different diagrams of perturbation theory. The corresponding matrix elements should be weighted with the amplitude of the component through which the final state is excited.

The above rules are also valid for those situations which cannot be treated in perturbation theory and where a full diagonalization is called for, for example, when the system displays a spurious state (see Sect. 2.5.3).

In what follows we discuss the energy of the $2p - 1h$–like excitations, the simplest modes which can display spuriosity. We distinguish between two types of states, namely,

$$|n_i = 1; m, 1\rangle, \quad \begin{cases} \omega_1 = \epsilon' - V\Omega, & \Lambda_1 = -\sqrt{\Omega}V \\ & (i = 1; m = 1, 2, \ldots, \Omega), \\ \omega_i = \epsilon', & \Lambda_i = 0 \\ & (i = 2, \ldots, \Omega; m = 1, 2, \ldots, \Omega), \end{cases}$$

(2.5.27)

and

$$|m', 1; m', -1; m, 1\rangle, \quad \epsilon' \quad (m, m' = 1, 2, \ldots, \Omega). \quad (2.5.28)$$

The physical states are to be written as

$$|qm\rangle = \sum_i \xi_{iqm} |n_i = 1; m, 1\rangle, \quad (2.5.29)$$

in keeping with the fact that (2.5.28) cannot be basis states according to rule I, but only intermediate states. The quantities ξ_{iqm} are the amplitudes of the physical state in the different components of the product basis of elementary excitations. Rule I eliminates most of the double counting of two-particle, one-hole states. The model space contains Ω "proper" states of the form $|n_i; m, 1\rangle$, in which case the odd particle is in the state $(m, 1)$, that is, $|n_1; m, 1\rangle$ $(\omega_1 = \epsilon' - V\Omega)$ and $|n_i; m, 1\rangle$ $(\omega_i = \epsilon', i = 2, \ldots, \Omega)$. However, there are only $\Omega - 1$ two-particle, one-hole states in

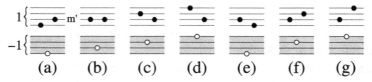

Figure 2.5.5 Schematic two-level model. Count of the states $|m, 1; m - 1; m', 1\rangle$ in the case of $j = 3/2$ and thus $\Omega = 2j + 1 = 4$. State (b) is not allowed because of the Pauli principle. The states ((a),(e)), ((c),(f)), and ((d),(g)) are pairwise identical, in keeping with the indistinguishability of the particles. Thus the states (a), (c), and (d) (equivalent (e), (f), (g)) exhaust the degrees of freedom of states of type (2.5.28). In other words, there are only $\Omega - 1 = 3$ two-particle one-hole states in which the odd particle is in the state $(m', 1)$. It is of notice that all substates m of levels $\sigma = \pm 1$ are degenerate. One displays them equally spaced only for didactical purposes.

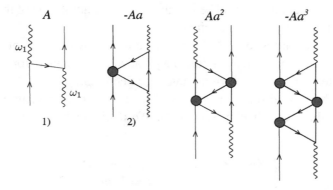

Figure 2.5.6 Contributions to the interaction of a fermion and a collective boson ω_i to order $1/\Omega^4$. The secular equation $E - E^{(0)} = A \sum_n a^n(-1)^n$ is given in terms of the quantities $A = 4\Omega V^2/(3\epsilon' - 2E)$ and $a = 2V/(3\epsilon' - 2E)$. The hatched circle stands for the four-point vertex (2.5.16) (see also Fig. 2.5.8 (I)).

which the odd particle is in the state $(m, 1)$ (Fig. 2.5.5). Therefore a spurious state remains in the spectrum based on elementary modes of excitation. In other words, allowing the quantum number m to run over all possible Ω-states, the model space contains Ω^2 states (one for each value of m), while the correct number is $\Omega(\Omega - 1)$. Thus the basis $|n_1 = 1; m, 1\rangle$ contains Ω spurious states. Its origin can be traced back to the violation of the Pauli principle.

To obtain the energy of $|qm\rangle$, we have to allow the states $|n_1 = 1; m, 1\rangle$ to interact through the vertices (2.5.16) and (2.5.26) and generate all the different perturbation theory diagrams (see rule II) except those containing bubbles (see rule III).

The different graphical contributions to the energy, calculated within the framework of the Brillouin–Wigner perturbation theory, are displayed in Fig. 2.5.6.

There is only one (diagonal) matrix element given by a single summation, which can be carried to all orders in the interaction vertices[89] and can be written as

$$X_{ii'} = A \sum_n (-1)^n a^n \delta(i, i')$$

$$= \frac{A}{1+a} \delta(i, i') \delta(n, 1) = -K(E)(\sqrt{\Omega}V)^2 \delta(i, i') \delta(i, 1), \quad (2.5.30)$$

where a and A are defined in the caption to the figure and

$$K(E) = \left(\tfrac{3}{2}\epsilon' - E + V\right)^{-1} \quad (2.5.31)$$

is the effective coupling strength. The associated secular equation

$$\left| (\omega_i - E)\delta(i, i') + X_{ii'} \right| = 0 \quad (2.5.32)$$

is equivalent to the dispersion relations ($i = 1$ and $i \neq 1$)

$$\frac{1}{K(E)} = \frac{\left(\sqrt{\Omega}V\right)^2}{\tfrac{1}{2}\epsilon' + \omega_i - E}. \quad (2.5.33)$$

Thus the energies of the system are determined by the equation

$$E = \frac{1}{2}\epsilon' + \omega_1 + \frac{\Omega V^2}{\tfrac{3}{2}\epsilon' - E + V}. \quad (2.5.34)$$

It admits the two solutions

$$E_{qm} = \begin{cases} \tfrac{3}{2}\epsilon' \\ \tfrac{1}{2}\epsilon' + \omega_1 + V = \tfrac{3}{2}\epsilon' - \Omega V + V, \end{cases} \quad (2.5.35)$$

which agree with the exact values.[90]

Because $\Lambda_i = 0$ for $i \neq 1$, there is no summation in (2.5.29) and

$$|qm\rangle = N_{qm}|n_1 = 1; m, 1\rangle, \quad (2.5.36)$$

where

$$1 = N_{qm}^2 \left(1 - \frac{\partial X_{11}}{\partial E}\right) = N_{qm}^2 \left(1 - \frac{\Omega V^2}{\left(\tfrac{3}{2}\epsilon' - E + V\right)^2}\right). \quad (2.5.37)$$

For $E_{qm} = \tfrac{1}{2}\epsilon' + \omega_1 + V$ we obtain

$$N_{qm}^2 = \frac{\Omega}{\Omega - 1}, \quad (2.5.38)$$

---

[89] Concerning the proper expansion parameter, see Eq. (2.5.85). See also Broglia (2020), App. A.
[90] The exact solutions can be obtained by noting that the operators $A^\dagger$, A, and $\tfrac{1}{2}N_0$ are generators of the SU_2 group (see Bortignon et al. (1977)).

while for $E_{qm} = \frac{3}{2}\epsilon'$ the state is nonnormalizable as the quantity in parentheses in (2.5.37) is either negative ($\Omega > 1$) or zero ($\Omega = 1$). The state defined by

$$|q, m\rangle = \sqrt{\frac{\Omega}{\Omega - 1}} |n_1 = 1; m, 1\rangle \qquad (2.5.39)$$

and

$$E_{qm} = \frac{1}{2}\epsilon' + \omega_1 + V = \frac{3}{2}\epsilon' - V(\Omega - 1) \qquad (2.5.40)$$

exhausts the inelastic sum rule in agreement with the exact results. Note that (2.5.39) is specifically excited in inelastic processes, as can be seen by direct inspection. The external inelastic field can act in two ways, exciting either a particle-hole pair or a phonon, with amplitudes

$$\langle m, 1; m', -1 | A_1^\dagger | 0 \rangle = \delta(m, m') \qquad (2.5.41)$$

and

$$\langle n_i = 1 | A_1^\dagger | 0 \rangle = \sqrt{\Omega} \delta(i, 1), \qquad (2.5.42)$$

respectively. The different graphical contributions to the inelastic scattering process are displayed in Fig. 2.5.7 and can again be summed to all orders in the interaction vertices, giving

$$\langle n_1 = 1; m, 1 | A_1^\dagger | m, 1 \rangle = \sqrt{\Omega} + \frac{\Lambda_1}{\frac{3}{2}\epsilon' - E_{qm} + V}. \qquad (2.5.43)$$

For $E_{qm} = \frac{3}{2}\epsilon'$ this quantity is equal to zero. Thus the corresponding states do not carry any inelastic strength, a feature which is closely related to the fact that they cannot be normalized and do not display correlation energy. On the other hand, the matrix element associated with (2.5.39) is

$$\langle qm | A^\dagger | m, 1 \rangle = \sqrt{\frac{\Omega}{\Omega - 1}} \frac{\Omega - 1}{\sqrt{\Omega}} = \sqrt{\Omega - 1}, \qquad (2.5.44)$$

a value which agrees with the exact answer.

The results (2.5.40) and (2.5.44) can be traced down to Pauli principle corrections. In fact, the state $|n_i = 1; m, 1\rangle$ has a nonvanishing matrix element, implying a single particle–vibration coupling vertex, with the state $|m, 1; m, -1; m, 1\rangle$. This component, which is spurious, is removed by the different graphs displayed in Figs. 2.5.6 and 2.5.7. *The presence of the odd particle* $(m, 1)$ *blocks the particle-hole excitation* $(m, 1; m, -1)$ *which was present in the uncoupled system. Thus the system increases its energy by a quantity* V. *The reduction of the inelastic amplitude from* $\sqrt{\Omega}$ *to* $\sqrt{\Omega - 1}$ *also indicates that there is one less particle-hole excitation responding to the external probe.*

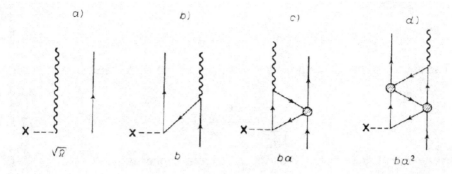

Figure 2.5.7 Graphical representation of the different terms contributing to the matrix element of the inelastic operator $\sqrt{\Omega}\,A^\dagger$ up to order $1/\Omega^3$. Note that the different contributions (b), (c), etc. have a one-to-one correspondence with the different contributions to E (see Fig. 2.5.6); $a = -2V/(3\epsilon' - 2E)$, $b = 2\Lambda_1/(3\epsilon' - 2E)$.

2.5.3 Spurious States

While the model space product of elementary modes of excitation discussed in the last section contains Ω^2 states, only $\Omega(\Omega - 1)$ are physically possible, the number of spurious states being Ω. On the other hand, the agreement between the exact and the nuclear-field-theoretical results shows that the effects of those spurious states are eliminated from all the matrix elements associated with physical observables.

In what follows we show that, in fact, the spurious states are isolated in an explicit way in nuclear field theory.[91] Their energy coincides with the initial unperturbed energy, while all physical operators have zero off-diagonal matrix elements between any physical state and a spurious state, in particular the unit operator, which measures the overlap of the two types of states.

For this purpose we use again a schematic model consisting of a number, Ω, of single-particle levels in which particles interact by means of a "monopole" force,

$$H = H_{sp} + H_{int}, \tag{2.5.45}$$

where

$$H_{sp} = \frac{1}{2} \sum_{m=1}^{\Omega} \epsilon_m \left(a_{m,1}^\dagger a_{m,1} - a_{m,-1}^\dagger a_{m,-1} \right) \tag{2.5.46}$$

and

$$H_{int} = -V A^\dagger A, \tag{2.5.47}$$

[91] Broglia et al. (1976).

with

$$A^\dagger = \sum_{m=1}^{\Omega} a_{m,1}^\dagger a_{m,-1}. \tag{2.5.48}$$

The energy of the ith phonon is determined by the RPA dispersion relation (see rule IV)

$$\sum_{m=1}^{\Omega} \frac{1}{\epsilon_m - \omega_i} = \frac{1}{V}. \tag{2.5.49}$$

The eigenfunction corresponding to the different modes is

$$|n_i = 1\rangle = \sum_m \frac{\Lambda_i}{\epsilon_m - \omega_i} a_{m,1}^\dagger a_{m,-1} |0\rangle. \tag{2.5.50}$$

The particle–vibration coupling constant is given by[92]

$$\Lambda_i = -\langle n_i = 1|H_{int}|m, 1; m', -1\rangle = \left[\sum_m \frac{1}{(\epsilon_m - \omega_i)^2} \right]^{-\frac{1}{2}} \delta(m, m'), \quad (2.5.51)$$

where $|n_i = 1\rangle$ denotes a state containing one phonon, while $|m, 1; m, -1\rangle$ is the eigenstate associated with a particle-hole excitation.

The other interaction to be included (rule II) is the four-point vertex which has the value

$$\langle m, 1; m', -1|H_{int}|m'', 1; m''', -1\rangle = -V\delta(m, m')\,\delta(m'', m'''). \tag{2.5.52}$$

The single-particle energies to be used in calculating the different graphs are $\frac{1}{2}\epsilon_m$, as the Hartree–Fock contribution (see rule IV) of H_{int} is zero.

Similarly to H_{int}, the "inelastic operator" has two different matrix elements, namely,

$$\langle n_i = 1|a_{m',1}^\dagger a_{m',-1}|0\rangle = \frac{\Lambda_i}{\epsilon_{m'} - \omega_i} \tag{2.5.53}$$

and

$$\langle m', 1; m'', -1|a_{m,1}^\dagger a_{m,-1}|0\rangle = \delta(m, m')\delta(m', m''). \tag{2.5.54}$$

In what follows we discuss again the system comprising an odd particle in the orbit $(m, 1)$, in addition to a single phonon excitation of the vacuum. According to rule I, initial and final states may involve both collective modes and particle modes, but not any particle configuration that can be replaced by a combination of

---

[92] The normalization of $|n_i = 1\rangle$ implies $\langle n_i' = 1|n_i = 1\rangle = 1 = \sum_m \Lambda_i^2/(\epsilon_m - \omega_i)^2$ and thus $\Lambda_i = [1/(\epsilon_m - \omega_i)^2]^{-1/2}$. In other words, the normalization is determined by the sum squared of the amplitudes, as the coupling matrix element does not depend on the energy. The HF solution can be obtained from the relation $[H, a_{m,1}^\dagger]|HF\rangle = \epsilon_{m,1}^{HF}|HF\rangle = \left(\frac{1}{2}\epsilon_m a_{m,1}^\dagger - VA^\dagger a_{m,-1}^\dagger\right)|HF\rangle = \frac{1}{2}\epsilon_m a_{m,1}^\dagger|HF\rangle$, as the state $(m, -1)$ is occupied in $|HF\rangle$. Thus $\epsilon_{m,1}^{HF} = \frac{1}{2}\epsilon_m$.

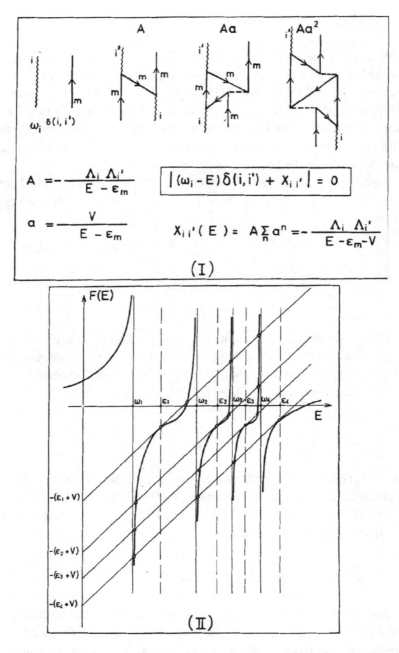

Figure 2.5.8 (I) Lower-order contributions to the energy matrix element between the basis states $|n_i = 1; m, 1\rangle$. The dashed line stands for the model bare interaction (see Eq. (2.5.52)). The quantity $X_{ii'}(E) = A \sum_n a^n = -\Lambda_i \Lambda_{i'}/(E - \epsilon_m - V)$, where $A = -\Lambda_i \Lambda_{i'}/(E - \epsilon_m)$ and $a = V/(E - \epsilon_m)$, is the matrix element iterated to all orders in $1/\Omega$. The secular equation of the problem is $|(\omega_i - E)\delta(i, i') + X_{ii'}| = 0$ and is equivalent to the dispersion relation (2.5.57). (II) Graphical solution of the dispersion relation (2.5.57) for the case $\Omega = 4$. The function $F(E) = \sum_i \Lambda_i^2/(\omega_i - E)$ is displayed as a continuous thick line, while the parallel lines $E - \epsilon_m - V$ have been drawn as thin continuous lines intersecting the ordinate axis at $-(\epsilon_m + V)$. The intersections between the two functions give the eigenvalues of the secular equation. For each value of ϵ_m there are $\Omega + 1$ roots, the root at $E = \epsilon_m$ being double.

collective modes. The exclusion of the states $|m, 1; m', 1; m', -1\rangle$ eliminates most of the double counting of two-particle, one-hole states. The Ω "proper" states of the form $|n_i = 1; m, 1\rangle$ are allowed. However, there are only $\Omega - 1$ (two-particle, one-hole) states in which the odd particle is in the state $(m, 1)$ (see Fig. 2.5.5). Therefore, a spurious state remains in the spectrum of the elementary modes of excitation. If we displace the zero-point energy of the odd system to $(1/2)\epsilon_m$, the unperturbed energy of the basis states $|n_i = 1; m, 1\rangle$ is ω_i.

The lower-order corrections to this energy which do not contain bubbles are drawn in Fig. 2.5.8 (I). Iterating these processes to infinite order, we obtain the secular equation

$$\left|(\omega_i - E)\delta(i, i') + X_{ii'}(E)\right| = 0, \tag{2.5.55}$$

where

$$X_{ii'} = -\frac{\Lambda_i \Lambda_{i'}}{E - \epsilon_m - V}. \tag{2.5.56}$$

The different contributions calculated within the framework of the Brillouin–Wigner perturbation theory are energy dependent and take into account renormalization effects of the states not explicitly included in the calculations. The dispersion relation fixing the energies E_{qm} of the physical states is

$$E - \epsilon_m - V = \sum_{i=1}^{\Omega} \frac{\Lambda_i^2}{\omega_i - E} = F(E). \tag{2.5.57}$$

There is one equation for each single-particle level because the monopole force cannot change the m-state of the odd particle. The relation (2.5.57) can be solved graphically as shown in Fig. 2.5.8 (II). The energy $E = \epsilon_m$ is always a root of (2.5.57), in fact a double root, since

$$\left[\frac{dF(E)}{dE}\right]_{E=\epsilon_m} = \sum_i \frac{\Lambda_i^2}{(\omega_i - \epsilon_m)^2} = 1, \tag{2.5.58}$$

and the line $E - \epsilon_m - V$ is at $45°$. The remaining intersections of this line and the function $F(E)$ give rise to $\Omega - 1$ additional roots denoted by (qm), whose energy E_{qm} agrees with the physical eigenvalues obtained from the exact solution of the model. The eigenvectors associated with the physical states (qm) are

$$|qm\rangle_F = \sum_i \xi_{iqm}|i; m, 1\rangle, \tag{2.5.59}$$

where

$$\xi_{iqm} = -N_{qm}\frac{\Lambda_i}{\omega_i - E_{qm}} = \langle i; m, 1|qm\rangle_F. \tag{2.5.60}$$

The normalization condition which determines N_{qm} is[93]

$$_F\langle qm|qm\rangle_F = 1 = \sum_{i,i'}\left(\delta(i,i') - \frac{\partial X_{ii'}}{\partial E}\right)\xi_{iqm}^*\xi_{i'qm}$$

$$= N_{qm}^2\left[\sum_i \frac{\Lambda_i^2}{(\omega_i - E_{qm})^2} - \frac{1}{(E_{qm} - \epsilon_m - V)^2}\sum_{i,i'}\frac{\Lambda_i^2\Lambda_{i'}^2}{(\omega_i - E_{qm})(\omega_{i'} - E_{qm})}\right]$$

$$= N_{qm}^2\left[\sum_i \frac{\Lambda_i^2}{(\omega_i - E_{qm})^2} - 1\right], \tag{2.5.61}$$

where the dispersion relation (2.5.57) has been utilized and where $X_{ii'}$ is the matrix element appearing in (2.5.55) and defined in (2.5.56). For $E_{qm} = \epsilon_m$ the factor multiplying N_{qm}^2 is zero (see Eq. (2.5.58)). Thus there are only $\Omega - 1$ states which can be normalized when solving the Hamiltonian (2.5.45) within the framework of nuclear field theory. The full spuriosity of the elementary-mode product basis is concentrated in a single state.[94]

The subscript F has been utilized above to indicate that we are dealing with the nuclear-field solution of the Hamiltonian (2.5.45) (for simplicity, it will not be used in the following). Note that these eigenvectors are expressed in terms of only allowed initial or final states (see rule I)

$$|i; m, 1\rangle \equiv a_{m,1}^\dagger|i\rangle, \tag{2.5.62}$$

which are assumed to form an orthonormal basis, in particular in deriving the relation (2.5.61). *This is equivalent to the basic assumption of nuclear field theory of the independence of the different modes of excitation, that is, in the present case,*

$$[\Gamma_i, a_{m,1}^\dagger] = 0. \tag{2.5.63}$$

Rules I–IV discussed in the last section give the proper mathematical framework to this ansatz, which has played a basic role in developing a unified theory of nuclear structure. The above discussion can be illuminated by utilizing a conventional treatment of the residual interaction. Expanding the states $|n_i = 1; m, 1\rangle$ in terms of particle and hole states, we can write, with the help of (2.5.50),

$$a_{m,1}^\dagger|n_i = 1\rangle = a_{m,1}^\dagger\sum_{m'\neq m}\frac{\Lambda_i}{\epsilon' - \omega_i}a_{m',1}^\dagger a_{m',-1}|0\rangle. \tag{2.5.64}$$

---

[93] This is in keeping with the fact that one is using Brillouin–Wigner perturbation theory (see, e.g., App. A of Bès et al. (1976b)), where the differences between the exact energy E and the unperturbed energy replace the differences between unperturbed energies in the denominators of the Rayleigh–Schrödinger expansion. Furthermore, in the Brillouin–Wigner treatment, the initial state never appears as an intermediate state, situations which are explicitly considered in the Rayleigh–Schrödinger treatment, and which in the Brillouin–Wigner one is taken into account through the use of the exact energy in the energy denominators. In the diagrams corresponding to transition matrix elements, they are taken into account, in addition, by the normalization coefficients ξ.

[94] Note that the mathematical relation $N^2 f(E) = 1$, N^2 being the norm of the state with energy E, implies that such state is spurious if $f(E) = 0$ or $f(E) < O$ (see Eq. (2.5.37) and subsequent discussion).

The overlap between the states $|n_i = 1; m, 1\rangle$ is thus given by

$$Z(i, i') = \langle i' | a_{m,1} a^\dagger_{m,1} | i \rangle$$

$$= \sum_{m' \neq m} \frac{\Lambda_i \Lambda_{i'}}{(\epsilon_{m'} - \omega_i)(\epsilon_{m'} - \omega_{i'})} = \delta(i, i') - \frac{\Lambda_i \Lambda_{i'}}{(\epsilon_m - \omega_i)(\epsilon_m - \omega_{i'})},$$

$$(2.5.65)$$

where the orthogonality relation

$$\sum_{m'} \frac{\Lambda_i \Lambda_{i'}}{(\epsilon_{m'} - \omega_i)(\epsilon_{m'} - \omega_{i'})} = \delta(i, i') \qquad (2.5.66)$$

of the RPA solutions in the even system has been utilized. Because of the non–orthogonality of the basis, the eigenvalues of the system are determined by the relation

$$|Z(E)(H - E)| = 0. \qquad (2.5.67)$$

This is fulfilled for

$$|H - E| = 0, \qquad (2.5.68)$$

which yields the $\Omega - 1$ physical roots, as well as for

$$|Z(E)| = 0. \qquad (2.5.69)$$

This solution corresponds to the spurious root $E_{qm} = \epsilon_m$. In fact[95],

$$\lim_{\delta \to 0} \sum_i \xi_{iqm}(E_{qm} = \epsilon_m + \delta) Z_{ii'} = \lim_{\delta \to 0} N_{qm}(E_{qm} = \epsilon_m + \delta)$$

$$\times \sum_i \frac{\Lambda_i}{\omega_i - (\epsilon_m + \delta)} \sum_{m \neq m'} \frac{\Lambda_i \Lambda_{i'}}{(\epsilon_{m'} - \omega_i)(\epsilon_{m'} - \omega_{i'})} = 0,$$

$$(2.5.70)$$

since

$$\sum_i \frac{\Lambda_i^2}{(\epsilon_m - \omega_i)(\epsilon_{m'} - \omega_i)} = \delta(m, m'). \qquad (2.5.71)$$

Note that this solution in terms of the overlap Z gives the exact answer in the present case, because of the simplicity of the model. In a general case which includes ground-state correlations, this may no longer be true.

We now calculate the one-particle stripping process leading to the odd system. This calculation illustrates the explicit concentration of the whole spuriosity into a

---

[95] Within the context of renormalization, one first calculates the expressions for a finite value of δ and then takes the limit.

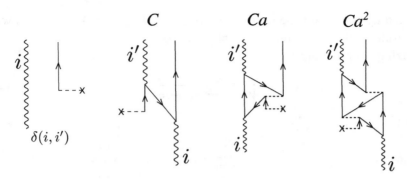

Figure 2.5.9 Lower-order contributions to the one-particle transfer reaction induced by $a^\dagger_{m,1}$. The result of iterating the different contributions to all orders in $1/\Omega$ is equal to $T_{qm}(i, i') = C \sum_n d^n = -\Lambda_i \Lambda_{i'} / \big((\omega_i - \epsilon_m)(E_{qm} - \epsilon_m - V)\big)$, $C = -\Lambda_i \Lambda_{i'} / \big((\omega_i - \epsilon_m)(E_{qm} - \epsilon_m)\big)$, $d = V / (E_{qm} - \epsilon_m)$.

single state which has zero correlation energy[96] and zero amplitude for the different physical processes exciting the $\Omega - 1$ physical states.

One has first to calculate the amplitude for the transition to a basis component $(n_i = 1; m, 1)$ including only those graphs in which all intermediate states are excluded from appearing as initial or final states. This exclusion reflects the fact that the diagonalization procedure has included all interaction effects that link these allowed states. The final amplitude for the transition to the state (qm) is obtained by summing the amplitudes $(n_i = 1; m, 1)$, each weighted by ξ_{iqm} given in Eq. (2.5.60).

The lower-order contributions to the one-particle transfer amplitude between the state $|n_i = 1\rangle$ and the state $|qm\rangle$ are displayed in Fig. 2.5.9. They can be summed up to all orders of $1/\Omega$, the result being equal to[97]

$$\langle qm | a^\dagger_{m,1} | n_i = 1 \rangle$$
$$= \sum_{i'} \xi_{i'qm} \left\{ \delta(i, i') - \frac{\Lambda_i \Lambda_{i'}}{(\omega_i - \epsilon_m)(E_{qm} - \epsilon_m)} \left[\frac{1}{1 - V/(E_{qm} - \epsilon_m)} \right] \right\}$$
$$= \sum_{i'} \xi_{i'qm} \left\{ \delta(i, i') - T_{qm}(i, i') \right\}$$
$$= -N_{qm} \left[\frac{\Lambda_i}{\omega_i - E_{qm}} - \frac{\Lambda_i}{(\omega_i - \epsilon_m)(E_{qm} - \epsilon_m - V)} \sum_{i'} \frac{\Lambda_{i'}^2}{\omega_{i'} - E_{qm}} \right]$$
$$= \frac{N_{qm}(E_{qm} - \epsilon_m)\Lambda_i}{(E_{qm} - \omega_i)(\omega_i - \epsilon_m)}. \tag{2.5.72}$$

---

[96] This is because the spurious state has zero phase space to correlate.

[97] For the definition of T_{qm}, see Fig. 2.5.9.

This quantity is zero for the spurious roots[98] (i.e., $E_{qm} = \epsilon_m$) and agrees with the exact result for the $\Omega - 1$ remaining physical roots.

Utilizing the relations

$$\frac{1}{V} = \sum_{m'} \frac{1}{\epsilon_{m'} - \omega_i} \qquad (2.5.73)$$

and

$$\frac{1}{V} = \sum_{m' \neq m} \frac{1}{\epsilon_{m'} - E_{qm}}, \qquad (2.5.74)$$

we obtain

$$\sum_{m' \neq m} \frac{1}{(\epsilon_{m'} - E_{qm})(\epsilon_{m'} - \omega_i)} = \frac{1}{(E_{qm} - \omega_i)(\epsilon_m - \omega_i)}. \qquad (2.5.75)$$

With the help of this relation we can derive the *one-particle transfer sum rule*. Note that (2.5.73) is the dispersion relation for the free phonon field. The second relation is, however, alien to the field theory results. Nevertheless, one can show that the solutions E_{qm} of (2.5.74) and of the nuclear field theory dispersion relation (2.5.57) are identical, except for the root $E_{qm} = \epsilon_m$. One can, therefore, utilize (2.5.74) as a mathematical relation without further justification in the present context. One obtains

$$\sum_{qm} \left| \langle qm | a^\dagger_{m,1} | n_i = 1 \rangle \right|^2 = \sum_{qm} \Lambda^2_{qm} \Lambda^2_i \sum_{m' \neq m} \frac{1}{(\epsilon_{m'} - E_{qm})(\epsilon_{m'} - \omega_i)}$$

$$\times \sum_{m'' \neq m} \frac{1}{(\epsilon_{m''} - E_{qm})(\epsilon_{m''} - \omega_i)}, \quad (2.5.76)$$

where

$$\Lambda_{qm} = -N_{qm}(E_{qm} - \epsilon_m) = \left[\sum_{m' \neq m} \frac{1}{(\epsilon_{m'} - E_{qm})^2} \right]^{-\frac{1}{2}}. \qquad (2.5.77)$$

Thus

$$\sum_{qm} \left| \langle qm | a^\dagger_{m,1} | n_i = 1 \rangle \right|^2 = \Lambda^2_i \sum_{m' \neq m} \frac{1}{(\epsilon_{m'} - \omega_i)^2} = 1 - \frac{\Lambda^2_i}{(\epsilon_m - \omega_i)^2}, \quad (2.5.78)$$

where use has been made of the orthogonality relation

$$\sum_{qm} \frac{\Lambda^2_{qm}}{(\epsilon_{m'} - E_{qm})(\epsilon_{m''} - E_{qm})} = \delta(m', m''), \qquad (m', m'' \neq m). \qquad (2.5.79)$$

---

[98] Making use of (2.5.61), one obtains $\lim_{\delta \to 0} [(E_{qm} - \epsilon_m) N_{qm}]_{E_{qm} = \epsilon_m + \delta} = \lim_{\delta \to 0} \left\{ \sqrt{2} \delta^{3/2} / [\sum_i \frac{\Lambda_i}{\omega_i - \epsilon_m}]^{1/2} \right\} = 0.$

The result (2.5.78) coincides with the exact result. Physically it means that the single-particle orbital $(m, 1)$ is blocked by the amount $\Lambda_i^2/(\epsilon_m - \omega_i)^2$, which is the probability that the phonon $(n_i = 1)$ is in the particle-hole configuration $(m, 1; m, -1)$, that is, with its particle in the orbital $(m, 1)$.

2.5.4 Applications

The nucleus $^{209}_{83}\text{Bi}_{126}$ can be viewed as a proton outside the double closed-shell nucleus $^{208}_{82}\text{Pb}_{126}$. It displays 12 states below 3 MeV. Within the framework of NFT, these states can be interpreted in terms of the elementary excitations of the ground state of ^{208}Pb, (gs), which plays the role of the "vacuum" state, namely, single-particle and $2p$–$1h$ states.

In what follows we discuss some aspects of this low-lying spectrum in terms of fermions (protons; a_j, $a_j^\dagger$), surface ($\Gamma^\dagger(\beta = 0\lambda)$), and pairing ($\Gamma^\dagger(\beta = 2\lambda)$ proton pair addition) vibrational modes.

The unperturbed two-particle, one-hole states of the closed-shell system can be written in terms of the free fields as

$$|n2\lambda, j; IM\rangle = [\Gamma_n^\dagger(2\lambda)a_j]_{IM}|0\rangle \tag{2.5.80}$$

and

$$|n0\lambda, j; IM\rangle = [\Gamma_n^\dagger(0\lambda)a_j^\dagger]_{IM}|0\rangle. \tag{2.5.81}$$

This constitutes the basis set of states $\{\alpha_i\}$. All other states give rise to the complementary Hilbert space $\{a_i\}$.

The elementary modes of excitation interact through the particle–vibration and four-point vertices displayed in Fig. 2.5.10, giving rise to the matrix elements

$$M_1(nj, n'j') \equiv \langle[\Gamma_{n'}^\dagger(0\lambda)a_{j'}^\dagger]_{IM}|h_{eff}(E)|[\Gamma_n^\dagger(0\lambda)a_j^\dagger]_{IM}\rangle, \tag{2.5.82}$$

$$M_2(nj, n'j') \equiv \langle[\Gamma_{n'}^\dagger(2\lambda)a_{j'}]_{IM}|h_{eff}(E)|[\Gamma_n^\dagger(2\lambda)a_j]_{IM}\rangle, \tag{2.5.83}$$

and

$$M_3(nj, n'j') \equiv \langle[\Gamma_{n'}^\dagger(2\lambda)a_{j'}]_{IM}|h_{eff}(E)|[\Gamma_n^\dagger(0\lambda)a_j^\dagger]_{IM}\rangle. \tag{2.5.84}$$

They are to be calculated by utilizing the graphical techniques of perturbation theory and the rules discussed in Sect. 2.5.2. There are two parameters on which to expand in carrying out a perturbative calculation. *The first one is the strength of the interaction vertices measured in terms of the average distance between single-particle levels. The second is* $1/\Omega$, *where* $\Omega = \sum_j (j + \frac{1}{2})$ *is the effective degeneracy of the valence shells* (in connection to this "standard" definition of Ω we refer to in footnote 87 of this chapter). These two parameters are in general

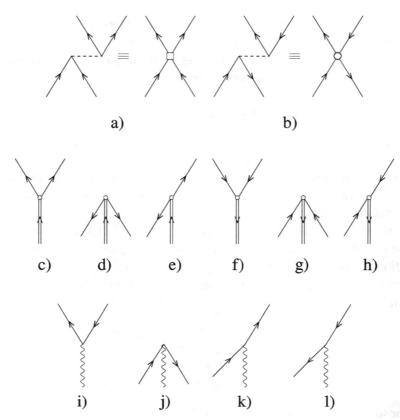

Figure 2.5.10 Interactions coupling the fermion fields with the pairing and surface vibrations. The different fermion and (quasi) boson free fields are particles, holes, pairing vibrations ($\beta = \pm 2$), and surface vibrations ($\beta = 0$), β being the transfer quantum number. The two possible four-point vertices are given in (a) and (b). They correspond to the pairing and particle-hole model bare interactions. In graphs (c)–(h) all possible couplings between the fermion fields (arrowed lines) and the pairing vibrational fields (double lines arrowed) are displayed. Graphs (i)–(l) are all the possible coupling vertices between the surface vibrations (wavy line) and the fermion fields. *Note that there is no direct coupling between the two (quasi) boson fields, as the field theory we are dealing with is linear in the different (quasi) boson field coordinates.*

connected through involved expressions. In the simple model discussed in Sect. 2.5.2, however, their relation is explicit and can be expressed as

$$\epsilon = \mathcal{O}(1), \quad \Lambda = \mathcal{O}\left(\frac{1}{\sqrt{\Omega}}\right) \quad \text{and} \quad V = \mathcal{O}\left(\frac{1}{\Omega}\right). \tag{2.5.85}$$

Another feature which determines the family of diagrams to select to a given order of perturbation is the number of internal lines which can be freely summed up. Each of these summations introduces a multiplicative factor Ω. Based on a wealth of detailed calculations for realistic distributions of levels, one can conclude that relations (2.5.85) are also valid in such cases.[99]

Returning to ^{209}Bi, this nucleus has been investigated by means of high-resolution anelastic process.[100] Through these experiments a septuplet of states of positive parity and lying around 2.6 MeV of excitation energy was identified, with spins ranging from $\frac{3}{2}^{+}$ to $\frac{15}{2}^{+}$.

In zeroth order these states can be interpreted in terms of a proton moving in the $h_{9/2}$ orbital coupled to the lowest octupole vibration of ^{208}Pb. The $\frac{3}{2}^{+}$ of this multiplet displays also a large parentage based on the proton pair addition mode of ^{208}Pb, and a proton hole moving in the $d_{3/2}$ orbital, as revealed by the (t, α) reaction[101] on ^{210}Po. The above results indicate that low-lying (two-particle, one-hole) types of states of ^{209}Bi are amenable to a simple description in terms of the basis states[102]

$$|j_1^{-1}, (\beta = 2), \lambda^{\pi}; IM\rangle \equiv |j_1^{-1} \otimes \lambda^{\pi}(^{210}\text{Po}); IM\rangle, \qquad (\lambda^{\pi} = 0^{+}, 2^{+}, 4^{+}),$$
$$(2.5.86)$$

and

$$|j_2, (\beta = 0), \lambda^{\pi}; IM\rangle \equiv |j_2 \otimes \lambda^{\pi}(^{208}\text{Pb}); IM\rangle, \qquad (\lambda^{\pi} = 3^{-}). \qquad (2.5.87)$$

Only the lowest states of each spin and parity λ^{π} are included in the basis states, while all the RPA solutions are included in the intermediate states. The quadrupole surface vibrational modes were allowed only as intermediate states. The single-hole and particle (proton) states j_1^{-1} and j_2, respectively, correspond to experimentally known levels around the $Z = 82$ ($N = 126$) shell closures. In what follows, the two $\frac{3}{2}^{+}$ states will be discussed.

The two basis states

$$|\alpha\rangle \equiv |d_{3/2}^{-1} \otimes \text{gs}(^{210}\text{Po}); 3/2^{+}\rangle \qquad (2.5.88)$$

and[103]

$$|\beta\rangle \equiv |h_{9/2} \otimes 3^{-}(^{208}\text{Pb}); 3/2^{+}\rangle \qquad (2.5.89)$$

---

[99] An alternative way to argue concerning the expansion parameter, in this case in connection with pairing vibrations, is to use the dimensionless quantity $x = 2G\Omega/D$ appropriate for a model made of two j-shells separated by an energy $D = 2\epsilon$ in which pairs of nucleons moving in time-reversal states interact through a pairing interaction of strength G. Phase transition takes place at $x \geq 1$, while situations away from phase transitions but still displaying consistent fluctuations typical of nuclei with low-lying collective vibrations correspond to $x \approx 0.5$. Assuming $D(\mathcal{O}(\epsilon)) = \mathcal{O}(1)$, one obtains $G(\mathcal{O}(V)) = \mathcal{O}(1/\Omega)$ and naturally $\Lambda = \mathcal{O}(1/\sqrt{\Omega})$, in keeping with the fact that the induced interaction can be written as $\Lambda^{2}/(\epsilon - \omega)$.

[100] Ungrin et al. (1971); see also Bohr and Mottelson (1975).

[101] Regarding this pickup reaction, the connection between theory and experiment will be carried out, in the present section, in terms of spectroscopic factors.

[102] Barnes et al. (1972).

[103] Although not likely, the reader is advised not to confuse the label of the state $|\beta\rangle$ with the transfer quantum number β used before.

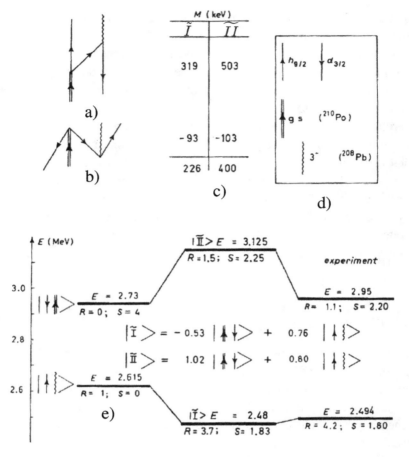

Figure 2.5.11 In (a), (b), and (c) we give the two graphical contributions and the corresponding numerical values to the matrix element $M(E) = \langle d_{3/2}^{-1} \otimes gs(^{210}\text{Po})|h_{eff}(E)|h_{9/2} \otimes 3^-(^{208}\text{Pb}); 3/2\rangle$ in lowest order in $1/\Omega$. The resulting wave functions $|\widetilde{\text{I}}\rangle$ and $|\widetilde{\text{II}}\rangle$ are displayed in (e) normalized according to (2.5.61). In (e) we also give the unperturbed and the renormalized theoretical energies of the levels. The (t, α) spectroscopic factor corresponding to the reaction $^{210}\text{Po}(t, \alpha)^{209}\text{Bi}$ is denoted by S, while $R = \dfrac{d\sigma(h_{9/2} \to J)}{d\sigma(gs(^{208}\text{Pb}) \to 3^-(^{208}\text{Pb}))}$ is the ratio of inelastic cross sections. In (d) we display the free fields, while in (e) we provide a summary of the results of the calculations in comparison with the data. The zeroth and order $1/\Omega$ contributions to the electromagnetic excitations are collected in (i) and (j). The value $0.577e^2b^3$ is the $B(E3; 0 \to 3)$ value associated with the 2.615 MeV state in ^{208}Pb. In (g) and (h) we give the zeroth and order $1/\Omega$ contributions to the spectroscopic factor associated with the $^{210}\text{Po}(t, \alpha)^{209}\text{Bi}$ reaction. Finally, in (f) we display the lowest contribution to the spectroscopic factor associated with the $^{208}\text{Pb}(^3\text{He}, d)$ reaction, which gives a measure of the ground state correlations of ^{208}Pb associated with the existence in its low-lying spectrum of an octupole (see Fig. 1.8.3) and a proton pair addition mode (see also Tables 2.5.1–2.5.3).

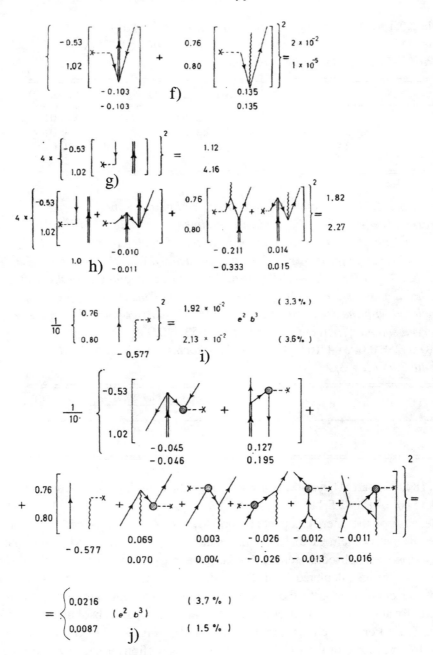

Table 2.5.1 *Observed single-particle strength (and corresponding errors) associated with the transfer reactions $^{210}Po(t,\alpha)^{209}Bi$ and $^{208}Pb(^3He,d)^{209}Bi$ (see Bortignon et al. (1977)).*

	$E_x(MeV)$	$S(t,\alpha)(2j+1)$	$S(^3He,d)$
$3/2^+$	2.49	$1.8 \pm 0.3\,(4)$	< 0.01
$3/2^+$	2.95	$2.2 \pm 0.3\,(4)$	< 0.01
$1/2^+$	2.43	$1.8\,(2)$	< 0.02
$11/2^-$	3.69	$10\,(12)$	< 0.05

Table 2.5.2 *Observed total inelastic cross section σ^{oct} (and corresponding errors) associated with the lowest octupole vibrational state of ^{208}Pb. It can be written in terms of that associated with a single magnetic substate σ' as $\sigma^{oct}_{3^-} = 7\sigma'$, that associated with the multiplet $(h_{9/2} \otimes 3^-)_{J^+}(J = 3/2 - 15/2)$ as $\sigma^{oct}_{3^-} = 70\sigma'$, in keeping with the fact that the $h_{9/2}$ state has 10 magnetic substates. Thus the strength associated with the 3/2 channel is $4/70 = 0.057$, to be compared with the observed summed (percentage) strength 0.053 ± 0.005 $(= (0.042 \pm 0.003) + (0.011 \pm 0.002))$ associated with the 2.45 MeV and the 2.95 MeV $3/2^+$ states.*

	$E_x(MeV)$	$\dfrac{\sigma(^{209}Bi(9/2^-;gs)\to{}^{209}Bi(3/2^+;E))}{\sigma(^{208}Pb(gs)\to(3^-;2.615\,MeV))}$
$3/2$	2.49	0.042 ± 0.003
$3/2$	2.95	0.011 ± 0.002

are 118 keV apart. They mix strongly through the couplings depicted by the graphs (a) and (b) of Fig. 2.5.11.

Because of the energy dependence of h_{eff} (Eqs. (2.5.82)–(2.5.84)), there is a different matrix element for each final stale. The diagonalization of the matrices was carried out self-consistently, that is, the energy denominators of the different graphs are to be calculated by utilizing the exact energies.[104] The corresponding graphical contributions to the single-particle spectroscopic amplitudes, and thus eventually absolute (t,α) differential cross sections[105], and absolute inelastic scattering (d,d') cross sections, are also collected in Fig. 2.5.11. To be noted is the very different ratio of the (d,d') and (t,α) cross sections associated with the two states. While $R_1 = B(E3;(\frac{3}{2})_1)/B(E3;(\frac{3}{2})_2))$ is approximately equal to 2.5, the ratio $R_2 = \sigma((t,\alpha);(\frac{3}{2})_2)/\sigma((t,\alpha);(\frac{3}{2})_1)$ is close to 1 (see Tables 2.5.1–2.5.3).

---

[104] For more details, see Bortignon et al. (1977); see also Bortignon et al. (1976).

[105] It is of note that in the present case, and at variance with the rest of the monograph, use will be made, for the sake of didactics, of the concept of spectroscopic factors (see end of Sect. 3.1 as well as Sect. 5.8).

Table 2.5.3 *Summary of NFT predictions concerning the structure of the two lowest $3/2^+$ states of ^{209}Bi, in comparison with the experimental data.*

	E_n(MeV)		$\frac{\sigma(h_{9/2}\to 3/2^+)}{\sigma(0^+\to 3^-)}(\%)$		$S(t,\alpha)$		$S(^3\text{He},d)$	
	Theory	Exp	Theory	Exp	Theory	Exp	Theory	Exp
3/2	2.480	2.49	3.76	4.2 ± 0.3	1.83	1.8 ± 0.3	0.02	< 0.01
3/2	3.125	2.95	1.56	1.1 ± 0.2	2.25	2.2 ± 0.3	10^{-5}	< 0.01

Because the component $|\beta\rangle$ carries the inelastic scattering strength, while the (t,α) reaction proceeds mainly through the component of type $|\alpha\rangle$, *the difference between R_1 and R_2 can be traced back to the corrections associated with the overcompleteness of the unperturbed basis states which give rise to rather different normalizations (see Sect. 2.5.3) of the two physical states $|\widetilde{I}\rangle$ and $|\widetilde{II}\rangle$ (see Fig. 2.5.11 (e)).*

Said differently, in a conventional two-state shell model calculation implying a single matrix, one would obtain

$$|\widetilde{I}\rangle = A|\alpha\rangle + B|\beta\rangle \tag{2.5.90}$$

and

$$|\widetilde{II}\rangle = -B|\alpha\rangle + A|\beta\rangle \tag{2.5.91}$$

with $A^2 + B^2 = 1$. The model would predict the value $(B/A)^2$ for both the ratio $R_1 = \sigma((d,d'); (3/2)_{\widetilde{I}})/\sigma((d,d'); (3/2)_{\widetilde{II}})$ and $R_2 = \sigma((t,\alpha); (3/2)_{\widetilde{II}})/\sigma((t,\alpha); (3/2)_{\widetilde{I}})$. The fact that $(R_1)_{th} = (0.0376/0.0156) = 2.41$ (against $(R_1)_{exp} = (0.042/0.011) = 3.8$) and $(R_2)_{th} = (2.25/1.83) = 1.22$ (against $(R_2)_{exp} = (2.2/1.8) = 1.22$) is a direct consequence of the overcompleteness of the basis which is taken care by NFT. This field theory provides a systematic procedure to deal with the spurious states, in the present case, associated with the overcompleteness of the basis ((2.5.88), (2.5.89)).

Within the framework of conventional shell model calculation ((2.5.90), (2.5.91)) the asymmetry between R_1 and R_2 can be related to the finite overlap[106] between the basis states $|\alpha\rangle$ and $|\beta\rangle$.

Summing up, most of the physics associated with the two $3/2^+$ states as revealed by an essentially "complete" set of experiments, namely, ^{209}Bi$(d,d')^{209}$Bi*, ^{210}Po$(t,\alpha)^{209}$Bi and ^{208}Pb$(^3\text{He},d)^{209}$Bi, is contained in the basis states, namely, in the states $|\alpha\rangle$ and $|\beta\rangle$, product of fermionic and (quasi-) bosonic elementary modes of excitation of simple representation. NFT provides the rules to diagonalize their coupling and calculate the associated spectroscopic amplitudes. These

---

[106] Making use of the experimental values displayed in Tables 2.5.1 and 2.5.2, this overlap is estimated to be ≈ 0.25.

amplitudes constitute the input to particle transfer and inelastic scattering computer codes needed to calculate the associated absolute differential cross sections,[107] that link theory with experiment.

2.6 Competition between ($\beta = 0$) and ($\beta = \pm 2$) ZPF

Particle-hole-like vibrations, as, for example, collective surface quadrupole vibrations, induce dynamical distortions of the mean field which virtually break the magnetic degeneracy of levels into twofold (Kramers) degenerate (Nilsson-like) levels and to a reduction of the size of the discontinuity at the Fermi surface typical of noninteracting Fermi systems (see Fig. 2.6.1 (f); see also Fig. 2.2.1 (c)–(e)). Pairing vibrations does similarly, through dynamical ($U_j V_j$) weighting factors (see Fig. 4.4.2), which are operative in an energy region[108] $\epsilon_F \pm E_{corr}$ ($\beta = \pm 2$) (see Fig. 2.6.1 (g); see also Figs. 2.2.1 (f), (g)). At the same time the dressing of nucleons by particle-hole and pairing vibrations leads to an effective ω-mass making nucleons heavier, thus approaching the centroid of the valence orbitals lying above and in the Fermi sea toward the Fermi surface. In Fig. 2.6.1 a schematic representation of the effects the interweaving of single-particle motion and collective vibrations has on pairing correlations is displayed.

Zero-point fluctuations induced by particle-hole-like vibrations and by pairing modes compete with each other for phase space, through the Pauli principle (see Fig. 2.6.2), thus eventually leading to a single ground state containing all of the dressed renormalized ZPF. The Pauli principle NFT diagrams shown in Figs. 2.6.2 (b) and (d) are at the basis of the stabilization of the ground state in general and of the competition between (as a rule quadrupole) deformations in 3D-space, which breaks single-particle degeneracy (Nilsson potential), and in gauge space, which thrives on large degeneracies.[109] It is also the reason why single open-shell nuclei are usually spherical. When tidal-like polarization effects in doubly open-shell nuclei become overwhelming, the nucleus makes use of a Jahn–Teller mechanism to profit at best and simultaneously of the quadrupole–quadrupole (alignment) and of the pairing (independent pair motion in Kramers degenerate levels) interactions, in other words, of potential energy (quadrupole deformation, localization) and of pairs of nucleons entangled with each other, over distances $\xi (\gg R_0)$ resulting in strongly overlapping entities and thus little sensitive to the orientation of the quadrupole deformed field.

---

[107] Making use, of course, of a set of optical parameters. Within this context, see footnote 26 of Ch. 1 and associated text of Sect. 1.2.

[108] E_{corr} is the correlation energy associated with the pair addition ($\beta = +2$) or with the pair removal modes ($\beta = -2$).

[109] Within this context see Bayman (1961); Bès and Sorensen (1969); Mottelson (1962); Bohr and Mottelson (1975), and refs. therein.

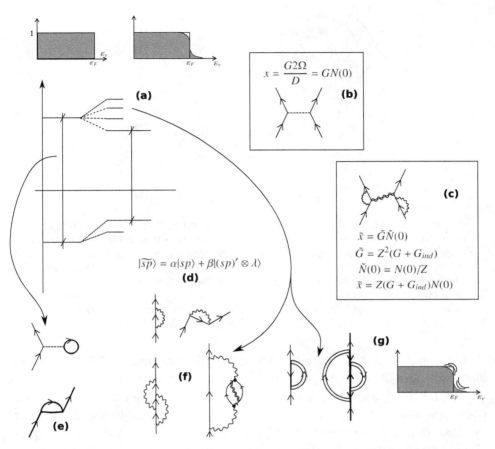

Figure 2.6.1 Schematic representation of some of the consequences the inter-weaving of the elementary modes of excitation with varied transfer quantum number ($\beta = 0, \pm 1, \pm 2$; (f), (g)) have in the (single-particle) spectrum ((a), (e)), in particular pair correlations (b, c), as measured by the (two-level) dimension-less parameter $x = G2\Omega/D = GN(0)$, product of the bare coupling constant G and the density of states (DOS) at the Fermi energy (ratio of the single-particle degeneracy $2\Omega = (2j+1)$ and the single-particle energy separation; see Högaasen-Feldman (1961); Broglia et al. (1968)). Coupling with surface modes (f) reduces the effective value of D, leading to an increase of $N(0)$ as measured by $1/Z$ but, at the same time decreases, through the breaking of the single-particle strength, the single-particle content (d) of each level (as measured by Z; see, e.g., Barranco et al. (2005), and refs. therein). The eventual increase of x, reflected by $\tilde{x}$, results from a delicate balance of the two effects. The increase in confine-ment kinetic energy is reflected by the dynamical smoothing of the Fermi energy through the coupling of single-particle states to $\beta = 0$ and $\beta = \pm 2$ vibrations, overwhelmed by the induced pairing interaction resulting from the exchange of collective ($\beta = 0$) vibrations between pairs of nucleons moving in time reversal states close to the Fermi energy.

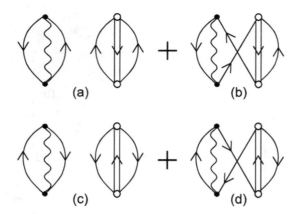

Figure 2.6.2 (a), (c) ZPF associated with $p - h$ and pairing vibrations (pair sub-straction and pair addition modes) make use of the same nucleon degrees of freedom to simultaneously, and independently, correlate $p - h$, $p - p$, and $h - h$ excitations, thus violating the Pauli principle. The NFT processes (b) and (d), which contribute to the correlation energy of the nucleus with opposite sign to that contributed by (a) and (c) (each unavoidable crossing of fermion lines con-tributes a minus sign), remove Pauli violating contributions to the corresponding order of perturbation in $1/\Omega$ (see Eq. (2.5.85) and related text).

It is of note that in the case of large-amplitude vibrations in 3D space associ-ated with Jahn–Teller effects and implying a number of level crossings (see Eq. (4.10.6)), pairing plays a central[110], symbiotic role. This is because it leads to a consistent reduction of the inertia, examples being provided by the exotic decay of ^{223}Ra (see Fig. 4.9.3; see also Fig. 1.1.1) and by the soft $E1$-mode (pygmy dipole resonance)[111] of ^{11}Li (Sect. 4.10.1, text following Eq. (4.10.5); see also Eq. (4.10.6)).

2.7 Coupling between Structure and Reaction Channels

In this section the unification of NFT of structure and reactions is further developed using as an example the light exotic two-neutron halo nucleus ^{11}Li. In particular, we dwell upon the variety of renormalization processes and associated formfactors needed to calculate one- and two-neutron transfer reactions. The use of the same elements in the eventual calculation of the corresponding polarization contribution to the optical potential[112], a subject not treated in the present monograph, is briefly mentioned.

---

[110] See also Matsuyanagi et al. (2013).
[111] See Broglia et al. (2019a,c).
[112] See footnote 26 in Ch 1.

2.7.1 Bare Particles and Hartree–Fock Field

Nucleon elastic scattering experiments at energies of tens of MeV can be accurately described in terms of an optical potential in which the real component is parameterized according to a (Woods–Saxon) potential[113]

$$U(r) = Uf(r), \tag{2.7.1}$$

$f(r)$ being a Fermi (sigmoidal) function of radius $R_0 = r_0 A^{1/3}$ ($r_0 = 1.2$ fm), diffusivity $a = 0.65$ fm, and strength

$$U = U_0 + 0.4E, \tag{2.7.2}$$

where

$$U_0 = V_0 + 30\frac{N-Z}{A} \text{ MeV}, \quad V_0 \approx -51 \text{ MeV}, \tag{2.7.3}$$

while E is the energy of the scattered particle $\epsilon_k = \hbar^2 k^2/2m$, measured from the Fermi energy, m being the nucleon mass. One can replace the k-dependence in (2.7.2) by the so-called k-mass[114]

$$m_k = m\left(1 + \frac{m}{\hbar^2 k}\frac{dU}{dk}\right)^{-1}, \tag{2.7.4}$$

where the energy-independent Woods–Saxon potential has a depth given by $\left(\frac{m}{m_k}\right)U_0 = U_0'$.[115] For the nucleons of the ^{11}Li core, namely, ^{9}Li, $m_k = m(1 + 0.4)^{-1} \approx 0.7m$. For the halo neutrons[116], $m_k/m = (1 + \mathcal{O} \times 0.4)^{-1}$, where $\mathcal{O}(= (R_0/R)^3)$ is the overlap between the core and the halo neutrons. Making use of the values $R_0 = 2.66$ fm and $R = 4.58$ fm (see Eq. (4.10.4)), one obtains $\mathcal{O} \approx 0.2$ (see (3.6.8)) and thus $m_k \approx 0.93\,m$.

---

[113] Cf. Bohr and Mottelson (1969), and refs. therein.

[114] See also Eq. (2.3.5). What in nuclear matter is called the k-mass and is a well-defined quantity, in finite systems like the atomic nucleus, in which linear momentum is not a conserved quantity, is introduced to provide a measure of the spatial nonlocality of the mean field and is defined for each state as the expectation value of the quantity inside the parentheses in Eq. (2.7.4), calculated making use of the corresponding single-particle wavefunction (see, e.g., Bernard and Giai (1981), in which case m_k is referred to as the nonlocality effective mass).

[115] See, e.g., Fig. 2.14 of Mahaux et al. (1985).

[116] Assuming a velocity-independent interaction v, the k-dependence of the mean field stems from the exchange (Fock) potential $U_x(\mathbf{r}, \mathbf{r}') = -\sum_i \varphi_i^*(\mathbf{r}')v(|\mathbf{r} - \mathbf{r}'|)\varphi_i(\mathbf{r})$ (linear in $\mathcal{O}$), while the central potential is written as $U(r) = \sum_i \int d\mathbf{r}' |\varphi_i(\mathbf{r}')|^2 v(|\mathbf{r} - \mathbf{r}'|)$, (independent of $\mathcal{O}$). It is of notice that the coupling between, e.g., the low-lying quadrupole vibration of the core (^{8}He) and a halo neutron is also linear in $\mathcal{O}$, i.e., $\langle H_c \rangle_{2^+(\text{(core)},\text{n (halo)})} = \beta_2 \left\langle \frac{R_0}{\sqrt{5}}\frac{\partial U}{\partial r} \right\rangle \mathcal{O}\langle j||Y^2||1/2 \rangle$, where $\langle j||Y^2||1/2\rangle \approx 0.7$ ($j = 5/2, 3/2$), and $\left\langle R_0 \frac{\partial U}{\partial r}\right\rangle \approx 1.44U_0 \approx -60$ MeV; see Brink and Broglia (2005), Eqs. (D20), p. 303, and (D26), p. 304. See also Fig. 2.7.1.

2.7.2 Physical Particles

Within the framework of the above scenario, the complexity of the many-body nuclear Hamiltonian

$$H = T + v \tag{2.7.5}$$

has been reduced to

$$H_{HF} = T + U(r) + U_x(\mathbf{r}, \mathbf{r}'). \tag{2.7.6}$$

In other words, and making use of the expression

$$H = T + v(|\mathbf{r} - \mathbf{r}'|) = H_{HF} + \left(v(|\mathbf{r} - \mathbf{r}'|) - (U(r) + U_x(|\mathbf{r} - \mathbf{r}'|))\right), \tag{2.7.7}$$

the many-body nuclear Hamiltonian has been approximated by the Hartree–Fock Hamiltonian by neglecting the term within parentheses. This approximation (adding an appropriate spin-orbit potential), although providing much insight into the nuclear structure, for example, concerning the sequence of single-particle levels[117] (and associated magic numbers), as well as in nuclear reactions, for example, nucleon-nucleus elastic phase shifts, disagrees with experiment in a number of points. In particular, it leads to a too low-level density of states around the Fermi energy and to infinite mean free paths, also for nucleons moving in states far removed from the Fermi energy, for example, deep hole states.

To move further, one has to go beyond independent particles as well as potential scattering motion. That is, one has to allow the particles moving in both bound and continuum states to interact with each other and among themselves through four-point vertices (see, e.g., diagrams (a) and (b) of Fig. 2.5.10) and by coupling to $\beta = \pm 2$ pairing and $\beta = 0$ surface mode (diagrams (c)–(h) and (i)–(l), respectively). Let us exemplify the consequences of the interaction between elementary modes of excitation in the case of ^{11}Li.

2.7.3 $^{11}_{3}Li_8$ Structure in a Nutshell

The sequence of single-particle levels for $^{10}_{3}Li_7$ associated with the mean field potential (2.7.3) implies that the distance between the last occupied neutron state $0p_{1/2}(\epsilon_{1/2-} = -1.2$ MeV) and the first empty one, $1s_{1/2}(\epsilon_{1/2+} = 1.5$ MeV), is 2.7 MeV (see Fig. 2.7.1; see also Fig. 1.1.2). In other words, within the framework of the Mayer and Jensen sequence of levels, in ^{10}Li, the $0s_{1/2}, 0p_{3/2}$ neutron orbitals are fully occupied, while $0p_{1/2}$ carries one neutron, making ^{10}Li a neutron closed-shell system ($N = 8$) with a hole in it.[118]

---

[117] Mayer and Jensen (1955). It is of note however the exceptions found in the case of light, halo nuclei, displaying parity inversion.

[118]   It is of note that in the characterization of nuclear single-particle levels in terms of number of nodes n, orbital angular momentum l, and total angular momentum j ($l, 1/2$) (i.e., (nlj)), while univoque for l and j, allows for the flexibility whether to include the node at the origin. In the first case, $n = 1$ is the first level of a given (l, j), in the second, $n = 0$.

There is experimental evidence[119] which testifies to the fact that the first unoccupied states of ^{10}Li are a virtual $1/2^+$ ($\epsilon_{1/2^+} \approx 0.2$ MeV) and a resonant $1/2^-$ ($\epsilon_{1/2^-} \approx 0.5$ MeV) state. According to NFT, this is a consequence of the self-energy renormalization of the bare $1s_{1/2}$ and $0p_{1/2}$ states through a mainly PO (polarization) and CO (correlation) process, respectively[120] (Fig. 2.7.1). Parity inversion leads to the melting of the $N = 8$ closed shell in favor of the new magic number $N = 6$.

While ^{10}Li is not bound, ^{11}Li displays a two-neutron separation energy $S_{2n} \approx$ 400 keV. This value is the result of a subtle bootstrap mechanism. Being at threshold and basically not feeling a centrifugal barrier, the $1/2^+$ and $1/2^-$ states are essentially not available for the short-range bare NN-pairing interaction[121], requiring a Cooper pair binding mechanism mediated by the exchange of long-wavelength collective modes. Parity inversion ($\epsilon_{1/2^-} - \epsilon_{1/2^+} \approx 0.3$ MeV) is the natural scenario of a very low-lying collective dipole mode for two main reasons, first, the presence of a dipole particle-hole transition $1/2^+ \rightarrow 1/2^-$, with energy <1 MeV. The second one is related to the fact that the neutron halo in ^{11}Li can hardly sustain multipole surface vibrations, for example, quadrupole vibrations, aside from displaying a very large radius as compared to that of the closed shell core $^9_3$Li$_6$ and thus a small overlap with it. This (halo) controlled phenomenon[122] has a twofold effect: (1) to screen the bare NN-$^1S_0$ short-range pairing interaction making it subcritical; (2) to screen the (repulsive) symmetry interaction, and (consequently), to allow for the presence, at low energies, of a consistent fraction of the TRK-sum rule.

In fact, a 1^- resonance carrying ≈ 8 percent of the TRK sum rule and with energy centroid $\lesssim 1$ MeV has been observed,[123] thus, the connotation of soft

---

[119] See footnote 30 of Ch. 5. See also Sect. 5.2.3.

[120] Barranco et al. (2001, 2020), CO (involving ground state correlation vertices), PO (particle moving around closed shells and polarizing the core); see Fig. 1.2.4.

[121] Bennaceur et al. (2000); Hamamoto and Mottelson (2003, 2004).

[122] Within this context, one is essentially forced to make a subtle extension of the statement according to which the single-particle motion is the most collective of all nuclear motions (Mottelson (1962)), emerging from the same properties of the nuclear interaction (both bare and induced) as collective motion does, and in turn at the basis of the detailed properties of each collective mode, acting as scaffolds and filters of the variety of embodiments. In fact, one has to add the characterization of "physical" to "single-particle motion" (i.e., dressed) to englobe in the above statement also the present situation (parity inversion). In other words, while the bare $s_{1/2}$ and $p_{1/2}$ orbitals could never lead to low-lying dipole strength, the corresponding clothed, physical states do so in a straightforward manner, the dressing effects being in this case of the same order of magnitude of the mean field effects. Consequently: "physical, dressed single-particle motion, is one of the most collective of nuclear motions," seems to be the right statement. Within this context, one can point to a tumbling-like chain of incipient phase transitions: parity-inversion → dipole (dynamic) distortion → pygmy dipole resonance → (single) BCS Cooper pair formation, similar to that observed in the phase transition between normal-superconducting phases in metals, spontaneously breaking of translational invariance (crystal)→ phonons → Cooper pairs →BCS condensation (Nambu (1991); see also Broglia et al. (2019a)).

[123] Zinser et al. (1997); Nakamura et al. (2006); Shimoura et al. (1995); Ieki et al. (1993); Sackett et al. (1993); Kobayashi et al. (1989); Kanungo et al. (2015); Aumann (2019). Within this context, see also Barranco et al. (2001), Fig. 2 (a) and (b).

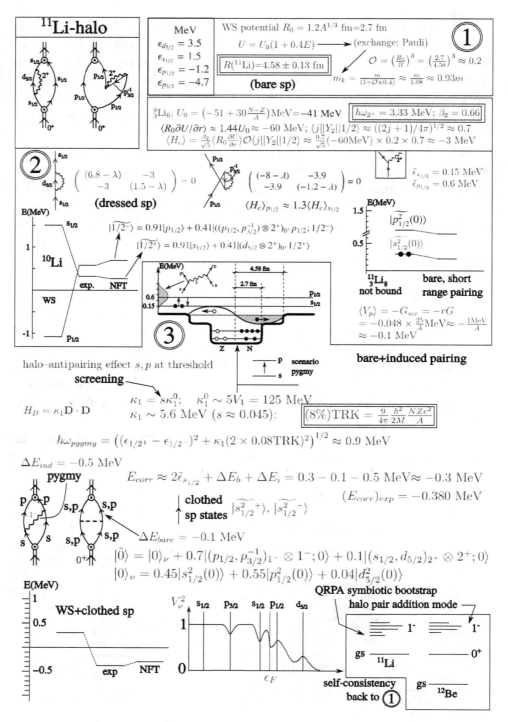

Figure 2.7.1 Parity inversion in ^{11}Li.

$E1$-mode (or pygmy dipole resonance (PDR)).[124] Exchanged between the heavily dressed neutrons moving in parity inverted states, it provides, together with the contribution of the strongly screened, bare NN-pairing interaction, the glue needed to bind the neutron halo Cooper pair to the ^{9}Li core. With some amount of experimental input[125], NFT allows to propagate dressing effects which not only renormalizes the mean field but overwhelms it, resulting in an overall account of the experimental findings. The discussion of the many-body effects at the basis of these phenomena is carried out in three steps, as schematically displayed in Fig. 2.7.1 and described below[126] (for details, see Sects. 3.6 and 5.2.3):

1. Starting with well-defined elements: Woods–Saxon (WS) potential, and the parameters characterizing the low-lying quadrupole vibration of the core ^{9}Li (**input**, double boxed quantities), calculate the single-particle levels and collective vibration (separable interaction) and determine the corresponding particle–vibration coupling vertices (strength and form factors). From the ratio of the WS radius (R_0) and of the observed one ($R(^{11}$Li) **input**), determine the overlap $\mathcal{O}$. Because $\mathcal{O} \ll 1$, the contribution of the exchange (Fock) potential to the empirical WS potential is small concerning the halo neutrons.[127] Consequently, the neutron halo k-mass m_k has a value close to the observed neutron mass m.

2. Making use of the above elements, one can dress the bare single-particle states, in particular the $s_{1/2}$ and $p_{1/2}$ states. Parity inversion ensues, with $1/2^+$ and $1/2^-$ at threshold. As a consequence, the $N = 8$ shell closure melts away, $N = 6$ becoming a new magic number, which testifies to the fact that the renormalization effects on single-particle motion of large-amplitude fluctuations can, in nuclei, be as important as static mean field effects. As a result, $^{10}_{3}$Li$_7$ is not bound. Adding one more neutron and switching on the bare pairing interaction (e.g., a contact force $V(r_{12}) = -4\pi V_0 \delta(\mathbf{r}_1 - \mathbf{r}_2)$ with constant matrix element[128] $G = 1.2$

---

[124] Broglia et al. (2019a).

[125] See, e.g., the corresponding discussion in Barranco et al. (2017).

[126] The information contained in this figure (see also Broglia et al. (2019b)) finds its origin going back to predictions made in 2001 (Barranco et al. (2001); see Fig. 3.6.2). Results from a *tour of force*–like experiment carried out at TRIUMF (Vancouver) by Tanihata et al. (2008) (see also Tanihata et al. (2013) and Fig. 7.1.2) provided evidence concerning the predicted phonon mediated pairing in nuclei (see Potel et al. (2010)). Figure 2.7.1 displays an analytic synthesis of a nuclear field theory solution – NFT, presented and discussed at the pedestrian level in Sect. 2.5 – of the structure of one- and two-neutron halo nuclei lying at the dripline, namely ^{10}Li (Barranco et al. (2020)) and ^{11}Li (Barranco et al. (2001)). Thus, of the unified interpretation of the results of almost two decades of experimental research carried out and being carried out at major laboratories around the world (TRIUMF (Vancouver; Tanihata et al. (2008)), ISOLDE (CERN; Jeppesen et al. (2006)), TRIUMF (Cavallaro et al. (2017)), etc.). Figs. 2.7.1, 2.7.2, and 2.7.3 condense also a number of facets about pygmy resonances in nuclei and of quantal dipole fluctuations mediated pairing, an intermediate (quasi) boson also found at the basis of the van der Waals interaction (see Broglia et al. (2019a)). Within this scenario, see Section 3.6, as well as the parallel carried out in connection with the role the van der Waals interaction plays in molecular systems in general, and proteins in particular (see Sect. 3.7 and App. 3.B).

[127] See footnote 116 of the present chapter.

[128] Brink and Broglia (2005), pp. 40–42.

fm$^{-3}V_0/A \approx (25/A)$ MeV), the screening $r = \frac{(M_j)\text{halo}}{(M_j)\text{core}} \approx \frac{2}{2j+1}\left(\frac{R_0}{R}\right)^3 \approx 0.048$ (see Eq. (3.6.3)) resulting from the poor overlap between halo and core neutrons leads to a value of the strength of the pairing interaction $(G)_{scr} = r \times G$ which is subcritical, and thus to an unbound system. In fact, $G_{scr} = 0.048 \times 25/A$ MeV ≈ 0.1 MeV and $\Delta E = 2\tilde{\epsilon}_{s_{1/2}} - G_{scr} \approx 0.3$ MeV $- 0.1$ MeV ≈ 0.2 MeV. Summing up, $G_{scr} < G_c$, G_c, being the minimum value of the pairing coupling constant leading to a bound state.[129]

3. The sloshing back and forth of the halo neutrons against the core nucleons moving in phase leads to a pygmy dipole mode which feels a strongly screened repulsive symmetry potential. In keeping with the fact that the strength of the bare dipole–dipole interaction $\kappa_1^0 \sim 1/R^2(^{11}\text{Li})$, the screening factor can be estimated as $((R(^{11}\text{Li})/\xi)^2 (R = 4.58$ fm, $\xi = 20$ fm (correlation length of the halo Cooper pair), $s \approx 0.052)$, the value obtained in Sect. 3.6 being $s = 0.043$. In other words, while it takes a quantity proportional to $5V_1 = 125$ MeV to separate protons from neutrons in the core, this value is reduced to $s \times 5V_1 = 5.4$ MeV $((V_1)_{scr} = sV_1 \approx 1$ MeV) for halo neutrons. This, as well as the small energy difference between the parity inverted levels, is at the basis of the fact that ≈ 8 percent of the Thomas–Reiche–Kuhn sum rule (**input**) is found at low energy ($\lesssim 1$ MeV). Another way to say the same thing is that $(V_1)_{scr} = sV_1$ is at the basis of the fact that the $\tilde{s}_{1/2} - \tilde{p}_{1/2}$ energy difference ($\Delta\tilde{\epsilon} \approx 0.45$ MeV) is only increased by a modest value ($\hbar\omega_{pygmy} \lesssim 1$ MeV $\approx 10^{21}$ Hz) and that, at the same time, the $E1$-strength remains essentially unchanged, not being shifted to higher energies by the GDR, typical values of the dipole strength in the case of nuclei lying along the stability valley being $\approx 10^{-4}B_W(E1)$ for low-energy single-particle transitions, while in the present case one finds a value close to $1B_W(E1)$.

---

[129] At the basis of superconductivity one finds the result obtained by Cooper (1956). He worked out the problem of two electrons interacting through an attractive interaction above a quiescent Fermi sea. Thus, all but two of the electrons are assumed to be noninteracting. The background of electrons enter the two-particle problem only through the Pauli principle by blocking states below the Fermi surface from participating in the remaining two-particle problem. If one measures the kinetic energy ϵ_k relative to its value at the Fermi surface only states with $\epsilon_k > 0$ are available to the interacting pair of electrons. Cooper found that a bound state of the pair always exists for arbitrarily weak coupling as long as the potential is attractive near the Fermi surface, a mechanism which implies the instability of the normal phase. Cooper pair binding is a rather remarkable result for the usual two-body problem. If one has only two particles coupled by an attractive interaction, they would not form bound states unless the attractive interaction exceeds a certain critical value. In other words, for a two-body system with an attractive interaction, a bound state exists for any value of the strength, provided the density of levels $N(E)$ ($E = |\epsilon - \epsilon_F|$) is different from zero around the Fermi energy ($N(0) \neq 0$). This is so for one- and two-dimensional systems, in which case $N(E) \sim E^{-1/2}$ and E^0 (constant) respectively, while for 3D-systems $N(E) \sim E^{1/2}$ (Fermi gas model). Cooper model reduces the three-dimensional to a lower-dimensional (essentially 2D) quantal system (Gor'kov (2012); see also Cohen-Tannoudji and Guéry-Odelin (2011)). In ^{11}Li the last two neutrons are very weakly bound. Consequently they move away from the nucleus core ^{9}Li, lowering in the process their relative momentum and forming a misty cloud or halo (Austin and Bertsch (1995)). One can thus view this system as the nuclear embodiment of a Cooper pair. The question then arises, why is there a critical value for the pairing interaction? The answer is likely related to spatial quantization and a not quiescent Fermi sea, features associated with the fact that the nucleus is a finite quantal many-body system.

The two halo neutrons dressed by the vibrations of the core (heavy arrowed lines in the lowest left corner of Fig. 2.7.1) and interacting through the bare NN-pairing force are not bound. Consequently, the pygmy resonance will fade away almost as soon as it is generated (essentially lasting the neutron transversal time $\approx 10^{-21}$ s in, e.g., a ^{9}Li(t, p) reaction), this, unless it is exchanged between the two neutrons, forcing them to jump from the configuration $s^2_{1/2}(0)$ at threshold $(2 \times \tilde{\epsilon}_{s_{1/2}} \approx 0.3$ MeV) into $p^2_{1/2}(0)$, also close to threshold $(2 \times \tilde{\epsilon}_{p_{1/2}} \approx 1.2$ MeV). In other words, one finds the dipole pygmy resonance acting as an intermediate boson which couples to the halo neutrons with a strength $\Lambda \approx 0.5$ MeV (QRPA calculation). Thus the corresponding $|gs(^{11}\text{Li})\rangle$ correlation energy $E_{corr} \approx -0.3$ MeV is mainly due to the pygmy dipole exchange process.

The resulting symbiotic halo pair addition mode of ^{11}Li can, in principle, be used as a building block of the nuclear spectrum, amenable to be moved around. A possible candidate for such a role is the first excited 0^+ state of ^{12}Be, together with the associated dipole state built on it and eventually other fragments of the associated $E1$-low energy strength.

To calculate the pygmy dipole resonance (PDR) of ^{11}Li, one needs to know the ground state of this nucleus (halo-pair addition mode) so as to be able to determine microscopically the occupation factors V^2_ν of the $0s_{1/2}, 0p_{3/2}, \epsilon s_{1/2}, \epsilon p_{1/2}, \epsilon d_{5/2}, \ldots$, and so on states. This is the basic input needed to work out the corresponding QRPA equations, whose diagonalization provides the energy, transition density, and $E1$-transition probability of the mode. But to do so, one is required to know the PDR. Having arrived to this point, the protocol schematically presented in Fig. 2.7.1 requires going back to **1** and repeating the whole procedure until convergence is achieved.

2.7.4 ^{11}Li$(p, p)^{11}$Li and Transfer Reaction Channels

Because NFT rules have, a priori, no limitations concerning whether or not the excitations studied lie in the continuum, or whether the single-particle motion displays asymptotic waves, it allows for a unified description of structure (s) and reactions (r) (NFT$_{(s+r)}$). An example of the above statement is provided by Fig. 2.7.2. Graph (a) describes one of the processes contributing to the elastic reaction ^{11}Li$(p, p)^{11}$Li as the system propagates in time, namely, a polarization contribution to the global (mean field) optical potential describing proton elastic scattering off ^{11}Li. In what follows we describe the processes taking place in the interval of time t_1–t_{11}, starting from the $t = 0$ situation in which a proton impinges[130] on ^{11}Li.

At time $\mathbf{t}_1$, the halo pair addition mode $|0_\nu\rangle$ couples to a pure, bare configuration $s^2_{1/2}(0)$. At time $\mathbf{t}_2$, due to the zero-point fluctuations associated with the quadrupole

---

[130] To be carried out in inverse kinematics, i.e., a beam of ^{11}Li on a ^{1}H target.

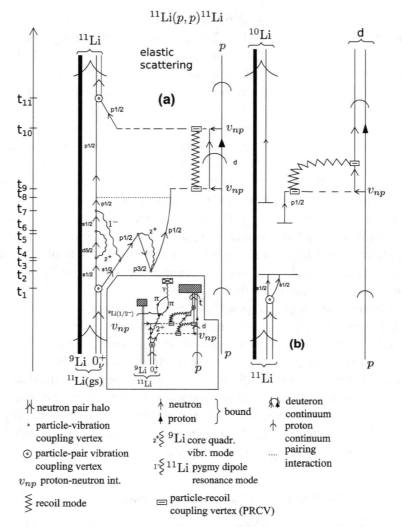

Figure 2.7.2 (a) $NFT_{(s+r)}$ diagram describing one of the processes contributing to the elastic reaction $^{11}Li(p,p)^{11}Li$ as the system propagates in time (polarization contribution to the global (mean field) optical potential). In the inset, a schematic NFT diagram describing the process $^{11}Li(p,t)^9Li(1/2^-)$ is displayed. A dashed open rectangle indicates the particle–recoil coupling vertex. A crossed box represents a γ-detector, while hatched rectangles stand for particle detectors. (b) Schematic NFT diagram describing the reaction $^{11}Li(p,d)^{10}Li$, i.e., same as in (a) up to time t_8 (the reason for which no details are repeated between t_2 and t_8). From there on, the deuteron continues to propagate to the detector, bringing to it, aside from nuclear structure information, the information resulting from its interaction with the recoil mode. After Broglia et al. (2016).

vibration of the ^{11}Li core, namely, ^{9}Li, the virtual state $((p_{1/2}, p_{3/2}^{-1})_{2^+} \otimes 2^+)_{0^+}$ is created. At time $\mathbf{t_4}$, one of the $s_{1/2}$ neutrons excites the quadrupole vibration of the core, reabsorbing it at time $\mathbf{t_6}$. As a result of this self-energy process, its energy is lowered to threshold, becoming a virtual state. The other $s_{1/2}$ neutron state excites at $\mathbf{t_3}$ the PDR and moves into the $p_{1/2}$ orbital, after which, due to the Pauli principle, it becomes exchanged[131] with the homologous $p_{1/2}$ of the $(p_{1/2}, p_{3/2}^{-1})_{2^+}$ configuration, an exchange process which is completed by time $\mathbf{t_5}$. As a result, the $p_{1/2}$ state undergoes a conspicuous repulsion, becoming a resonant state.[132] At time $\mathbf{t_7}$ the $s_{1/2}$ neutron absorbs the PDR and moves into a $p_{1/2}$ state. At time $\mathbf{t_8}$ it interacts, through the bare $^1S_0$-(pairing) interaction, with the other $p_{1/2}$ neutron. Before the two $p_{1/2}$ neutrons couple to the $|0_\nu\rangle$ state, one of them is picked up at time $\mathbf{t_9}$ under the action of the proton–neutron interaction v_{np}, by the projectile (proton), to form a virtual deuteron,[133] the recoil effect associated with the new mass partition being taken care of by the particle-recoil coupling vertex and associated recoil mode (jaggy line). At time $\mathbf{t_{10}}$, under the effect of v_{np}, the neutron of the virtual deuteron is transferred back to the virtual ^{10}Li nucleus to form again ^{11}Li and thus the original mass partition, as testified by the fact that the recoil phonon (jaggy line initiated at time $\mathbf{t_9}$) is reabsorbed at a second particle–recoil coupling vertex. The resulting $(p_{1/2}^2(0))$ configuration couples, at time $\mathbf{t_{11}}$, to the $|0_\nu\rangle$ state, leading to the $|^{11}\text{Li(gs)}\rangle$ state. The system has thus returned to the entrance channel configuration, namely, that of ^{11}Li $+ p$, which propagates to $t \rightarrow +\infty$, the proton eventually bringing the nuclear structure information to the detector.

In other words, the real and imaginary nonlocal and ω-dependent processes shown in Fig 2.7.2 (a) contribute to the polarization component of the optical potential[134] and to its A-dependence. It is of note that Fig. 2.7.2 exemplifies

---

[131] The (inevitable) crossing between the two $p_{1/2}$ arrowed lines; see Fig. 2.7.1, diagram at the center of the panel. indicated by a (double circled) number 2, and the corresponding one in the upper left corner (^{11}Li-halo); see also diagrams (a) and (b) of Fig. 1.2.4.

[132] It is of note that the fact that to follow the dressing of the $p_{1/2}$ neutron by the quadrupole vibration of the core ^{9}Li, one has to go backward in time in a N-like trajectory testifies to the fact that the hole states play, in nuclei, the role antiparticles do in field theory, in particular, in QED.

[133] In the time interval t_9–t_{10}, the two partners of the halo Cooper pair can, in principle, be at a relative distance of more than a nuclear diameter.

[134] Concerning the optical potential, we refer to footnote 26 of Ch. 1. In relation with the program of NFT$_{(s+r)}$ one can mention that Landau felt that Feynman diagrams have an independent basic importance, because of the possibility of relating them directly to physical observables. Feynman diagrams allow to describe processes where one set of particles with given energies, momenta, angular momenta, go in and another set (or the same) comes out. At the basis of this approach one finds vertex processes and dispersion relations. Now, vertex processes can simply mean the variety of processes connecting the incoming particles with the outgoing ones. In other words, within the present framework the processes taking place between times t_2–t_{11} (Figs. 2.7.2 (a) and (b)) and t_1–t_2 (Fig. 2.7.3 (b)) (Landau (1959); ter Haar (1969)). See also Sect. 5.3.5.

the elements needed to extend and formalize NFT rules of structure so as to be able to deal, on equal footing, or, more properly, in a unified fashion, with reactions.

As schematically shown in Fig. 2.7.1 (see also Sect. 3.6.1), to each nuclear structure process displayed in diagram (a) of Fig. 2.7.2 corresponds a specific equation, amplitude, and so on, and thus a number with appropriate units. This is also so regarding the reaction aspects of the diagram, and thus reaction amplitudes, formfactors, and absolute differential cross sections. Within this context we refer to Fig. 4.1.2 (see also App. 7.A) in connection with the inset to Fig. 2.7.2 (a), as well as to graphs (a) and (b) of Fig. 2.7.3.

In Fig. 2.7.2 (b) one assumes the same processes to take place as in (a) up to time t_8 (the reason for which no details are repeated between t_2 and t_8). From there on the deuteron continues to propagate to the detector, and the effect of the particle–recoil coupling vertex is to be worked out and the corresponding outgoing distorted waves modified accordingly. Likely, the halo neutron in ^{10}Li will break up before the system can be recorded by the particle detector. Summing up, in the center of the mass reference frame, both p and ^{11}Li display asymptotic states in entrance as well as in exit channels in case (a), and only in the entrance channel in case (b), while in the exit channel, only ^{10}Li(^{9}Li+n) and the deuteron do so.

Additional examples of the NFT diagrams of structure and reactions are given in Fig. 2.7.3. In (a) a contribution associated with the reaction ^{11}Li(p, t)^{9}Li(gs) is shown, while in (b) one associated with the population of the first excited $1/2^-$ (2.69 MeV) state of ^{9}Li is displayed. We refer to Figs. 2.7.4 and 2.7.5 for a compact graphical representation of this last process, which provided evidence of phonon mediated pairing interaction in nuclei as theoretically predicted.[135]

Concerning the question of the Pauli principle in reaction processes, for example, between the incoming proton and the collective modes of the core (^{9}Li), we refer to Fig. 2.7.6.

2.8 Characterization of an Open-Shell Nucleus: ^{120}Sn

In this section we aim to give an overall view of the versatility of (NFT)$_{\text{ren (s+r)}}$ to describe both structure and the results of experimental probes which provide an essentially "complete" characterization of an atomic nucleus or, better, of a small island of closely related nuclei (isotopes). Details of the structure and reaction calculations, techniques, and computer codes employed to achieve such characterization are provided in the following chapters.

In keeping with the fact that $^{120}_{50}$Sn$_{70}$ is a typical example of a single open-shell superfluid nuclei, it has been studied extensively with a variety of probes, both

---

[135] Barranco et al. (2001); Tanihata et al. (2008); Potel et al. (2010); Tanihata et al. (2013); Beceiro-Novo et al. (2015).

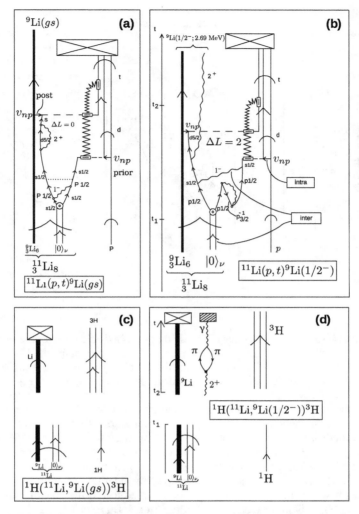

Figure 2.7.3 NFT$_{(s+r)}$ diagram describing a contribution to the reaction amplitude (a) ^{11}Li$(p, t)^9$Li(gs), (b) ^{11}Li$(p, t)^9$Li$(1/2^-)$, (c) ^{1}H(^{11}Li,^{9}Li(gs))^{3}H, and (d) ^{1}H(^{11}Li,^{9}Li$(1/2^-)$)^{3}H. These last two processes make it concrete, through inverse kinematics, the gedankenexperiments (a) and (b). Time is assumed to run upward. A single arrowed line represents a fermion (proton) (π) or neutron (ν), a double arrowed line two correlated nucleons. In the present case are two correlated (halo) neutrons (halo-neutron pair addition mode $|0\rangle_\nu$). A heavy arrowed line represents the core system $|^9$Li(gs)$\rangle$. A standard pointed arrow refers to structure, while "round" arrows refer to reaction. A wavy line represents (particle-hole) collective vibrations, like the low-lying quadrupole mode of ^{9}Li, or the dipole pygmy resonant state, which, together with the bare pairing interaction (horizontal dotted line), binds the neutron halo Cooper pair to the core. A short horizontal arrow labels the proton–neutron interaction v_{np} responsible for the single-particle transfer processes, represented by a horizontal dashed line. A dashed open square indicates the particle–recoil coupling vertex. The jagged line represents the recoil mode resulting from the mismatch between the relative center of mass coordinates associated with the mass partitions ^{11}Li+p, ^{10}Li+d (virtual), and ^{9}Li+t. The γ-detector is represented by a hatched box, the particle detector by a crossed rectangle. After Broglia et al. (2016).

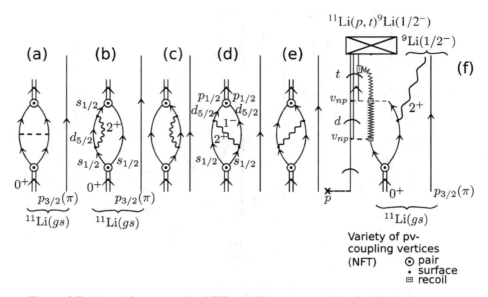

Figure 2.7.4 (a–e) Lowest-order NFT$_{(s+r)}$ diagrams associated with the processes contributing to the binding of the neutron halo Cooper pair (double arrowed line) of ^{11}Li to the core ^{9}Li through the bare pairing interaction (dashed line) as well as the exchange of the core quadrupole phonon and of the soft dipole mode of ^{11}Li (wavy line). Single arrowed lines describe the nucleon independent-particle motion of neutrons ($s_{1/2}$, $d_{5/2}$, etc.) as well as of the $p_{3/2}(\pi)$ proton considered as a spectator. (a) Bare pairing interaction, four-point vertex (horizontal dotted line). (b, c) Self-energy, effective mass polarization (PO) process dressing the $s_{1/2}(\nu)$ single-particle state (a similar diagram, but corresponding to correlation (CO) processes dressing the $p_{1/2}$ state is not shown, see Fig. 2.7.2). (d, e) Vertex correction (induced pairing interaction) renormalizing the vertex with which the pair addition mode couples to the fermions (dotted open circle). (f) NFT diagram describing the reaction ^{11}Li$(p, t)^9$Li$(1/2^-)$ populating the first excited state of ^{9}Li, the dashed horizontal line starting with a cross stands for the (p, t) probe. The successive transfer of the two halo neutrons (^{11}Li(gs)$+p \rightarrow{}^{10}$Li$+d \rightarrow{}^9$Li$(1/2^+) + t$) is shown, in keeping with the fact that this process provides the largest contribution to the absolute differential cross section. The jagged line represents the recoil mode carrying to the outgoing particle the effect of the momentum mismatch associated with the transfer process (recoil).

elastic and anelastic (Coulomb excitation and subsequent γ-decay as well as inelastic scattering), and also one- and two-particle transfer reactions. The corresponding absolute differential cross sections and transition probabilities involve as targets and residual systems the island of superfluid nuclei [118,119,120,121,122]Sn.

A theoretical description of the variety of observables has been carried out solving the Nambu–Gorkov equation to propagate the different (NFT)$_{ren (s)}$ processes, which dress the single-particle states and renormalize the bare pairing

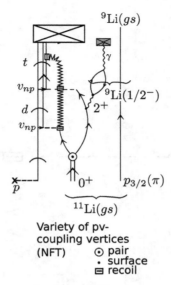

Variety of pv-
coupling vertices
(NFT)   ⊙ pair
         • surface
         ⊟ recoil

Figure 2.7.5 Gedankenexperiment: NFT$_{(s+r)}$ γ-ray coincidence between ^{1}H(^{11}Li,^{9}Li)^{3}H and ^{9}Li(gs)+γ (E2; 2.69 MeV) decay process. The virtual quadrupole phonon associated with self-energy and vertex correction processes becomes real through the action of the ^{1}H,^{3}H (p, t) external field. Thus it is not only that recoil modes are "measured" by particle detectors in connection with outgoing particles which have asymptotic wavefunctions, but also the quadrupole vibration, whose eventual γ-decay (see Fig. 5.3.2 (II)) can be measured by the γ-detector. The apparent bubble process made out of a proton particle-hole component of the quadrupole mode does not contradict NFT rule III (Sect. 2.5). In fact, the initial vertex (solid dot) corresponds to a nuclear PVC, while the second vertex corresponds to the proton excitation to the electromagnetic field, leading to γ-decay.

interaction, as well as the spectroscopic and the transition amplitudes.[136] With the help of the computer codes SINGLE and COOPER (App.7.A), tailored to propagate the (NFT)$_{ren (s)}$ spectroscopic amplitude content to the detector, one- and two-nucleon transfer absolute differential cross sections were worked out. In Fig. 2.8.1, theory is compared to experiment in terms of renormalized energies, absolute differential cross sections, and electromagnetic transition probabilities.

In the ***upper left*** part of this figure, the cartoon representation displayed in Fig. 2.1.1 is here used to schematically illustrate[137] anelastic processes, Coulomb excitation, and the quadrupole γ-decay of ^{119}Sn (***middle left*** (a) experiment, (b) theory, (c) standard deviation between theory and experiment calculated as a function of

---

[136] See Idini (2013); Idini et al. (2012); Idini et al. (2015), and refs. therein; see also Broglia et al. (2016).

[137] While in Fig. 2.1.1 one refers to the excitation of an octupole mode, in the present case we deal with a quadrupole vibration.

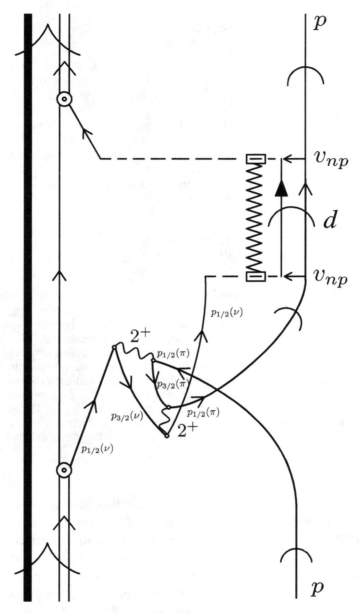

Figure 2.7.6 In keeping with standard direct reaction praxis, neither in Fig. 2.7.2 nor in 2.7.3 is antisymmetrization carried out between the impinging proton and the protons of ^{11}Li. Within the present discussion (^{11}Li(p, p)^{11}Li), an example of such processes corresponds to the exchange of a proton participating in the quadrupole vibration of the core, with the projectile, as shown in the figure. Such a process will not only be two orders higher in perturbation in the particle–vibration coupling vertex than the original one shown in graph (a) of Fig. 2.7.2. It will also be strongly suppressed by the square of the overlap between a proton moving in the continuum, and a $p_{1/2}$ proton of the ^{9}Li core. After Broglia et al. (2016).

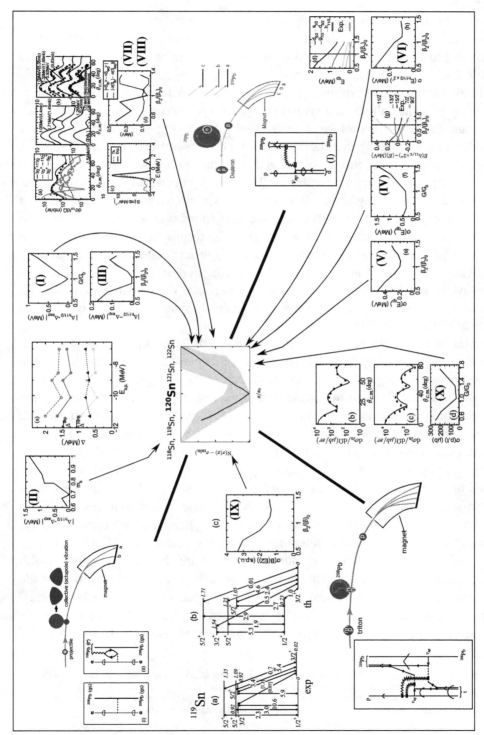

Figure 2.8.1 Characterization of ^{120}Sn. For the color version of this figure, refer to cambridge.org/nuclearcooperpair.

β_2 and displayed normalized with respect to the experimental value $(\beta_2)_0$ (**IX**)). This value, together with those also labeled by roman numerals, referred to below, was used to draw the center boxed nuclear structure landscape (see Fig. 2.4.1 and Table 2.4.1).

In the **bottom left** part of the figure, a schematic representation of an experimental setup to measure two-nucleon transfer processes is given. Also, a $(\text{NFT})_{\text{ren (s+r)}}$ diagram describing successive transfer is displayed (Fig. 2.1.3). In the **bottom middle** part of the figure, the theoretical absolute differential cross sections $d\sigma_{2n}/d\Omega$ (continuous curves) associated with the reactions $^{120}\text{Sn}(p, t)^{118}\text{Sn}(\text{gs})$ and $^{122}\text{Sn}(p, t)^{120}\text{Sn}(\text{gs})$ are displayed ((b) and (c)), in comparison with the experimental data (solid dots). The theoretical cross sections were recalculated as a function of the pairing strength G and the corresponding standard deviation with respect to the experimental value determined. The results are shown in (d) (**X**) as a function of G, normalized with respect to the (equivalent) value G_0 of v_{14} (Argonne $^1S_0$ bare NN-interaction; only the values corresponding to $^{120}\text{Sn}(p, t)^{118}\text{Sn}$ are displayed). It is of note that the corresponding relative standard deviation σ/L for $G = G_0$ ($L \equiv \sigma_{2n}$ being the experimental cross section), is quite small $(70\mu\text{b}/2250\mu\text{b} \approx 0.03).$[138]

In the **upper middle** part of the figure and under the label (a), the state-dependent pairing gap of the valence orbitals of ^{120}Sn are displayed, in comparison with the experimental findings ($\Delta^{exp} \approx 1.45$ MeV, arrow left). The gap associated with the lowest quasiparticle state $h_{11/2}$ calculated as a function of m_k (different Skyrme interactions) as well as of G/G_0 and $\beta_2/(\beta_2)_0$ have been used to work out the corresponding standard deviations with respect to the experimental findings and are displayed to the right ((**I**) and (**III**)) and the left (**II**) of (a).

In the **upper right** part of the figure, absolute differential cross sections and strength functions associated with the one-particle transfer processes $^{120}\text{Sn}(d, p)^{121}\text{Sn}$ are displayed. In (a), the absolute differential cross sections associated with the low-lying states $h_{11/2}, d_{3/2}, s_{1/2}$, and $d_{5/2}$ are shown (theory: continuous curve; data: solid dots). In (b) (left), the calculated $^{121}\text{Sn}(5/2^+)$ absolute differential cross sections (continuous curves) are shown and compared with the experimental data (right, solid dots; also given here are DWBA fits used in the analysis of the experimental data). In (c) the calculated strength function associated with the $5/2^+$ state (dashed curve) is compared to the data (continuous curve), while in (d) the differences between the centroid and width of the experimental

---

[138] This result is related with the fact that $\sigma \sim |\alpha_0|^2$, α_0 being the BCS order parameter, that is, the number of Cooper pairs participating in the BCS condensate which measures the deformation in gauge space. Because the state $|BCS\rangle$ describing phase-correlated, independent pair motion (Cooper pair condensate) is a coherent state displaying off-diagonal long-range order (ODLRO), it is not surprising that

$$|\alpha|^2 = \left|\sum_{\nu>0} U_\nu V_\nu\right|^2 = \left|\langle BCS|P^\dagger|BCS\rangle\right|^2 = |\langle BCS|P|BCS\rangle|^2 \text{ plays the role of a physical,}$$

non-energy-weighted sum rule (Potel et al. (2017)). See Sect. 7.2.1.

and calculated $d_{5/2}$ strength function are shown as a function of the ratio $\beta_2/(\beta_2)_0$ in terms of solid and dashed curves (**(VII)** and **(VIII)**). In the **middle right** part of the figure, a cartoon representation of a setup to measure one-nucleon transfer reactions is displayed. Also shown is a (NFT)$_{\text{ren (s+r)}}$ diagram describing the process (see Fig. 2.1.2).

In (d) (**lower right** part of the figure), the lowest quasiparticle energy values of each spin and parity are displayed as a function of $\beta_2/(\beta_2)_0$ in comparison with the data. The root-mean-square deviation between the experimental and theoretical levels as a function of $\beta_2/(\beta_2)_0$ and of G/G_0 is shown in (e) and (f), respectively (**(V)**, **(IV)**). In (g) the experimental energies of the members of the $h_{11/2} \otimes 2^+$ multiplet are compared with the theoretical values calculated as a function of the ratio $\beta_2/(\beta_2)_0$. Finally, in (h), the root-mean-square deviation between the experimental and theoretical energies of the members of the $h_{11/2} \otimes 2^+$ multiplet shown in (g) are given as a function of $\beta_2/(\beta_2)_0$ (**VI**).

Summing up, the nuclear structure landscape is well funneled, and theory provides an overall account of the data when the physical values of β_2, G, and m_k are used.[139]

2.9 Summary

In Fig. 2.8.1 we report on the (NFT)$_{\text{ren (s+r)}}$ analysis of a rather complete set of data aimed at characterizing the superfluid nucleus ^{120}Sn, namely, (**1**) ^{120}Sn$(p, t)^{118}$Sn(gs), ^{122}Sn$(p, t)^{120}$Sn(gs); (**2**) ^{120}Sn$(p, d)^{119}$Sn, ^{121}Sn$(p, d)^{120}$Sn; (**3**) ^{119}Sn$(\alpha, \alpha')^{119}$Sn (γ-decay); (**4**) quasiparticle energies; (**5**) quasiparticle lifetimes (strength functions); (**6**) state-dependent pairing gaps; and (**7**) $B(E2)$ transition probabilities.

Arbitrarily forcing the particle–vibration coupling (PVC) strength, the strength of the bare pairing force, and the value of the k-mass to depart from their "physical" values, one can test the robustness of the NFT$_{\text{ren (s+r)}}$ picture of ^{120}Sn given, and of the funneled character of the associated nuclear structure and reaction landscape.

In a very real sense, this, namely, the results collected in Fig. 2.8.1, is a nucleus, that is, the summed experimental and theoretical structural information accessed through asymptotic states, outcome of probing the system with a "complete" array of experiments interpreted with the help of a unified structure and reaction theoretical scheme as, for example, that provided by (NFT)$_{\text{ren (s+r)}}$.

2.A NFT$_{(s+r)}$: Linear Theory

NFT is linear in the particle–vibration coupling vertices associated with the variety of modes (see, e.g., Fig. 2.5.10). This is also true concerning its extension to

---

[139] For details, see Idini et al. (2015) and Broglia et al. (2016); see also Fig. 2.4.1.

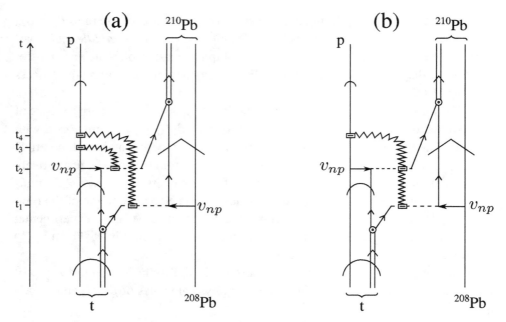

Figure 2.A.1 NFT$_{(s+r)}$ diagram describing the successive transfer contribution to the reaction ^{208}Pb$(t, p)^{210}$Pb(gs), that is, the population of the lowest energy, monopole pair addition mode of ^{208}Pb. Concerning the different symbols used, we refer to Figs. 2.1.3 and 2.7.2, in particular concerning the recoil mode (jagged line) and the associated particle-recoil coupling vertex (dashed open rectangle). (a) Of the variety of contributions associated with the different sequence of events taking place at times t_1, t_2, t_3, and t_4, one has chosen that in which the information of a mass $2m$ transfer process to be carried out to the detector is conveyed to the outgoing particle at $t_3(t_4)$ by the recoil process associated with the nucleon transfer at $t_2(t_1)$; diagram (b) is used only for simplicity when one is interested in discussing aspects of the (s+r)-diagram other than recoil.

describe reaction processes, as can be seen from graph (a) of Fig. 2.A.1. For simplicity, this diagram and similar ones are drawn as displayed in Fig. 2.A.1 (b) (see, e.g., Fig. 2.1.3 (I)). However, in all cases, the effects of recoil are properly taken into account following (a), as shown, for example, in Fig. 4.1.2 as well as in Sect. 6.1 within the framework of second-order DWBA and in Sect. 6.5 (Eq. (6.5.9)) in the semiclassical approximation.

2.B NFT and Reactions

Nuclear field theory was systematically developed to describe nuclear structure processes. This fact did not prevent the translation into this graphical language of expressions which embodied the transition amplitude of a variety of reaction processes, in particular, second-order (in v_{np}) transition amplitudes associated with two-nucleon transfer reactions.[140]

The new feature to be considered regarding transfer processes and not encountered either in structure or in elastic or anelastic processes is the graphical representation of recoil effects, that is, a physical phenomenon associated with the change in the coordinate of relative motion reflecting the difference in mass partition between entrance (also intermediate, if present) and exit channels. It is a phenomenon which mixes in a nonanalytically separable fashion structure and reaction degrees of freedom.

Nuclear structure (intrinsic) processes do not affect the center of mass of the nucleus, with a proviso. In fact, the shell model potential violates translational invariance of the total nuclear Hamiltonian and thus single-particle excitations can be produced by a field proportional to the total center-of-mass coordinate. The translational invariance can be restored by including the effects of the collective field generated by a small displacement α of the nucleus. Such a displacement, for example, in the x-direction, gives rise to a coupling which can be written as

$$H_c = \kappa \alpha F, \qquad (2.B.1)$$

where

$$F = -\frac{1}{\kappa} \frac{\partial U}{\partial x} \qquad (2.B.2)$$

and

$$\kappa = \int \frac{\partial U}{\partial x} \frac{\partial \rho_0}{\partial x} d\tau = -A \left\langle \frac{\partial^2 U}{\partial x^2} \right\rangle, \qquad (2.B.3)$$

corresponding to a normalization of α such that $\langle F \rangle = \alpha$. It is of note that both κ and U are negative for attractive fields.[141]

The spectrum of normal modes generated by the field coupling (2.B.1), namely, by a Galilean transformation of amplitude α, contains an excitation mode with zero energy for which zero-point fluctuations diverge in just the right way to restore translational invariance to leading order[142] in α. In fact, while the zero-point fluctuations

$$\lim_{\omega_\alpha \to 0} \left(\frac{\hbar \omega_\alpha}{2C_\alpha} \right)^{1/2} = \lim_{\omega_\alpha \to 0} \left(\frac{\hbar^2}{2D_\alpha \hbar \omega_\alpha} \right)^{1/2}, \qquad (2.B.4)$$

diverge, the inertia remains finite and equal to $D_\alpha = Am$, as expected, C_α being the restoring force constant. The additional dipole roots include, in particular, the

---

[140] Broglia (1975); Broglia and Winther (2004).

[141] Bohr and Mottelson (1975), p. 444.

[142] Within this context, see Sect. 1.4.3 in connection with gauge invariance restoration.

isoscalar dipole modes associated with $\hat{D} = \sum_{i=1}^{A} r_i^3 Y_{1\mu}(\hat{r}_i)$, which can be viewed as a nonisotropic compression mode.[143]

The operators leading to transformations associated with the change in the coordinates of relative motion (recoil effects) in nuclear reaction processes are Galilean operators ($\sim \exp\left(\mathbf{k}_{\alpha\beta} \cdot (\mathbf{r}_\beta - \mathbf{r}_\alpha)\right)$). Their action (on, e.g., the entrance channel distorted waves), as that of (2.B.1) on the shell model ground state, can be graphically represented in terms of NFT diagrams (or eventual extensions of them). In Figs. 2.1.2 and 2.1.3 as well as 2.7.2–2.7.6, they are drawn in terms of jagged lines. How does one calculate such couplings? Let us elaborate on this point.

When one states that the small displacement α of the nucleus leads to a coupling (2.B.1), one means a coupling between the single-particle and the collective displacement of the system as a whole. When one talks about the spectrum of normal modes associated with such a coupling, one refers to the harmonic approximation (RPA). Thus, to the solutions of the dispersion relation[144],

$$-\frac{2\kappa}{\hbar} \sum_i \frac{|F|_i^2 \omega_i}{\omega_i^2 - \omega_a^2} = 1, \tag{2.B.5}$$

where the sum is over dipole particle-hole excitations. This dispersion relation can be represented graphically through the diagrams shown in Fig. 1.2.3 (inset). In particular, α acting on the vacuum creates the collective mode. This can also be seen by expressing α in second quantization, namely (see Eq. (1.1.11)),

$$\alpha = \sqrt{\frac{\hbar\omega_\alpha}{2C_\alpha}}(\Gamma_\alpha^\dagger + \Gamma_\alpha), \tag{2.B.6}$$

where $\sqrt{\hbar\omega_\alpha/2C_\alpha} = \sqrt{\frac{\hbar^2}{2D_\alpha}\frac{1}{\hbar\omega_\alpha}}$ is the zero-point amplitude of the collective (displacement) mode. Now, none of the above arguments lose their meaning in the case in which there is a root with $\omega_\alpha = 0$, also in keeping with the fact that inertia remains finite. In Figs. 2.7.2–2.7.6 we do something similar to what is done in Fig. 1.2.3 (inset). The dot, which in this figure represents the particle–vibration coupling, is replaced by a small dashed open square, which we label "particle–recoil coupling vertex" (see Fig. 2.7.2). It constitutes a graphical mnemonic for counting the degrees of freedom that are at play, in this case, the coordinates of relative motion. Also is the fact that in connection with the appearance of such vertices one has to calculate matrix elements of precise formfactors which involve the recoil phases.

Concerning the actual calculation of a particle-mode vertex in which $\omega_a \to 0$, an empirical way out is that of a coarse-grained-like symmetry restoration. In this

[143] See, e.g., Colò et al. (2000).
[144] Bohr and Mottelson (1975), Eq. (6-244); Brink and Broglia (2005), Sect. 8.3.1.

case, κ is adjusted in such a way that the lowest solution of Eq. (2.B.5), although being smaller than the rest of them, remains finite.[145]

2.B.1 Transfer

We now consider, within the framework of the semiclassical[146] approximation, a general reaction

$$a + A \rightarrow b + B, \tag{2.B.7}$$

in which the nucleus a impinges on the nucleus A in the entrance channel $\alpha(a, A)$ and where the two nuclei in the exit channel β, namely, b and B, may differ from those in α, by the transfer of one or more nucleons.

In the center-of-mass system, the total Hamiltonian may be written as

$$\begin{aligned} H &= T_{aA} + H_a + H_A + V_{aA} \\ &= T_{bB} + H_b + H_B + V_{bB}, \end{aligned} \tag{2.B.8}$$

where T_{aA} is the kinetic energy of the relative motion in channel α,

$$T_{aA} = -\frac{\hbar^2}{2m_{aA}} \nabla^2_{aA}, \quad (m_{aA} = \frac{m_a m_A}{m_a + m_A}), \tag{2.B.9}$$

and similarly for T_{bB}. Assuming the nuclei in (2.B.7) to be heavy ions, one can solve the time-dependent Schrödinger equation

$$i\hbar \frac{\partial \Psi}{\partial t} = H\Psi \tag{2.B.10}$$

with the initial condition that a and A are in their ground states and that the relative motion is described as a narrow wavepacket of rather well-defined impact parameter and velocity. Because of the quantal nature of the process under consideration, we study the quantal description in the limit of small wavelength of relative motion (semiclassical approximation). One expands Ψ on the channel wavefunctions

$$\Psi_\beta(t) = \Psi^b_m(\xi_b)\Psi^B_n(\xi_B)e^{i\delta_\beta}, \tag{2.B.11}$$

---

[145] Within this context we refer to Bohr and Mottelson (1975), p. 446. With no coupling H_c (Eq. (2.B.1)), the ZPF $\alpha_0^{(0)}$ of the nuclear CM are small ($\sim A^{-2/3}$). Thus it is possible to tune κ so as to make the ZPF associated with the lowest root large as compared to $\alpha_0^{(0)}$, but still compatible with the ansatz at the basis of RPA (small-amplitude harmonic vibrations). It is of note that a similar requirement, i.e., that of eliminating the contribution of the $\omega_\alpha \rightarrow 0$ (Goldstone mode), is also operative in connection with the calculation of the GDR in RPA (QRPA) (see Sect. 1.5), as well as in restoring gauge invariance in connection with the BCS solution of the pairing Hamiltonian (Sect. 1.4.3; see also Fig. 3.1.2).

[146] Assuming Eq. (2.B.7) to describe a transfer process induced by a heavy ion reaction with bombarding energy lying somewhat above the Coulomb barrier, the associated de Broglie reduced wavelength $\lambda = \hbar/\sqrt{2mE}$ is much smaller than the nuclear dimension, and the relative motion can be described in terms of classical trajectories. On the other hand, the particle transfer process is described fully quantum mechanically. The coupling between the intrinsic (structure) nucleon motion and the relative motion associated with the recoil effect is dealt with in terms of narrow wavepackets. See, e.g., Broglia and Winther (2004), and refs. therein.

where Ψ^b and Ψ^B describe the structure of the two nuclei and satisfy the equations

$$H_b \Psi^b_m(\xi_b) = E^b_m \Psi^b_m(\xi_b) \tag{2.B.12}$$

and

$$H_B \Psi^B_n(\xi_B) = E^B_n \Psi^B_n(\xi_B), \tag{2.B.13}$$

while ξ_b and ξ_B denote the intrinsic coordinates. The phase δ_β is defined by

$$\delta_\beta = \frac{1}{\hbar} \left\{ m_\beta \mathbf{v}_\beta(t)(\mathbf{r}_\beta - \mathbf{R}_\beta(t)) - \int_0^t \left(U_\beta(R_\beta(t')) - \frac{1}{2} m_\beta v_\beta^2(t') \right) dt' \right\}. \tag{2.B.14}$$

The index β labels both the partition of nucleons into b and B and the quantal states of the two nuclei. The quantity U_β is the ion–ion potential in this channel. It is equal to the expectation value of $V_\beta = V_{bB}$ in the channel β. The distance between the centers of mass of the two systems is denoted by

$$\mathbf{r}_\beta = \mathbf{r}_{bB} = \mathbf{r}_b - \mathbf{r}_B. \tag{2.B.15}$$

The quantity $\mathbf{R}_\beta$ and its derivative $\mathbf{v}_\beta = \dot{\mathbf{R}}_\beta$ describe the motion of the centers of mass of the wavepackets and satisfy the corresponding classical equation of motion,

$$m_\beta \dot{\mathbf{v}}_\beta = -\nabla_\beta U_\beta(\mathbf{R}_\beta). \tag{2.B.16}$$

The phase factor $e^{i\delta_\beta}$ in the channel wavefunction is essentially a Galilean transformation where an additional phase (related to the Q-value) has been added to eliminate, as far as possible, the diagonal matrix elements of the coupled equations. Using the notation $E_\beta = E^b_m + E^B_n$ and inserting the ansatz

$$\Psi = \sum_\beta c_\beta((r_\beta - R_\beta), t) \Psi_\beta(t) e^{-iE_\beta t/\hbar} \tag{2.B.17}$$

in Eq. (2.B.10), one obtains, assuming narrow wavepackets, the product of an amplitude $a_\beta(t)$ and a shape[147] $\chi_\beta(\mathbf{r} - \mathbf{R}_\beta(t), t)$, $(c_\beta = a_\beta \chi_\beta)$,

$$i\hbar \sum_\beta \dot{a}_\beta(t) \langle \Psi_\xi | \Psi_\beta \rangle_{\mathbf{R}_{\xi\gamma}} e^{-iE_\beta t/\hbar}$$
$$= \sum_\gamma \langle \Psi_\xi | V_\gamma - U_\gamma(r_\gamma) | \Psi_\gamma \rangle_{\mathbf{R}_{\xi\gamma}} a_\gamma(t) e^{-iE_\gamma t/\hbar}, \tag{2.B.18}$$

where the subindex on the matrix elements indicates that the integration over the degree of freedom of the two nuclei, the average center-of-mass coordinate

---

[147] See Broglia and Winther (2004), p. 294.

$\mathbf{r}_{\beta\gamma} = (\mathbf{r}_\beta + \mathbf{r}_\gamma)/2$, should be identified with the average classical coordinate, that is,

$$\mathbf{r}_{\beta\gamma} \rightarrow \mathbf{R}_{\beta\gamma} = \frac{1}{2}(\mathbf{R}_\beta + \mathbf{R}_\gamma), \tag{2.B.19}$$

and the functions $\langle \Psi_\xi | V_\gamma - U_\gamma(r_\gamma) | \Psi_\gamma \rangle_{\mathbf{R}_{\xi\gamma}}$ are the formfactors. The coupled equations (2.B.18) can be written in a more compact way by an orthogonalization procedure, which makes use of the *adjoint channel wavefunctions*

$$\omega_\xi = \sum_\gamma g_{\xi\gamma}^{-1} \Psi_\gamma, \tag{2.B.20}$$

where g^{-1} is the inverse of the overlap matrix

$$g_{\xi\gamma} = \langle \Psi_\xi | \Psi_\gamma \rangle, \tag{2.B.21}$$

that is,

$$\sum_\xi g_{\gamma\xi} g_{\xi\beta}^{-1} = \sum_\xi g_{\gamma\xi}^{-1} g_{\xi\beta} = \delta(\gamma, \beta). \tag{2.B.22}$$

With this definition,

$$(\omega_\xi, \Psi_\beta) = \delta(\xi, \beta), \tag{2.B.23}$$

which takes care of nonorthogonality. Making use of the above relations, one can rewrite (2.B.18) in the form

$$i\hbar \dot{a}_\beta(t) = \sum_\gamma \langle \omega_\beta | V_\gamma - U_\gamma | \Psi_\gamma \rangle_{\mathbf{R}_{\beta\gamma}} e^{i(E_\beta - E_\gamma)t/\hbar} a_\gamma(t). \tag{2.B.24}$$

That is, the proper transfer (tunneling) equations are obtained from (2.B.18) by a basis orthogonalization process.[148] Solving these coupled equations with the condition that at $t = -\infty$ the system is in the ground state of a and A (entrance channel α), that is, $a_\gamma(-\infty) = \delta_{\gamma,\alpha}$, one can calculate the differential cross section

$$\frac{d\sigma_{\alpha\rightarrow\beta}}{d\Omega} = P_{\alpha\rightarrow\beta} \sqrt{\left(\frac{d\sigma_\alpha}{d\Omega}\right)_{el} \left(\frac{d\sigma_\beta}{d\Omega}\right)_{el}}, \tag{2.B.25}$$

where $P_{\alpha\rightarrow\beta}$ is the absolute value squared of the transition amplitude $|a_\beta(t = +\infty)|^2$. It gives the probability that the system at $t = +\infty$ is in the final channel. The quantities $(d\sigma/d\Omega)_{el}$ are the (semiclassical) elastic cross sections.

---

[148] Cf. with Eq. (4.6.1) and following discussion, in connection with the derivation of the Cooper pair tunneling Hamiltonian (Cohen et al. (1962)) between two weakly coupled superconductors (Josephson junction).

We now solve the coupled equations in first-order perturbation theory. For this purpose we insert $\delta(\gamma, \alpha)$ at the place of $a_\gamma(t)$, obtaining

$$
\begin{aligned}
a_\beta(t) &= \frac{1}{i\hbar} \int_{-\infty}^{t} \langle \omega_\beta | V_\alpha - U_\alpha | \Psi_\alpha \rangle_{\mathbf{R}_{\beta\alpha}(t)} \exp^{i(E_\beta - E_\alpha)t'/\hbar} dt' \\
&= \frac{1}{i\hbar} \int_{-\infty}^{t} dt' \langle \Psi_\beta | V_\alpha - U_\alpha | \Psi_\alpha \rangle_{\mathbf{R}_{\beta\alpha}(t)} \exp^{i(E_\beta - E_\alpha)t'/\hbar},
\end{aligned}
\tag{2.B.26}
$$

where the expansion

$$
\omega_\beta = \Psi_\beta - \langle \Psi_\alpha | \Psi_\beta \rangle_{\mathbf{R}_{\beta\alpha}(t)}
\tag{2.B.27}
$$

has been used, and the ansatz made, so that the global optical potentials (U: real part), and standard nucleon–nucleon interactions V fulfill the relation

$$
\langle \Psi_\alpha | V_\alpha - U_\alpha | \Psi_\alpha \rangle = 0.
\tag{2.B.28}
$$

Let us consider for simplicity the one-particle transfer reaction[149]

$$
a(= b + 1) + A \rightarrow b + B(= A + 1).
\tag{2.B.29}
$$

Making use of (2.B.11), that is,

$$
\Psi_\alpha = \Psi^a \Psi^A e^{i\delta_\alpha}
\tag{2.B.30}
$$

and

$$
\Psi_\beta = \Psi^b \Psi^B e^{i\delta_\beta},
\tag{2.B.31}
$$

one can write

$$
\begin{aligned}
\langle \Psi_\beta | V_\alpha - U_\alpha | \Psi_\alpha \rangle_{\mathbf{R}_{\alpha\beta}} &= \langle \Psi^b \Psi^B | (V_\alpha - U_\alpha) e^{i\sigma_{\alpha\beta}} | \Psi^a \Psi^A \rangle_{\mathbf{R}_{\alpha\beta}} e^{i\gamma_{\alpha\beta}} \\
&= \langle \phi^{B(A)}(S^B(n), r_{1A}), U(r_{1b}) e^{i\sigma_{\alpha\beta}} \phi^{a(b)}(S^a(n), r_{1b}) \rangle_{\mathbf{R}_{\alpha\beta}} e^{i\gamma_{\alpha\beta}}.
\end{aligned}
\tag{2.B.32}
$$

To obtain the above relation, we have separated the difference $\delta_{\alpha\beta} = \delta_\alpha - \delta_\beta$ between the phases δ_α and δ_β into a part $\gamma_{\alpha\beta}$ which only depends on time and is related to the effective Q-value of the reaction process, and a phase $\sigma_{\alpha\beta}$ which also depends on the center-of-mass coordinate of the transferred particles. That is,

$$
\gamma_{\alpha\beta}(t) = \frac{1}{\hbar} \int_0^t dt' \left\{ U_\alpha(R_\alpha(t')) - \frac{1}{2} m_\alpha v_\alpha^2(t') - U_\beta(R_\beta(t')) + \frac{1}{2} m_\beta v_\beta^2(t') \right\}
$$
$$
+ \mathbf{k}_{\alpha\beta}(t)(\mathbf{R}_\alpha - \mathbf{R}_\beta),
\tag{2.B.33}
$$

where $\mathbf{k}_{\alpha\beta}$ is the average wave vector

$$
\mathbf{k}_{\alpha\beta} = \frac{1}{2\hbar} \left(m_\alpha \mathbf{v}_\alpha(t) + m_\beta \mathbf{v}_\beta(t) \right).
\tag{2.B.34}
$$

---

[149] Concerning the question of nonorthogonality in one-particle transfer processes, the first contribution arises in second order of perturbation theory, within the framework of DWBA (Thompson and Nunes (2009)).

Similarly,

$$\sigma_{\alpha\beta} = \mathbf{k}_{\alpha\beta}(t) \cdot (\mathbf{r}_\beta - \mathbf{r}_\alpha). \tag{2.B.35}$$

The phase σ is characteristic of transfer processes since the dynamical variables $\mathbf{r}_\alpha$ and $\mathbf{r}_\beta$ are identical for elastic and inelastic scattering. It arises from the change in the center-of-mass coordinate taking place when mass is transferred from one system to an other. It gives rise to the recoil effect. Summing up, the one-particle transfer amplitude reads

$$\left(a_\beta(t = +\infty)\right)^{(1)} = \int_{-\infty}^{\infty} \langle \phi^{B(A)}(S^B(n), \mathbf{r}_{1A}), U_{1b}(r_{1b}) e^{\sigma_{\alpha\beta}} \phi^{a(b)}(S^a(n), \mathbf{r}_{1b}) \rangle_{\mathbf{R}_{\alpha\beta}}$$
$$\times \exp\left\{ i \left[(E_\beta - E_\alpha)t'/\hbar + \gamma_{\alpha\beta} \right] \right\}. \tag{2.B.36}$$

The dimensionless phases $\delta_{\alpha\beta}$, $[\sigma_{\alpha\beta} + \gamma_{\alpha\beta}]$ play a similar role in the determination of transfer processes reaction amplitudes, as δ_l does in connection with elastic scattering cross sections. In fact, $\delta_{\alpha\beta}$ determines the shift between incoming and outgoing waves and thus the interference process which is at the basis of the absolute value of the transfer differential cross section. In other words, the reaction part of the elastic and one-nucleon-transfer reaction cross section is embodied in δ_l and $\delta_{\alpha\beta}$, respectively. The nuclear structure part is contained in the reduced mass μ and potential U in the case of elastic scattering and in the single-particle content and radial wavefunctions (formfactors), mean field potential U_{1b}, and Q-value phase in the transfer case. Within the diagrammatic representation of particle-transfer reaction theory, the recoil phase is represented by a jagged line. Similar to δ_l, $\delta_{\alpha\beta}$ and $\sigma_{\alpha\beta}$ cannot be measured directly but can, in principle, be inferred from the absolute differential cross section.[150] In other words, the jagged line does not display asymptotic behavior, representing in all cases a virtual process.

Let us conclude this section by making some comments concerning two nucleon transfer processes

$$\alpha \equiv a(= b + 2) + A \rightarrow b + B(= A + 2) \equiv \beta, \tag{2.B.37}$$

and associated sum rules. For such reactions, the overlap appearing in Eq. (2.B.27) contributes with an amplitude $\langle \Psi_\beta | \mathbb{1} | \Psi_\gamma \rangle \langle \Psi_\gamma | V_\alpha - U_\alpha | \Psi_\alpha \rangle$, where $\mathbb{1}$ is the unit operator, while $\gamma \equiv f(= b + 1) + F(= A + 1)$ denotes the mass partition of the intermediate channel. The above expression indicates that a consistent description of two-nucleon transfer reactions in a nonorthogonal basis involves three strongly interweaved reaction channels[151], namely, α, γ, and β. Consequently, it has to be

---

[150] For more details, see Broglia and Winther (2004), in particular Sect. V.4, p. 308 and subsequent ones, and refs. therein.

[151] This is also testified by the fact that their formal expression can be shifted around by changing representation, among post-post, prior-prior, and prior-post; see, e.g., Eq. (2.B.8) as well as Fig. 6.5.2.

calculated, at least, up to second order of perturbation theory. Thus the need to calculate also $a^{(2)}(t)$, that is, to add to simultaneous, also successive transfer and nonorthogonality corrections.

Regarding the sum rules we consider, for simplicity, the simultaneous transfer amplitude (2.B.26) (see also Sect. 6.5, Eq. (6.5.4)). That is,

$$a^{(1)}(t = +\infty) = \frac{1}{i\hbar} \int_{-\infty}^{\infty} dt \, \exp^{\left[\frac{i}{\hbar}(E^{bB} - E^{aA})t + \gamma_{\beta\alpha}(t)\right]}$$
$$\times \langle \phi^{B(A)}(\mathbf{r}_{1A}, \mathbf{r}_{2A}) | U(r_{1b}) | e^{i\sigma_{\beta,\alpha}} \phi^{a(b)}(\mathbf{r}_{1b}, \mathbf{r}_{2b}) \rangle_{\mathbf{R}_{\alpha\beta}(t)}, \quad (2.B.38)$$

where[152]

$$\sigma_{\beta,\alpha} = \frac{1}{\hbar} \frac{2m_n}{m_A} (m_{aA}\mathbf{v}_{aA}(t) + m_{bB}\mathbf{v}_{bB}(t)) \cdot (\mathbf{r}_\alpha - \mathbf{r}_\beta) \quad (2.B.39)$$

takes care of recoil effects, the phase factor $e^{i\sigma_{\beta,\alpha}}$ being a generalized Galilean transformation associated with the mismatch between entrance and exit channels, while the phase

$$\gamma_{\beta\alpha}(t) = \frac{1}{\hbar} \int_0^t dt' \left\{ U_\beta(\mathbf{R}_\beta(t')) - \frac{1}{2}m_\beta v_\beta^2(t') - U_\alpha(\mathbf{R}_\alpha(t')) + \frac{1}{2}m_\alpha v_\alpha^2(t') \right\}$$
$$+ \frac{1}{2\hbar} (m_\alpha \mathbf{v}_\alpha(t) + m_\beta \mathbf{v}_\beta(t)) \cdot (\mathbf{R}_\beta(t) - \mathbf{R}_\alpha(t)) \quad (2.B.40)$$

is related to the effective Q-value of the reaction.

The rate of change of the formfactor $\langle \phi^{B(A)}, U(r_{1b})e^{i\sigma_{\alpha\beta}} \phi^{a(b)} \rangle$ with time is slow, being completely overshadowed by the rapidly varying phase factor $\exp\left[\frac{i}{\hbar}(E^{bB} - E^{aA})t + \gamma_{\beta\alpha}(t)\right]$. Similar relations concerning recoil and Q-value effects can be obtained from the amplitudes associated to successive and to nonorthogonality terms, that is, $a^{(2)}$ and $a^{(NO)}$ (see Ch. 6, Sect. 6.5).

Summing up, to compare two-nucleon transfer cross sections on equal structural footing, one has to eliminate the kinematical oscillating phase, which can completely distort the "intrinsic" (reduced matrix element) value of the two-nucleon cross section. And for that, one has to work on each of the three amplitudes to extract at best the phases which couple relative and intrinsic motion (see also Sect. 2.2.1).

Let us make a parallel with the sum rule associated with electromagnetic decay (Coulomb excitation, γ-decay). The absolute transition probability for absorption (emission) of a photon is measured in sec^{-1}:

$$T(E1; I_1 \rightarrow I_2) = (1.59 \times 10^{15}) \times (E)^3 \times B(E1; I_1 \rightarrow I_2), \quad (2.B.41)$$

---

[152] See Eqs. (13)–(15), p. 310; Broglia and Winther (2004).

where E is the energy of the transition and $B(E1) = \langle I_2||\mathcal{M}(E1)||I_1\rangle / \sqrt{3}$ is[153] the reduced transition probability.[154] The TRK-sum rule is written as

$$S(E1) = \langle 0|[[H, \mathcal{M}(E1)]], \mathcal{M}(E1)]|0\rangle/2. \qquad (2.B.42)$$

In this case the Q-value dependence of the observed absolute transition probabilities can be eliminated analytically (E^3 dependence), in keeping with the fact that the mass partition ($a + A \rightarrow a + A^*$) does not change between entrance and exit channels or, equivalently, the coordinate of relative motion $\mathbf{R}_{\alpha\alpha'}(t)$ is always that describing the relative center of mass position of the system in the entrance channel.

Expressed differently, and returning to the expression of the first-order (simultaneous) two-nucleon transfer amplitude $a^{(1)}(t = +\infty)$, one can only devise empirical protocols to try to extract the γ- and σ-phase dependence from it and set differential absolute two-nucleon transfer cross sections $d\sigma/d\Omega \sim |a|^2$ on equal footing regarding kinematics, so as to be able to compare intrinsic, reduced transition probabilities (structure), that is, extract the structure information contained in, for example,[155]

$$\phi^{B(A)}(\mathbf{r}_{1A}, \mathbf{r}_{2A}) = \langle \mathbf{r}_{1A}, \mathbf{r}_{2A}|\Gamma_1^\dagger(\beta = +2)|\tilde{0}\rangle, \qquad (2.B.43)$$

as well as in

$$\phi^{B(A)}(\mathbf{r}_{1A}, \mathbf{r}_{2A}) = \langle \mathbf{r}_{1A}, \mathbf{r}_{2A}| \left[a_k^\dagger a_k^\dagger\right]_0 |0\rangle, \qquad (2.B.44)$$

namely, in the RPA pair addition mode representation and in the pure two-particle configuration $|j_k^2(0)\rangle$ describing two nucleons moving in time-reversal states around the close shell system $|0\rangle$. *If it were possible to disentangle the γ- and σ-dependence of $a^{(1)}$ (as well as that of $a^{(2)}$ and $a^{(NO)}$) from its formfactor dependence, the comparison between the quantities[156] $\sum_{n,k} |c_k^{(n)}|^2$ and $\sum_{n,j} \left|\sum_k X_k^{(n)}\delta(j,k) - \sum_i Y_i^{(n)}\delta(j,i)\right|^2$ could eventually be phrased in terms of exact sum rules. This not being the case, one has to deal with approximate TNTR sum rules. With this proviso in mind, such sum rules are quite useful (see Sect. 7.2.2; also Sect. 2.2.1).*

---

[153] $\mathcal{M}(E1) = e \sum_k \left(\left(\frac{N-Z}{A} - t_Z\right) r Y_{1\mu}\right)_k$.

[154] Bohr and Mottelson (1969), p. 382.

[155] Where $\Gamma_n^\dagger(\beta = +2) = \sum_k X_k^n \left[a_k^\dagger a_k^\dagger\right]_0 - \sum_i Y_i^n \left[a_i^\dagger a_i^\dagger\right]_0$ is the RPA pair creation addition mode acting on a correlated closed-shell system $|\tilde{0}\rangle$.

[156] See, e.g., Broglia and Riedel (1967).

2.C NFT Vacuum Polarization

The role zero-point fluctuations play in the nuclear ground state, that is, in the NFT vacuum, can be clarified by relating them to the polarization of the QED vacuum. Let us briefly dwell on the "reality" of such a phenomenon by recalling the fact that to the question of Rabi of whether the polarization of the QED vacuum could be measured[157], in particular the change in charge density felt by the electrons of an atom, for example, the electron of a hydrogen atom, due to virtual creation and annihilation of electron–positron pairs, Lamb gave a quantitative answer, both experimentally and theoretically.[158] The corresponding correction (Lamb shift) implies that the $2s_{1/2}$ level lies higher than the $2p_{1/2}$ level by about 1000 MHz.

In connection with the discussion of Feynman of vacuum polarization, where a field produces a pair, the subsequent pair annihilation producing a new field, namely, a closed loop, he implemented in his space-time trajectories Wheeler's idea of electrons going backward in time (positrons). Such trajectories would be like an **N** in time, that is, electrons which would back up for a while and go forward again. Being connected with a minus sign, these processes are associated with the Pauli principle in the self-energy of electrons (see Fig. 2.4.2 (f)). The divergences affecting such calculations could be renormalized by first computing the self-energy diagram in second order and finding the answer which is finite but contains a cutoff to avoid a logarithmic divergence.[159] Expressing the result in terms of the experimental mass, one can take the limit (cutoff $\rightarrow \infty$) which now exists (see also Sect. 5.3.3).

In the nuclear case, for example, Skyrme effective interactions give rise to particle–vibration coupling vertices which, because of the contact character of these interactions, may lead to divergent zero-point energies[160], unless a cutoff is introduced.[161] The Gogny force, being of finite range, does not display such problems. The question still remains concerning the stability of the results to be

---

[157] Pais (1986), pp. 450, 451; Pais (2000), pp. 255–267.

[158] Lamb and Retherford (1947); Kroll and Lamb (1949).

[159] See Bethe (1947); Feynman (1961); Weinberg (1996a), p. 578; see also Mehra (1996), p. 295, and Bjorken and Drell (1998).

[160] See Hellemans et al. (2013); Pastore et al. (2015), and refs. therein.

[161] The question emerges which are the provisos to be taken in the use of effective forces to higher orders in the PVC. Within this context, see Mahaux et al. (1985) and also Broglia et al. (2016) and Barranco et al. (2017) concerning the implementation of renormalization in both configuration and 3D spaces within the framework of NFT. In a nutshell, the bare mean field exists, but its properties cannot be measured (not any more than the bare electron mass in renormalized quantum electrodynamics), and it corresponds to a set of parameters of a Fermi-like function which ensure that the dressed states reproduce all of the experimental findings, for both structure and reactions.

obtained by carrying out a complete summation over both collective and noncollective contributions. Likely, one ensures it by going to higher orders in the oyster diagrams (see Figs. 1.8.1 (a) and (b), 1.8.3 (a), 2.4.2 (a), 2.5.1, and 2.6.2, as well as Sect. 6.3). The fermion exchange between two of these diagrams (Pauli principle) eliminates noncollective contributions, leading eventually to convergent results.

3

Collective Pairing Modes

I wrote down the trial ground state as a product of operators – one for each pair state – acting on the vacuum ... $\Psi_0 = \prod_{\mathbf{k}}(U_{\mathbf{k}} + V_{\mathbf{k}}b_{\mathbf{k}}^\dagger)|0\rangle$.... Since the pair creation operators $b_{\mathbf{k}}^\dagger$ commute for different $\mathbf{k}$'s ... Ψ_0 represents uncorrelated occupancy of the various pair states.... The operators $b_{\mathbf{k}}^\dagger = c_{\mathbf{k}\uparrow}^\dagger c_{\mathbf{k}\downarrow}^\dagger$, being a product of two fermions (quasiparticle) creation operators, do not satisfy Bose statistics, since $b_{\mathbf{k}}^{\dagger 2} = 0$.

J. R. Schrieffer

3.1 Structure and Reactions

The low-energy properties of the finite, quantal, many-body nuclear system, in which nucleons interact through the strong force of strength $v_0(\approx -100 \text{ MeV})$ and range $a(\approx 0.9 \text{ fm})$, are controlled, in first approximation, by independent particle motion. This is a consequence of the fact that nucleons display a sizable value of the quantal zero-point (kinetic) energy of localization ($\hbar^2/ma^2 \approx 50 \text{ MeV}$) as compared to the absolute value of the strength of the NN-potential[1] $|v_0| = 100 \text{ MeV}$. The corresponding ground state $|HF\rangle = \Pi_i a_i^\dagger|0\rangle$ describes a step function in the probability of the occupied ($\epsilon_i \leq \epsilon_F$) and empty ($\epsilon_k > \epsilon_F$) states, displaying a sharp discontinuity at the Fermi surface. Pushing the system, it reacts with an inertia Am, sum of the nucleon masses (App. 2.B). Setting it into rotation, assuming the density $\rho(r) = \sum_i |\langle \mathbf{r}|i\rangle|^2$ ($|i\rangle = a_i^\dagger|0\rangle$) to be spatially deformed, it responds

---

[1] The corresponding ratio $q = \left(\frac{\hbar^2}{ma^2}\right)\frac{1}{|v_0|}$ is known as the quantality parameter and was first used in connection with the study of condensed matter (de Boer (1948, 1957); de Boer and Lundbeck (1948); Nosanow (1976)). In connection with nuclear physics, it is discussed by Mottelson (1998), who notes that its value $q = 0.5$ testifies to the validity of independent particle motion. Questions like the one posed in connection with localization and long mean free path were already discussed by Lindemann (1910) in connection with the study of the stability or less of crystals. The generalization to aperiodic crystals, like proteins (Schrödinger (1944)), was carried out in Stillinger and Stillinger (1990). Its possible application to the atomic nucleus is discussed in App. 3.A.

with the rigid moment of inertia. This is because the single-particle orbitals are solidly anchored to the mean field (Fig. 3.3.2).

Pairing acting on nucleons moving in time-reversal states ν, $\bar{\nu}$ ($\nu \equiv (nlj)$), in configurations of the type $((l)_{L=0}^2, (s)_{S=0}^2)$, and lying close to the Fermi energy, alter the above picture in a conspicuous way.[2] Within an energy range of the order of the absolute value of the pair correlation energy[3] $|E_{corr}| (\approx 2\text{MeV})$ centered around ϵ_F ($|E_{corr}|/\epsilon_F \ll 1$), the role of independent particles is taken over by pairs of nucleons, correlated over distances $\xi \approx \hbar v_F/(\pi \Delta)$ (≈ 14 fm; $\Delta = 1.2$ MeV), and being equally phased (see Sect. 3.4). In other words, and bringing the argument to a previous step, one can posit that since the creation operators $a_i^\dagger$ commute for different i's, $|HF\rangle$ represents uncorrelated occupancy of the different single-particle states. Similarly, because the pair operators $P_\nu^\dagger$ commute for different $\nu's$, $|BCS\rangle$ represents uncorrelated occupancy of the various pair states. The independent-particle (normal density) and, phase-correlated, independent Cooper[4] pair (abnormal density) parallel suggested above is summarized in Fig. 3.4.3.

For intrinsic[5] nuclear excitation energies and rotational frequencies[6] sensibly smaller than $|E_{corr}/2|$ and than $\hbar\omega_{rot} \approx 0.5$ MeV, respectively, as well as distances of closest approach (D_0) in Cooper pair transfer processes $D_0 \lesssim \xi$, the system can be described in terms of gauge phase–correlated, independent pair motion, each pair contributing a phase ϕ (see Eq. (3.4.2)) This is a consequence of the fact that the kinetic energy of (Cooper) pair confinement ($\hbar^2/(2m\xi^2) \approx 10^{-2}$ MeV) is much smaller than the value of the pair binding energy $|E_{corr}|$, leading essentially to binding by deconfinement[7] and implying that each pair of entangled nucleon partners behaves as an entity[8] of mass $2m$ and intrinsic structure $L = S = 0$. That is, under these circumstances, Cooper pairs behave like (quasi) bosons[9], and the

---

[2] Bohr et al. (1958); see also Broglia and Zelevinsky (2013), and refs. therein.

[3] In BCS, $E_{corr} \approx -\frac{N(0)}{2}\Delta^2$, where $N(0) = \frac{g}{2}$ is the density of nuclear states at the Fermi energy and for one spin orientation, $g_i = i/16$ MeV^{-1} ($i = N, Z$) being the result of an empirical estimate which takes surface effects into account (Bohr and Mottelson (1975), Eq. (2-47); Bortignon et al. (1998), Eq. (7.15), Brink and Broglia (2005) pp. 63, 64), while Δ is the pairing gap, 2Δ being the binding energy of the Cooper pair.

[4] Cooper (1956).

[5] As opposed to collective excitations, which do not alter the temperature of the system, as all the energy is concentrated in a single mode.

[6] Coriolis force acts oppositely on each member of a Cooper pair. When the difference in rotational energy between superfluid and normal rotation becomes about equal to the correlation energy, the nucleon moving opposite to the collective rotation becomes so retarded in its revolution period with respect to the partner nucleon that eventually it can no longer correlate with it and "align" its motion (and spin) with the rotational motion, becoming again a pair of fermions and not participating in the BCS condensate. This happens for a (critical) angular momentum $I_c \approx (120 \times |E_{corr}|)^{1/2} \approx 20\hbar$, corresponding to a rotational frequency $\hbar\omega_c \approx 0.5$ MeV (see Bohr and Mottelson (1975); Broglia and Zelevinsky (2013), Brink and Broglia (2005) p. 128 and refs. therein).

[7] See also footnote 42 in Chapter 4.

[8] The ratio $q_\xi = \frac{\hbar^2}{2m\xi^2}\frac{1}{|E_{corr}|} \approx 0.02$ provides a generalized quantality parameter. It testifies to the stability of nuclear Cooper pairs in superfluid nuclei.

[9] See footnote 46 in Chapter 3.

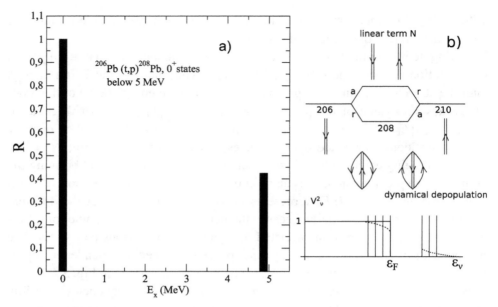

Figure 3.1.1 (a) Ratio of the absolute $L = 0$ differential cross sections $d\sigma(E_x, \theta = 59°)/d\sigma(gs, \theta = 59°)$ $(= (0.05 \text{ mb/sr})/(0.12 \text{ mb/sr}))$ below 5 MeV for the reaction $^{206}\text{Pb}(t, p)^{208}\text{Pb}$ at the second maximum ($\theta = 59°$; Bjerregaard et al. (1966)). Of note are the large experimental errors of the corresponding angular distributions associated with the poor statistics of the cross section at the first maximum $\theta = 5°$. This is the reason why the maximum at 59° was preferred to report the ratio of the cross sections. (b) Schematic representation of the pairing vibrational spectrum around ^{208}Pb (see also Fig. 1.3.2). Also shown is a cartoon representation of the softening of the sharp mean field Fermi surface due to the ZPF of the pairing vibrational modes. The label, a and r indicate the pair addition and pair removal modes, respectively. It is to be noted that a linear term in N has been added to the binding energy to make the values associated with ^{206}Pb ($N = 124$) and ^{210}Pb ($N = 128$) equal, in an attempt to emphasize a harmonic picture for the two-phonon state. Concerning the anharmonicities of the modes, see the last paragraph of Sect. 3.5. After Potel et al. (2013a).

single-particle orbits on which they correlate become dynamically detached from the mean field, leading to a BCS condensate. It is, however, very different from a standard condensate of real bosons. Within this context, see Sect. 3.4 (Figs. 3.4.1 and 3.4.2) and App. 4.A.

Cooper pairs exist also in situations in which the environmental conditions are above critical, for example, in metals at room temperature, in closed-shell nuclei as well as in deformed open-shell nuclei at high values of the angular momentum.[10] However, in such circumstances, they break essentially as soon as they are generated (pairing vibrations). While these pair addition and substraction fluctuations

---

[10] See, e.g., Shimizu and Broglia (1990), and refs. therein.

have little effect in condensed matter systems with the exception of at[11] $T \approx T_c$ (critical temperature), they play an important role in normal (nonsuperfluid) nuclei, in particular around closed shells[12] (Fig. 3.1.1), and specially in the case of light, highly polarizable, exotic halo nuclei.[13] From this vantage point, one can posit that it is not so much, or, at least not only, the superfluid phase which is abnormal in the nuclear case but the normal state around the closed-shell systems, displaying a consistent component of dynamical abnormal density. The importance of pairing vibrations is also found in connection with the self-energy of nucleons moving around closed shells[14] as well as concerning the binding energy of the closed-shell systems.[15]

Furthermore, of note is the role pairing vibrations play in the transition between superfluid and normal nuclear phases (cf. Fig. 3.1.2) as a function of the rotational frequency (angular momentum) as emerging from the experimental studies of high spin states carried out by, among others, Garrett and collaborators.[16] While the (dynamic) pairing gap associated with pairing vibrations leads to $\approx 20\%$ increase of the static pairing gap for low rotational frequencies, it becomes the overwhelming contribution above the critical frequency.[17]

The central role played by pairing vibrations within the present circumstances is to restore particle-number conservation, another example after that provided by the quantality parameter and by its generalization to pair motion of the fact that potential functionals are, as a rule, best profited by special arrangements of particles (spontaneous symmetry breaking), while fluctuations favor symmetry.[18] Within this context, see Eqs. (1.4.55) and (1.4.59).

A number of methods allow one to go beyond BCS mean-field approximation or its generalization, known as the Hartree–Fock–Bogolyubov approximation (HFB). Generally referred to as number projection methods[19](NP), they make use of a variety of techniques, such as the Generator Coordinate Method, as

[11] See Schmidt (1968), Schmid (1969), Abrahams and Woo (1968); concerning superfluid ^{3}He, see Wölfe (1978). We make reference to systems of dimensions much larger than the correlation length, the situation being different for zero-dimensional superconducting particles of small dimension compared to ξ (see, e.g., Tinkham (1980), Ch. 7, in particular Sect. 7-2), a connotation which can also be ascribed to the nucleus.

[12] See Sects. 3.5 and 3.6; see also Bohr (1964), Bès and Broglia (1966), Högaasen-Feldman (1961), Schmidt (1972), Barranco et al. (2001), Potel et al. (2013a), Potel et al. (2014), Schmidt (1964).

[13] See Potel et al. (2013a), and refs. therein. Also see Potel et al. (2013b) in connection with the closed-shell system ^{132}Sn.

[14] See, e.g., Bès and Broglia (1971a); Flynn et al. (1971).

[15] Baroni et al. (2004).

[16] Garrett (1985); Garrett et al. (1986); see also Shimizu et al. (1989), Barranco et al. (1987), and Chapter 6 of Brink and Broglia (2005).

[17] Shimizu et al. (1989); Shimizu and Broglia (1990); Shimizu (2013); Dönau et al. (1999); Shimizu et al. (2000).

[18] Anderson and Stein (1984); Anderson (1976).

[19] See Allaart et al. (1974); Ring and Schuck (1980); Egido (2013); Vaquero et al. (2013); Robledo and Bertsch (2013); see also Frauendorf (2013); Ring (2013); Heenen et al. (2013), and refs. therein.

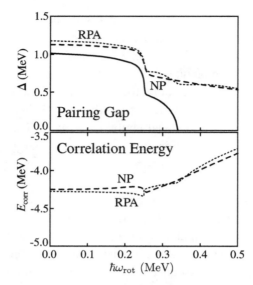

Figure 3.1.2 Pairing gap calculated taking into account the correlation associated with pair vibrations in the RPA approximation ($\Delta = (\Delta_{BCS}^2 + \frac{1}{2}G^2 S_0(RPA))^{1/2}$) (upper panel) and RPA correlation energy (lower panel) for neutrons in ^{164}Er as a function of the rotational frequency (Shimizu et al. (1989); Shimizu (2013)). Both quantities are in MeV (dash-dotted curves). The value of the static (mean-field) pairing gap Δ, which vanishes at $\hbar\omega_{rot} = 0.34$ MeV, is also displayed in the upper panel (continuous curve). The results of the number-projection (NP) calculations are shown as dotted curves. The non-energy-weighted sum rule S_0 (RPA) $= \sum_{n \neq AGN} \left[\langle n|P|0 \rangle + \langle n|P^\dagger|0 \rangle \right]_{RPA}^2$, $P^\dagger = \sum_{\nu > 0} a_\nu^\dagger a_{\bar\nu}^\dagger$, describes the contribution of pairing fluctuations to the effective (RPA) gap and is intimately associated with projection in particle number. It is of note that $\sum_{n \neq AGN}$ means that the divergent contribution from the zero-energy mode (Anderson–Goldstone–Nambu mode; see, e.g., Broglia et al. (2000), and refs. therein), associated with the lowest ($\hbar\omega_0$) solution of the $H = H_p + H_p''$, is to be excluded (see Sect. 1.4.3, as well as Brink and Broglia (2005), App. J).

well as protocols like variation after projection. The advantages of NP methods over the RPA is to lead to smooth functions for both the correlation energy and the pairing gap at the pairing phase transition between normal and superfluid phases.

In Fig. 3.1.3 we display the excitation function associated with the reaction ^{122}Sn$(p, t)^{120}$Sn(J^π), populating the low-energy states of the single open-shell superfluid nucleus ^{120}Sn. The angle selected to report the value of the absolute differential cross sections, that is, 5°, corresponds to the first, and largest, peak of the absolute $L = 0$ differential two-nucleon transfer cross section. Essentially all the strength in the $L = 0$ channel is concentrated in the ground state, the strongest

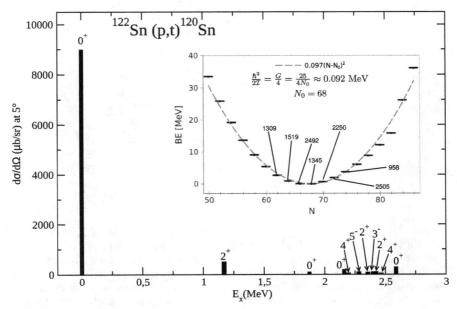

Figure 3.1.3 Excitation function associated with the reaction ^{122}Sn$(p,t)^{120}$ Sn(J^{π}). The absolute experimental values of $d\sigma(J^{\pi})/d\Omega|_{5°}$ are given as a function of the excitation energy E_x (after Guazzoni et al. (2011)). In the inset the neutron pairing rotational band between magic numbers $N = 50$ and $N = 82$ is displayed (see also the caption to Fig. 1.4.5), the absolute $^{A+2}_{50}$Sn$_{N+2}$ (p,t) $^{A}_{50}$Sn$_N$(gs) experimental cross sections taken from Guazzoni et al. (1999, 2004, 2006, 2008, 2011, 2012) are also given in μb (number connected by lines to the corresponding levels of the pairing rotational band). After Potel et al. (2013a). For the color version of this figure, refer to cambridge.org/nuclearcooperpair.

0^+-excited state carrying a cross section of the order of 3% of that of the ground state. Within this context, the difference with the results displayed in Fig. 3.1.1 is apparent.[20]

In the inset to Fig. 3.1.3 a quantity closely related to the Sn-isotope binding energy is displayed (boldface levels), namely, $B(^{50+N}$Sn$_N)$-8.124N MeV+46.33 MeV, resulting from the substraction of the contribution of the single nucleon addition to the nuclear binding energy (linear function in N) obtained by a linear fitting of the binding energies of all the Sn-isotopes. Also displayed is the parabolic fit to these energies, a quantity to be compared with $E_N = (\hbar^2/2\mathfrak{I})(N - N_0)^2$, namely,

---

[20] While this "distortion" of the (t,p) excitation function is useful to emphasize the parallels between vibrational and rotational bands in 3D and in gauge spaces, it has to be used with care concerning the parallel with Cooper pair tunneling between weakly coupled superconductors (Sect. 4.6 and, in particular, Sect. 7.3). This is also in keeping with the fact that at the basis of nuclear BCS, one finds an interdisciplinary connection (see Bohr et al. (1958)).

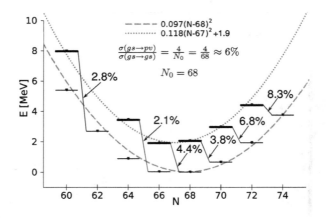

Figure 3.1.4 The weighted average energies ($E_{exc} = \sum_i E_i\sigma_i / \sum_i \sigma_i$) of the excited 0^+ states below 3 MeV in the Sn isotopic chain are shown on top of the ground state pairing rotational band. Also indicated is the percentage of cross section for two-neutron transfer to excited states, normalized to the cross sections populating the ground states. The estimate of the ratio of cross sections displayed on top of the figure was obtained making use of the single j-shell model (Brink and Broglia (2005), App. H). After Potel et al. (2013b). For the color version of this figure, refer to cambridge.org/nuclearcooperpair.

the energy associated with the members of the pairing rotational band. The difference with the spectrum of pair addition and substraction modes displayed in Fig. 3.1.1 b) (see also Fig. 1.3.2) is again evident. Concerning the parallel, one can draw between the 3D- and pairing-vibrations and -rotations (see Fig. 3.1.5).

3.1.1 Pairing Rotational and Vibrational Bands Probed with One- and Two-Nucleon Transfer

A simple estimate of the pairing rotational band moment of inertia is provided by the single j-shell model[21], namely, $(\hbar^2/2\mathcal{J})=G/4 \approx 25/(4N_0)$ MeV. This estimate turns out to be rather accurate (Figs. 3.1.3 and 3.1.4). On the other hand, one is reminded of the fact that we are discussing properties which specifically characterize a coherent state[22], namely, $|BCS\rangle$ (see Eq. (1.4.39)).

Also reported in Fig. 3.1.3 (inset) are the integrated values of the measured absolute two-neutron transfer cross sections, quantities which are reproduced by the theoretical predictions within experimental errors (Figs. 3.1.7 and 7.2.1). In principle, one could have expected a sensible constancy of these cross sections (transitions) as the pairing rotational model implies a common intrinsic deformed state in gauge space, namely, $|BCS(N_0 = 82)\rangle$ (see Eq. (1.4.65)). On the other

[21] See, e.g., Brink and Broglia (2005), App. H and refs. therein.
[22] See Sect. 7.2; see also Potel et al. (2017).

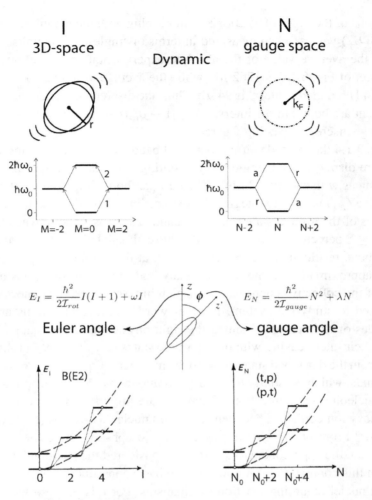

Figure 3.1.5 Parallel between dynamic and static deformations in 3D and in gauge space for the nuclear finite many-body system. In the first case, the angular momentum **I** and the Euler angles are conjugate variables. In the second are particle number N and gauge angle. While the fingerprint of static (quadrupole and gauge) deformations are quadrupole and pairing rotational bands, vibrational bands are the expression of such phenomena in nondeformed systems.

hand, the number of Cooper pairs α'_0 which defines deformation in gauge space is rather small (≈ 6) and thus subject to conspicuous fractional fluctuations ($\Delta\alpha'_0/\alpha'_0 \approx \sqrt{6}/6 \approx 0.4$). Because $\sigma \sim \alpha'^2_0$, fluctuations in σ of the order of 100 percent can be expected.

In keeping with the analogies discussed in connection with Figs. 3.1.5 and 3.3.2 between pairing and quadrupole rotational bands, we note that in the electromagnetic decay of these last bands, one expects, in the case of heavy nuclei, fractional fluctuations of the order of $\sqrt{200}/200$, in keeping with the magnitude of the associated $B(E2)$ values, when measured in terms of single-particle units. Within this context, the average value of the absolute experimental cross sections displayed in the inset of Fig. 3.1.3 is 1762 μb, while the average difference between experimental and predicted values[23] is 94 μb. Thus, the discrepancies between theory and experiment are bound in the interval $0 \leq |1 - \sigma_{th}(i \rightarrow f)/\sigma_{exp}(i \rightarrow f)| \leq 0.09$, the average discrepancy being 5 percent.

In Fig. 3.1.4 the excited pairing rotational bands based on 0^+ pairing vibrational modes are displayed as a single band, resulting from the average value of the 0^+ excited states with energy ≤ 3 MeV. The best parabolic fit is shown. Also given are the relative (p, t) absolute integrated cross sections normalized to the corresponding values of the ground state rotational band. The cross talk between bands is in all cases ≤ 8 percent, the single j-shell value estimate being[24] 6 percent.

The above results underscore the fact that, at the basis of an operative coarse-grained approximation to the nuclear many-body problem, one finds a judicious choice of the collective coordinates,[25] which in turn is closely connected with the probe used to study the system. In other words, pairing vibrations are elementary modes of excitation containing the right physics to restore gauge invariance through their interweaving with quasiparticle states (Eqs. (1.4.59)–(1.4.62)). They project from the deformed state ($\alpha'_0 \neq 0$) the members (Eq. (1.4.65)) of pairing rotational bands, which, similarly to pairing vibrational bands, are specifically excited in two-nucleon transfer reactions. Examples are provided by Cooper pair tunneling in heavy ion collisions between superfluid nuclei, as well as (t, p), (p, t), etc., reactions.[26] From other fields of physics, the Josephson[27] effect between weakly coupled metallic superconductors is rightly considered the paradigmatic example.

Within this context one can now take the basic consequence of pairing condensation in nuclei regarding reaction mechanisms. For this purpose let us consider a *gedankenexperiment* in which the superfluid target and the projectile can at best come in weak contact. Because $\left(\hbar^2/2m\xi^2\right)/|E_{corr}| \approx 10^{-2}$, Cooper pairs in superfluid nuclei behave as (potentially) very extended entities of mean square radius ($\xi \gg R_0$) and mass $2m$, even in the case in which the $^1S_0$, NN-potential vanishes in the zone between the weakly overlapping (normal) densities of the two interacting nuclei. One then expects Cooper pair transfer to be observed. One also expects the associated absolute differential cross section to be of the same order of magnitude

[23] Potel et al. (2013b).

[24] Brink and Broglia (2005), App. H.

[25] In this connection, we quote from S. Weinberg: "You can use any degrees of freedom that you like to describe a physical system, but if you choose the wrong ones, you will be sorry" (Weinberg, 1983).

[26] See, e.g., Yoshida (1962); Glendenning (1965); Bohr (1964); Bayman (1971); Broglia et al. (1973); Hansen (2013); Potel et al. (2013a), and refs. therein.

[27] Josephson (1962).

as one-nucleon transfer ones and to be dominated by successive transfer. These expectations have been confirmed experimentally.[28] The above parlance, being at the basis of the Josephson effect, reflects both one of the most solidly established results in the study of BCS pairing and the workings of a paradigmatic probe of spontaneous symmetry breaking phenomena.

Because, away from the Fermi energy ($\gtrsim \epsilon_F \pm \Delta$), pair motion becomes essentially independent particle motion, one-particle transfer reactions like (d, p) and (p, d) can be used together with (t, p) and (p, t) processes as valid tools to cross check pair correlation predictions, in particular, to shed light on the origin of pairing in nuclei and concerning the relative importance of the bare- and induced-pairing interaction. For this purpose, one should be able to reproduce the absolute differential cross sections within a 10 percent error.

While the calculation of two-nucleon transfer spectroscopic amplitudes and differential cross sections are, a priori, more involved to be worked out than those associated with one-nucleon transfer reactions, the former are, as a rule, more "intrinsically" accurate than the latter ones. This is because, in the case of two-nucleon transfer reactions, the quantity (order parameter α_0') which expresses the collectivity of the members of a pairing rotational band reflects the properties of a coherent state ($|BCS\rangle$). In other words, it results from the sum over many two-nucleon transfer spectroscopic amplitudes ($\sqrt{j_\nu + 1/2}\, U_\nu' V_\nu'$), all phased in the same way (phase correlated).

There is a further reason which confers $\alpha_0' = \sum_j (j + 1/2) U_j' V_j'$ a privileged position with respect to the single-particle spectroscopic amplitudes (U_j', V_j'). It is the fact that $\alpha_0' = e^{2i\phi} \sum_j (j + 1/2) U_j V_j = e^{2i\phi} \alpha_0$ defines a privileged orientation in gauge space, α_0 being the order parameter referred to the laboratory system which makes an angle ϕ in gauge space with respect to the intrinsic system to which α_0' is referred.[29] In other words, the quantities α_0' measure the deformation of the superfluid nuclear system in gauge space.

Similar arguments can be used regarding the excitation of pairing vibrations in terms of Cooper pair transfer around closed shells as compared to one-particle

---

[28] Both in heavy ion reaction between superfluid nuclei (Montanari et al. (2014); see also Fig. 4.6.1 and Sect. 7.3) and light ions (d, p) and (t, p) reaction. See, e.g., Cavallaro et al. (2017); $d\sigma(^9\text{Li}(d, p)^{10}\text{Li}(1/2^-))/d\Omega|_{\theta_{max}} \approx 0.8$ mb/sr, as compared to Tanihata et al. (2008); $d\sigma(^{11}\text{Li}(p, t)^9\text{Li}(1/2^-))/d\Omega|_{\theta_{max}} \approx 1$ mb/sr, Fortune et al. (1994); $^{10}\text{Be}(t, p)^{12}\text{Be(gs)}$ ($\sigma = 1.9 \pm 0.5$ mb, $4.4° \leq \theta_{cm} \leq 54.4°$) as compared to Schmitt et al. (2013). $^{10}\text{Be}(d, p)^{11}\text{Be}(1/2^+)$ ($\sigma = 2.4 \pm 0.013$ mb, $5° \leq \theta_{cm} \leq 39°$) in the case of light nuclei around closed ($N = 6$) shell, and Bassani et al. (1965) $^{120}\text{Sn}(p, t)^{118}\text{Sn(gs)}$ ($\sigma = 3.024 \pm 0.907$ mb, $5° \leq \theta_{cm} \leq 40°$) as compared to Bechara and Dietzsch (1975) $^{120}\text{Sn}(d, p)^{121}\text{Sn}(7/2^+)$ ($\sigma = 5.2 \pm 0.6$ mb, $2° \leq \theta_{cm} \leq 58°$).

[29] It is of note that the square root of $\alpha_0 = e^{-2i\phi}\alpha_0'$, normalized to a proper volume element V, that is, $\Psi = e^{-i\phi}\sqrt{n_S}$, where $n_S = \alpha_0'/V$ is the density of superconducting electrons, constitutes the order parameter at the basis of the Ginzburg–Landau theory of superconductivity (Ginzburg and Landau (1950)), called by Ginzburg the Ψ-theory of superconductivity (Ginzburg (2004)), published seven years before BCS.

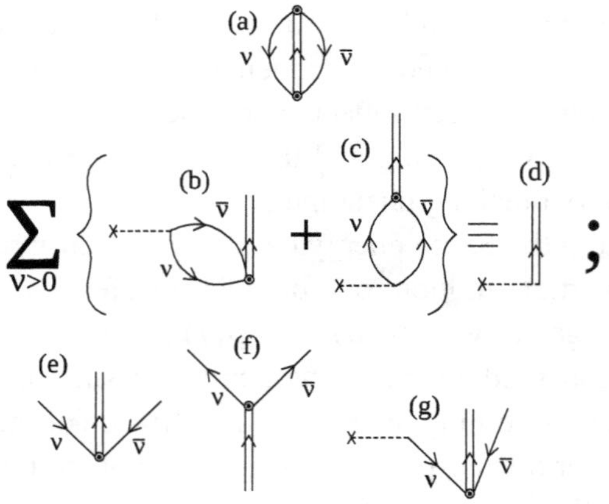

Figure 3.1.6 NFT diagrams associated with one- and two-particle transfer from a closed shell associated with ZPF of pair addition modes. (a) ZPF associated with the virtual excitation of a pair addition mode and two uncorrelated holes. (b) Two-particle transfer filling a two-hole state associated with the ZPF. (c) Two-particle transfer to time-reversal states lying above the Fermi energy. These processes receive contribution from all $(\nu, \bar{\nu})$ pairs (sum over $\nu > 0$), leading to (d), the direct excitation of the collective pair addition mode. The relation (b) + (c) $\equiv$ (d) is the NFT graphical representation of the random phase approximation (RPA) dispersion relation used to calculate the properties of the pair addition mode in the harmonic approximation (Sect. 3.5). The backward- and forward-going RPA amplitudes are displayed in (e) and (f), respectively. (g) One-particle transfer proceeding through the filling of a hole associated with the ZPF.

transfer. As seen from Fig. 3.1.6 (c)–(b), the random phase approximation (RPA) amplitudes X_ν^a and Y_ν^a are coherently summed over pairs of time-reversal states to give rise to the spectroscopic amplitudes associated with the direct excitation of the pair addition mode displayed in (d). Because of the (dispersion) relation (b)+(c)$\equiv$(d), the X_ν- and Y_ν-amplitudes are, in the Cooper pair transfer process, correlated, the situation being not so in the case of one-particle transfer (see, e.g., Fig. 2.4.2 (e), (f); see also Fig. 3.1.6 (g)). In other words, while the (renormalized) spectroscopic amplitude and (renormalized) radial form factor associated with a one-nucleon transfer process depend on a single renormalized energy and associated wavefunction, the corresponding quantities for Cooper pair transfer depend on a distribution of (renormalized) states and wavefunctions of levels around the Fermi energy, leading to a coherent state.[30]

We write now, in somewhat more detail, the spectroscopic amplitudes (B-coefficients) for the two-nucleon transfer processes mentioned above, namely, one for the case in which A and $B(= A + 2)$ are members of a pairing rotational band

---

[30] We note that, despite that one is dealing with the connection between structure and direct transfer reactions, no mention has been made of spectroscopic factors in relation with one-particle transfer processes, let alone when discussing two-particle transfer processes.

Table 3.1.1 *Two-nucleon transfer spectroscopic amplitudes associated with the reactions* $^{112}Sn(p,t)^{110}Sn(gs)$, *and* $^{124}Sn(p,t)^{122}Sn(gs)$. [a] *Quantum numbers of the two-particle configurations* $(nlj)^2_{J=0}$ *coupled to angular momentum* $J = 0$. [b], [d] $\langle BCS|T_\nu|BCS\rangle = \sqrt{(2j_\nu+1)/2}\,U_\nu(A)V_\nu(A+2)$ $(A + 2 = 112$ *and* 124 *respectively*), *where* $T_\nu = [a^\dagger_\nu a^\dagger_\nu]^0/\sqrt{2}\,(\nu \equiv nlj)$ (cf. Potel et al. (2011, 2013a,b)) [c] *two-nucleon transfer spectroscopic amplitudes calculated making use of initial and final state wavefunctions obtained by diagonalizing a* v_{low-k}, *that is, a renormalized, low-momentum interaction derived from the CD–Bonn nucleon–nucleon potential (see Guazzoni et al. (2006), and refs. therein).* [e] *Two-neutron overlap functions obtained making use of the shell-model wavefunctions for the ground state of* ^{122}Sn *and* ^{124}Sn *calculated with the code NuShell (Brown and Rae, 2007). The wavefunctions were obtained starting with a G-matrix derived from the CD–Bonn nucleon–nucleon interaction (Machleidt et al. (1996)). These amplitudes were used in the calculation of* $^{124}Sn(p,t)^{122}Sn$ *absolute cross sections carried out by I. J. Thompson (2013).*

	$^{112}Sn(p,t)^{110}Sn$(gs)		$^{124}Sn(p,t)^{122}Sn$(gs)	
nlj [a]	BCS[b]	V^c_{low-k}	BCS[d]	NuShell[e]
$1g_{7/2}$	0.96	−1.1073	0.44	0.63
$2d_{5/2}$	0.66	−0.7556	0.35	0.60
$2d_{3/2}$	0.54	−0.4825	0.58	0.72
$3s_{1/2}$	0.45	−0.3663	0.36	0.52
$1h_{11/2}$	0.69	−0.6647	1.22	−1.24

and a second one for the case in which they are members of a pairing vibrational band. That is,

$$B((nlj)^2(0)) = \langle BCS(N+2)|\frac{[a^\dagger_{nlj}a^\dagger_{nlj}]^0_0}{\sqrt{2}}|BCS(N)\rangle \qquad (3.1.1)$$

$$= \sqrt{j+1/2}\,U_{nlj}(N)V_{nlj}(N+2)$$

and

$$B((nlj)^2(0)) = \langle(N_0+2)(gs)|\frac{[a^\dagger_{nlj}a^\dagger_{nlj}]^0_0}{\sqrt{2}}|N_0(gs)\rangle \qquad (3.1.2)$$

$$= \begin{cases} \sqrt{j_k+1/2}\ X^a(n_k l_k j_k) & (\epsilon_{j_k} > \epsilon_F) \\ \sqrt{j_i+1/2}\ Y^a(n_i l_i j_i) & (\epsilon_{j_i} \le \epsilon_F), \end{cases}$$

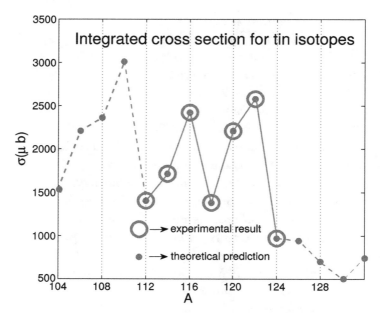

Figure 3.1.7 Absolute value of the two-nucleon transfer cross section $^{A}_{50}\mathrm{Sn}_N(p,t)^{A-2}_{50}\mathrm{Sn}_{N-2}(\mathrm{gs})$ ($102 \le A \le 132$, i.e., from (final $(A-2)$)) closed-shell $^{100}_{50}\mathrm{Sn}_{50}$ to (initial A) closed-shell $^{132}_{50}\mathrm{Sn}_{82}$ isotopes (see Potel et al. 2013a, 2013b); see also Sect. 7.2) calculated making use of RPA and BCS spectroscopic amplitudes (Eqs. (3.1.1) and (3.1.2), respectively; in connection with this last one, see Tables 3.1.1 and 7.2.1) and taking into account successive and simultaneous transfer in second-order DWBA, properly corrected for nonorthogonality contributions, in comparison with the experimental data (Guazzoni et al. (1999, 2004, 2006, 2008, 2011, 2012)). For the color version of this figure, refer to cambridge.org/nuclearcooperpair.

respectively, where here the X and Y coefficients are the forward-going and backward-going RPA amplitudes of the pair addition mode. For actual numerical values, see Tables 3.1.1 and 3.5.2–3.5.5. Making use of these spectroscopic amplitudes to calculate the successive and simultaneous transfer amplitudes, correcting both of them for nonorthogonality contributions, makes the above picture the quantitative probe of Cooper pair correlations in nuclei[31], as can be seen from the variety of examples shown in the present monograph, in particular, those shown in Figs. 3.1.7, 4.5.4, and 7.1.5. In the first one, we display the absolute cross sections associated with the $^{A}_{50}\mathrm{Sn}_N(p,t)^{A-2}_{50}\mathrm{Sn}_{N-2}(\mathrm{gs})$ reactions studied between the double closed-shell systems $^{132}_{50}\mathrm{Sn}_{82}$ and $^{100}_{50}\mathrm{Sn}_{50}$, that is, from pairing vibrational through pairing rotational to pairing vibrational nuclei again. Examples of the population of pairing vibrational states are displayed in Figs. 4.5.4 and 7.1.5.

---

[31] Bayman and Chen (1982); Thompson (1988); Thompson and Nunes (2009); Broglia and Winther (2004); Potel et al. (2013a); see also Götz et al. (1975); Broglia et al. (1977).

Summing up, one will use throughout the present monograph, except when explicitly mentioned, absolute cross sections as the link between spectroscopic amplitudes and experimental observations.[32]

3.2 Renormalization and Spectroscopic Amplitudes

As a result of the interweaving of single-particle and collective motion, the nucleons acquire a state-dependent self-energy $\Delta E_j(\omega)$. Consequently, the single-particle potential which was already nonlocal in space (exchange potential, related to the Pauli principle) becomes also nonlocal in time. A possible technique to make it local is that of the effective mass approximation. In it one describes the single-particle motion in terms of a local potential given by $U'(r) = (m/m^*)U(r)$. The quantity $m^* = m_k m_\omega / m$ is the effective nucleon mass, m_k being the so-called k-mass associated with the Fock potential and $m_\omega = m(1 + \lambda)$ being the ω-mass (nonlocality in time, as implied by the relation $\Delta\omega\Delta t \geq 1$). The quantity $\lambda = -\partial\Delta E(\omega)/\partial\hbar\omega$ is the so-called mass enhancement factor. It reflects the ability with which vibrations dress single particles. In other words, it measures the probability with which a nucleon moving at $t = -\infty$ in a "pure" orbital j can be found at a later time in a $2p - 1h$ like (doorway state) $|j'L; j\rangle$, L being the multipolarity of a vibrational state. Within this context, the discontinuity taking place at the Fermi energy in the dressed-particle picture ($Z_\omega = (m/m_\omega)$) reflects the associated single-particle occupancy probability.[33]

It is of note that dressed particles imply also an induced pairing interaction v_p^{ind} which renormalizes the bare NN-$^1S_0$ interaction v_p^{bare} and results from the exchange of the dressing vibrations between pairs of nucleons moving in time-reversal states close to the Fermi energy.[34] *In other words, fluctuations in the normal density $\delta\rho$ and associated particle–vibration coupling vertices (time-dependent HF δU) lead to abnormal (pairing) density.* Whether this results in a dynamic or static phenomenon depends on whether the parameter (see Fig. 3.5.7[35])

$$x' = G'N'(0),\qquad (3.2.1)$$

product of the effective pairing strength,

$$G' = Z_\omega^2(v_p^{bare} + v_p^{ind}),\qquad (3.2.2)$$

---

[32] It is of note that concerning nuclear structure information, it is also contained in the optical potential used in the calculation of the corresponding cross sections. See footnote 26 in Chapter 1.

[33] See, e.g., Brink and Broglia (2005), Chapter 9, and refs. therein.

[34] See, e.g., Figs. 4.8.3 and 7.E.1 (II) (b),(d),(e).

[35] Brink and Broglia (2005), App. H, Sect. H4 and refs. therein; Barranco et al. (2005).

and of the renormalized density of levels $N'(0)$ is considerably smaller (larger) than $\approx 1/2$. The quantity G' is the sum of the bare and induced pairing interaction, renormalized by the degree of single-particle content of the levels in which nucleons correlate. The quantity

$$N'(0) = Z_\omega^{-1} N(0) = (1 + \lambda) N(0) \tag{3.2.3}$$

is the similarly renormalized density of levels at the Fermi energy. From the above relations one obtains

$$x' = Z_\omega (v_p^{bare} + v_p^{ind}) N(0). \tag{3.2.4}$$

Typical values of $Z_\omega \approx 0.7$, while for nuclei along the stability valley, the bare and the induced pairing contributions are about equal, thus, according to Eq. (3.2.2) $G' \approx G = v_p^{bare}$, as in the case of a nonrenormalized situation. On the other hand, the physics is radically different, particles being a consistent fraction of the time in excited states coupled to collective vibrations and pairing acquiring a state dependence.

All of the above many-body, ω-dependent effects which imply in many cases a coherent sum of amplitudes, together with the corresponding renormalizations of the single-particle radial wavefunctions (formfactors) not discussed within the present framework, are not simple to capture in a spectroscopic factor,[36] neither in connection with one-particle transfer nor regarding two-nucleon transfer processes.[37]

3.3 Quantality Parameter

The quantality parameter[38] is defined as the ratio of the quantal kinetic energy of localization (confinement) and the potential energy (cf. Fig. 3.3.1 and Table 3.3.1). Fluctuations, quantal or classical, favor symmetry: gases and liquids are homogeneous. Potential energy, on the other hand, prefers special arrangements:

---

[36] In keeping with the fact that $m_k \approx 0.6 - 0.7m$ and that $m^* \approx m$, as testified by the satisfactory fitting standard Woods–Saxon potentials, provides for the valence orbitals of nucleons of mass m around closed shells, one obtains $m_\omega \approx 1.4 - 1.7m$. Thus $Z_\omega \approx 0.6 - 0.7$. It is still an open question how much of the observed single-particle depopulation can be ascribed to hard core effects, which shift the associated strength to high momentum levels (see Dickhoff and Van Neck (2005); Jenning (2011); Kramer et al. (2001); Barbieri (2009); Schiffer et al. (2012); Duguet and Hagen (2012); Furnstahl and Schwenk (2010)). An estimate of such an effect of about 20 percent will not qualitatively alter the long wavelength estimate of Z_ω given above. Arguably, a much larger depopulation through hard core effects remains an open problem within the overall picture of elementary modes of nuclear excitation and of medium polarization effects. Within this context remains an open question the role the renormalization of the radial dependence of the single-particle wavefunctions (formfactors), due to many-body effects, can play in the determination of spectroscopic factors. Discussions with Augusto Macchiavelli on this subject are gratefully acknowledged.

[37] See Barranco et al. (2005, 1999).

[38] Nosanow (1976); de Boer (1957); de Boer (1948); de Boer and Lundbeck (1948); Mottelson (1998).

Table 3.3.1 *Zero-temperature phase for a number of systems of mass M (M_n: nucleon mass), the first four depending on atomic interactions (range Å, strength eV), the last one referring to the atomic nucleus.*
[a] *Delocalized (condensed),* [b] *localized,* [c] *non-Newtonian solid (cf., e.g., Bertsch (1988), De Gennes (1994)), that is, systems which react elastically to sudden solicitations and plastically under prolonged strain.* [d] *Paradigm of quantal, strongly fluctuating, finite many-body systems. Delocalization or less does not seem to depend much on whether one is dealing with fermions or bosons (Mottelson (1998), and refs. therein; cf. also Ebran et al. (2012, 2013, 2014a, 2014b).*

Constituent	M/M_n	a (cm)	v_0 (eV)	q	Phase ($T = 0$)
^{3}He	3	2.9 (−8)	8.6 (−4)	0.19	liquid[a]
^{4}He	4	2.9 (−8)	8.6 (−4)	0.14	liquid[a]
H_2	2	3.3 (−8)	32 (−4)	0.06	solid[b]
^{20}Ne	20	3.1 (−8)	31 (−4)	0.007	solid[b]
Nucleons	1	9 (−14)	100 (+6)	0.5	liquid[a,c,d]

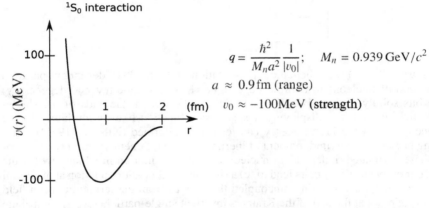

$$q = \frac{\hbar^2}{M_n a^2}\frac{1}{|v_0|}; \quad M_n = 0.939\,\text{GeV}/c^2$$

$$a \approx 0.9\,\text{fm (range)}$$

$$v_0 \approx -100\,\text{MeV (strength)}$$

Figure 3.3.1 Schematic representation of the bare NN-interaction acting among nucleons displayed as a function of the relative coordinate $r = |\mathbf{r}_1 - \mathbf{r}_2|$, used to estimate the quantality parameter q, ratio of the zero-point fluctuations (ZPF) of confinement and the potential energy.

atoms like to be at specific distances and orientations from each other (spontaneous breaking of translational and of rotational symmetry reflecting the homogeneity and isotropy of empty space[39]).

---

[39] As already stated in footnote 1 of this chapter, within such a general context, the physics embodied in the quantality parameter is closely related to that which is at the basis of the classical Lindemann criterion (Lindemann (1910)) to measure whether a system is ordered (e.g., a crystal) or disordered (e.g., a melted system) (Bilgram (1987); Löwen (1994); Stillinger and Stillinger (1990); Stillinger (1995)). The above

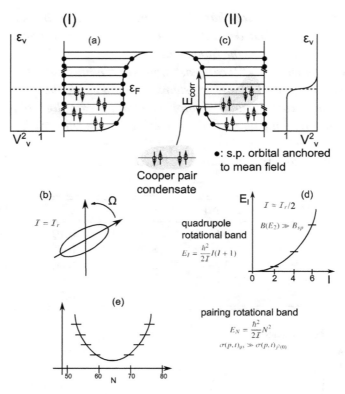

Figure 3.3.2 **(I)** (a) Schematic representation of "normal" (independent-particle) motion of nucleons in twofold degenerate (Kramers, time-reversal degeneracy) orbits solidly anchored to the mean field (solid dots at the ends of the single-particle levels) and displaying a sharp, step-function-like discontinuity in the occupancy at the Fermi energy can lead to a deformed (Nilsson (1955)) rotating nucleus with a rigid moment of inertia $\mathcal{I}_r$ (b). **(II)** Schematic representation of phase-correlated, independent nucleon Cooper pair motion in which few (of the order of five to eight) pairs lead to (c) a sigmoidal occupation function at the Fermi energy and, having mainly uncoupled themselves from the fermionic mean field (no solid dot at the end of the Kramers invariant single-particle levels), contribute in a reduced fashion to (d) the moment of inertia of quadrupole rotational bands leading to $\mathcal{I} \approx \mathcal{I}_r/2$ (cf. Belyaev (1959, 2013); Bohr and Mottelson (1975), and refs. therein). (e) Pairing rotational bands in gauge space, an example of which is provided by the ground states of the superfluid Sn isotopes.

When q is small, quantal effects are small, and the lower state for $T < T_c$ will be a solid (crystalline structure), T_c denoting the critical temperature. For sufficiently

statement concerning the competition between potential energy and fluctuations is also valid for the generalized Lindemann parameter (Stillinger and Stillinger (1990); Zhou et al. (1999)) used to provide similar insight into inhomogeneous finite systems like proteins (aperiodic crystals; Schrödinger (1944); see also Ehrenfest's theorem (Basdevant and Dalibard (2005), p. 138; see also App. 3.A).

large values of q ($>$ 0.15) the system will display particle delocalization and, likely, be amenable, within some approximation, to a mean field description (Figs. 1.2.2 and 3.3.2). In fact, the step *delocalization* $\rightarrow$ *mean field* is neither automatic nor guaranteed, in any case, neither for all properties nor for all levels of the system.

As already stated, independent particle motion can be viewed as the most collective of all nuclear properties, reflecting the effect of all nucleons on a given one resulting in a macroscopic effect. Consequently, it should be possible to calculate the mean field in an accurate manner, arguably, as accurately as one can calculate collective vibrations, e.g., quadrupole vibrations. But this does not mean that one knows how to correctly calculate the energy and associated dynamical deformation parameter of each single state of the quadrupole response function. Within this context one may find, through mean field approximation, a good description for the energy of the valence orbitals of a nucleus, but for specific levels (e.g., the $d_{5/2}$ level[40] of the isotopes $^{119-120}$Sn; see Fig 5.2.3). It is not said that including particle–vibration coupling corrections, a process which on average makes theory come closer to experiment[41], single specific quasiparticle energies will satisfactorily agree with the data.[42] Cases like this one constitute a fruitful experience concerning the intricacies of the many-body problem in general, and of the nuclear one (finite many-body system, FMBS) in particular, where spatial quantization plays a central role. In other words, one is dealing with a self-bound, strongly interacting, finite many-body system generated from collisions originally associated with a variety of astrophysical events and thus with the coupling and interweaving of different scattering channels and resonances, a little like the Hoyle monopole resonance ($\alpha + \alpha + \alpha \rightarrow ^{12}$C). Within the anthropomorphic scenario, such phenomena are found in the evolution of the universe to eventually allow for the presence of organic matter and, arguably, life on earth[43] more likely than to make mean field approximation, also renormalized mean field theory, an "exact" description of nuclear structure and reactions for every single nuclear level.

3.4 Generalized Quantality Parameter

Within the framework of independent particle motion, but allowing for the dressing of particles, one obtains a sigmoidal distribution around the Fermi energy reflecting the ω-dependence of $Z_\nu(\omega)$. In the case of the Sn-isotopes where the density of

---

[40] Idini et al. (2015).

[41] See, e.g., Bohr and Mottelson (1975); Bohr and Broglia (1977); Hamamoto (1977); Bès et al. (1977); Barnes et al. (1972); Reich et al. (1974); Hamamoto and Siemens (1976); Bortignon et al. (1977); Bernard and Giai (1981); Mahaux et al. (1985); Bès and Broglia (1971a); Flynn et al. (1971); Bortignon et al. (1976); Bès et al. (1988); Barranco and Broglia (1987); Vinh Mau (1995); Barranco et al. (2001); Orrigo and Lenske (2009), and refs. therein.

[42] See Tarpanov et al. (2014).

[43] See, e.g., Rees (2000); Meißner (2015), and refs. therein.

levels is $N(0) \approx 4\text{MeV}^{-1}$ and the energy of the lowest $\beta = 0$ collective vibration ($\lambda^\pi = 2^+$) is $\hbar\omega_2 \approx 1.3$ MeV, one expects the width of the distribution to be $\approx 2-3$ MeV around the Fermi energy. Allowing nucleons not only to emit and reabsorb $\beta = 0$ vibrational modes but also to exchange them with the other nucleons, as well as to interact through v_p^{bare}, one is confronted with a many-body problem like the one discussed in Sect. 3.2 for $x \gg 1/2$, which can be solved at profit within the BCS framework. It leads, again for Sn-isotopes, to a sigmoidal occupation probability extending over an energy range also of ≈ 3 MeV ($\approx 2\Delta$) around the Fermi energy.[44]

Despite the similitudes in occupancy of levels lying around the Fermi energy between the two physical situations discussed above, the associated $L = 0$ (gs)$\rightarrow$(gs) two-nucleon transfer cross sections are quite different. In the first case, it is proportional to $(Z_{\nu_1}\sqrt{\sigma_{\nu_1}})^2 \approx (1/2) \times \sigma_{\nu_1}$ and to $(\sum_{\nu>0} U_\nu' V_\nu' \sqrt{\sigma_\nu})^2 \approx (\alpha_0')^2\bar{\sigma} \approx 50 \times \bar{\sigma}$ in the second. The quantity σ_{ν_1} is the cross section associated with the lowest $j_{\nu_1}^2(0)$ two-particle configuration. The summation $\sum_{\nu>0}$ extends over the orbitals around ϵ_F within a range $\approx 2\Delta$, and $\bar{\sigma}$ is the average of the cross sections σ_ν associated with the valence configurations $j_\nu^2(0)$.

In connection with the Cooper pair (BCS solution), one should, more correctly, use the wavefunction (1.4.65) to calculate the two-nucleon transfer cross section. In keeping with the definition $c_\nu' = V_\nu'/U_\nu'$ and the phasing (1.4.41), it is found that

$$\sum_{\nu>0} c_\nu' P_\nu^\dagger |0\rangle_F = e^{2i\phi} \sum_{\nu>0} c_\nu P_\nu^\dagger |0\rangle_F, \qquad (3.4.1)$$

leading to

$$|\Psi_{N_1}\rangle \sim \left(e^{2i\phi} \sum_{\nu>0} c_\nu P_\nu^\dagger\right)^{N_1/2} |0\rangle_F, \qquad (3.4.2)$$

that is, independent pair motion, each Cooper pair contributing a phase ϕ to $|N\rangle$.

Note that in a metallic superconductor as described by BCS, there are of the order of 10^6 Cooper pairs around the Fermi surface. Because the phase ϕ is common to so many pairs, its effects can be felt on the macroscopic scale. *By the same token, the quantal behavior of the BCS condensate must also be the same as the quantal behavior of a single Cooper pair,* a feature which reflects itself in superfluid nuclei, although in this case one is talking of only six to eight Cooper pairs.

---

[44] Within this context, see discussion following Eq. (1.4.53) (see also the discussion in connection with Fig. 2-1 of Tinkham (1980)).

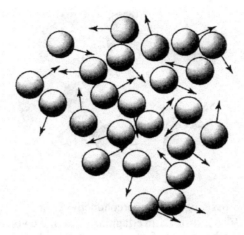

Figure 3.4.1 A system of localized pairs (Schafroth pairs; Schafroth (1958); Schafroth et al. (1957); see also Schrieffer (1964)), in which the size of the pair bound state is small compared to the average spacing between pairs.

If one finds, as one does[45], that the tunneling energy is a function of the relative phase, it will adjust so as to minimize the energy through tunneling of electrons. The generalized quantality parameter, ratio of the kinetic energy of confinement ($T_\xi = \hbar^2/(2m_e\xi^2) \approx 4 \times 10^{-5}$meV; $\xi \approx 10^4$Å) of the electrons forming the Cooper pairs undergoing tunneling, and the Cooper pair correlation (binding) energy E_{corr}(≈ 3 meV, $\Delta \approx 1.4$ meV, Pb), are much smaller than 1 ($q_\xi \approx 10^{-5}$), a result which implies that for $T < T_c$ ($T_c \approx 7.1$ K, critical temperature of Pb), the electron Cooper pair partners are solidly anchored to each other. Said differently, the intrinsic Cooper pair motion is frozen because of the pairing gap. Consequently, if under such conditions ($T < T_c$) one observes a current across the unbiased junction between the two superconductors, it must be a current of carriers of mass $2m_e$ and charge $2e$. If one electron partner of a Cooper pair tunnels, the other entangled electron does it also. *It is of note that in the present case, this result emerges not because of the large role played by potential effects but because of the very large dimensions and low relative intrinsic momentum of the Cooper pair partners.*

BCS condensation, that is, condensation of strongly overlapping, very extended, weakly bound Cooper pairs, corresponds to ordering in occupying momentum space and not space-like condensation of strongly bound clusters which undergo Bose condensation (Figs. 3.4.1 and 3.4.2, respectively). In BCS condensation, the inner, intrinsic structure of the pair, that is, the fact that it is made out of fermions entangled in time-reversal states, is the characterizing feature at the

---

[45] Josephson (1962); Anderson and Rowell (1963).

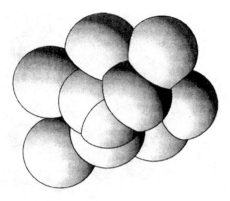

Figure 3.4.2 At the basis of BCS superconductivity, one finds condensation of very extended ($\approx 10^4$ Å), strongly overlapping (within the volume of a given pair the centers of $\approx 10^6$ other pairs is found), weakly bound ($\approx 2\Delta \approx$ meV) Cooper pairs corresponding to ordering in occupying momentum space, and not space-like condensation of strongly bound pairs which undergo Bose condensation (cf. Schrieffer (1964)).

basis of ODLRO. Cooper pairs (see Eqs. (1.4.48) and (3.4.1)) are not bosons, as $(P_\nu^\dagger)^2 = 0$ testifies (see also App. 4.A). However, it is true that, under certain circumstances, namely, $T < T_c$, $q < 1/\xi$, and $d \approx D_0 \leq \xi$ in the case of heavy ion reactions between superfluid nuclei, where ξ is the correlation length and d is the width of the Josephson junction (D_0 being the distance of closest approach), Cooper pairs show properties which strongly resemble those of bosons.[46]

We conclude this section by summarizing in[47] Fig. 3.4.3 the central aspect of the parallel between independent-particle and independent-pair motion discussed above.

3.5 Two-Nucleon Spectroscopic Amplitudes Associated with Pairing Vibrational Modes in Closed-Shell Systems: The ^{208}Pb Case

The solution of the pairing Hamiltonian leads, in the case of closed-shell systems and within the harmonic approximation (RPA), to pair addition (a) pair removal (r) two-particle, two-hole correlated modes. The associated creation operator have been defined in Eqs. (1.3.7)–(1.3.9) for the pair addition mode. Similarly, in the

---

[46] Ter Haar (1995a, Ch. 4.9; 1995b); Ehrenfest and Oppenheimer (1931).

[47] In connection with Fig. 3.4.3, the estimate $R = 8/k_F$ was carried out with the help of the Fermi gas model (see, e.g., Bohr and Mottelson (1969)). The Fermi momentum is written as $k_F \approx (3\pi^2 A/2V)^{1/3} \approx (\frac{3\pi^2}{2}\rho(0))^{1/3}$. It is of note that this expression leads to $\rho(0) \approx 0.17$ fm^{-3} and to the Fermi momentum $k_F \approx 1.36$ fm^{-1}. $A^{1/3} \approx 5$ for medium heavy nuclei.

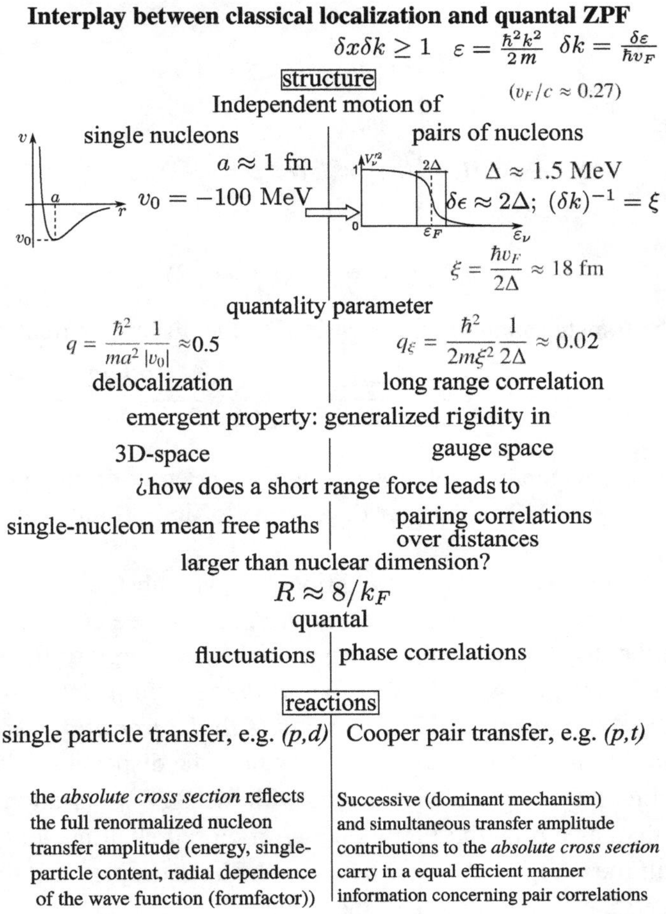

Interplay between classical localization and quantal ZPF

$$\delta x \delta k \geq 1 \quad \varepsilon = \frac{\hbar^2 k^2}{2m} \quad \delta k = \frac{\delta \varepsilon}{\hbar v_F}$$

structure

Independent motion of $(v_F/c \approx 0.27)$

single nucleons | pairs of nucleons

$a \approx 1 \text{ fm}$ | $\Delta \approx 1.5 \text{ MeV}$

$v_0 = -100 \text{ MeV}$ | $\delta \epsilon \approx 2\Delta; \ (\delta k)^{-1} = \xi$

$$\xi = \frac{\hbar v_F}{2\Delta} \approx 18 \text{ fm}$$

quantality parameter

$$q = \frac{\hbar^2}{ma^2} \frac{1}{|v_0|} \approx 0.5 \qquad q_\xi = \frac{\hbar^2}{2m\xi^2} \frac{1}{2\Delta} \approx 0.02$$

delocalization | long range correlation

emergent property: generalized rigidity in

3D-space | gauge space

¿how does a short range force leads to

single-nucleon mean free paths | pairing correlations over distances

larger than nuclear dimension?

$$R \approx 8/k_F$$

quantal

fluctuations | phase correlations

reactions

single particle transfer, e.g. *(p,d)* | Cooper pair transfer, e.g. *(p,t)*

the *absolute cross section* reflects the full renormalized nucleon transfer amplitude (energy, single-particle content, radial dependence of the wave function (formfactor)) | Successive (dominant mechanism) and simultaneous transfer amplitude contributions to the *absolute cross section* carry in a equal efficient manner information concerning pair correlations

Figure 3.4.3 Classical localization and zero-point fluctuations associated with independent-particle (normal density) and phase-correlated, independent-pair (abnormal density) motion.

case of the pair removal mode,

$$\Gamma_r^\dagger(n) = \sum_i X_n^r(i)\Gamma_i^\dagger + \sum_k Y_n^r(k)\Gamma_k,$$

X, Y fulfilling

$$\sum X^2 - Y^2 = 1$$

and

$$\Gamma_k^\dagger = a_k^\dagger a_{\bar{k}}^\dagger, \quad (\epsilon_k > \epsilon_F).$$

Similarly,

$$\Gamma_i^\dagger = a_{\bar{i}} a_i, \quad (\epsilon_i \leq \epsilon_F).$$

The relations

$$[H, \Gamma_a^\dagger(n)] = \hbar W_n (\beta = +2)$$

and

$$[H, \Gamma_r^\dagger(n)] = \hbar W_n (\beta = -2),$$

where β is the transfer quantum number, lead to the dispersion relations

$$\frac{1}{G(\pm 2)} = \sum_k \frac{(\Omega_k/2)}{2\epsilon_k \mp W_n(\pm 2)} + \sum_i \frac{(\Omega_i/2)}{2\epsilon_i \pm W_n(\pm 2)}, \qquad (3.5.1)$$

n labeling the corresponding solutions in increasing order of energy. In the above equation, $\Omega_j = j + 1/2$ is the pair degeneracy of the orbital with total angular momentum j.

For the case of the (neutron) pair addition and pair substraction modes of ^{208}Pb the above equations are graphically solved in Fig 3.5.1 (see also Table 3.5.1). The minimum of the dispersion relation defines the Fermi energy of the system under study. This is in keeping with the fact that *in the case in which* $W_1(\beta = +2) = W_1(\beta = -2) = 0$, *the situation corresponding to the transition between normal and superfluid phases*[48], the energy value at which the dispersion relation touches for the first time the energy axis coincides with the BCS λ variational parameter. It is of note that, as a rule, the Fermi energy of closed-shell nuclei is empirically defined as half the energy difference between the last occupied and the first empty single-particle state.[49] Making use of the values[50]

$$\begin{cases} E_{corr}(+2) = BE(208) + BE(210) - 2BE(209) = 1.248\,\text{MeV} \\ E_{corr}(-2) = BE(208) + BE(206) - 2BE(207) = 0.640\,\text{MeV}, \end{cases} \qquad (3.5.2)$$

one obtains $W_1(-2) + W_1(+2) = (BE(208) - BE(206)) - (BE(210) - BE(208))$ $= 14.11 - 9.115 = 4.995\,\text{MeV}$. *One notes that in the above calculations, all energies differences are positive.* In particular (see Table 3.5.1),

$$\epsilon_i < \epsilon_F \Rightarrow \epsilon_F - \epsilon_i = -|\epsilon_F| + |\epsilon_i| = |\epsilon_i| - |\epsilon_F| > 0$$

and

$$\epsilon_k > \epsilon_F \Rightarrow \epsilon_k - \epsilon_F = -|\epsilon_k| + |\epsilon_F| = |\epsilon_F| - |\epsilon_k| > 0.$$

---

[48] See footnote 31 Ch. 1.

[49] Cf., e.g., Mahaux et al. (1985).

[50] It is of note that in the present section, the quantities E_{corr} are defined in such a way that their value is positive.

Table 3.5.1 *Empirical energies of the valence single-particle levels of* ^{208}*Pb.*

In the upper part the occupied levels ($\epsilon_i \leq \epsilon_F$) *are displayed, while in the lower part the empty levels are shown* ($\epsilon_k > \epsilon_F$). *Of note is that* $\epsilon_{p_{1/2}} - \epsilon_{g_{9/2}} = 3.41\ MeV$ *is the single-particle gap associated with* $N = 126$ *shell closure.*

Orbit	ϵ_j	$\epsilon_{p_{1/2}} - \epsilon_k \equiv \lvert\epsilon_k\rvert - \lvert\epsilon_{p_{1/2}}\rvert$
$0h_{9/2}$	-10.62	3.47
$1f_{7/2}$	-9.50	2.35
$0i_{13/2}$	-8.79	1.64
$2p_{3/2}$	-8.05	0.90
$1f_{5/2}$	-7.72	0.57
$2p_{1/2}$	-7.15	0.00
$\epsilon_F = -5.825$ MeV		$\epsilon_k - \epsilon_{g_{9/2}} \equiv \lvert\epsilon_{g_{9/2}}\rvert - \lvert\epsilon_k\rvert$
$1g_{9/2}$	-3.74	0.00
$0i_{11/2}$	-2.97	0.77
$0j_{15/2}$	-2.33	1.41
$2d_{5/2}$	-2.18	1.56
$3s_{1/2}$	-1.71	2.03
$1g_{7/2}$	-1.27	2.47
$2d_{3/2}$	-1.23	2.51

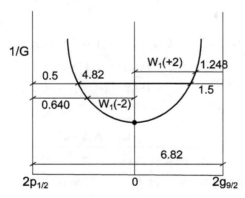

Figure 3.5.1 The right-hand side of the RPA pairing vibrational dispersion relation for neutrons in the case of the closed-shell system ^{208}Pb in the region between the two neighboring shells ($p_{1/2}$ and $g_{9/2}$). All quantities are in MeV. For each value of G there is a straight horizontal line, which is divided by the curve in three sections. The first one, from the left, corresponds to the pairing correlation energy of the nucleus ^{206}Pb (two correlated neutron hole states), while the last segment, to the right, measures the pairing correlation energy of ^{210}Pb (two correlated neutrons above closed shell), and the intermediate segment measures the energy of the two-phonon (correlated $(2p - 2h)$) pairing vibrational state of ^{208}Pb.

Thus,

$$\begin{cases} 2(\epsilon_F - \epsilon_{p_{1/2}}) = W_1(-2) + E_{corr}(-2) > 0 \\ 2(\epsilon_{g_{9/2}} - \epsilon_F) = W_1(+2) + E_{corr}(+2) > 0. \end{cases}$$

From Fig. 3.5.1 and Table 3.5.1 one can then write

$$2 \times (-5.825 - (-7.15))\,\text{MeV} = 2.650\,\text{MeV} = W_1(-2) + 0.640\,\text{MeV}$$

and

$$2 \times (-3.74\,\text{MeV} - (-5.825)\,\text{MeV}) = 4.17\,\text{MeV} = W_1(+2) + 1.248\,\text{MeV}.$$

Consequently,

$$W_1(-2) = 2.01\,\text{MeV} \quad \text{and} \quad W_1(+2) = 2.92\,\text{MeV},$$

leading to

$$W_1(+2) + W_1(-2) = 4.93\,\text{MeV}. \tag{3.5.3}$$

3.5.1 Pair Removal Mode

In Fig. 3.5.2, a graphical representation of the forward-going RPA amplitude of the pair removal mode is shown. Its expression for $n = 1$ is

$$X_1^r(i) = \frac{\frac{1}{2}\Omega_i^{1/2}\Lambda_1(-2)}{2(\epsilon_F - \epsilon_i) - W_1(-2)},$$

where

$$\begin{aligned} 2 \times (\epsilon_F - \epsilon_i) - W_1(-2) &= 2 \times (\epsilon_F - \epsilon_i) - 2 \times (\epsilon_F - \epsilon_{p_{1/2}}) + E_{corr}(-2) \\ &= 2 \times (\epsilon_{p_{1/2}} - \epsilon_i) + E_{corr}(-2) \\ &= 2 \times (|\epsilon_i| - |\epsilon_{p_{1/2}}|) + E_{corr}(-2). \end{aligned}$$

Thus,

$$X_1^r(i) = \frac{\frac{1}{2}\Omega_i^{1/2}\Lambda_1(-2)}{2(|\epsilon_i| - |\epsilon_{p_{1/2}}|) + E_{corr}(-2)}.$$

Making use of the empirical value of $E_{corr}(-2)$ worked out above, one obtains

$$X_1^r(i) = \frac{\frac{1}{2}\Omega_i^{1/2}\Lambda_1(-2)}{2(|\epsilon_i| - |\epsilon_{p_{1/2}}|) + 0.640\,\text{MeV}}. \tag{3.5.4}$$

In Fig. 3.5.3 we display the graphical process associated with the backward-going ($n = 1$) RPA amplitude,

$$Y_1^r(k) = \frac{\frac{1}{2}\Omega_k^{1/2}\Lambda_1(-2)}{2(\epsilon_k - \epsilon_F) + W_1(-2)}.$$

Figure 3.5.2 NFT representation of the forward-going RPA amplitude of the pair-removal mode (double downward-going arrowed line) describing a two-correlated hole state (single downward-going arrowed line for each hole with quantum numbers collectively labeled i).

Figure 3.5.3 Same as Fig. 3.5.2, but for the backward-going amplitudes.

Making use of

$$2 \times (\epsilon_F - \epsilon_{p_{1/2}}) - E_{corr}(-2) = W_1(-2),$$

one can write

$$2 \times (\epsilon_F - \epsilon_{p_{1/2}}) + 2 \times (\epsilon_k - \epsilon_F) - E_{corr}(-2) = 2 \times (\epsilon_k - \epsilon_F) + W_1(-2),$$

leading to

$$2 \times (|\epsilon_{p_{1/2}}| - |\epsilon_k|) - E_{corr}(-2) = 2 \times (|\epsilon_{p_{1/2}}| - |\epsilon_{g_{9/2}}|)$$
$$+ 2 \times (|\epsilon_{g_{9/2}}| - |\epsilon_k|) - E_{corr}(-2).$$

Thus,

$$Y_1^r(k) = \frac{\frac{1}{2}\Omega_k^{1/2}\Lambda_1(-2)}{2(|\epsilon_{g_{9/2}}| - |\epsilon_k|) + 2(|\epsilon_{p_{1/2}}| - |\epsilon_{g_{9/2}}|) - E_{corr}(-2)}.$$

With the help of $2 \times (|\epsilon_{p_{1/2}}| - |\epsilon_{g_{9/2}}|) - E_{corr}(-2) = 6.82\text{MeV} - 0.640\text{MeV} = 6.18\text{MeV}$, one obtains

$$Y_1^r(k) = \frac{\frac{1}{2}\Omega_k^{1/2}\Lambda_1(-2)}{2(|\epsilon_{g_{9/2}}| - |\epsilon_k|) + 6.18\,\text{MeV}}. \tag{3.5.5}$$

The above expressions of $X_1^r(i)$ and $Y_1^r(k)$ contain the experimental values of the two-hole correlation energies (0.640 MeV). *Because (see Fig. 3.5.1) the associated values of G do not lead to the observed correlation energy of the pair addition mode (1.248 MeV), we prefer to choose a single value of G and use the resulting $E_{corr}(-2)$ (=0.5 MeV) and $E_{corr}(+2)$ (=1.5 MeV) correlation energies to calculate* *the corresponding X, Y amplitudes* for both the lowest removal and lowest addition pairing modes. Making use of

$$2 \times (|\epsilon_{p_{1/2}}| - |\epsilon_{g_{9/2}}|) = 6.82 \, \text{MeV} \quad \text{and} \quad 2 \times (|\epsilon_{p_{1/2}}| - |\epsilon_{g_{9/2}}|) - E_{corr}(-2)$$

$$= (6.82 - 0.5) \, \text{MeV} = 6.32 \, \text{MeV},$$

one can write

$$X_1^r(i) = \frac{\frac{1}{2}\Omega_i^{1/2}\Lambda_1(-2)}{2(|\epsilon_i| - |\epsilon_{p_{1/2}}|) + 0.5 \, \text{MeV}} \tag{3.5.6}$$

$$Y_1^r(k) = \frac{\frac{1}{2}\Omega_k^{1/2}\Lambda_1(-2)}{2(|\epsilon_{g_{9/2}}| - |\epsilon_k|) + 6.32 \, \text{MeV}}. \tag{3.5.7}$$

Tables 3.5.2 and 3.5.3 contain the amplitudes of the pair removal mode of ^{208}Pb ($\Gamma_r^\dagger(1) = \sum X_1^r(i)\Gamma_i^\dagger + \sum Y_1^r(k)\Gamma_k$), that is, of the two-neutron correlated hole state describing $|^{206}\text{Pb (gs)}\rangle = \Gamma_r^\dagger(1)|0\rangle$.

It is of note that the coupling strength $\Lambda_1(-2)$ with which the pair removal mode couples to the two single-particle (-hole) states is calculated by normalizing the amplitudes (1) Tamm Dancoff, TD implying no ground state correlations[51], i.e., to assume that $Y \equiv 0$) $\sum_i A^2(i) = 1.5549 \, \text{MeV}^{-2}$ and thus $\Lambda_1(-2) = 0.802$ MeV, ($\sum X(i)_{TD}^2 = 1$); (2) RPA, $\Lambda_1^2(-2) \times (\sum_i A^2(i) - \sum_k B^2(k)) = \Lambda_1^2(-2) \times 1.45073 = 1$. Thus $\Lambda_1(-2) = 0.830 \, \text{MeV}$.

The above results show that there is a few percentage difference between the two values of Λ (TD and RPA), as well as for the corresponding X amplitudes (Table 3.5.2). Nonetheless, ground state correlations as expressed by the Y amplitudes (Table 3.5.3) give rise to a 53 percent increase in the $^{206}\text{Pb}(t, p)^{208}\text{Pb(gs)}$ absolute

Table 3.5.2 *Forward-going RPA amplitudes $X_1^r(i)$ of the lowest pair removal mode of ^{208}Pb (i.e., $|gs\,(^{206}\text{Pb})\rangle$ state), cf. Table XVI Broglia et al. (1973)*

Units		MeV	MeV^{-1}	RPA	TD
nlj	Ω_i	$|\epsilon_i| - |\epsilon_{p_{1/2}}|$	$A(i) = \frac{\frac{1}{2}\Omega_i^{1/2}}{2(|\epsilon_i|-|\epsilon_{p_{1/2}}|)+0.5\,\text{MeV}}$	$X_1^r(i)$	$X_1^r(i)$
$2p_{1/2}$	1	0	1	0.83	0.80
$1f_{5/2}$	3	0.57	0.528	0.44	0.42
$2p_{3/2}$	2	0.90	0.307	0.25	0.25
$0i_{13/2}$	7	1.64	0.350	0.29	0.28
$1f_{7/2}$	4	2.35	0.192	0.16	0.15
$0h_{9/2}$	5	3.47	0.150	0.12	0.12

---

[51] See footnote 31 of Chapter 1.

Table 3.5.3 *Same as Table 3.5.2, but for the backward amplitudes $Y_1^r(k)$ of the lowest energy pair removal mode*

Units		MeV	MeV^{-1}	RPA
nlj	Ω_k	$|\epsilon_{g9/2}| - |\epsilon_k|$	$B(k) = \dfrac{\frac{1}{2}\Omega_i^{1/2}}{2(|\epsilon_{g9/2}|-|\epsilon_k|)+6.23\,\text{MeV}}$	$Y_1^r(i)$
$1g_{9/2}$	5	0	0.179	0.15
$0i_{11/2}$	6	0.77	0.158	0.13
$0j_{15/2}$	8	1.41	0.156	0.13
$2d_{5/2}$	3	1.56	0.093	0.08
$3s_{1/2}$	1	2.03	0.046	0.04
$1g_{7/2}$	4	2.47	0.090	0.07
$2d_{3/2}$	2	2.51	0.063	0.05

cross section, from 0.34 mb to 0.52 mb to be compared with experimental data $\sigma = 0.68 \pm 0.24$ mb (see Fig. 4.5.4).

3.5.2 Pair Addition Mode

In Fig. 3.5.4 the X-amplitude of the pair addition mode is shown (NFT diagram). The associated expression ($n = 1$),

$$X_1^a(k) = \frac{\frac{1}{2}\Omega_k^{1/2}\Lambda_1(+2)}{2(\epsilon_k - \epsilon_F) - W_1(+2)},$$

can, making use of the relation

$$2 \times (\epsilon_k - \epsilon_F) - W_1(+2) = 2 \times (\epsilon_k - \epsilon_F) - 2 \times (\epsilon_{g9/2} - \epsilon_F) + E_{corr}(+2)$$
$$= 2 \times (\epsilon_k - \epsilon_{g9/2}) + E_{corr}(+2)$$
$$= 2 \times (|\epsilon_{g9/2}| - |\epsilon_k|) + E_{corr}(+2),$$

be expressed as

$$X_1^a(k) = \frac{\frac{1}{2}\Omega_k^{1/2}\Lambda_1(+2)}{2(|\epsilon_{g9/2}| - |\epsilon_k|) + E_{corr}(+2)}.$$

Similarly (Fig. 3.5.5),

$$Y_1^a(i) = \frac{\frac{1}{2}\Omega_i^{1/2}\Lambda_1(+2)}{2(\epsilon_F - \epsilon_i) + W_1(+2)}$$

Figure 3.5.4 Same as Fig. 3.5.2, but for the pair addition mode.

Figure 3.5.5 Same as Fig. 3.5.3, but for the pair addition mode.

can be written, with the help of the relation

$$2 \times (\epsilon_F - \epsilon_i) + W_1(+2) = 2 \times (\epsilon_F - \epsilon_i) - 2 \times (\epsilon_{g_{9/2}} - \epsilon_F) - E_{corr}(+2)$$
$$= 2 \times (\epsilon_{p_{1/2}} - \epsilon_i) + 2 \times (\epsilon_{g_{9/2}} - \epsilon_{p_{1/2}}) - E_{corr}(+2)$$
$$= 2 \times (|\epsilon_i| - |\epsilon_{p_{1/2}}|) + 2 \times (|\epsilon_{p_{1/2}}| - |\epsilon_{g_{9/2}}|) - E_{corr}(+2),$$

as

$$Y_1^a(i) = \frac{\frac{1}{2}\Omega_i^{1/2}\Lambda_1(+2)}{2(|\epsilon_i| - |\epsilon_{p_{1/2}}|) + 2\Delta\epsilon_{sp} - E_{corr}(+2)}. \qquad (3.5.8)$$

Making use of $E_{corr}(+2) = 1.5$ MeV (cf. Fig. 3.5.1) and

$$2 \times \Delta\epsilon_{sp} = 2 \times (|\epsilon_{p_{1/2}}| - |\epsilon_{g_{9/2}}|) = 6.82 \, \text{MeV}, \qquad (3.5.9)$$

one can write $2\Delta\epsilon_{sp} - E_{corr}(+2) = (6.82 - 1.5)$ MeV=5.32 MeV, leading to

$$\begin{cases} X_1^a(k) = \dfrac{\frac{1}{2}\Omega_k^{1/2}\Lambda(-2)}{2(|\epsilon_{g_{9/2}}| - |\epsilon_k|) + 1.5 \, \text{MeV}} \\[2mm] Y_1^a(i) = -\dfrac{\frac{1}{2}\Omega_i^{1/2}\Lambda(+2)}{2(|\epsilon_i| - |\epsilon_{p_{1/2}}|) + 5.32 \, \text{MeV}}. \end{cases} \qquad (3.5.10)$$

The corresponding numerical values are displayed in Tables 3.5.4 and 3.5.5, while in Fig. 3.5.6 we display a schematic summary of the graphical solution of the dispersion relations.

Let us conclude this section by noting that while the harmonic (RPA) description of the pair vibrational modes of ^{208}Pb provides a fair picture of the two-neutron transfer spectroscopic amplitudes, in keeping with the collective character of these (coherent) states, conspicuous anharmonicities in the multiphonon spectrum have been observed and calculated.[52] Within the framework of Fig. 3.1.5, we

---

[52] Cf., e.g., Flynn et al. (1972a); Lanford and McGrory (1973); Bortignon et al. (1978); Clark et al. (2006).

Table 3.5.4 *Forward-going RPA amplitudes associated with the pair addition mode of* ^{208} *Pb.*
Cf Table XVI, Broglia et al. (1973). In the TD approximation (amplitudes not shown) the particle-pair addition coupling constant is calculated from the normalization of the amplitudes C(k). That is, $\sum_k C^2(k) = 0.903$ *MeV*$^{-2}$ *and* $\Lambda_1(+2) = (0.903)^{-1/2}$ *MeV =1.052 MeV.*

Units		MeV		MeV^{-1}	
nlj	Ω_k	$\|\epsilon_{g9/2}\| - \|\epsilon_k\|$	$C(k) = \dfrac{\frac{1}{2}\Omega_k^{1/2}}{2(\|\epsilon_{g9/2}\|-\|\epsilon_k\|)+1.5\,\text{MeV}}$		$X_1^a(k)$
$1g_{9/2}$	5	0	0.745		0.82
$0i_{11/2}$	6	0.77	0.403		0.44
$0j_{15/2}$	8	1.41	0.327		0.36
$2d_{5/2}$	3	1.56	0.187		0.21
$3s_{1/2}$	1	2.03	0.090		0.10
$1g_{7/2}$	4	2.47	0.155		0.17
$2d_{3/2}$	2	2.51	0.108		0.12

Table 3.5.5 *Same as Table 3.5.4, but for the backward-going amplitude.*
$\sum_i D^2(i) = 0.079$ *and* $\Lambda_1^2(+2)(\sum_k C^2(k) - D^2(i)) = \Lambda_1^2(+2)(0.903 - 0.079)$
MeV$^{-2} = 0.824$*MeV*$^{-2}$; $\Lambda_1(+2) = (0.824)^{-1/2}$*MeV, thus* $\Lambda_1(+2) = 1.102$ *MeV.*

Units		MeV		MeV^{-1}	
nlj	Ω_i	$\|\epsilon_i\| - \|\epsilon_{p1/2}\|$	$D(i) = \dfrac{\frac{1}{2}\Omega_i^{1/2}}{2(\|\epsilon_i\|-\|\epsilon_{p1/2}\|)+5.32\,\text{MeV}}$		$Y_1^a(i)$
$2p_{1/2}$	1	0	0.094		-0.1
$1f_{5/2}$	3	0.57	0.134		-0.15
$2p_{3/2}$	2	0.90	0.099		-0.11
$0i_{13/2}$	7	1.64	0.154		-0.17
$1f_{7/2}$	4	2.35	0.100		-0.11
$0h_{9/2}$	5	3.47	0.091		-0.10

schematically emphasize in Fig. 3.5.7 the relative importance of dynamic and static pairing distortions, in comparison with the corresponding quantities in the case of quadrupole surface distortions in 3D space.[53] These results underscore the major role pairing vibrations play in nuclei around closed shells, while those shown in Fig. 3.1.2 emphasize their importance in gauge invariance restoration in systems far away from closed shells.

[53] For details cf. Bès and Broglia (1977); Broglia et al. (1968); Bès et al. (1988); Barranco et al. (1987); Shimizu et al. (1989); Shimizu (2013); Vaquero et al. (2013), and refs. therein.

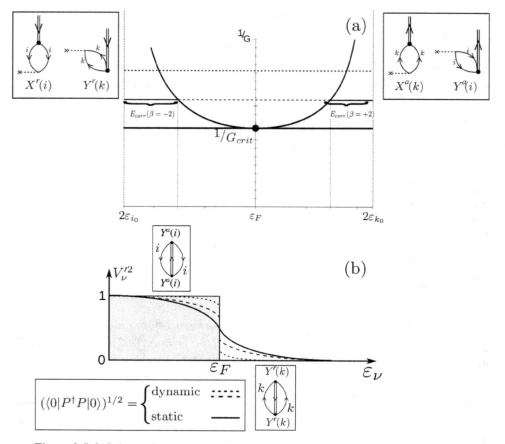

Figure 3.5.6 Schematic representation of the quantal phase transition taking place as a function of the pairing coupling constant in a closed-shell nucleus. (a) Dispersion relation associated with the RPA diagonalization of the Hamiltonian $H = H_{sp} + H_p$ for the pair addition and pair removal modes. In the insets are shown the two-particle transfer processes exciting these modes, which testify to the fact that the associated zero-point fluctuations (ZPF), which diverge at $G = G_{crit}$, blur the distinction between occupied and empty states typical of closed shell nuclei. (b) Occupation number associated with the single-particle levels. For $G < G_{crit}$, there is a dynamical depopulation (population) of levels $i(k)$ below (above) the Fermi energy. For $G > G_{crit}$, the deformation of the Fermi surface becomes stable, although with a nonvanishing dynamic component (see Fig. 3.1.2).

3.6 Halo Pair Addition and Pygmy Dipole Modes: A New Mechanism to Break Gauge Invariance

Pairing correlations are intimately connected with particle number violation and thus spontaneous breaking of gauge invariance, as testified by the order parameter

$$P^\dagger = \sum_{\nu>0} a_\nu^\dagger a_{\bar\nu}^\dagger$$

$$x' = \frac{2G\Omega'}{D} = GN(0)$$

$x'>1$ $x'>1$

$$\alpha_0 = <P^\dagger> = \frac{\Delta}{G} \approx 7$$

$$\alpha_{dyn} = \frac{<PP^\dagger>^{1/2} + <P^\dagger P>^{1/2}}{2}$$

$$\approx \frac{1}{2}\left(\frac{E_{corr}(A+2)}{G} + \frac{E_{corr}(A-2)}{G}\right) \approx 10$$

$$\frac{\alpha_0}{\alpha_{dyn}} \approx 0.7$$

$$\frac{\beta_2}{(\beta_2)_{dyn}} \approx 3-6$$

Figure 3.5.7 Relative importance of dynamic and static pairing distortion (α_{dyn} and α_0, respectively) associated with closed-shell and open-shell nuclei, calculated in terms of a two-level model, as compared with similar quantities for the case of quadrupole surface degrees of freedom (β_2-values). The parameter x' (product of the effective pairing strength $G' = Z_\omega^2(v_p^{bare} + v_p^{ind})$ and of the effective density of levels at the Fermi energy $N'(0) = Z_\omega^{-1}N(0) = Z_\omega^{-1}(2\Omega/D) = 2\Omega'/D = 2\Omega/D'$; $\Omega' = Z_\omega^{-1}\Omega$, $D' = Z_\omega D$), measures the relative importance of the single-particle gap $D' = Z_\omega D$ and of the pair correlation $G'\Omega$ (see Sect. 3.2; see also Brink and Broglia (2005), App. H, Sect. H.4). It is of note that in referring to α_0, one actually has in mind α_0' ($\alpha_0 = e^{-2i\phi}\alpha_0'$, α_0' being a real quantity). We nonetheless avoid the use of α_0', in keeping with the use of primes to refer to renormalized quantities, both here and in Sect. 3.2. For the color version of this figure, refer to cambridge.org/nuclearcooperpair.

$\langle BCS|P^\dagger|BCS \rangle = \alpha_0$. In the nuclear case, dynamical breaking of gauge symmetry is similarly important to that associated with static distortions. The fact that the average single-particle potential acts as an external field is one of the reasons for the existence of a critical value G_c of the pairing strength G to bind Cooper pairs in nuclei. Spatial quantization in finite systems at large and in nuclei in particular is intimately connected with the paramount role the surface plays in these systems.[54] Another consequence of this role is the fact that in nuclei, an important fraction (≈ 50 percent in the case of nuclei lying along the stability valley and even more,

[54] See Bohr and Mottelson (1975); see also Broglia (2002), and refs. therein.

up to 80 percent, for light halo nuclei) of Cooper pair binding is due to the exchange of collective vibrations between the pair partners[55], the rest being associated with the bare NN-interaction in the $^1S_0$ channel.

The study of light exotic nuclei lying along the neutron drip line have revealed a novel aspect of the interplay between shell effects and induced pairing interaction. It has been found that there are situations in which spatial quantization screens, essentially completely, the bare nucleon–nucleon paring interaction. This happens in the case in which the nuclear valence orbitals are s, p-states at threshold.[56] An example of situations of this type is provided by some of the $N = 7$ isotones, in particular $^{10}_{3}$Li and, to some extent, $^{11}_{4}$Be, nuclei which display *"parity inversion"* in the sequence of single-particle levels $1p_{1/2}$ and $2s_{1/2}$ as compared to the Mayer–Jensen prediction (Fig. 1.1.2). In what follows we discuss the (unbound) nucleus ^{10}Li (see also Sect. 5.2.3), in connection with the (bound) two-neutron halo system ^{11}Li.

The $N = 7$ isotone ^{10}Li displays a virtual $s_{1/2}$ and a resonant $p_{1/2}$ state.[57] The binding provided by a contact pairing interaction $V_\delta(|\mathbf{r} - \mathbf{r}'|)$ (δ-force) to a pair of fermions moving in time–reversal states in a single j-shell[58] is given by the matrix element

$$M_j = \langle j^2(0)|V_\delta|j^2(0)\rangle = -\frac{(2j+1)}{2}V_0 I(j) \approx -\frac{(2j+1)}{2}V_0\frac{3}{R^3}. \qquad (3.6.1)$$

It is of note that $G = V_0 I(j)$ ($\approx 25/A$ MeV≈ 2.3 MeV ($A = 11$)). The ratio of the above matrix element associated with the two halo neutrons of ^{11}Li and with an hypothetical normal nucleus of mass $A = 11$ is

$$r = \frac{(M_j)_{halo}}{(M_j)_{core}} = \frac{2}{(2j+1)}\left(\frac{R_0}{R}\right)^3. \qquad (3.6.2)$$

The quantities $R_0 = 1.2A^{1/3}$fm $= 2.7$fm ($A = 11$), and $R = \sqrt{\frac{5}{3}}\langle r^2\rangle^{1/2}_{^{11}\text{Li}} = \sqrt{\frac{5}{3}}$ (3.55 ± 0.1) fm $=(4.58 \pm 0.13)$ fm are the radius of a nucleus of mass $A = 11$ (systematics) and the measured (mean square) radius[59] of ^{11}Li, respectively.[60] The

---

[55] See Barranco et al. (1999); Terasaki et al. (2002); Brink and Broglia (2005); Saperstein and Baldo (2013); Avdeenkov and Kamerdzhiev (2013); Lombardo et al. (2013); Barranco et al. (2001); Potel et al. (2010); Pankratov et al. (2011); Idini et al. (2015); Barranco et al. (2005).

[56] Pairing antihalo effect (Bennaceur et al. (2000); Hamamoto and Mottelson (2003, 2004)).

[57] See Barranco et al. (2020); Moro et al. (2019), and refs. therein, as well as footnotes 30, 31, and 32 of Chapter 5.

[58] Cf., e.g., Eq. (2.12) of Brink and Broglia (2005).

[59] In other words, a contact interaction has an effective strength which scales with $R^{-3} \sim A^{-1}$. In the case of halo nuclei, the radius is larger as a rule than that expected from systematics. In particular, in the case of ^{11}Li, the value of the radius corresponds, assuming $r_0 = 1.2$ fm, to an effective mass number $A_{eff} \approx 56$.

[60] See footnote 70 of this chapter.

quantity j is the effective angular momentum of a single j-shell which can accommodate eight neutrons ($j \sim k_F R_0 \approx 1.36\,\text{fm}^{-1} \times 2.7\,\text{fm} \approx 3.7$). One thus obtains

$$r = 0.048. \tag{3.6.3}$$

Consequently, the bare NN-nucleon pairing interaction is expected to become strongly screened, the resulting effective G-value ($A \approx 11$)

$$G_{scr} = r \times G = 0.048 \times 25\text{MeV}/A \approx 1\text{MeV}/A \approx 0.1\,\text{MeV} \tag{3.6.4}$$

becoming subcritical and thus unable to bind the halo Cooper pair ($2\tilde{\epsilon}_{s_{1/2}} = 0.3$ MeV; see Fig. 2.7.1) to the ^{9}Li core.

Further insight into this question can be gained making use of the multipole expansion of a general interaction

$$v(|\mathbf{r}_1 - \mathbf{r}_2|) = \sum_\lambda V_\lambda(r_1, r_2) P_\lambda(\cos\theta_{12}). \tag{3.6.5}$$

Because the function P_λ drops from its maximum at $\theta_{12} = 0$ in an angular distance $1/\lambda$, particles 1 and 2 interact through the component λ of the force, only if $r_{12} = |\mathbf{r}_1 - \mathbf{r}_2| < R/\lambda$, where R is the mean value of the radii $\mathbf{r}_1$ and $\mathbf{r}_2$. Thus, as λ increases, the effective force range decreases. For a force of range much greater than the nuclear size, only the lowest λ (long-wavelength) terms are important. At the other extreme, a δ-function force has coefficients $V_\lambda(r_1, r_2) \left(= \frac{(2\lambda+1)}{4\pi r_1^2}\delta(r_1 - r_2) \right)$ that increases with λ. In the case of ^{11}Li(gs), one is thus confronted with the need to accept a long-range, low-λ pairing interaction as responsible for the binding of the dineutron, halo Cooper pair to the ^{9}Li core. This is equivalent to saying an induced pairing interaction arising from the exchange of vibrations with low λ-value.

3.6.1 Cooper Pair Binding: A Novel Embodiment of the Axel–Brink Hypothesis

In what follows we discuss a possible novel test of the Axel–Brink hypothesis.[61] Within the s, p subspace, the most natural low-multipolarity, long-wavelength vibration is the dipole mode. From systematics, the centroid of these vibrations is found at $\hbar\omega_{GDR} \approx 100$ MeV/R, R being the nuclear radius.[62] Thus, in the

---

[61] The color of an object can be determined in two ways: by illuminating it with white light and seeing which wavelength it absorbs or by heating it up and seeing the wavelength it emits. In both cases, one is talking about dipole radiation. To describe the deexcitation process of hot nuclei requires the knowledge of the photon interactions with excited states. The common assumption, known as the Axel–Brink hypothesis, has been that each excited state of a nucleus carries a giant dipole resonance (GDR) on top of it and that the properties of such resonances are unaffected by any excitation of the nucleus (Brink (1955); Lynn (1968), p. 321; Axel (1962); cf. also Bertsch and Broglia (1986); Bortignon et al. (1998); Martin et al. (2017), and refs. therein).

[62] See Bohr and Mottelson (1975); Bortignon et al. (1998); Bertsch and Broglia (2005), and refs. therein.

case of ^{11}Li, one expects the centroid of the *giant dipole resonance* (GDR) carrying ≈ 100 percent of the energy-weighted sum rule (EWSR) at $\hbar\omega_{GDR} \approx 100$ MeV/4.6 ≈ 22 MeV. Such a high-frequency mode can hardly be expected to give rise to anything but polarization effects. On the other hand, there exists experimental evidence which testifies to the presence in ^{11}Li of a well-defined dipole resonance with centroid at $\lesssim 1$ MeV and carrying ≈ 8 percent of the EWSR.[63] The existence of this *pygmy (dipole) resonance* (PDR) which can be viewed as a consequence of the existence of a low-lying particle hole state associated with the transition $s_{1/2} \rightarrow p_{1/2}$ testifies, arguably, to the *coexistence*[64] of two states with rather different radii in the ground state, one closely connected with the ^{9}Li core (≈ 2.5 fm), the second with the diffuse halo (≈ 4.6 fm), namely, displaying a large radial (isotropic) deformation (neutron skin) and thus able to induce a conspicuous (radial, isotropic)[65] inhomogeneous damping to the dipole mode.

The importance of this mechanism is underscored by the fact that ^{11}Li, displaying a neutron excess $(N - Z)/A \approx 0.45$ as compared to the value of 0.21 in the case of ^{208}Pb, is able to bring down by tens of MeV a consistent fraction (≈ 8 percent) of the TRK-sum rule, a consequence of the fact that the nucleus is most sensitive to changes in density (saturation phenomena). In the case of the halo of ^{11}Li, we are confronted with nuclear structure phenomena in a medium displaying a density of ≈ 4 percent of saturation density (Sect. 4.10; see also Eq. (4.3.6)).

Before proceeding, let us estimate the overlap $\mathcal{O}$ between the two "ground states." Making use of a schematic expression for the single-particle radial wavefunctions,[66]

$$\mathcal{R} = \sqrt{3/R_0^3}\ \Theta(r - R_0),\tag{3.6.6}$$

where

$$\Theta = 1 \quad (r \leq R_0); \quad 0 \quad (r > R_0),$$

leading to

$$\int_0^\infty dr\, r^2 \mathcal{R}^2(r) = \frac{3}{R_0^3}\int_0^{R_0} dr^3/3 = 1,\tag{3.6.7}$$

---

[63] See footnote 123 Chapter 2.

[64] Within this context, one can mention similar situations concerning the coexistence of spherical and quadrupole deformed states (cf., e.g., Wimmer et al. (2010); Federman and Talmi (1965); Federman and Talmi (1966); Dönau et al. (1967), and refs. therein; cf. also Bohr and Mottelson (1963)), typically of nuclei with $N \approx Z$. Surface quadrupole inhomogeneous damping has modest consequences in bringing dipole strength at low energies as compared with radial (isotropic) deformations in ^{11}Li. This is understood in terms of the plasticity displayed by the atomic nucleus regarding quadrupole deformations (low-lying collective 2^+ surface vibrations, fission, exotic decay (cf. Barranco et al. (1988); Barranco et al. (1989); Bertsch (1988); Bertsch et al. (1987)), and of the little tolerance to both compressibility and rarification displayed by the same system and connected with saturation properties (see also Broglia et al.(2019a, 2019b)).

[65] See Sect. 4.10.1, in particular footnote 164 of Chapter 4.

[66] Bohr and Mottelson (1969).

one can work out the overlap $\mathcal{O}$ between the two halo neutrons and the core nucleons. That is,

$$
\mathcal{O} = |\langle \mathcal{R}_{halo}|\mathcal{R}_{core}\rangle|^2 = \left(\sqrt{\frac{3}{R_0^3}} \sqrt{\frac{3}{R^3}} \int_0^\infty dr\, r^2 \Theta(r-R)\Theta(r-R_0) \right)^2
$$

$$
= \left(\sqrt{\frac{3}{R_0^3}} \sqrt{\frac{3}{R^3}} \int_0^{R_0} dr^3/3 \right)^2 = (R_0/R)^3 = 0.20,
$$

(3.6.8)

where use has been made of $\Theta(r-R)\Theta(r-R_0) = \Theta(r-R_0)$, $R_0 = 1.2A^{1/3}$ fm $=$ 2.7 fm$(A = 11)$ and $R = (4.58 \pm 0.013)$ fm. Because of the small value of this overlap, one can posit that the $E_x \lesssim 1$ MeV $(\Gamma \approx 0.5$ MeV) soft $E1$-mode of ^{11}Li is a bona fide dipole resonance based on an exotic, unusually extended $|0_\nu\rangle$ state of radial dimensions equivalent, according to systematics, to a system of effective A-mass number about five times that of the actual system $(A \approx (4.6/1.2)^3 \approx 60)$, thus consistent with the connotation of PDR.

It is of note that the small values of r and of $\mathcal{O}$ have essentially the same origin. On the other hand, they have apparently rather different physical consequences. In fact, the first makes the bare pairing interaction strength G subcritical, while the second one screens the repulsive symmetry potential $V_1(\approx +25$ MeV$)$[67], that is, the energy price one has to pay to separate protons from neutrons. This effect allows for a consistent fraction of the dipole Thomas–Reiche–Kuhn sum rule, that is, of the $J^\pi = 1^-$ energy-weighted sum rule (EWSR), to come low in energy $(p_{1/2}-s_{1/2}$ transition) from the value $E_{GDR} \approx (100/R)$ MeV and, acting as an intermediate boson between the two halo neutrons, glue them to the ^{9}Li core. *Summing up, the halo antipairing effect $G_{scr} = r \times G \ll G < G_{crit}$ also triggers $(\mathcal{O}V_1 \ll V_1)$, the virtual presence of a "gas" of dipole (pygmy) bosons which, exchanged between the two halo neutrons (cf. Fig. 3.6.1), overcompensates the reduction of the bare pairing interaction, leading to the binding of the halo Cooper pair to the core (anti-(halo antipairing effect)). It can thus be stated that the halo of ^{11}Li and the pygmy dipole resonance[68] built on top of it constitute a pair of symbiotic states (see also Chapter 7).*

Let us further elaborate on these issues. Making use of the relation $\langle r^2 \rangle^{1/2} \approx (3/5)^{1/2} R$ between mean square radius and the radius, one may write

$$
\langle r^2 \rangle_{^{11}Li} \approx \frac{3}{5} R_{eff}^2(^{11}Li),
$$

(3.6.9)

---

[67] See, e.g., Bortignon et al. (1998), Eq. (3.48), and refs. therein.

[68] In which the two halo neutrons oscillate out of phase with respect to nucleons of the core ^{9}Li moving in phase (see Broglia et al. (2019a)).

with

$$R_{eff}^2(^{11}\text{Li}) = \left(\frac{9}{11}R_0^2(^9\text{Li}) + \frac{2}{11}\left(\frac{\xi}{2}\right)^2\right),$$ (3.6.10)

where

$$R_0(^9\text{Li}) = 2.5\text{fm}$$ (3.6.11)

is the ^{9}Li radius ($R_0 = r_0 A^{1/3}$, $r_0 = 1.2$fm), while ξ is the correlation length of the Cooper pair neutron halo. An estimate of this quantity is provided by the relation[69]

$$\xi = \frac{\hbar v_F}{\pi |E_{corr}|} \approx 20\,\text{fm},$$ (3.6.12)

in keeping with the fact that in ^{11}Li, $(v_F/c) \approx 0.16$ and $E_{corr} \approx -0.5$ MeV (see App. 7.C). Consequently,

$$R_{eff}(^{11}\text{Li}) \approx 4.8\,\text{fm},$$ (3.6.13)

and $\langle r^2 \rangle_{^{11}\text{Li}}^{1/2} \approx 3.7$ fm, to be compared with the experimental value[70] $\langle r^2 \rangle_{^{11}\text{Li}}^{1/2} = 3.55 \pm 0.1$fm. It is of note that this experimental value implies the radius $R(^{11}\text{Li}) = \sqrt{5/3\langle r^2\rangle_{^{11}\text{Li}}} = 4.58 \pm 0.13$ fm.

We now proceed to the calculation of the centroid of the dipole pygmy resonance of ^{11}Li in RPA, making use of the separable interaction[71]

$$H_D = -\kappa_1 \vec{D} \cdot \vec{D},$$ (3.6.14)

where $\vec{D} = \vec{r}$ and

$$\kappa_1 = \frac{-5V_1}{AR^2},$$ (3.6.15)

$V_1 = 25$ MeV being the symmetry potential energy. The RPA dispersion relation is[72]

$$W(E) = \sum_{k,i} \frac{2(\epsilon_k - \epsilon_i)|\langle \tilde{i}|F|k\rangle|^2}{(\epsilon_k - \epsilon_i)^2 - E^2} = \frac{1}{\kappa_1}.$$ (3.6.16)

Making use of this relation as well as (see Fig. 2.7.1) $\tilde{\epsilon}_{p_{1/2}} - \tilde{\epsilon}_{s_{1/2}} \approx 0.45$MeV for the value of $\epsilon_{v_k} - \epsilon_{v_i}$, and that the EWSR associated with the ^{11}Li pygmy resonance is ≈ 8 percent of the total Thomas–Reiche–Kuhn sum rule[73]

---

[69] The associated generalized quantality parameter being $q_\xi = \frac{\hbar^2}{2m\xi^2}\frac{1}{|E_{corr}|} \approx 0.1$ (see also App. 7.C).

[70] Kobayashi et al. (1989).

[71] For details, we refer to Bortignon et al. (1998), Sect. 3.2.3.

[72] See (3.30), p. 55, of Bortignon et al. (1998).

[73] The Thomas–Reiche–Kuhn sum rule (Bohr and Mottelson (1975); Bortignon et al. (1998))
    TRK=$\frac{9}{4\pi}\frac{\hbar^2 e^2}{2m}\frac{NZ}{A} = 14.8\frac{NZ}{A}e^2$ fm^2 MeV has a value of 32.3 e^2 MeV fm^2 for $^{11}_3$Li$_8$. Assuming a systematic behavior, the centroid of the giant dipole resonance is expected at $\hbar\omega_D \approx 80/A^{1/3}$ MeV ≈ 36 MeV, leading to the *ratio* ($\equiv (32.3e^2$ MeV fm$^2)/(36$MeV$) \approx 0.9e^2$ fm^2). The $E1$–single–particle

$$\sum_n |\langle 0|F|n\rangle|^2 (E_n - E_0) = \frac{\hbar^2}{2m} \int d\mathbf{r} |\vec{\nabla} F|^2 \rho(r), \tag{3.6.17}$$

which, for $F = r$, has the value[74] $\hbar^2 A/2m$, one can write the numerator of Eq. (3.6.16) as

$$2 \times 0.08 \times \frac{\hbar^2 A}{2m} = \frac{1}{\kappa_1}[(0.45\text{MeV})^2 - (\hbar\omega_{pygmy})^2], \tag{3.6.18}$$

and thus

$$(\hbar\omega_{pygmy})^2 = (0.45\text{MeV})^2 - 2 \times 0.08 \times \frac{\hbar^2 A}{2m}\kappa_1, \tag{3.6.19}$$

where[75]

$$\kappa_1 = -\frac{5V_1}{A(\xi/2)^2}\left(\frac{2}{11}\right) = -\frac{125\text{MeV}}{A\,100\,\text{fm}^2}\left(\frac{2}{11}\right) \approx -0.021\,\text{fm}^{-2}\,\text{MeV}. \tag{3.6.20}$$

The ratio in parentheses reflects the fact that only 2 out of 11 nucleons slosh back and forth in an extended configuration with little overlap with the other nucleons. The quantity

$$\kappa_1^0 = -\frac{5V_1}{A R_{eff}^2(^{11}\text{Li})} \approx 0.49\,\text{MeV}\,\text{fm}^{-2} \tag{3.6.21}$$

is the "standard" self-consistent dipole strength.[76] The screening factor $s = (\kappa_1/\kappa_1^0) = (R_{eff}^2 \times (2/11))/(\xi/2)^2 = 0.043$ is very close in magnitude to the ratio $r(= 0.048$, Eq. (3.6.3)) and has a similar physical origin. It is of note that $(V_1)_{scr} = sV_1 \approx 1$ MeV. Making use of (3.6.20), one obtains

$$-2 \times 0.08 \frac{\hbar^2 A}{2m}\kappa_1 \approx 0.74\text{MeV}^2 \approx (0.86\text{MeV})^2. \tag{3.6.22}$$

Consequently,

$$\hbar\omega_{pygmy} = \sqrt{(0.45)^2 + (0.86)^2}\text{MeV} \approx 1\,\text{MeV}, \tag{3.6.23}$$

(Weisskopf) unit can be written as (Bohr and Mottelson (1969), p. 389, Eq. (3C-38))
$B_W(E1) \approx ((1.2)^2/4\pi)(3/4)^2 A^{2/3}(e_{E1})^2\,\text{fm}^2 = 0.32(e_{E1})^2\,\text{fm}^2$, e_{E1} being the effective dipole charge equal to $(N/A)e = 0.73e$ for the protons of ^{11}Li, and $-(Z/A)e = 0.27$ for the neutrons. Making use of the average value, one can write $\bar{B}_W(E1) \approx 0.1e^2\,\text{fm}^2$. Thus 8%× ratio /$\bar{B}_W(E1) \approx 0.072/0.1 \approx 0.7$. In other words, about one single-particle unit is associated with the eventual γ-decay of the PDR of ^{11}Li.

[74] Cf. Bertsch and Broglia (2005), p. 53.
[75] See Bortignon et al. (1998).
[76] Cf. Bohr and Mottelson (1975).

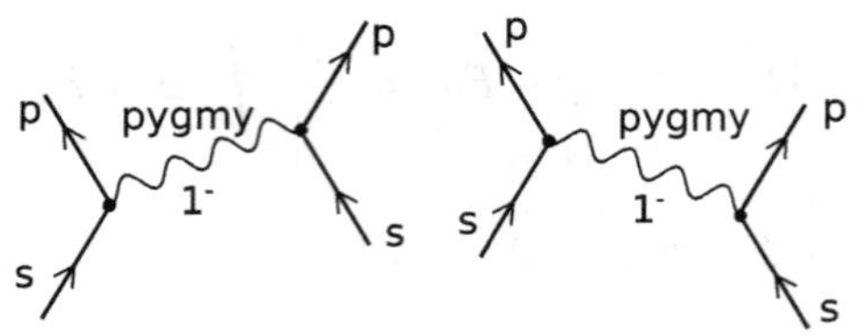

Figure 3.6.1 Diagrammatic representation of the exchange of a collective 1^- pygmy resonance between pairs of nucleons moving in the time-reversal configurations $s^2_{1/2}(0)$ and $p^2_{1/2}(0)$. It is of note that both these configurations can act as initial states, the figure showing only one of the two possibilities. Consequently, the energy denominator to be used in the simple estimate (3.6.25) is the average value $DEN = (DEN_1 + DEN_2)/2 = -\hbar\omega_{pygmy}$, where $DEN_1 = \Delta\epsilon - \hbar\omega_{pygmy}$ and $DEN_2 = -\Delta\epsilon - \hbar\omega_{pygmy}$, while $\Delta\epsilon = \epsilon_{s_{1/2}} - \epsilon_{p_{1/2}}$.

in overall agreement with the experimental findings.[77] It is of note that the centroid of the pygmy dipole resonance (microscopically) calculated in QRPA with the help of a separable dipole interaction is[78] $\approx (0.6\,\text{MeV} + 1.6\,\text{MeV})/2 \approx 1.1\,\text{MeV}$.

Let us now estimate the binding energy which the exchange of the pygmy resonance between the two neutrons of the halo Cooper pair of ^{11}Li can provide. The associated particle–vibration coupling[79] is $\Lambda = \left(\partial W(E)/\partial E\big|_{\hbar\omega_{pygmy}}\right)^{-1/2}$. Note the use in what follows of a dimensionless dipole single–particle field $F' = F/R_{eff}(^{11}\text{Li})$. This is in keeping with the fact that one wants to obtain a quantity with energy dimensions ($[\Lambda] = \text{MeV}$) and that κ_1 has been introduced through the Hamiltonian H_D with the self-consistent value normalized in terms of $R^2_{eff}(^{11}\text{Li})$ (Eq. (3.6.21)). One then obtains

$$
\begin{aligned}
\Lambda &= \left\{ 2\hbar\omega_{pygmy} \frac{2 \times 0.08(\frac{\hbar^2 A}{2m})/R^2_{eff}}{\left[(\tilde{\epsilon}_{p_{1/2}} - \tilde{\epsilon}_{s_{1/2}})^2 - (\hbar\omega_{pygmy})^2\right]^2} \right\}^{-1/2}, \\[2mm]
&= \left\{ 2\text{MeV} \frac{0.16(\hbar^2 A/2m)(1/4.8)^2\,\text{fm}^2}{\left[(0.45\,\text{MeV})^2 - (1\text{MeV})^2\right]^2} \right\}^{-1/2}, \\[2mm]
&= \left(\frac{3\,\text{MeV}^2}{(0.8)^2\,\text{MeV}^4} \right)^{-1/2} \approx 0.5\,\text{MeV}.
\end{aligned}
\tag{3.6.24}
$$

The value of the induced interaction matrix elements is then given by (Fig. 3.6.1),

$$
M_{ind} = \frac{2\Lambda^2}{DEN} \approx -\frac{2\Lambda^2}{\hbar\omega_{pygmy}} \approx -0.5\,\text{MeV},
\tag{3.6.25}
$$

---

[77] See footnote 123 of Chapter 2.

[78] See Barranco et al. (2001), in particular Fig. 2a, where the doubled peaked ($\approx$0.6 MeV and $\approx$1.6 MeV) $dB(E1)/dE$ strength function is displayed.

[79] Cf., e.g., Brink and Broglia (2005), Eq. (8.42), p. 189.

the factor of 2 arising from the two time-ordering contributions. The resulting correlation energy is thus $E_{corr} = 2\tilde{\epsilon}_{s_{1/2}} - G_{scr} + M_{ind} = (0.3 - 0.1 - 0.5)$ MeV $\approx$ -0.3 MeV, in overall agreement with the experimental[80] findings (-0.380 MeV). It is of note that in this estimate the (subcritical) effect of the screened bare pairing interaction has also been used (see Eq. (3.6.4))[81], as well as the theoretical value[82] $\tilde{\epsilon}_{1/2} = 0.15$ MeV (Fig. 2.7.1).

This schematic model[83] has been implemented with microscopic detail[84] within the framework of a field theoretical description of the interweaving of collective vibrations and single-particle motion and is also discussed within the context of single-particle (Chapter 5) and two-particle (Chapter 7) transfer processes. Here we provide a summary of the theoretical findings.

In Fig. 3.6.2 (**I**), the lowest single-particle neutron virtual and resonant states of ^{10}Li are indicated.[85] The position of the levels $s_{1/2}$ and $p_{1/2}$ determined making use of mean-field theory is shown (left hatched area and thin horizontal line, respectively). The coupling of a single-neutron (upward pointing arrowed line) to a vibration (wavy line) calculated making use of NFT Feynman diagrams (schematically depicted also in terms of either solid dots (neutron) or open circles (neutron hole) moving in a single-particle level around or in the ^{9}Li core (gray circle)), leads to conspicuous shifts in the energy centroid of the $s_{1/2}$ and $p_{1/2}$ resonances (shown by thick horizontal lines to the right) and eventually to an inversion in their sequence. In Fig. 3.6.2 (**II**) the processes contributing to binding the halo neutron system ^{11}Li are displayed. One starts with the dressed mean field picture in which two neutrons (solid dots) coupled to angular momentum and parity 0^+ move in time-reversal states around the core ^{9}Li (hatched area) in the $s_{1/2}$ virtual state leading to an unbound $s_{1/2}^2(0)$ configuration. The associated spatial structure of the uncorrelated pair is shown in (**a**). The exchange of vibrations between the two neutrons displayed in the upper part of the figure leads to an induced pairing interaction that, added to the subcritical bare nucleon–nucleon Argonne potential[86] (see boxed inset), leads to a bound state $|\tilde{0}\rangle$. The corresponding wavefunction is displayed in (**b**), together with the spatial structure of the resulting Cooper pair. It is of note that a large fraction of the induced interaction arises from the exchange of

---

[80] Bachelet et al. (2008); Smith et al. (2008).

[81] That new physics, namely, a novel mechanism to (dynamically) violate gauge invariance finds, to express itself, a scenario of a barely bound Cooper pair at the drip line (half-life 8.75 ms), seems to confirm a recurrent expectation that truly new complex phenomena appear at the border between rigid order and randomness (see De Gennes (1994)).

[82] See footnotes 30 and 31 of Chapter 5.

[83] See also Broglia et al. (2019b).

[84] Cf. Barranco et al. (2001); see also Potel et al. (2010).

[85] Barranco et al. (2001). For a more detailed (NFT) study of ^{10}Li, see Barranco et al. (2020) (see also Sect. 5.2.3). See also Moro et al. (2019).

[86] Wiringa et al. (1984).

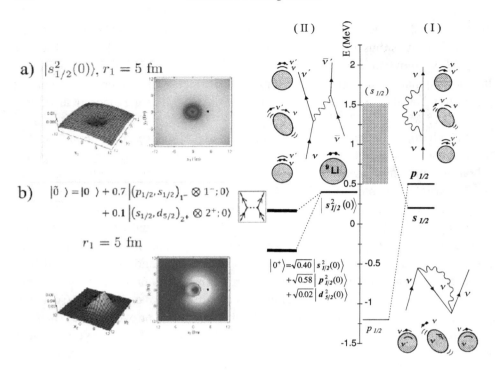

Figure 3.6.2 In (I) and (II) the NFT processes renormalizing the single-particle motion (^{10}Li) and leading to the effective interaction, sum of the bare (horizontal dotted line, inset) and induced (wavy line) interactions which bind the two-neutron halo to the core ^{9}Li, thus leading to the $|^{11}$Li$\rangle$ ground state, are displayed. It is of note that the odd $p_{3/2}(\pi)$ proton, considered as a spectator, is not shown. In (a) and (b) the spatial structure of the pure $|s^2_{1/2}(0)\rangle$ configuration and that of the two-neutron halo $|\tilde{0}\rangle$ Cooper pair is displayed. The modulus squared of the wave function $|\Psi_0(\mathbf{r}_1, \mathbf{r}_2)|^2 = |\langle \mathbf{r}_1, \mathbf{r}_1 | 0^+\rangle|^2$ describing the motion of the two halo neutrons around the ^{9}Li core is shown as a function of the Cartesian coordinates of particle 2, for fixed values of the position of particle 1 ($r_1 = 5$ fm) represented by a solid dot, while the core radius (^{9}Li) is shown as a solid circle. The numbers appearing on the z-axis of the three-dimensional plots displayed on the left side of the figure are in units of fm^{-2}. For the color version of this figure, refer to cambridge.org/nuclearcooperpair.

the pygmy dipole resonance between the two halo neutrons. Within this scenario one can posit that the ^{11}Li PDR can hardly be viewed but in symbiosis with the ^{9}Li halo neutron pair addition mode, and vice versa, and furthermore, that the two halo neutrons of ^{11}Li provide the first example of a Van der Waals Cooper pair, the first of its type in nuclei (Fig. 3.7.1). For further details, see Chapter 7 as well as reference.[87]

[87] Barranco et al. (2001).

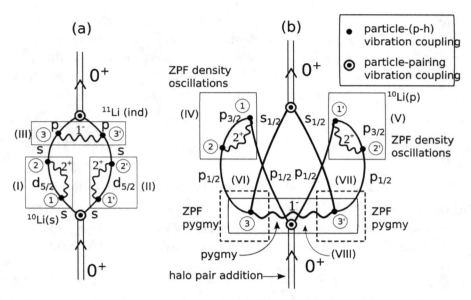

Figure 3.7.1 NFT Feynman diagrams describing the binding of the halo Cooper pair through pygmy, that is, producing the symbiotic mode involving the pair addition mode and the PDR. The single-particle states $s_{1/2}$ and $p_{1/2}$ are labeled in (a) s and p for simplicity. The different particle–vibration coupling vertices (either with the quadrupole (2^+) or with the pygmy (1^-) modes drawn as wavy lines) are denoted by a solid dot and numbered in increasing time sequence so as to show that diagram (b) emerges from (a) through time ordering. The motions of the neutrons are drawn in terms of continuous solid curves. In keeping with the fact that the occupation of the single-particle states is neither 1 nor 0, we treat these states as quasiparticle states, for simplicity. Thus no arrow is drawn on them. Diagram (a) emphasizes the self-energy renormalization of the state $s_{1/2}$ lying in the continuum and which, through its clothing with the quadrupole mode, is brought down, becoming a virtual ($\widetilde{\epsilon}_{s_{1/2}} \approx 0.2$ MeV) state (see (I) and (II)), while (III) contributes to the induced pairing interaction through pygmy (see also Fig. 3.6.1). The "eagle" diagram (b) contains ((IV) and (V)) Pauli corrections which push the bound state $p_{1/2}$ into a resonant state in the continuum ($\widetilde{\epsilon}_{p_{1/2}} \approx 0.5$ MeV). In other words, processes (I), (II), (IV), and (V) are at the basis of parity inversion and of the appearance of the new magic number $N = 6$. Processes (III), (VI) and (VII) are associated with the pygmy ZPF, while (VIII) contributes to the induced pairing interaction through pygmy (van der Waals–like process).

3.7 Nuclear van der Waals Cooper Pair

The atomic van der Waals (dispersive, retarded) interaction which, like gravitation, acts between all atoms and molecules, also nonpolar, can be written for two systems placed at a distance R as (see App. 3.B)

$$\Delta E = -\frac{6 \times e^2 \times a_0^5}{R^6} = -\frac{16}{(R/a_0)^6} \left(\frac{e^2}{2a_0} \times 0.75 \right), \qquad (3.7.1)$$

where a_0 is the Bohr radius. One recognizes in the term in parentheses the transition energy $2P \to 1S$ of the hydrogen atom. A possible qualitative nuclear parallel (see Fig. 3.7.1) can be established, making the correspondences of the term in parentheses with $\tilde{\epsilon}_{p_{1/2}} - \tilde{\epsilon}_{s_{1/2}} \approx 0.45$ MeV, R with $R_{eff} \approx 4.6$ fm and a_0 with $R_0(^{11}\text{Li}) \approx 3$ fm, leading to $\Delta E \approx -0.6$ MeV $(=M_{ind})$. Thus,

$$E_{corr} = 2\tilde{\epsilon}_{s_{1/2}} - G_{scr} + \Delta E = 0.3\,\text{MeV} - 0.1\,\text{MeV} - 0.6\,\text{MeV} \approx -0.4\,\text{MeV},$$

which is not inconsistent with the experimental finding (-0.380 MeV).

3.8 Renormalized Coupling Constants ^{11}Li: Resumé

The fact that the screening factors r and s (see Eqs. (3.6.3) and (3.6.2); also (3.6.21) and subsequent paragraph) essentially coincide within numerical approximations is in keeping with the fact that both quantities are closely related to the overlap[88]

$$\mathcal{O} \approx \left(\frac{R_0}{R}\right)^3 \approx \left(\frac{2.7\,\text{fm}}{4.6\,\text{fm}}\right)^3 \approx 0.2, \tag{3.8.1}$$

a quantity that has two main effects concerning the mechanism which is at the basis of much of the nuclear structure of the halo exotic nucleus ^{11}Li: (1) it makes subcritical the screened bare NN-pairing interaction $G_{scr} = r \times G < G_c$ ($G_{scr} = 1$ MeV/A); (2) it screens the symmetry potential drastically, reducing the energy price one has to pay to separate protons from delocalized neutrons. In this way it permits a consistent chunk ($\approx 8\%$) of the TRK sum rule[89] to become essentially degenerate with the ground state $((V_1)_{scr} = 1$ MeV), allowing for the first nuclear example of a van der Waals Cooper pair and a novel mechanism to break dynamically gauge invariance, namely, dipole–dipole fluctuating fields associated with the exchange of the pygmy dipole resonance between the halo neutrons of ^{11}Li. As a result, we are in the presence of a new, (composite) elementary mode of nuclear excitation: a halo pair addition mode carrying on top of it a low-lying collective soft $E1$-vibration. This symbiotic mode can be studied through two-particle transfer reactions, eventually in coincidence with γ-decay, in particular, making use of the reaction $^9\text{Li}(t, p)^{11}\text{Li}(f)$ for

$|f\rangle$: ground state $(L = 0)$ and pygmy $(L = 1; E_x \lesssim 1$ MeV$)$.

Similarly, but in this case in connection with the reaction $^{10}\text{Be}(t, p)^{12}\text{Be}(f)$ for

$|f\rangle$: first excited 0^+ state $(E_x = 2.24$ MeV$)$,

as well as pygmy $(L = 1)$ on top of it, arguably the 1^- state at $E_x = 2.70$ MeV is part of this mode.[90]

---

[88] One can equivalently use $(R_0/R_{eff})^3 \approx (2.7\,\text{fm}/(4.8\,\text{fm}))^3 \approx 0.18$.

[89] See footnote 123 of Chapter 2.

[90] Iwasaki et al. (2000).

3.A Lindemann Criterion and Connection with Quantality Parameter

The original Lindemann criterion[91] compares the atomic fluctuation amplitude $\langle \Delta r^2 \rangle^{1/2}$ with the lattice constant a of a crystal. If this ratio, which is defined as the disorder parameter Δ_L, reaches a certain value, fluctuations cannot increase without damaging or destroying the crystal lattice. The results of experiments and simulations show that the critical value of Δ_L for simple solids is in the range of 0.10–0.15, relatively independent of the type of substance, the nature of the interaction potential, and the crystal structure.[92] Applications of this criterion to an inhomogeneous finite system like a protein in its native state (aperiodic crystal)[93] require evaluation of the generalized Lindemann parameter[94]

$$\Delta_L = \frac{\sqrt{\sum_i \langle \Delta r_i^2 \rangle / N}}{a'}, \qquad (3.A.1)$$

where N is the number of atoms and a' the most probable nonbonded near-neighbor distance, $\mathbf{r}_i$ being the position of atom i, $\Delta r_i^2 = (\mathbf{r}_i - \langle \mathbf{r}_i \rangle)^2$, and $\langle \rangle$ denoting configurational averages at the conditions of measurement or simulations (e.g., biological, in which case $T \approx 310$ K, PH ≈ 7, etc.[95]). The dynamics as a function of the distance from the geometric center of the protein is characterized by defining an interior (int) Lindemann parameter,

$$\Delta_L^{int}(r_{cut}) = \frac{\sqrt{\sum_{i, r_i < r_{cut}} \langle \Delta r_i^2 \rangle / N}}{a'}, \qquad (3.A.2)$$

which is obtained by averaging over the atoms that are within a chosen cutoff distance, r_{cut}, from the center of mass of the protein.

Simulations and experimental data for a number of proteins, in particular Barnase, Myoglobin, Crambin, and Ribonuclease A, indicate 0.14 as the critical value distinguishing between solid-like and liquid-like behavior, and $r_{cut} \approx 6$ Å. As can be seen from Table 3.A.1, the interior of a protein, under physiological conditions, is solid-like ($\Delta_L < 0.14$), while its surface is liquid-like ($\Delta_L > 0.14$). The beginning of thermal denaturation in the simulations appears to be related to the melting of its interior (i.e., $\Delta_L^{int} > 0.14$) so that the entire protein becomes liquid-like. This is also the situation of the denatured state of a protein under physiological conditions.[96]

---

[91] Lindemann (1910).

[92] Bilgram (1987); Löwen (1994); Stillinger (1995).

[93] Schrödinger (1944).

[94] Stillinger and Stillinger (1990).

[95] Fluctuations, classical (thermal) or quantal, imply a probabilistic description. While one can only predict the odds for a given outcome of an experiment in quantum mechanics, probabilities themselves evolve in a deterministic fashion (Born (1948)).

[96] See, e.g., Rösner et al. (2017).

Table 3.A.1 *The heavy-atom* $\Delta_L(\Delta_L^{int})$ *value, for four proteins at 300 K. After Zhou et al. (1999).*

	$\Delta_L(\Delta_L^{int}(6\ \text{Å}))(300\ \text{K})$			
	MD simulations			X-ray data
Proteins	Barnase	Myoglobin	Crambin	Ribonuclease A
All atoms	0.21 (0.12)	0.16 (0.11)	0.16 (0.09)	0.16 (0.12)
Backbone atoms only	0.16 (0.10)	0.12 (0.09)	0.12 (0.08)	0.13 (0.10)
Side chain atoms only	0.25 (0.14)	0.18 (0.12)	0.19 (0.10)	0.19 (0.13)

3.A.1 Lindemann ("Disorder") Parameter for a Nucleus

An estimate of $\sqrt{\sum_i \langle \Delta r_i^2 \rangle / A}$ in the case of nuclei considered as a sphere of nuclear matter of radius R_0 is provided by the "spill out" of nucleons due to quantal effects. That is,[97] $\sqrt{\quad} \approx 0.69 \times a_0$, where a_0 is of the order of the nuclear diffusivity (≈ 0.65 fm). The average internucleon distance can be estimated to be $2 \times r_0 \approx 2.4$ fm. Thus,

$$\Delta_L = \frac{0.69 a_0}{2.4 \text{ fm}} \approx 0.19. \tag{3.A.3}$$

While it is difficult to compare among them crystals, aperiodic finite crystals, and atomic nuclei, arguably, the above value indicates that a nucleus is liquid-like.

3.B The Van Der Waals Interaction

Historically one can distinguish two contributions to the van der Waals interaction[98]:

1. **dispersive** retarded contribution[99], emerging from the dynamical dipole–dipole, as well as from higher multipolarities, interaction associated with the quantum mechanical zero-point fluctuations (ZPF) of the ground state of the two interacting atoms or molecules[100]
2. **inductive**, implying the polarization of one molecule in the permanent dipole or quadrupole field of the other molecule[101]

---

[97] Bertsch and Broglia (2005); see, e.g., Chapter 5. See also the paragraph following Eq. (4.10.3).
[98] Let us think mainly of nonpolar (NP) molecules.
[99] Dispersion: variation of a quantity, e.g., spatial separation of white light (rainbow), as a function of frequency (cf., e.g., Israelachvili (1985), p. 65).
[100] These forces act between all atoms and molecules, even nonpolar, totally neutral ones.
[101] Debye (1920, 1921).

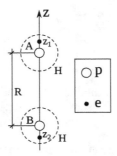

Figure 3.B.1 Planar configuration assumed for two hydrogen atoms at a relative distance R, where p stands for proton and e for electron.

Only the first one is a bona fide van der Waals interaction. In fact, with the advent of quantum mechanics, it was very early recognized[102] that for most molecules, interactions of type 2 are small compared with interactions of type 1, that is, the interaction corresponding to the mutual polarization of one molecule in the rapidly changing field – due to the instantaneous configuration of electrons and nuclei – of the other molecule.[103]

3.B.1 Van der Waals Interaction between Two Hydrogen Atoms

For large values of the internuclear distance $r_{AB} = R$, the exchange phenomenon is unimportant (Pauli principle), and one can take as the unperturbed wavefunction for a system of two hydrogen atoms the simple product of two hydrogenlike wavefunctions,

$$\Psi^0 = u_{1sA}(1)u_{1sB}(2). \tag{3.B.1}$$

The perturbation for this wavefunction arises from the potential energy terms

$$H' = \frac{e^2}{r_{12}} + \frac{e^2}{r_{AB}} - \frac{e^2}{r_{A2}} - \frac{e^2}{r_{B1}}, \tag{3.B.2}$$

corresponding to the variety of Coulomb interactions involving electrons and protons. Let us assume for simplicity that we are dealing with a one-dimensional problem, in which case one can write (Fig. 3.B.1)

$$\mathbf{r}_{12} = (R + z_1 + z_2)\,\hat{\mathbf{z}},$$
$$\mathbf{r}_{AB} = R\,\hat{\mathbf{z}},$$
$$\mathbf{r}_{A2} = (R + z_2)\,\hat{\mathbf{z}},$$
$$\mathbf{r}_{B1} = (R + z_1)\,\hat{\mathbf{z}}. \tag{3.B.3}$$

---

[102] London (1930).
[103] Pauling and Wilson (1963), p. 384; Born (1969), p. 471.

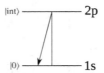

Figure 3.B.2 Schematic representation of the virtual process associated with (3.B.14), the intermediate, virtual state being the $2p$ state.

Because all these distances are much larger than the radius of the atom ($a_0 \approx$ 0.529 Å, Bohr radius), the expression (3.B.2) can be calculated making use of a Taylor expansion. One obtains

$$r_{12}^2 = (R + z_1 + z_2)^2 = R^2 \left[1 + 2\frac{(z_1 + z_2)}{R} + \frac{(z_1 + z_2)^2}{R^2} \right], \qquad (3.B.4)$$

which leads to

$$\frac{e^2}{r_{12}} = \frac{e^2}{R \left[1 + \frac{2(z_1+z_2)}{R} + \frac{(z_1+z_2)^2}{R^2} \right]^{1/2}} \approx \frac{e^2}{R} \left[1 - \frac{(z_1 + z_2)}{R} - \frac{(z_1 + z_2)^2}{2R^2} \right].$$
$$(3.B.5)$$

Similarly,

$$r_{A2}^2 = \left(R^2 + 2Rz_2 + z_2^2 \right) = R^2 \left(1 + 2\frac{z_2}{R} + \frac{z_2^2}{R^2} \right) \qquad (3.B.6)$$

and

$$r_{B1}^2 = R^2 \left(1 + 2\frac{z_1}{R} + \frac{z_1^2}{R^2} \right), \qquad (3.B.7)$$

leading to

$$-\frac{e^2}{r_{A2}} = -\frac{e^2}{R} \left(1 - \frac{z_2}{R} - \frac{z_2^2}{2R^2} \right) \qquad (3.B.8)$$

and

$$-\frac{e^2}{r_{B1}} = -\frac{e^2}{R} \left(1 - \frac{z_1}{R} - \frac{z_1^2}{2R^2} \right). \qquad (3.B.9)$$

Finally,

$$\frac{e^2}{r_{AB}} = \frac{e^2}{R}. \qquad (3.B.10)$$

With the exception of the cross term of (3.B.5), there is complete cancellation between the different contributions to (3.B.2). Thus,

$$H' = -\frac{\mathbf{D}_1 \cdot \mathbf{D}_2}{R^3}, \qquad (3.B.11)$$

where

$$\mathbf{D}_i = ez_i\,\hat{\mathbf{z}} \tag{3.B.12}$$

is the dipole moment operator associated with electron i. Because $R \gg z_i$, one can diagonalize the interaction Hamiltonian (3.B.11) perturbatively. In keeping with the fact that a single-particle quantum state displaying a given parity (and in the present case, angular momentum $(-1)^\ell = \pi$) cannot sustain a permanent dipole moment, in particular,

$$\int d\tau u_{1s}(z) \times z \times u_{1s}(z) = \int d\tau (u_{1s}(z))^2 z = \int d\tau \rho(z) z = 0, \tag{3.B.13}$$

the lowest perturbative correction to (3.B.1) is of second order. The associated energy correction is given by the relation

$$\Delta E_z^{(2)} = -\sum_{int} \frac{\langle 0|H'|int\rangle\langle int|H'|0\rangle}{E_{int} - E_0}. \tag{3.B.14}$$

Because we are concerned with the $1s \rightarrow 2p$ transition (Fig. 3.B.2), the intermediate state is

$$\Psi^{int} = u_{2pA}(1)u_{2pB}(2), \tag{3.B.15}$$

and thus

$$D_{en} = E_{int} - E_0. \tag{3.B.16}$$

One can then write

$$\begin{aligned}
\Delta E_z^{(2)} &= -\frac{|\langle 0|H'|0\rangle|^2}{D_{en}} = -\frac{e^4}{R^6}\frac{|\langle 0|z_1^2 z_2^2|0\rangle|^2}{D_{en}}, \\
&\approx -\frac{e^4}{R^6}\frac{\int d\tau_1\,d\tau_2 u_{1s}^2(1)u_{1s}^2(2)z_1^2 z_2^2}{D_{en}}, \\
&= -\frac{e^4}{R^6}\frac{\int d\tau_1 \rho(z_1)z_1^2 \int d\tau_2 \rho(z_2)z_2^2}{D_{en}}, \\
&= -\frac{e^4}{R^6}\frac{\bar{z}_1^2 \bar{z}_2^2}{D_{en}} \approx -\frac{e^4}{R^6}\frac{a_0^2 a_0^2}{\frac{e^2}{2a_0}} = -\frac{2e^2 a_0^5}{R^6}.
\end{aligned} \tag{3.B.17}$$

This result corresponds to the z-degree of freedom of the system (two H atoms at a distance $R \gg a_0$). One has then to multiply the above result by 3 to take into account the x and y degrees of freedom. Thus,

$$\Delta E^{(2)} = -\frac{6e^2 a_0^5}{R^6}. \tag{3.B.18}$$

Let us now calculate the van der Waals interaction between two H atoms at a distance of the order of 10 times the summed radii of the two atoms ($\approx 2a_0 \approx 1\text{Å}$), that is, for $R \approx 10\text{Å}$,

$$\Delta E^{(2)}_{H-H}(10\ \text{Å}) \approx - \frac{6 \times 14.4\ \text{eV Å}(0.529\ \text{Å})^5}{(10\ \text{Å})^6}$$
$$\approx -3.6 \times 10^{-6}\ \text{eV} = -3.6\,\mu\text{eV}. \qquad (3.\text{B}.19)$$

Making use of the relation

$$1\ \text{eV} = 2.42 \times 10^{14}\ \text{Hz}, \quad (1\text{Hz=s}^{-1}), \qquad (3.\text{B}.20)$$

one obtains

$$|\Delta E^{(2)}_{H-H}(10\ \text{Å})| \approx 3.6 \times 10^{-6} \times 2.42 \times 10^{14}\ \text{Hz} \approx 9 \times 10^8\ \text{Hz} \approx 10^3\ \text{MHz},$$
$$(3.\text{B}.21)$$

a quantity which can be compared with the Lamb shift (1058 MHz; see Fig. 5.2.4). It is of note that $|\Delta E^{(2)}_{H-H}(2.5\ \text{Å})| \approx 15\ \text{meV/part} \approx 0.35\ \text{kcal/mole}$ (1meV/part≈ 0.02306 kcal/mole), a value of the order of $kT/2$, that is, one-half of the thermal energy under biological conditions ($T \approx 300$ K, $kT \approx 0.6$ kcal/mole)[104], in keeping with the role played by the Van der Waals interaction in the folding and stability of proteins.

---

[104] Huang (2005).

4

Phase-Coherent Pair Transfer

A superconductor has rather perfect internal phase order.... The importance of
the Josephson effect ... is that ... it can pin down the order parameter.

P. W. Anderson

4.1 Simultaneous versus Successive Cooper Pair Transfer in Nuclei

Cooper pair transfer is commonly thought to be tantamount to simultaneous transfer. In this process a nucleon goes over through the NN-interaction v; the second one does it making use of the correlations with its partner (cf. Figs. 4.1.1 and 6.5.1 (I)). Consequently, in the independent-particle limit, simultaneous transfer should not be possible (see Sect. 6.5.1). Nonetheless, it remains operative. This is because, in this limit, the particle transferred through v does it together with a second one which profits from the nonorthogonality of the wavefunctions describing the single-particle motion in target and projectile (Figs. 4.1.2 and 6.5.1 (II)). This is the reason why this (nonorthogonality) transfer amplitude has to be treated on equal footing with the previous one representing, within the nonorthogonal basis employed, a natural contribution to simultaneous transfer. In fact $(T^{(1)} - T^{(1)}_{NO})$ is the correct, two-nucleon transfer amplitude to lowest order (first) in v. The resulting cancellation is quite conspicuous in actual nuclei (see Figs. 4.5.2 (b), 4.5.3, and Fig. 4.5.4 for examples in open-shell (superfluid) and closed-shell (normal) nuclei, respectively). This is in keeping with the fact that Cooper pairs are weakly correlated systems ($2\Delta \ll \epsilon_F$), and the reason why the successive transfer process in which v (e.g., the mean field U in the post-post representation[1]) acts twice is the dominant mechanism in pair transfer reactions. While this mechanism seems antithetical to the transfer of correlated fermions pairs, it probes, in the nuclear case, the

---

[1] Bayman and Chen (1982); Potel et al. (2013a). See also Pinkston and Satchler (1982).

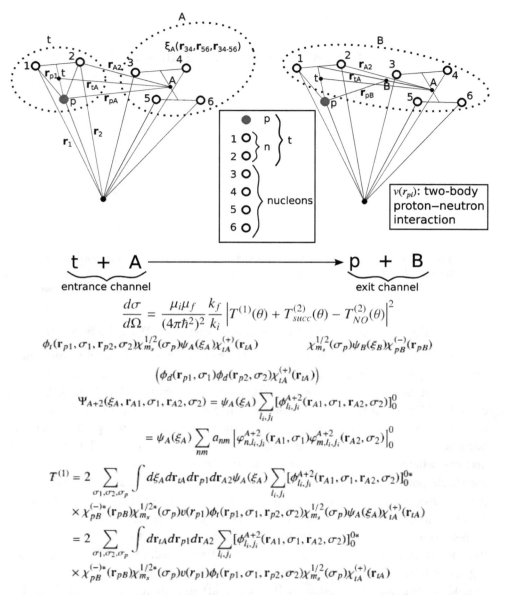

Figure 4.1.1 Contribution of simultaneous transfer, in first-order DWBA, to the reaction $A(t, p)B(\equiv A + 2)$. The nucleus A is, for simplicity, assumed to contain four nucleons, the triton being composed of two neutrons and one proton. The set of coordinates used to describe the entrance and exit channels are shown in the upper part (boldface vectors represent the coordinates used to describe the relative motion, while the intrinsic coordinates ξ_A are $\mathbf{r}_{34}$, $\mathbf{r}_{56}$, and $\mathbf{r}_{34-56}$). In the lower part of the figure, the simultaneous two-nucleon transfer amplitude is written (see Potel et al. (2013b)). It is of note that of all the relative motion coordinates, only those describing the relative motion of (t, A) and of (p, B), have asymptotic values, being those associated with distorted waves.

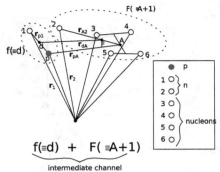

$$f(\equiv d) \quad + \quad F(\equiv A+1)$$

$$\underbrace{}$$

intermediate channel

$$\chi_{m_s}^{1/2}(\sigma_p)\phi_d(\mathbf{r}_{p1},\sigma_1)\psi_A(\xi_A)\varphi_{l_f,j_f,m_f}^{A+1}(\mathbf{r}_{A2},\sigma_2)$$

$$G(\mathbf{r}_{dF},\mathbf{r}'_{dF}) = i\sum_l \sqrt{2l+1}\,\frac{f_l(k_{dF},r_<)g_l(k_{dF},r_>)}{k_{dF}r_{dF}r'_{dF}}\left[Y^l(\hat{r}_{dF})Y^l(\hat{r}'_{dF})\right]_0^0$$

$$T_{succ}^{(2)} = 2\sum_{l_i,j_i}\sum_{l_f,j_f,m_f}\sum_{\substack{\sigma_1\sigma_2\\ \sigma'_1\sigma'_2}}\int d\xi_A d\mathbf{r}_{dF}d\mathbf{r}_{p1}d\mathbf{r}_{A2}\chi_{pB}^{(-)*}(\mathbf{r}_{pB})\chi_B^*(\xi_B)v(\mathbf{r}_{p1})\phi_d(\mathbf{r}_{p1})\varphi_{l_f,j_f,m_f}^{A+1}(\mathbf{r}_{A2},\sigma_2)$$

$$\times \chi_{m_s}^{1/2}(\sigma_p)\Psi_A(\xi_A)\frac{2\mu_{dF}}{\hbar^2}\int d\xi'_A d\mathbf{r}'_{dF}d\mathbf{r}'_{p1}d\mathbf{r}'_{A2}G(\mathbf{r}_{dF},\mathbf{r}'_{dF})$$

$$\times \chi_{tA}^{(+)}(\mathbf{r}_{tA})\psi_A^*(\xi'_A)v(\mathbf{r}'_{p2})\phi_d(\mathbf{r}'_{p1})\varphi_{l_f,j_f,m_f}^{A+1}(\mathbf{r}'_{A2},\sigma'_2)$$

$$= 2\sum_{l_i,j_i}\sum_{l_f,j_f,m_f}\sum_{\substack{\sigma_1\sigma_2\\ \sigma'_1\sigma'_2}}\int d\mathbf{r}_{dF}d\mathbf{r}_{p1}d\mathbf{r}_{A2}\chi_{pB}^{(-)*}(\mathbf{r}_{pB})v(\mathbf{r}_{p1})\phi_d(\mathbf{r}_{p1})\left[\varphi_{l_f,j_f,m_f}^{A+2}(\mathbf{r}_{A1},\sigma_1,\mathbf{r}_{A2},\sigma_2)\right]_0^0$$

$$\times \frac{2\mu_{dF}}{\hbar^2}\int d\mathbf{r}'_{dF}d\mathbf{r}'_{p1}d\mathbf{r}'_{A2}G(\mathbf{r}_{dF},\mathbf{r}'_{dF})\chi_{tA}^{(+)}(\mathbf{r}'_{tA})v(\mathbf{r}'_{p2})\phi_d(\mathbf{r}'_{p1},\sigma'_1)\phi_d(\mathbf{r}'_{p2},\sigma'_2)\varphi_{l_f,j_f,m_f}^{A+1}(\mathbf{r}'_{A2},\sigma'_2)$$

$$T_{NO}^{(1)} = 2\sum_{l_i,j_i}\sum_{l_f,j_f,m_f}\sum_{\substack{\sigma_1\sigma_2\\ \sigma'_1\sigma'_2}}\int d\xi_A d\mathbf{r}_{dF}d\mathbf{r}_{p1}d\mathbf{r}_{A2}\chi_{pB}^{(-)*}(\mathbf{r}_{pB})\chi_B^*(\xi_B)v(\mathbf{r}_{p1})\phi_d(\mathbf{r}_{p1})\varphi_{l_f,j_f,m_f}^{A+1}(\mathbf{r}_{A2},\sigma_2)$$

$$\times \chi_{m_s}^{1/2}(\sigma_p)\Psi_A(\xi_A)\frac{2\mu_{dF}}{\hbar^2}\int d\xi'_A d\mathbf{r}'_{dF}d\mathbf{r}'_{p1}d\mathbf{r}'_{A2}$$

$$\times \chi_{tA}^{(+)}(\mathbf{r}_{tA})\psi_A^*(\xi'_A)\phi_d(\mathbf{r}'_{p1})\mathbb{1}\varphi_{l_f,j_f,m_f}^{A+1}(\mathbf{r}'_{A2},\sigma'_2)$$

$$= 2\sum_{l_i,j_i}\sum_{l_f,j_f,m_f}\sum_{\substack{\sigma_1\sigma_2\\ \sigma'_1\sigma'_2}}\int d\mathbf{r}_{dF}d\mathbf{r}_{p1}d\mathbf{r}_{A2}\chi_{pB}^{(-)*}(\mathbf{r}_{pB})v(\mathbf{r}_{p1})\phi_d(\mathbf{r}_{p1})\left[\varphi_{l_f,j_f,m_f}^{A+2}(\mathbf{r}_{A1},\sigma_1,\mathbf{r}_{A2},\sigma_2)\right]_0^0$$

$$\times \frac{2\mu_{dF}}{\hbar^2}\int d\mathbf{r}'_{dF}d\mathbf{r}'_{p1}d\mathbf{r}'_{A2}\chi_{tA}^{(+)}(\mathbf{r}'_{tA})\phi_d(\mathbf{r}'_{p1},\sigma'_1)\phi_d(\mathbf{r}'_{p2},\sigma'_2)\varphi_{l_f,j_f,m_f}^{A+1}(\mathbf{r}'_{A2},\sigma'_2)$$

Figure 4.1.2 Successive and nonorthogonality contributions to the amplitude describing two-nucleon transfer in second-order DWBA, entering in the expression of the absolute differential cross section $d\sigma/d\Omega = \frac{\mu_i\mu_f}{(4\pi\hbar^2)^2}\frac{k_f}{k_i}\left|T^{(1)} + T_{succ}^{(2)} - T_{NO}^{(2)}\right|^2$. Concerning $T^{(1)}$, we refer to Fig. 4.1.1. In the upper part of the figure the coordinates used to describe the intermediate channel $d + F(\equiv A + 1)$ are given (boldface vectors represent the coordinates used to describe the relative motion, while the intrinsic coordinates ξ_A are $\mathbf{r}_{34}$, $\mathbf{r}_{56}$, and $\mathbf{r}_{34-56}$). In the lower part of the figure, the expressions corresponding to the (t, p) process are displayed (Potel et al. (2013b)). Schematically, the three contributions $T^{(1)}$, $T_{succ}^{(2)}$, and $T_{NO}^{(2)}$ to the transfer amplitude can be written as $\langle pB|v|tA\rangle$, $\sum\langle pB|v|dF\rangle\langle dF|v|tA\rangle$, and $\sum\langle pB|v|dF\rangle\langle dF|\mathbf{1}|tA\rangle$, respectively, where v is the proton–neutron interaction and $\mathbf{1}$ the unit operator. Within this context, while $T_{NO}^{(2)}$ receives contributions from the intermediate (virtual) closed $(d + F)$ channel, as $T_{succ}^{(2)}$ does, it is first order in v, as $T^{(1)}$ is.

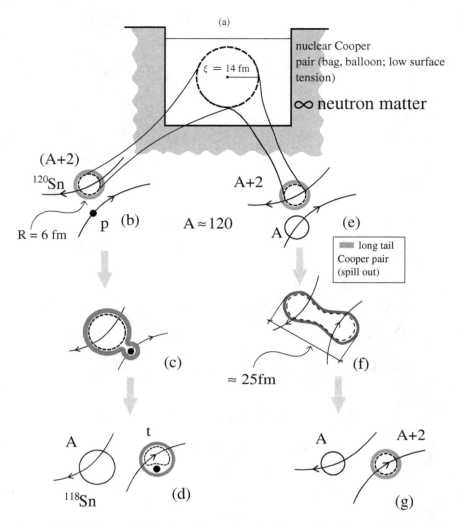

Figure 4.1.3 The correlation length associated with a nuclear Cooper pair is of the order of $\xi \approx \hbar v_F/(\pi \Delta) \approx 14$ fm ($\Delta \approx 1.2$ MeV, $v_F/c \approx 0.27$). (a) In neutron matter at densities of the order of the surface density of finite nuclei, the $NN\text{-}^1S_0$ short-range force, eventually renormalized by medium polarization effects, makes pairs of nucleons moving in time-reversal states to correlate over distances larger than nuclear dimensions (dashed circle). How can one get evidence, in the laboratory, for such an extended object? (b) Hardly when the Cooper bag (balloon) is introduced in a superfluid nucleus, which, acting as a strong external field, leads to spatial quantization (related to angle correlations; see Sect. 4.11) and confines the Cooper pair with some spillout (tail of Cooper pair, gray, shaded area; see Bertsch and Broglia (2005), p. 88) as testified by (c,d) two-nucleon transfer process (e.g., (p, t) reaction), in which case the absolute cross section can change in a conspicuous fashion, in going from pure two-particle (uncorrelated configurations) to long-tail Cooper pair spillouts. This effect is expected to become stronger by allowing pair transfer between similar superfluid nuclei (see Sect. 7.3), in which case the Cooper pair can move in the combined nuclear system resulting from the two heavy ions in weak contact, and at the same time profit from the same type of correlations (superfluidity) as resulting from similar pair mean fields (e), (f), (g) (see, e.g., von Oertzen (2013); von Oertzen and Vitturi (2001), and refs. therein).

same pairing correlations as simultaneous transfer does.[2] This is because nuclear Cooper pairs (quasi-bosons) are quite extended objects, the two nucleons being (virtually) correlated over distances ξ larger than typical nuclear dimensions (see Fig. 4.1.3). In a two-nucleon transfer process this virtual property becomes real, in the sense that the presence of (normal) density over regions larger than that of the dimensions of each of the interacting nuclei allows for Cooper pair (abnormal density) manifestation over distances of the order of ξ.

Within this context, let us refer to the Josephson effect, associated with the Cooper pair tunneling across a thin barrier separating two metallic superconductors. Because the probability of one-electron-tunneling is small, (conventional) simultaneous tunneling associated with that probability quantity square would hardly be observed. Nonetheless, Josephson currents are standard measures in low-temperature laboratories.[3] The same arguments related to the large value of the correlation length is operative in explaining the fact that Coulomb repulsion is rather weak between partners of Cooper pairs which are, in average, at a distance ξ ($\approx 10^4$ Å) much larger than the Wigner–Seitz radius r_s typical of metallic elements (≈ 1–2Å). Consequently, it can be overcompensated and turned into an effective (weakly) attractive force by the long-range electron–phonon pairing.

Similarly, in widely extended light halo nuclei, the short-range bare pairing interaction may become subcritical due to the poor overlap between the halo neutrons and the core ones (cf. Sect. 3.6). The fact that such systems are nonetheless bound, although weakly, testifies to the important role the exchange of collective vibrations between halo nucleons have in binding the associated halo Cooper pair.

The above arguments are at the basis of the fact that second-order DWBA theory which add both successive and nonorthogonality contributions to the simultaneous transfer amplitudes, are needed in a quantitative description of the experimental findings.

4.2 One- and Two-Nucleon Transfer Probabilities

In what follows we define transfer probabilities P_1 and P_2 for one- and two-nucleon transfer. For this purpose let us remind some useful relations. In particular that of the differential reaction cross section

$$\frac{d\sigma}{d\Omega} = |f(\theta)|^2, \tag{4.2.1}$$

where

$$f(\theta) = \frac{1}{k} \sum_l (2l + 1) e^{i\delta_l} \sin \delta_l P_l(\cos \theta), \tag{4.2.2}$$

---

[2] Within this context, note, however, the possible interplay in reactions between heavy ions at energies below the Coulomb barrier, between the distance of closest approach and the correlation length ξ (see Sect. 7.3 as well as Eqs. (4.9.5)–(4.9.7)).

[3] Cf., e.g., Rogalla and Kes (2012), and refs. therein.

δ_l being the partial wave l phase shift. Let us now use for simplicity the results associated with hard sphere scattering[4] in the low and high energy limit, R being the radius of the sphere. Making use of the fact that in the case under discussion the phase shifts δ_l are related to the regular and irregular spherical Bessel functions,

$$\tan \delta_l = \frac{j_l(kR)}{n_l(kR)}, \tag{4.2.3}$$

and that $\sin^2 \delta_l = \tan^2 \delta_l/(1 + \tan^2 \delta_l)$, one can write in the case in which $kR \ll 1$, that is, in the low-energy, long-wavelength regime

$$\tan \delta_l \approx \frac{-(kR)^{2l+1}}{(2l+1)[(2l-1)!!]^2}, \tag{4.2.4}$$

implying that one can ignore essentially all δ_l with $l \neq 0$. Thus, $\tan \delta_0 = -kR$, $\sin^2 \delta_0 = (kR)^2$ and one can write

$$\frac{d\sigma}{d\Omega} = \frac{\sin^2 \delta_0}{k^2} = R^2. \tag{4.2.5}$$

Consequently,

$$\sigma_{tot} = \int \frac{d\sigma}{d\Omega} d\Omega = 4\pi R^2 \quad (kR \ll 1), \tag{4.2.6}$$

a cross section which is four times the geometric cross section πR^2, namely, the area of the disc of radius R that blocks the propagation of the incoming (plane) wave, and has the same value as that of a hard sphere. Because $kR \ll 1$ implies long-wavelength scattering, it is not surprising that quantal effects are important, so as to overwhelm the classical picture.[5]

Let us now consider the high energy limit $kR \gg 1$. The total cross section is, in this case, given by

$$\sigma_{tot} = \int |f_l(\theta)|^2 d\Omega = \frac{1}{k^2} \int_0^{2\pi} d\phi \int_{-1}^{1} d(\cos\theta) \sum_{l=1}^{kR} \sum_{l'=1}^{kR} (2l+1)(2l'+1)$$

$$\times e^{i\delta_l} \sin \delta_l \times e^{-i\delta_{l'}} \sin \delta_{l'} \times P_l P_{l'} = \frac{4\pi}{k^2} \sum_{l=1}^{kR} (2l+1) \sin^2 \delta_l. \tag{4.2.7}$$

---

[4] Cf., e.g., Sakurai (1994).

[5] Effects related with Heisenberg's indeterminacy principle. The original German word *Unbestimmtheitsprinzip* ("indefiniteness" or "indeterminacy principle") is sometimes incorrectly translated as "uncertainty principle." This is misleading, since it suggests that the electron or the nucleon actually has a definite position and momentum of which one is uncertain. In fact, the quantum formalism simply does not allow the ascription of a definite position and momentum simultaneously (see Leggett (1987)).

Making use of the relation

$$\sin^2 \delta_l = \frac{\tan^2 \delta_l}{1 + \tan^2 \delta_l} = \frac{[j_l(kR)]^2}{[j_l(kR)]^2 + [n_l(kR)]^2} \approx \sin^2 \left(kR - \frac{\pi l}{2} \right) \qquad (4.2.8)$$

and the fact that so many l-values contribute to (4.2.7), one can replace $\sin^2 \delta_l$ by its average value 1/2. Because the number of terms of the sum is roughly kR, the same being true for the average value of $(2l + 1)$, one can thus write

$$\sigma_{tot} = \frac{4\pi}{k^2} (kR)^2 \frac{1}{2} = 2\pi R^2, \qquad (kR \gg 1), \qquad (4.2.9)$$

which, in this short-wavelength limit, is not the geometric cross section either. In fact, (4.2.9) can be split into two contributions each of value πR^2. One due to reflection in which it can be shown that there is no interference among contributions from different l-values. A second one (coherent contribution in the forward direction) called shadow because for hard-sphere scattering at high energies, waves with impact parameter less than R must be deflected. Consequently, behind the scatterer there must be zero probability for finding the scattered particle and a shadow must be generated.[6] The quantity

$$\lambdabar = \frac{\lambda}{2\pi} = \frac{h}{2\pi p} = \frac{\hbar}{p} = \frac{1}{k} = \frac{\hbar}{\sqrt{2mE}} \qquad (4.2.10)$$

is the reduced de Broglie wavelength[7] for a massive particle ($E = p^2/2m$). For a proton of energy $E_p \approx 20$ MeV, typical of beams used in ^{120}Sn$(p, t)^{118}$Sn(gs) and ^{120}Sn$(p, d)^{119}$Sn(j) reactions[8] $\lambdabar \approx 1$ fm, to be compared with the value $R \approx 6$ fm

---

[6] In terms of wave mechanics, this shadow is due to the destructive interference between the original wave (which would be there even if the scatterer was absent) and the newly scattered wave. Thus, one needs scattering in order to create a shadow. This contribution is intimately related to the optical theorem (Sakurai (1994))

$$\sigma_{tot} = \frac{4\pi}{k} \Im[f(\theta = 0, k)] = \frac{4\pi}{k} [f_{shad}(\theta = 0, k)] = \frac{4\pi}{k^2} \sum_l (2l + 1) \sin^2 \delta_l,$$

to which it provides its physical interpretation. In fact, there are two independent ways of measuring σ_{tot}, namely: (1) by integrating the differential cross section $d\sigma/d\Omega = |f(\theta)|^2$ moving around the detector and (2) measuring the attenuation of the incoming beam. Both procedures should give the same result. One then identifies $(4\pi/k)f(\theta = 0, k)$ with the attenuation arising from the interference of the elastic wave with the incoming wave. It is of note that in (4.2.7), the factor $(\pi/k^2)(2l + 1) = \pi \lambdabar^2 (2l + 1)$ is the area of a ring with radius $b = (l + 1/2)\lambdabar$ and width $\lambdabar$ due to quantal indeterminacy. Thus

$$\sigma_{tot} = 2\pi(R + \lambdabar/2)^2 \quad (kR \gg 1).$$

---

[7] De Broglie (1925).
[8] Note that the reduced wavelength of a photon ($p = E/c$) of the same energy ($E = 20$ MeV) is $\lambdabar(= \lambda/2\pi = \hbar/p = \hbar c/E) \approx 10$ fm (cf. Table 2.1 of Satchler (1980), p. 22.)

of the radius of ^{120}Sn. Consequently, we are in a situation of type (4.2.9), that is,

$$\sigma_{tot} = 2\pi (6)^2 \text{ fm}^2 \approx 2.3 \text{ b}. \tag{4.2.11}$$

Because typical values of the absolute one-particle cross section associated with the (p, d) reaction mentioned above are few mb (see, e.g., Fig. 5.2.3, right panel) one can use, for order of magnitude estimate purposes,

$$P_1 \approx \frac{5.35 \text{ mb}}{2.3 \text{ b}} \approx 10^{-3}, \tag{4.2.12}$$

as typical probability for such processes. *Consequently, one may argue that the probability for a pair of nucleons to simultaneously tunnel in, for example, the (p, t) process mentioned above is $(P_1)^2 \approx 10^{-6}$, as near impossible as no matter.*

Within this context we note that the integrated gs $\rightarrow$ gs absolute cross section $\sigma(^{120}\text{Sn}(p, t)^{118}\text{Sn(gs)}) \approx 2.25 \pm 0.338$ mb, quantity representative of the population of the members of the Sn-isotopes ground state rotational band (see Figs. 3.1.3, 3.1.7, and 7.2.1),[9] a fact which implies[10] that the empirical two-nucleon transfer probability is of the order of $P_2 \approx 10^{-3}$, that is, $P_2 \approx P_1$.

It is to be noted that perplexities similar to the one expressed above, also emerged in connection with the study of Cooper pair tunneling between weakly coupled superconductors. In fact, objections were raised[11] in connection with the prediction of Josephson[12] that there should be a contribution to the current through an insulating barrier between two superconductors which would behave like direct tunneling of condensed pairs. This is in keeping with the fact that having the single electron a rather small probability of getting through the insulating junction, the "classical" estimate of simultaneous pair tunneling will not be observable. Let alone the fact that the (local) pairing interaction vanishes inside the barrier.[13]

4.3 Phase Correlation and Enhancement Factor

When one turns on, in an open-shell atomic nucleus like $^{120}_{50}\text{Sn}_{70}$, a pairing interaction of strength larger or equal than critical, the system moves into a Cooper pair regime.[14] This fact has quantitative (but not qualitative) consequences concerning the one-particle transfer mechanism, and regards the size of the mismatch between

---

[9] Concerning the marked fluctuations observed in these cross sections, see Sect. 3.1.1, second paragraph.

[10] See footnote 28 of Chapter 3 and related text.

[11] Bardeen (1962, 1961) (see also Pippard (2012); Cohen et al. (1962); McDonald (2001)).

[12] Josephson (1962); Anderson (1964b).

[13] Anderson (1970); see also Broglia (2020), p. xxi–xxiii, in connection with the nuclear case.

[14] Regime which is conditioned by the "external" mean field, in other words, regime (abnormal density) which expresses itself, provided there is nucleon (normal) density available. It is of note that pairing in turn may help extend the range over which normal nuclear density can reach, as in the case of the neutron halo nucleus ^{11}Li defining (lying at) the neutron drip line.

the relative motion incoming (e.g., $p + {}^{120}$Sn(gs)) and outgoing ($d + {}^{119}$Sn(gs)) trajectories (Q-value and recoil effect). This is in keeping with the fact that one has to break a Cooper pair to populate a single quasiparticle state. From a structure point of view, the depletion of the occupation probability measured in a (p, d) process is correlated with the corresponding increase in occupation observed in (d, p) (U^2, V^2 factors). Aside from the detailed quantitative values, this is quite similar to what is observed in dressed single-particle states in normal nuclei (see Fig. 1.2.5), the single-particle sum rule involving both the $(A - 1)$ and $(A + 1)$ aside from the A systems (see Sect. 5.8). Concerning the phase coherence of the pair correlated wavefunction, it has no consequence for one-particle transfer process, in keeping with the fact that $|e^{i\phi} \sqrt{P_1}|^2 = P_1$. A further reminder that not one, but two-particle transfer is the specific probe of pairing in nuclei.

In fact, qualitative, and not only quantitative differences are found concerning pair transfer when the Fermi system, also the renormalized one (Fig. 1.2.4), changes from the normal into the superconducting state, a result of the fact that in this phase transition, more or less the same set of (phase disordered) renormalized many-body states become superposed with a fixed phase relation (see Eq. (3.4.2)) described by the $|BCS\rangle_{\mathcal{K}'} = \prod_{\nu>0} \left(U_\nu' + e^{-2i\phi} V_\nu' a_\nu^\dagger a_{\bar{\nu}}^\dagger \right) |0\rangle$ state which displays ODLRO.[15] As a consequence, in the calculation of the Cooper pair transfer probability, one has to add phased amplitudes before one takes modulus squared (Josephson (1962)). That is,

$$P_2 = \left| \frac{1}{\sqrt{2}} \left(U_\nu' \sqrt{P_1} + e^{-2i\phi} V_\nu' \sqrt{P_1} \right) \right|^2$$

$$= \frac{P_1}{2} \left(1 + 2U_\nu' V_\nu' \cos 2\phi \right) \approx P_1. \tag{4.3.1}$$

In other words, it is like interference in optics, with phase-coherent mixing. We note that above it was assumed $U_\nu' V_\nu'|_{\epsilon_F} \approx 1/2$ and $\cos 2\phi \approx 1$. Within this context, see Sect. 5.5.1.

In connection with pair transfer experimental probes involving intrinsic (as opposite to relative motion) energy exchange $< 2\Delta$ and relative distances of closest approach $D_0 \lesssim \xi$, the partners nucleons of the transferred Cooper pair are correlated over distances of the order of 15–20 fm (Fig. 4.3.1 (b)), the intrinsic motion being frozen because of the pairing gap, a situation quite different from that of a normal system characterized by independent nucleon motion, where the relative distance between nucleons is ≈ 2.4 fm (Fig. 4.3.1 (a)).

---

[15] See footnote 47 of Chapter 1.

To the extent that the mean field acting as an "external" field allows particle density to be present, the properties of phase-correlated independent Cooper pair motion will explicit themselves. And thus it is a physical condition which is assumed fulfilled each time one makes use of the physical picture provided by Fig. 4.3.1 (b). In other words, inside ^{120}Sn all (6–8) Cooper pairs will be found within a volume of radius $R_0 \approx 6$ fm, in a similar way in which the transferred Cooper pair can be distributed over two nuclear volumes during the contact time in, for example, a Sn + Sn heavy ion reaction (Fig. 4.5.1). In this phenomenon, the long-range induced pairing interaction (exchange of collective vibrations) contributes in the case of nuclei lying along the stability valley, let alone in the case of light halo nuclei like ^{11}Li (Fig. 4.3.2).[16]

The interest of the picture shown in Fig. 4.3.1 (b) can also be exemplified by referring to the fact that the moment of inertia of heavy deformed nuclei is considerably smaller than the rigid moment of inertia, but still larger than the irrotational one[17] ($5\mathfrak{J}_{irrot} \lesssim \mathfrak{J} \lesssim \mathfrak{J}_r/2$). Even confined within the mean field of the nucleus, the small but finite number of phase-correlated pairs having the "intrinsic," infinite-matter-like tendency will, to some extent, average out the different orientations of the rotating system and react to it in terms of an effective inertia smaller than the one related to the independent-particle motion in a deformed potential (rigid moment of inertia). However, constrained by such a potential, they cannot fully profit of BCS condensation ODLRO). To have irrotational flow, one has to have $\Delta \gg \hbar\omega_0$. Summing up, $\mathfrak{J}_{irrot} = \delta^2 \mathfrak{J}_{rig} \approx \delta^2 \times 2\mathfrak{J}_{exp} \approx (2/10) \times \mathfrak{J}_{exp} \approx \mathfrak{J}_{exp}/5$, where[18] $\delta \approx 0.95\,\beta_2$, typical values of β_2 being ≈ 0.3.

Within this context we note that the (approximate) form of the (local) pair wavefunction can be written as[19]

$$F(r) \approx \Delta N(0) \frac{\sin k_F r}{k_F r} \exp\left(-\frac{\sqrt{2}\,r}{\xi}\right), \qquad (4.3.2)$$

where Δ is the pairing gap and $N(0)$ is the density of levels at the Fermi energy for one spin orientation. For $r \lesssim \xi$ the pair wavefunction is approximately proportional to that of two particles at the Fermi energy moving freely in a relative s-wave state. In a typical metallic superconductor ξ is of the order of 1 μm (10^4 Å), much larger than the interelectron spacing (≈ 2Å). Note that relative to the total correlation (binding) energy of a metal (nucleus), the correlation energy ($E_{corr} = (-1/2)N(0)\Delta^2$) associated with Cooper pairing is very small, $\approx 10^{-7}$

---

[16] In this case, $R(^{11}$Li$) = 4.58 \pm 0.13$ fm, the "effective" Wigner–Seitz radius being $(r_0)_{eff} \approx 4.58$
      fm$/(11)^{1/3} \approx 2.1$ fm.

[17] Bohr and Mottelson (1975); Belyaev (1959, 2013).

[18] See Bohr and Mottelson (1975), p. 75.

[19] Cf. Leggett (2006), p. 185; for the nonlocal nuclear version, cf. Broglia and Winther (1983).

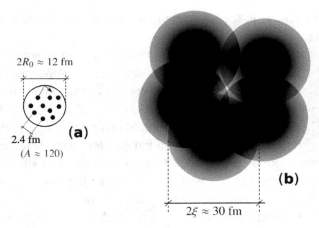

$2R_0 \approx 12$ fm

2.4 fm **(a)**

$(A \approx 120)$

(b)

$2\xi \approx 30$ fm

Figure 4.3.1 (a) Schematic representation of independent-particle motion and (b) independent-pair motion. In the first case, nucleons (fermions) move independently of each other, their relative distance being $2r_0 (\approx 2.4$ fm). Switching on the pairing interaction (bare plus induced) leads to Cooper pair formation in which the correlation length (mean square radius) is ξ. Pairs of nucleons moving in time-reversal states close to the Fermi energy will tend to recede from each other, lowering their relative momentum ($r_0 \rightarrow \xi$), thus boosting the stability of the system, provided that the external mean field allows for it, or better, if there is (normal) nucleon density available to do so, something controlled to a large extent by the single-particle potential. From this point on, and at least for the levels lying close to the Fermi surface, one cannot talk about particles but about Cooper pairs (unless one does not intervene the system with an external field, e.g., (p, d), and provides the energy, angular, and linear momentum needed to break a pair). The picture displayed in (b) is likely to be the one that is operative in the case of, e.g., two nuclei of Sn at a relative (CM) distance somewhat larger than $2R_0$ (≈ 12 fm), but still allowing for (weak) contact. The pair field associated with Cooper pairs will extend from one to the other partner of the heavy ions participating in the reaction through the weakly overlapping interaction region, allowing two nucleons to correlate over a distance ξ and, eventually, in a reaction like $^{A+2}Sn + ^{A'}Sn \rightarrow$ $^{A}Sn(gs) + ^{A'+2}Sn(gs)$, allow for the successive transfer of two nucleons correlated over distances larger than nuclear dimensions. For the color version of this figure, refer to cambridge.org/nuclearcooperpair.

($\approx 10^{-3}$). Arguably, the most important consequence of this fact is the exponentially large mean square radius and thus very small value of the relative momentum associated with Cooper pairs. In other words, the typical scenario for a very small value of the quantum localization kinetic energy and thus of the generalized quantality parameter (see Sect. 3.4 and App. 7.C), implying that the two partners of the Cooper pair, are anchored to (entangled with) each other. This phenomenon is at the basis of the emergence of new elementary modes of excitation and associated tunneling mechanism. They correspond to pairing vibrations for single Cooper pairs, pairing rotations for few ones and to supercurrents, and Josephson currents for macroscopic amounts of them. Common to all of them, when probed with

Cooper pair tunneling, is the associated successive transfer of the pair partners (entanglement).

The situation of very extended Cooper pairs sounds, in principle, very different in the case of condensed matter (e.g., low-temperature superconductors) than in atomic nuclei, in keeping with the fact that nuclear Cooper pairs are, as a rule, subject to an overwhelming external (mean) field ($2\Delta \approx 2.4$ MeV $\ll |U(r \approx R_0)| \approx |V_0/2| \approx 25$ MeV), and thus confined within dimensions $R_0 \ll \xi$. But even in this case one can posit that the transition from independent-particle to phase-correlated independent-pair motion, implies that Cooper pair partners show a tendency to recede from each other. Let us clarify this point for the case of a single pair, for example, ^{210}Pb(gs). It is true that allowing the pair of neutrons to correlate in the valence orbitals leads to a pair wavefunction which is angle correlated ($\Omega_{12} \approx 0$), as compared to, for example, the pure $j^2(0)(j = g_{9/2})$ configuration[20] (Sect. 4.11). On the other hand, the correlated pair addition mode (Tables 3.5.4 and 3.5.5) will display a sizable spill out as compared to the pure two particle state, and thus a tendency to lower the relative momentum between Cooper pair partners.[21] This is also connected with the fact that close to ≈ 40 percent of the pairing matrix elements is contributed by the induced pairing interaction resulting from the exchange of long-wavelength, low-lying collective modes, the other ≈ 60 percent resulting from the bare nucleon–nucleon $^1S_0$ pairing interaction.[22]

The situation described above becomes likely clearer, even if extreme, in the case of ^{11}Li. In this case, the Fermi momentum is $k_F \approx 0.8$ fm^{-1}, the radius $R \approx 4.58$fm being much larger than $R_0 = 2.7$ fm expected from systematics. Furthermore essentially all of the correlation energy ($E_{corr} \approx -0.5$ MeV)[23] is associated with the exchange of the pygmy dipole resonance between the halo neutrons.[24] It is of note that in this case, as already stated, renormalization effects due to the clothing of single-particle states by vibrations, are as strong as mean field effects.

Again in this case $s^2_{1/2}(0)$ and $p^2_{1/2}(0)$ are not correlated in Ω_{12}, while the Cooper state probability density displays a clear angular correlation (see Fig. 3.6.2 (II) (a)

---

[20] Bertsch et al. (1967); Ferreira et al. (1984); Matsuo (2013); Kubota et al. (2020), and refs. therein.

[21] Within this context, of note is the 53 percent increase in the absolute two-nucleon transfer cross section for the reaction ^{206}Pb$(t, p)^{208}$Pb(gs) (see Fig. 4.5.4), observed by changing from a TD to a RPA description of the pair removal mode $|^{206}$Pb$(gs)\rangle$ (see Table 3.5.2 and associated text), change which leads to an increase in the neutron spillout (see Fig. 6.4.3), boosting at the same time successive transfer.

[22] In carrying out the above arguments, the values of $|E_{corr}|/BE(A) \approx \left|\frac{-3\,\text{MeV}}{10^3\,\text{MeV}}\right| \approx 3 \times 10^{-3}$ and $\xi = \frac{\hbar v_F}{\pi \Delta} \approx 14$ fm $((\frac{v_F}{c}) \approx (k_F)_{\text{fm}^{-1}}/5 \approx 0.27$; $\Delta \approx 1.2$ MeV; $N(0) \approx 4$ MeV^{-1}), typical for superfluid nuclei lying along the stability valley, were used ($A \approx 120$) (see Sect. 4.8.4).

[23] $|E_{corr}|/\epsilon_F \approx 0.380/13 \approx 3 \times 10^{-2}$, $\xi \approx 20$ fm ($v_F/c \approx 0.2(k_F)_{\text{fm}^{-1}} \approx 0.16$; Eq. (7.C.6) and footnote 35 of Chapter 7.

[24] See Barranco et al. (2001); see also Broglia et al. (2019a).

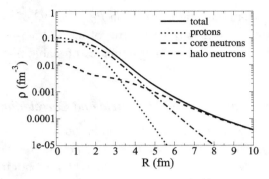

Figure 4.3.2 The nuclear density associated with ^{11}Li, as resulting from the microscopic NFT calculations which are at the basis of the results displayed in Figs. 3.6.2 and 7.1.2 (Barranco et al. (2001)). The contribution arising from the protons and the neutrons of the core (^{9}Li) are displayed with dotted and dash-dotted curves, respectively, while that associated with the two halo neutrons is shown in terms of a dashed curve. The sum of these contributions labeled tot (total) is drawn with a continuous curve. After Broglia et al. (2019a).

and (b)). Nonetheless, the average distance between the partners of the neutron halo Cooper pair, is considerably larger than that associated with the ^{9}Li core nucleons as a consequence of the induced (soft $E1$-mode exchange) pairing interaction, and as testified by the following numerical estimates (see also Fig. 4.3.2):

$$\text{a)} \quad R(^{11}\text{Li}) = 4.58 \pm 0.13 \,\text{fm} \quad (V = (4\pi/3)\, R^3 = 402.4 \,\text{fm}^3), \quad (4.3.3)$$

$$\text{b)} \quad R_0(^{11}\text{Li}) = 2.7 \,\text{fm} \quad (V = 82.4 \,\text{fm}^3), \quad (4.3.4)$$

$$\text{c)} \quad R_0(^{9}\text{Li}) = 2.5 \,\text{fm} \quad (V = 65.4 \,\text{fm}^3), \quad (4.3.5)$$

and associated mean distance between nucleons,

$$\text{a)} \quad \left(\frac{337 \,\text{fm}^3}{2}\right)^{1/3} \approx 5.5 \,\text{fm} \quad ((402.4 - 65.4)\,\text{fm}^3), \quad (4.3.6)$$

$$\text{b)} \quad \left(\frac{82.4 \,\text{fm}^3}{11}\right)^{1/3} \approx 1.96 \,\text{fm}, \quad (4.3.7)$$

$$\text{c)} \quad \left(\frac{65.4 \,\text{fm}^3}{9}\right)^{1/3} \approx 1.94 \,\text{fm}. \quad (4.3.8)$$

The above quantities are to be compared with the definition[25]

$$d = \left(\frac{\frac{4\pi}{3} R^3}{A}\right)^{1/3} = \left(\frac{4\pi}{3}\right)^{1/3} \times r_0 \approx 1.93 \,\text{fm}, \quad (4.3.9)$$

[25] Brink and Broglia (2005), App. C.

consistent with the standard parameterization $R_0 = r_0 A^{1/3}$ of the nuclear radius. It is written in terms of the radius r_0 (= 1.2 fm) of the Wigner–Seitz cell associated with each nucleon, and derived from systematics of stable nuclei lying along the stability valley.

4.3.1 Interplay between Mean Field and Correlation Length

In Fig. 4.3.1 one displays a schematic representation of the basis for two possible *gedankenexperiment* situations: (**a**) (*independent-particle motion*), system which can be probed in a (p, t) reaction leading insight into noninteracting pairs of nucleons moving in time-reversal states and confined by a mean field potential; (**b**) (*phase-correlated independent-pair motion*), system probed in, for example, two-nucleon transfer process induced by a heavy ion reaction between superfluid nuclei (see Sect. 7.3). In this case it is assumed that pairs of nucleons moving in time-reversal states interact through an effective pairing interaction v_p^{eff}, sum of a (short-range) bare NN-$^1S_0$ potential (v_p^{bare}) and an induced (v_p^{ind}) pairing component, which is long range. As a result they become weakly bound with an energy 2Δ ($\ll \epsilon_F$), condensing in the BCS state and giving rise to the nuclear abnormal density in which the average distance between the pair partners is considerably larger than the relative distance associated with normal density ($\approx 2r_0$). Expressed differently, and in keeping with the parallel with the BCS description of low-temperature superconductivity found at the basis of nuclear BCS[26], one can assume that pairs of correlated nucleons moving in time-reversal states close to the Fermi energy, will tend to recede from each other. For example, in the case of Sn, Cooper pair partners are expected to be at the antipodes[27], in keeping with the fact that $\xi \approx 14$ fm, and $R_0 \approx 6$ fm.

This expectation is not confirmed by studies of the Cooper pair wavefunction[28] or, better, its modulus squared, which indicated that in going from situation (a) to situation (b), Cooper pair partners come close to each other, if nothing else because of angular correlation (see Sect. 4.11), a result which is also valid for systems with two nucleons outside closed shells.

In hindsight, one can posit that this result has to be interpreted with care in connection with the experiments associated with the specific probe[29] of

---

[26] Bohr et al. (1958).

[27] While not profiting completely from the latitude given by Cooper's mean square radius ξ, they do the best in the strong "external" mean field they are subject to, eventually displaying a nonnegligible amount of spillout.

[28] See footnote 20 of this chapter.

[29] On the other hand, to learn about the structure of the Cooper pair subject to the single-particle potential acting as a strong external field, one can carry out electron scattering experiments or (p, pn) knockout reactions at high energies, something that is also true for ^{11}Li, not particularly because of confinement, but because of spatial quantization, in keeping with the fact that the components $|s_{1/2}^2(0)\rangle$ and $|p_{1/2}^2(0)\rangle$ of $|gs(^{11}\text{Li})\rangle$ have similar amplitudes, and of the symmetry relations of the spherical harmonics discussed in

Cooper pair correlations, namely, two-nucleon transfer reactions, in which case the closest quantity to be observable is the two-nucleon transfer formfactor (Sect. 7.4). The soundness of picture (b) (Fig. 4.3.1) gets also support from the fact that one- and two-particle transfer absolute cross sections have similar values.[30]

Within this context we note that the fact that $^9_3Li_6$ is well bound ($N = 6$ parity-inverted closed shell), $^{10}_3Li_7$ is not while $^{11}_3Li_8$ is again bound, indicates that we are confronted with a pairing phenomenon. Allowing the two neutrons moving outside $N = 6$ closed shell to correlate in the configurations $j^2(0)$ ($s^2_{1/2}, p^2_{1/2}, d^2_{5/2} \ldots$) through a short-range bare pairing interaction, for example, the v_{14} Argonne NN-potential[31], does not lead to a bound state. The system lowers the relative momentum of the pair by exchanging the low-lying dipole vibration (soft $E1$ mode, pygmy resonance)[32] of the associated diffuse system becoming, eventually, bound, ever so weakly ($S_{2n} = 380$ MeV). The radius of the resulting system ($R(^{11}Li)=4.58 \pm 0.13$ fm) corresponds, in the parameterization $R_0 = 1.2A^{1/3}$ fm, to an effective mass number $A \approx 60$. So undoubtedly the system has swelled in moving from $A = 9$ single closed-shell system to the $A = 11$ two-nucleon pair correlated outside the closed shell, in a manner that goes beyond the $1.2A^{1/3}$ (fm) expected dependence. Although the correlation length of the neutron Cooper pair is restricted to a spherical volume of diameter $2 \times R(^{11}Li) \approx 9.2$ fm, half of the estimated value of the correlation length $\xi \approx 20$ fm (see App. 7.C), it is almost double as large as $2 \times R_0(^{11}Li) \approx 5.4$ fm. Consequently, the function ($|\Psi_0(\mathbf{r}_1, \mathbf{r}_2)|^2$) displayed in Fig. 3.6.2 (II) b) should be read with care.

It will be surprising if the afore mentioned bootstrap-like mechanism (see Fig. 2.7.1), namely, that of profiting from low, unstable, nuclear densities to generate transient medium polarization effects to stabilize a Cooper pair halo system, was a unique property of ^{11}Li. In fact, one can expect situations of s and p states at threshold eventually leading to a symbiotic halo Cooper pair with a small value of S_{2n}, also in connection with nuclear excited states.

Within the bootstrap ansatz of symbiotic Cooper pair binding, we introduce a generalization of the Axel–Brink[33] hypothesis based on well established experimental results, namely, the fact that the line shape and thus also the percentage of EWSR per energy interval as well as the decay properties of the GDR will

---

Sect. 4.11, in particular in the text following Eq. (4.11.11). Within this scenario, see the exhaustive discussion in Kubota et al. (2020); see also Hansen and Jonson (1987); Bertsch and Esbensen (1991); Esbensen et al. (1997); Matsuo et al. (2005); Hagino and Sagawa (2005); Matsuo (2006); Hagino et al. (2007); Sagawa and Hagino (2007); Hagino and Sagawa (2007); Matsuo (2013).

[30] See footnote 28 in Chapter 3. See also Eq. (4.2.12) and following text, as well as Sect. 7.3.

[31] Wiringa et al. (1984).

[32] Broglia et al. (2019a).

[33] Namely, the hypothesis that on top of each excited level of the nuclear spectrum is built a GDR equal to that built on the ground state; Axel (1962); Brink (1955).

reflect the adiabatic (inhomogeneous damping) and dynamic (motional narrowing) deformation properties of the state on which the GDR is built upon.[34]

In the case of halo nuclei, the afore mentioned generalization is both qualitative and quantitative. A sensible fraction of the TRK sum rule is found almost degenerate with the ground state. From the "elastic" antenna-like response typical of the high-energy GDR ($\hbar\omega_{GDR} \approx 80\text{MeV}/A^{1/3}$) one is now confronted with a "plastic," low-energy ($\lesssim 1$ MeV, $\Gamma \approx 0.5$ MeV) pygmy dipole resonance (PDR) or soft $E1$-mode[35] where pairing, that is, Cooper pairs, play an important role. Regarding the consequences this phenomenon has for the $L = 1$ induced interaction between nucleons, one moves from dipole–dipole (static moment interactions, effective charges) to dispersive (retarded) contributions, emerging essentially from quantum mechanical ZPF.[36] As already expressed, it is this second one which dominates the van der Waals interaction (Sect. 3.7) and, similarly, it is one which can lead to an almost resonant (PDR) gluing of Cooper pair halos (see Eq. 4.8.19). Similar to the resonant mechanism, phonons in this case, found at the basis of superconductivity in metals (see Eq. (4.8.64) and following discussion).

The energy centroid, the width and the percentage of the TRK sum rule (EWSR) of dipole resonances can be strongly affected by nuclear dynamical fluctuations and deformations which, in turn, depend on pairing, and thus on Cooper pairs. Because of angular momentum conservation, such phenomenon is restricted in lowest order, to fluctuations and deformations of quadrupole and monopole type. In the case of ^{11}Li, to an isotropic radial deformation, build up by the two less bound neutrons. This highly extended, low-momentum neutron halo, vibrating out of phase against the nucleons of the core (^{9}Li), gives rise to the soft $E1$-mode (PDR).[37] It can be viewed as the tailored glue which, exchanged between the halo neutrons, binds the resulting neutron halo Cooper pair to the core.[38]

The challenges faced in trying to learn about the physical basis of pairing in nuclei are comparable to those encountered to extract a collective vibration from a background much larger than the signal, as it was the case in the discovery of the GDR in hot nuclei.[39,40] In trying to observe pairing effects in nuclei close to the

---

[34] Le Tourneaux (no. 11, 1965); Bohr and Mottelson (1975); Bortignon et al. (1998), and refs. therein. See also Dattagupta (1987).

[35] See footnote 123 of Chapter 2.

[36] Within this context, see Bohr and Mottelson (1975), Sect. 6-5 f p. 432.

[37] Broglia et al. (2019a), and refs. therein.

[38] Barranco et al. (2001).

[39] Snover (1986); Gaardhøje (1985); Gaardhøje et al. (1986). See also Bortignon et al. (1998), Figs. 1.4 and 6.8, and refs. therein.

[40] A statement which becomes more than an analogy in connection with the nuclear analogue to the (ac) Josephson effect proposed in Potel et al. (2021), which can eventually be tested by extracting a dipole γ-signal of Josephson frequency $\nu_J = Q_{2n}/h$ (or better, that of the somehow higher peaked γ-strength

ground state, one has the advantage to start with the system at zero temperature for free. On the other hand, one needs to substract the very large, state-dependent effects of the "external" mean field, a challenge not second to that faced by condensed matter practitioners to study low-temperature superconductivity in general, and the Josephson effect in particular. Within this context, ^{11}Li provides a textbook example of the fact that, given the possibility[41], nucleon pair partners recede from each other, lowering in the process the momentum of relative motion and thus the confinement kinetic energy[42], allowing to extend the limits of stability of nuclear species through a subtle long-range pairing mechanism.

Concerning the fact that the role of pairing increases for weakly bound nuclei far from stability and the fact that some nuclei turn out to be bound only due to pairing see Zelevinsky and Volya (2004).[43]

4.4 Correlations between Nucleons in Cooper Pair Tunneling

Let us call x_1 and x_2 the coordinates of the Cooper pair partners. Let us furthermore assume they can only take two values: 0 when they are bound to the target nucleus, 1 when they have tunneled and become part of the outgoing particle (see Fig. 4.4.1).

The correlation between the two nucleons is measured by the value[44]

$$\langle x_1 x_2 \rangle - \langle x_1 \rangle \langle x_2 \rangle = \int d\gamma \, P_2 \times 1 \times 1 - \int d\gamma \, P_1 \times 1 \int d\gamma' \, P_1' \times 1 = P_2 - P_1 P_1',$$

(4.4.1)

$d\gamma$ being the differential volume in phase space, normalized with respect to the corresponding standard deviations, that is, with respect to

$$\sigma_{x_1} \sigma_{x_2} = \left[\left(\langle x_1^2 \rangle - \langle x_1 \rangle^2 \right) \left(\langle x_2^2 \rangle - \langle x_2 \rangle^2 \right) \right]^{1/2}.$$

(4.4.2)

function, energy shifted by the γ-phase space volume element), from a continuum plus discrete lines background.

[41] Namely, the presence of normal density, a feature which in the present case goes hand in hand with the presence of abnormal density (single Cooper pair).

[42] That is, from $\hbar^2/(m(2r_0)^2) \approx 7$ MeV ($r_0 = 1.2$ fm) to $\hbar^2/(m(2R(^{11}\text{Li}))^2) \approx 0.5$ MeV ($R(^{11}\text{Li}) \approx 4.6$ fm; within this context, it is of note that the overlap between radial wavefunctions of the halo neutrons and of those of the core is $\mathcal{O} = (R_0/R)^3 \approx 0.2$; see Fig. 2.7.1; see also Broglia et al. (2019b)). While the system loses in this way a consistent fraction of the bare, short-range pairing interaction through screening, it opens up for long-range pairing contributions which, ever so weak, can still bind the halo Cooper pair to the core. In other words, symbiotic binding by deconfinement (note that the correlation length expected to be operative in the case of a two-neutron transfer process in a heavy ion collision is estimated to be $\xi \approx 20$ fm; see Eq. (3.6.12)). In the case of superconducting metals, the correlation length is $\xi \approx 10^4$ Å, a distance to be compared to the Wigner–Seitz radius of ≈ 2Å. This variation leads to a decrease of the Coulomb repulsion by almost four orders of magnitude. From $U_c = e^2/r = 14.4 \times$ eV$\times$Å/(2Å) = 7.2 eV to $U_c = 14.4$ eVÅ/(10^4Å) = 1.4 meV.

[43] Within this context and regarding a powerful method to treat pairing, based on the quasispin symmetry, see Volya et al. (2001) and Zelevinsky and Volya (2003).

[44] Basdevant and Dalibard (2005).

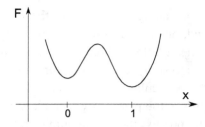

Figure 4.4.1 A schematic representation of nucleon tunneling between target and projectile. The free energy $F = U - TS$ which for the zero-temperature situation under consideration coincides with the potential energy as a function of the nucleon coordinate x. For $x = 0$ the nucleon is assumed to be bound to the target system. For $x = 1$ the nucleon has undergone tunneling, becoming bound to the outgoing particle. In other words x_1 jumps from the value 0 to the value 1 in the tunneling process $(x, 0 \rightarrow 1)$, the same for the coordinate of the second nucleon.

Making use of the fact that

$$\langle x_1^2 \rangle = \int d\gamma \, P_1 \times 1^2 = P_1, \qquad (4.4.3)$$

and

$$\langle x_1 \rangle = \int d\gamma \, P_1 \times 1 = P_1, \qquad (4.4.4)$$

one can calculate the function which measures the correlations between nucleons 1 and 2, namely,

$$Corr = \frac{\langle x_1 x_2 \rangle - \langle x_1 \rangle \langle x_2 \rangle}{\sqrt{(\langle x_1^2 \rangle - \langle x_1 \rangle^2)(\langle x_2^2 \rangle - \langle x_2 \rangle^2)}} = \frac{P_2 - P_1 P_1'}{\sqrt{(P_1 - P_1^2)(P_1' - P_1'^2)}}. \qquad (4.4.5)$$

Because both nucleons are identical and thus interchangeable, $P_1 = P_1'$. Thus

$$Corr = \frac{P_2 - P_1^2}{P_1 - P_1^2}. \qquad (4.4.6)$$

Making use of the empirical values

$$P_1 \approx P_2 \approx 10^{-3} \qquad (4.4.7)$$

leads to

$$Corr = \frac{10^{-3} - 10^{-6}}{10^{-3} - 10^{-6}} \approx 1. \qquad (4.4.8)$$

In other words, within the Cooper pair motion regime, nucleon partners are solidly anchored (entangled) to each other: if one nucleon goes over, the other does it also. This is so in spite of the very liable and fragile (intrinsic) structure of the nuclear Cooper pairs ($2\Delta/\epsilon_F \ll 1$). An example of such a scenario is provided

$$
\text{Order parameter} \quad (\langle \tilde{0} | PP^\dagger | \tilde{0} \rangle)^{1/2} = \begin{cases} \alpha_0 = \Sigma_{v>0} \, U'_v V'_v \\ \alpha_{dyn} = \Sigma_{v>0} \, U_v^{eff} V_v^{eff} \end{cases}
$$

pairing vibrations

$$
(V_v^{eff})^2 = 2 Y_a^2(j_v)/\Omega_v; \qquad (U_v^{eff})^2 = 1 - (V_v^{eff})^2
$$

$$
\left. \begin{array}{l} X_a(j_v) \\ Y_a(j_v) \end{array} \right\} = \frac{(\sqrt{\Omega_j/2})\Lambda_a}{2|E_j| \mp W_a}
$$

pairing rotations

$$
\left. \begin{array}{l} U'_v \\ V'_v \end{array} \right\} = \frac{1}{\sqrt{2}} \left(1 \pm \frac{\epsilon_v}{\sqrt{\epsilon_v^2 + \Delta^2}} \right)^{1/2}
$$

Figure 4.4.2 Order parameter associated with static and dynamic pair correlations.

by ^{11}Li. In fact, if one picks up a neutron from ^{11}Li (^{11}Li(p,d)^{10}Li), the other one breaks up essentially instantaneously, ^{10}Li being unbound (Sect. 5.2.3). Despite this, the probability associated with the reaction ^{1}H(^{11}Li,^{9}Li(gs))^{3}H is given by[45]

$$
P_2 = \frac{(5.7 \pm 0.9)\text{ mb}}{2\pi((4.58 \pm 0.13)\text{ fm})^2} \approx (4.3 \pm 1)10^{-3}. \tag{4.4.9}
$$

This value is of the order of that associated with the breakup process mentioned above[46] and reported in Table 4.4.1.

The fact that $P_2 \approx P_1$ contained in the relation (4.3.1) is applicable both for static and dynamic pairing modes is in keeping with the fact that nuclei dynamic spontaneous breaking of gauge invariance is of similar importance as the static one[47] (see Figs. 3.5.7 and 4.4.2).

4.5 Pair Transfer

In the semiclassical approximation, the second-order two-nucleon transfer amplitude can be written as (see Eq. (6.5.3)),

$$
a(t = +\infty) = a^{(1)}(\infty) - a^{(NO)}(\infty) + \tilde{a}^{(2)}(\infty), \tag{4.5.1}
$$

[45] See Fig. 7.1.2, Sect. 7.1.2 regarding the experimental values and Table 7.1.2 concerning the theoretical ones of $\sigma(^{11}$Li $\to^9$ Li (gs)), and see Kobayashi et al. (1989) concerning the experimental value of $R(^{11}$Li). See also the paragraph following Eq. (3.6.13).

[46] It is of note that the value given in Eq. (4.4.9) essentially coincides with that shown in the $l = 0$, column **1** of Table 7.1.2.

[47] See also Fig. 4 of Potel et al. (2013b).

Table 4.4.1 *Probabilities p_l associated with the reaction $^1H(^{11}Li,^{10}Li(gs))^2H$ calculated with the same bombarding conditions as those associated with $^1H(^{11}Li,^9Li(gs))^3H$ (see also Table 7.1.2).*

l	p_l
0	1.02×10^{-3}
1	2.40×10^{-3}
2	1.26×10^{-2}
3	1.84×10^{-2}
4	6.13×10^{-3}
5	1.39×10^{-3}
6	2.89×10^{-4}
7	5.04×10^{-5}
8	6.51×10^{-6}
9	5.87×10^{-7}

where $a^{(1)}(\infty) = a^{(1)}_{sim}$ and $\tilde{a}^{(2)}(\infty) = a^{(2)}_{succ}$ describe the simultaneous and successive transfer amplitudes, respectively. In the **independent-particle limit**, these amplitudes fulfill the relations[48]

$$a^{(1)}_{sim} = a^{(1)}_{NO} \tag{4.5.2}$$

and

$$a^{(2)}_{succ} = a^{(1)}_{one-part} \times a^{(1)}_{one-part}, \tag{4.5.3}$$

with

$$a + A \rightarrow f + F \rightarrow b + B, \tag{4.5.4}$$

corresponding to the product of two single nucleon transfer amplitudes. On the other hand, in the **strong correlation limit** one can write, making use of the post-prior representation

$$a^{(2)}_{succ} = \tilde{a}^{(2)}_{succ} - a^{(1)}_{NO}, \tag{4.5.5}$$

a relation which fulfills

$$\lim_{E_{corr} \rightarrow \infty} a^{(2)}_{succ} = 0. \tag{4.5.6}$$

---

[48] See Sect. 6.5, also Potel et al. (2013a).

That is, the transfer process is, in this case, associated with simultaneous transfer. Actual nuclei are close to the independent-particle limit ($E_{corr} \approx$ (2–3 MeV) $\ll$ $\epsilon_F (\approx 36$ MeV)). Successive transfer is thus expected to be the major contribution to pair transfer processes. But successive transfer seems to break the pair *right? Wrong. Why?* Let us see below.

4.5.1 Cooper Pair Dimensions

At the basis of the relations used to estimate the dimensions of a Cooper pair (correlation length; mean square radius) one finds

$$\delta x \delta p \geq \hbar \quad \delta \epsilon \approx |E_{corr}|, \tag{4.5.7}$$

where

$$\epsilon = \frac{p^2}{2m}; \quad \delta \epsilon = \frac{p \delta p}{m} \approx v_F \delta p, \tag{4.5.8}$$

and thus

$$\delta \epsilon \approx |E_{corr}| \approx v_F \delta p, \tag{4.5.9}$$

leading to

$$\xi \approx \delta x \approx \frac{\hbar}{\delta p} \approx \frac{\hbar v_F}{|E_{corr}|} \quad \text{(correlation length).} \tag{4.5.10}$$

In what follows we use, for normal systems,

$$\xi = \frac{\hbar v_F}{\pi |E_{corr}|}, \tag{4.5.11}$$

and, in the case of open-shell superfluid nuclei,[49]

$$\xi = \frac{\hbar v_F}{\pi \Delta}. \tag{4.5.12}$$

For Sn-isotopes, a typical value of the pairing gap is $\Delta \approx 1.2$ MeV. Making use of $k_F = 1.36$ fm^{-1} associated with nuclei lying along the stability valley, and thus $(v_F/c \approx 0.2(k_F)_{\text{fm}^{-1}} \approx 0.27)$,

$$\xi \approx 14 \text{ fm.} \tag{4.5.13}$$

Consequently, successive and simultaneous transfer can feel equally well the pairing correlations giving rise to long-range order.[50] This virtual property can become real in, for example, a pair transfer between two superfluid nuclei[51], in

---

[49] So called Pippard correlation length. See Schrieffer (1964), pp. 18 and 34; Annett (2013), p. 62.

[50] See footnote 47 of Chapter 1; see also, e.g., Potel et al. (2017), and refs. therein.

[51] See von Oertzen (2013); von Oertzen and Vitturi (2001); Weiss (1979); for a more recent experimental and theoretical study, see Montanari et al. (2014, 2016); Potel et al. (2021).

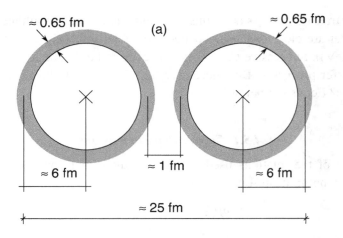

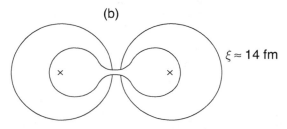

Figure 4.5.1 (a) Schematic representation of two Sn-isotopes (radius $R_0 \approx 6$ fm) at the distance of closest approach in a heavy ion collision; (b) single Cooper pair in which each nucleon partner is in a different nucleus. The gray areas indicate the nuclear diffusivity. After Potel et al. (2013b).

particular two tin isotopes (see Figs. 4.1.3 and 4.5.1; see also Sect. 7.3 with regard to the reaction between superfluid Sn- and Ni-isotopes).

Objection

What about $v_{pairing}(= G)$ becoming zero, for example, between the two nuclei?

Answer

$$\frac{d\sigma(a(= b + 2) + A \rightarrow b + B(= A + 2))}{d\Omega} \sim |\alpha_0|^2, \quad (4.5.14)$$

$$\alpha_0 = \langle BCS(A + 2)|P^\dagger|BCS(A)\rangle = \sum_{v>0} U_v(A)V_v(A + 2). \quad (4.5.15)$$

That is, pair transfer does not depend explicitly but only functionally on G.

Objection

Relation (4.5.15) is only valid for simultaneous transfer, *right? Wrong.*

Answer

The order parameter can also be written as

$$
\begin{aligned}
\alpha_0 &= \sum_{\nu,\nu'>0} \langle BCS|a_\nu^\dagger|int(\nu')\rangle\langle int(\nu')|a_{\bar\nu}^\dagger|BCS\rangle \\
&\approx \sum_{\nu,\nu'>0} \langle BCS(A+2)|a_\nu^\dagger\alpha_{\nu'}^\dagger|BCS(A+1)\rangle\langle BCS(A+1)|\alpha_{\nu'}a_{\bar\nu}^\dagger|BCS(A)\rangle \\
&= \sum_{\nu,\nu'>0} \langle BCS(A+2)|V_\nu(A+2)\alpha_{\bar\nu}\alpha_{\nu'}^\dagger|BCS(A+1)\rangle \\
&\quad\times \langle BCS(A+1)|\alpha_{\nu'}U_\nu(A)\alpha_{\bar\nu}^\dagger|BCS(A)\rangle \approx \sum_{\nu>0} V_\nu(A+2)U_\nu(A), \quad (4.5.16)
\end{aligned}
$$

where the (inverse) quasiparticle transformation relation $a_\nu^\dagger = U_\nu\alpha_\nu^\dagger + V_\nu\alpha_{\bar\nu}$ was used.[52] Examples of two-nucleon spectroscopic amplitudes involving superfluid targets, namely, those associated with the reactions ^{112}Sn$(p,t)^{110}$Sn(gs) and ^{124}Sn$(p,t)^{122}$Sn(gs) are given in Table 3.1.1 (see also Table 7.2.1). Making use of some of these amplitudes (first column of Table 3.1.1) and of global optical parameters (Table 7.2.2), the two-nucleon transfer absolute differential cross section of the reaction ^{112}Sn$(p,t)^{110}$Sn(gs) at a bombarding energy of $E_p = 26$ MeV, was calculated making use of the software COOPER based on second-order DWBA and taking into account successive and simultaneous transfer properly corrected for nonorthogonality (cf. Chapter 6 and App. 7.A). It is compared with experimental data in Fig. 4.5.2 (a). The corresponding absolute integrated cross sections are 1301 μb and 1309 $\pm$ 200 μb, respectively. The largest contribution to the cross section arise from successive transfer, the cancellation between simultaneous and nonorthogonality amplitudes being important (Fig. 4.5.2 (b)). The above provides a typical example of results of a systematic study of two-nucleon transfer reactions in terms of absolute cross sections[53] (Figs. 3.1.7 and 7.2.1).

Making use of two-nucleon spectroscopic amplitudes worked out within the framework of an extended shell model calculation (Table 3.1.1, second column) one obtains very similar results to those displayed in Fig. 4.5.2 (a). In Fig. 4.5.3 we report results similar to those displayed in Fig. 4.5.2, but for the case of the reaction ^{124}Sn$(p,t)^{122}$Sn(gs) calculated within second-order DWBA making use of the BCS spectroscopic amplitudes (Table 3.1.1 third column).

---

[52] See, e.g., Brink and Broglia (2005), App. G.
[53] Potel et al. (2013a,b).

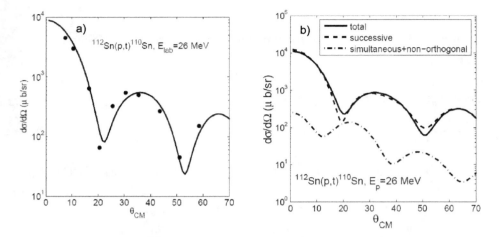

Figure 4.5.2 (a) Absolute differential cross section associated with the reaction $^{112}\text{Sn}(p,t)^{110}\text{Sn}(\text{gs})$ calculated with the software COOPER in comparison with the experimental data (Guazzoni et al. (2006)). (b) Details of the different contributions to the total absolute (p,t) differential cross section.

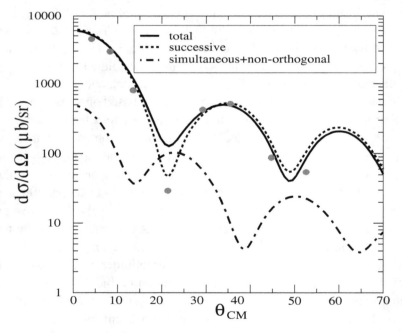

Figure 4.5.3 Absolute differential cross section associated with the reaction $^{124}\text{Sn}(p,t)^{122}\text{Sn}(\text{gs})$ calculated making use of second-order DWBA, taking into account nonorthogonality corrections and the two-nucleon spectroscopic amplitudes resulting from BCS (see Table 3.1.1, third column; for details, see Potel et al. (2013a), Potel et al. (2013b)) in comparison with experimental data (Guazzoni et al. (2011)). For the color version of this figure, refer to cambridge.org/nuclearcooperpair.

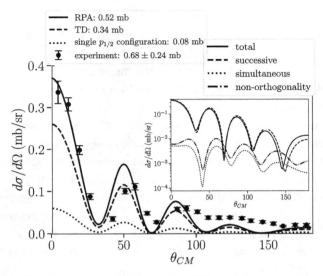

Figure 4.5.4 Absolute two-nucleon transfer differential cross section associated with the ^{206}Pb$(t, p)^{208}$Pb(gs) transfer reaction, that is, the annihilation of the pair removal mode of ^{208}Pb in comparison with the data (Bjerregaard et al. (1966)). The theoretical cross sections were calculated making use of the spectroscopic amplitudes given in Tables 3.5.2 and 3.5.3 and of global optical parameters as reported in the reference above. Both RPA and TD amplitudes were used in the calculations as well as a pure configuration $p_{1/2}^2(0)$.

Let us now discuss an example of two-nucleon transfer around a closed-shell nucleus displaying well defined collective pairing vibrational modes. We refer, in particular, to the pair removal mode of ^{206}Pb, that is, to the reaction, ^{206}Pb$(t, p)^{208}$Pb(gs). Making use of the spectroscopic amplitudes displayed in Tables 3.5.2 and 3.5.3 and of global optical parameters, the associated absolute differential cross section was calculated with the software COOPER. It is displayed in Fig. 4.5.4 in comparison with the experimental findings. In the same figure, the absolute total differential cross section is compared with that associated with the TD (Tamm-Dancoff) description of ^{206}Pb(gs), that is, setting the pairing ground state correlations to zero ($\sum_i X_r^2(i) = 1$, $Y_r(k) \equiv 0$, see Table 3.5.2). In this case, theory underpredicts observation by about a factor of 2. Also given in Fig. 4.5.4 is the predicted cross section associated with the pure configuration $|p_{1/2}^{-2}(0)\rangle$.

It is of note that within the effective reaction mechanism described in Sect. 6.4 pairing correlations increase the value of $\Omega_0 (\approx 0.97)$. As a consequence the $l = n = 0$ two-neutron system gives a consistently larger contribution to the two-nucleon transfer process than those associated with $n = 1$ and 2, that is those proportional to Ω_1 and Ω_2 whose values are 0.25 and 0.06, respectively (cf. Eq. (6.4.13)). All these features boost the effective absolute two-nucleon transfer cross section. While the results displayed in Fig. 4.5.4 were calculated making use of the full formalism of second-order DWBA (Figs. 4.1.1 and 4.1.2) the simplified expressions given in Eqs. (6.4.1–6.4.3) and (6.4.13–6.4.16) were useful to gain physical insight into the effective two-nucleon transfer formfactors $u_{LSJ}^{J_i J_f} f(R)$. In fact, these functions, multiplied by the factors D_0, provided a simple parameterization to account for the absolute two-nucleon transfer differential cross sections. *On the other hand, they were a source of misunderstanding concerning the reaction mechanism of two-nucleon transfer.*[54]

Collective surface vibrations of closed-shell nuclei can be viewed as correlated particle-hole excitations. The phase coherence existing between the different RPA amplitudes of the corresponding wavefunction leads to a decrease of the average distance (angular correlation) between the particle and the hole, as compared with pure *ph*-configurations, similar to what happens in pair addition (substraction) modes[55], in which case are the *pp* (*hh*) which angular correlate with each other. It has been argued that this is the reason why both pairing and surface vibrations of closed-shell nuclei display enhanced (t, p) cross sections as compared to pure configurations.[56] In the triton, the two neutrons lie rather close to each other.

In the case of collective surface vibrations of closed-shell nuclei, like the octupole vibration of ^{208}Pb ($E_x = 2.65$ MeV), the specific probe is not two-nucleon transfer, but Coulomb excitation or inelastic scattering, the wavelength of the γ-ray associated with the corresponding electromagnetic decay being $\lambda \approx 460$ fm. This is about two orders of magnitude larger than nuclear dimensions. Within this scenario, whether the particle is angle correlated with the hole or less in the collective (RPA) state, as compared to a pure $p - h$ state can hardly be of any relevance to explain the enhancement of the absolute transition probability ($B(E3) = 32$ Bsp) of the particle-hole correlated state, one of the largest of the whole mass table. Not only this, it sets a question mark on the validity of the argument as applied to Cooper pair transfer and pairing correlations. This question is taken up in Sects. 4.9–4.12.

4.6 Weak Link between Superconductors

Two-nucleon transfer reactions involving superfluid nuclei display some similarities with Cooper pair tunneling between weakly coupled superconductors, in particular when discussing heavy ion reactions, but not only.[57] Within this context it is useful to remind the basic elements of the pair tunneling which is at the basis of the Josephson effect. In this section we essentially reproduce the description of

---

[54] For more (historic) details, see Broglia (2020), Overview.

[55] Concerning this parallel, see Fig. 4.9.3; see also Fig. 7 of Barranco et al. (2019b).

[56] Bertsch et al. (1967).

[57] Goldanskii and Larkin (1968); Dietrich (1970); Dietrich et al. (1971); Hara (1971); Kleber and Schmidt (1971); Weiss (1979); von Oertzen and Vitturi (2001); von Oertzen (2013); Broglia and Winther (2004).

the tunneling of Cooper pairs between two weakly coupled superconductors to be found in[58], likely the best physical presentation of the Josephson effect.[59]

One starts with the many-body Hamiltonian[60]

$$H = H_1 + H_2 + \sum_{kq} T_{kq}(a_{k\uparrow}^\dagger a_{q\uparrow} + a_{-q\downarrow}^\dagger a_{-k\downarrow}) + hc \qquad (4.6.1)$$

where H_1 and H_2 are the separate Hamiltonians of the two superconductors on each side of the junction, T_{kq} being the (exponentially) small tunneling matrix element from state k on one side to state q on the other.

One can arrive to (4.6.1) by first finding sets of single-particle wavefunctions for each side separately, in the absence of the potential of the other system. *Then one eliminates the nonorthogonality effects by perturbation theory (cf. the similarity with the arguments used in Sect. 4.1 to introduce the two-nucleon transfer amplitudes in (second order) DWBA; see also App. 2.B, Eq. (2.B.23)).* A nuclear embodiment of such strategy[61] is worked out in Chapter 6 (see in particular Sect. 6.2.9 and Sect. 6.5) and was implemented in the code COOPER.[62]

At the basis of the Josephson effect one find $P_2 \approx P_1$ which, in the nuclear case implies $\sigma_{2p} \approx \sigma_{1p}$ (see Sect. 4.3; also footnote 28 of Chapter 3). A remarkable finding valid also whether only one of the two interacting systems belongs to a pairing vibrational band, or is superfluid. This is because of the importance of pairing vibrations in nuclei as compared to superconductors. Similar results are found in the case in which both systems are superfluid, namely, in the case of one- and two-nucleon transfer reactions in a heavy ion collision between superfluid nuclei (see Chapter 7).[63] It is likely that the results obtained in this case at bombarding energies of few MeV below the Coulomb barrier, come closest to the tunneling phenomena observed between two weakly coupled superconductors.

Let us now calculate the second-order expression of (4.6.1) in the case in which the gaps of the two weakly linked superconductors are different. Making use of

---

[58] Anderson (1964b).

[59] Josephson (1962).

[60] Cohen et al. (1962).

[61] Because pairing vibrations in nuclei are quite collective, leading to effective U and V occupation factor (cf. Fig. 4.4.2) (see also Potel et al. (2013b)), the nuclear and the condensed matter expressions for Cooper tunneling processes are very similar. Of course, no supercurrent is expected between nuclei. However, the systems ^{120}Sn(gs), ^{119}Sn(j), ^{118}Sn(gs) form an ensemble of weakly coupled Fermi superfluids, with different (average) number of particles (N, $N-1$, $N-2$), to which essentially all the BCS techniques, including those of the present section, can be applied.

[62] Cf. App. 7.A; cf. also Broglia and Winther (2004).

[63] Note that in the case of a heavy ion collision between two superfluid nuclei at energies below the Coulomb barrier, one compares gs→gs with gs→qp, summed over a number of qp states (see Sect. 7.3).

relations presented in[64] Sects. 1.4.2 and 4.5 one can write, for $T = 0$,

$$\Delta E_2 = -2 \sum_{kq} |T_{kq}|^2 \frac{|V_k U_q + V_q U_k|^2}{E_k + E_q}. \tag{4.6.2}$$

With the help of

$$2U_k V_k^* = \frac{\Delta_k}{E_k}, \quad 2U_q V_q^* = \frac{\Delta_q}{E_q}, \tag{4.6.3}$$

and

$$|U_k|^2 - |V_k|^2 = \frac{\epsilon_k}{E_k}, \quad |U_q|^2 - |V_q|^2 = \frac{\epsilon_q}{E_q}, \tag{4.6.4}$$

where

$$E = \sqrt{\epsilon^2 + \Delta^2}, \tag{4.6.5}$$

and[65]

$$\Delta_k = \Delta_1 e^{i\phi_1}, \quad \Delta_q = \Delta_2 e^{i\phi_2}, \tag{4.6.6}$$

one can write for the numerator of Eq. (4.6.2),

$$\begin{aligned} NUM &= \left(V_k U_q + V_q U_k\right)\left(V_k^* U_q^* + V_q^* U_k^*\right) \\ &= \left\{V_k^2 U_q^2 + V_q^2 U_k^2\right\} + \left[(U_k^* V_k)(U_q V_q^*) + (U_q^* V_q)(U_k V_k^*)\right]. \end{aligned} \tag{4.6.7}$$

It is of note that, for simplicity, throughout this section

$$V^2 \equiv |V|^2. \tag{4.6.8}$$

With the help of (4.6.3) the expression in the squared bracket in (4.6.7) can be written as

$$[\,] = \frac{1}{4E_k E_q}\left(\Delta_k^* \Delta_q + (\Delta_k^* \Delta_q)^*\right) = \frac{1}{4E_k E_q} 2\Re(\Delta_k^* \Delta_q). \tag{4.6.9}$$

Making use of the relations

$$\begin{aligned} \left(U_k^2 - V_k^2\right)\left(U_q^2 - V_q^2\right) &= U_k^2 U_q^2 - U_k^2 V_q^2 - V_k^2 U_q^2 + V_k^2 V_q^2 \\ &= -\left(U_k^2 V_q^2 + V_k^2 U_q^2\right) + \left(U_k^2 U_q^2 + V_k^2 V_q^2\right), \end{aligned} \tag{4.6.10}$$

and

$$\begin{aligned} 1 &= \left(U_k^2 + V_k^2\right)\left(U_q^2 + V_q^2\right) = U_k^2 U_q^2 + U_k^2 V_q^2 + V_k^2 U_q^2 + V_k^2 V_q^2 \\ &= \left(U_k^2 V_q^2 + V_k^2 U_q^2\right) + \left(U_k^2 U_q^2 + V_k^2 V_q^2\right), \end{aligned} \tag{4.6.11}$$

---

[64] See also App. G of Brink and Broglia (2005).
[65] Note the difference in gauge phasing used in this section to follow that of Anderson (1964b) as compared with that used in the rest of the monograph.

one obtains,

$$1 - \left(U_k^2 - V_k^2\right)\left(U_q^2 - V_q^2\right) = 2\left(U_k^2 V_q^2 + V_k^2 U_q^2\right), \tag{4.6.12}$$

that is, twice the expression written in curly brackets in (4.6.7). Consequently

$$\{ \ \} = \frac{1}{2}\left(1 - \left(U_k^2 - V_k^2\right)\left(U_q^2 - V_q^2\right)\right) = \frac{1}{2}\left(1 - \frac{\epsilon_k \epsilon_q}{E_k E_q}\right). \tag{4.6.13}$$

Thus, the sum of (4.6.9) and (4.6.13) leads to,

$$NUM = \frac{1}{2}\left(1 - \frac{\epsilon_k \epsilon_q}{E_k E_q} + \frac{\Re(\Delta_q^* \Delta_k)}{E_k E_q}\right) \tag{4.6.14}$$

and

$$\Delta E_2 = -\sum_{kq} \frac{|T_{kq}|^2}{E_k + E_q}\left(1 - \frac{\epsilon_k \epsilon_q}{E_k E_q} + \frac{\Re(\Delta_q^* \Delta_k)}{E_k E_q}\right). \tag{4.6.15}$$

In what follows, we concentrate in the gauge phase-dependent term.

With the help of (4.6.6) one can write

$$\Delta_k \Delta_q^* = \Delta_1 \Delta_2 e^{i(\phi_1 - \phi_2)} = \Delta_1 \Delta_2\left(\cos(\phi_1 - \phi_2) + i\sin(\phi_1 - \phi_2)\right). \tag{4.6.16}$$

Thus

$$\Re \Delta_k \Delta_q^* = \Delta_1 \Delta_2 \cos(\phi_1 - \phi_2), \tag{4.6.17}$$

where $\Re$ stands for real part. Making use of

$$\sum_k \rightarrow N_1 \int d\epsilon_1, \quad \sum_q \rightarrow N_2 \int d\epsilon_2 \tag{4.6.18}$$

where N_1 and N_2 are the density of levels of one spin orientation at the Fermi energy one obtains[66]

$$\Delta E_2 \approx -N_1 N_2 \Delta_1 \Delta_2 \langle |T_{kq}|^2 \rangle \cos(\phi_1 - \phi_2) \int_{-\infty}^{\infty}\int_{-\infty}^{\infty} \frac{d\epsilon_1 d\epsilon_2}{E_1 E_2 (E_1 + E_2)}$$

$$\approx -2\pi N_1 N_2 \langle |T_{kq}|^2 \rangle \cos(\phi_1 - \phi_2) \times \left(\pi \frac{\Delta_1 \Delta_2}{\Delta_1 + \Delta_2}\right). \tag{4.6.19}$$

Replacing the phase difference in (4.6.19) by the gauge-invariant phase difference $\gamma = (\phi_1 - \phi_2) - \frac{2e}{\hbar}\int_1^2 \mathbf{A}\cdot d\boldsymbol{\ell}$, where $\mathbf{A}$ is the magnetic vector potential ($\nabla \times \mathbf{A} = \mathbf{B}$), and the line integral is carried across the barrier of thickness

---

[66] It is of notice that ΔE_2 is bilinear in the density of levels (see Potel et al. (2017)).

d ($\int_1^2 d\ell = d$), one can calculate the supercurrent[67] whose maximum value for two identical superconductors is[68]

$$J_0 = \left(\frac{\pi \Delta}{2e}\right) \times \frac{1}{R_b} = \frac{\pi}{4} \frac{V_{equiv}}{R_b} = \frac{\pi}{4} I_N, \qquad (4.6.20)$$

where

$$\frac{1}{R_b} = \frac{4\pi e^2}{\hbar} N^2 \langle |T|^2 \rangle. \qquad (4.6.21)$$

The quantity R_b is the junction's resistance multiplied by the unit area (S, i.e., Ohm $\times S$) when both metals are in the normal state, and $V_{equiv} = 2\Delta/e$ is the minimum voltage at which a normal current I_N, upon the breaking of Cooper pairs, starts to flow.[69] It is, within a factor $\pi/4$, equal to the critical Josephson current J_0. This result is a consequence of $P_2 \approx P_1$ (see Eq. (4.3.1)), and provides experimental validation to the prediction.

As it emerges from Fig. 4.6.1, in which a bias of value $V \gtrsim 2\Delta/e$ is applied to the barrier, transferred Cooper pairs are broken and quasiparticle excitations created –thus the labeling (S-Q)– resulting in the flow of a normal (dissipative) current of carriers $q = e$. For $T = 0$, one is in presence of processes connecting a ground state (S) with ground and excited states (Q). The importance of this fact in connection with the Josephson-like junction transiently formed in heavy ion reactions becomes apparent in Sect. 7.3.

For Pb and Sn, the gaps are 1.4 meV and 0.7 meV, leading to eV_{equiv} (V_{equiv}) $\approx$ 2.8 meV (2.8 mV) and 1.4 meV (1.4 mV), respectively. Assuming R_b to be of the order of $1\Omega \times S$, implies maximum values of the Josephson supercurrent $J = J_0 \sin \gamma$, of the order of $J_0 \approx 2$ mA, as experimentally observed.

It is suggestive that the expression (4.6.19) is formally similar to that of the ion-ion potential acting between two heavy ions in weak contact, namely, at a distance a diffusivity away from th grazing distance r_g. In this case the role of the reduced gap is played by a quantity closely related to the reduced radius of curvature[70]

---

[67] $J_0 = \frac{1}{d} (\delta(\Delta E_2)/\delta \mathbf{A}) =$

$\left(4\pi e^2/\hbar\right) N_a N_A \langle |T|^2 \rangle \times (\pi(\Delta_a \Delta_A/e(\Delta_a + \Delta_A))) \sin\left((\phi_1 - \phi_2) - \frac{2e}{\hbar}\int_1^2 \mathbf{A} \cdot d\ell\right)$, where one has

divided by the thickness of the barrier d, so as to obtain a supercurrent per unit area, i.e., J_0/S. Thus $\langle |T|^2 \rangle$ is assumed to be an energy squared per unit area.

[68] Tinkham (1980), Chapter 6, Eq. (6-4) and subsequent discussion; see also Anderson (1964b), Eq. (11) and following discussion.

[69] Giaever (1973).

[70] Broglia and Winther (2004), p.114, Eq. (40), $U_{aA}^N(r) = -V_0/(1 + \exp(\frac{r - R_0}{a}))$, $V_0 = 16\pi \gamma R_{aA} a$,

$R_0 = R_a + R_A + 0.29$ fm, which for two ^{120}Sn nuclei ($R = 5.883$ fm) leads to $R_0 \approx 12.1$ fm and $V_0 = 83.8$ MeV. For energies somewhat above the Coulomb barrier, the grazing distance (Eq. (25), p. 128) of the above reference is $r_g = r_B - \delta \approx 12.8$ fm ($r_B \approx 13.3$ fm, $\delta \approx 0.5$ fm). Thus $(1 + \exp(\frac{r_g + a - R_0}{a})) \approx 9.26$.

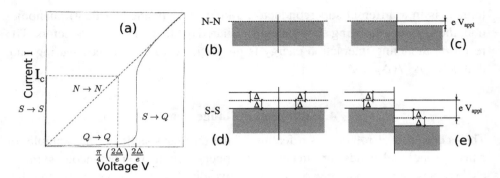

Figure 4.6.1 (**a**) Current-voltage characteristic curves for tunneling between two metals across a barrier (thin oxide layer of 10–30Å). It is assumed that both metals are equal and that they can become superconductors at low temperature (e.g., Al-Al$_2$O$_3$-Al, Sn-SnO$_x$-Sn). (**b, c**) **N-N** indicates the current resulting from the *tunneling of single electrons* between the metals, both of them in the normal phase. **S-Q** indicates the single-electron tunneling between two superconductors (**S-S**) when the barrier is subject to a potential difference (direct current bias) $V \gtrsim 2\Delta/e$, in which case Cooper pairs are broken and quasiparticles (**Q**) become excited. (**d, e**) **S-S** labels the (super) current associated with the *tunneling of Cooper pairs* through the unbiased barrier ($V = 0$, no potential loss), that is, the (dc) Josephson effect. The critical (dc) Josephson current is equal to $\pi/4$ ($\approx$ 80 percent) of the **N-N** normal (single electron carrier) current at the so-called equivalent voltage $V_{equiv} = 2\Delta/e$. (b)–(e); ϵ_F: Fermi energy, dashed horizontal line; occupied levels, gray areas; Δ, pairing gap of metal; V_{appl}, applied (direct current) bias.

$$U_{aA}^N(r_g + a) \sim -\gamma \frac{R_a R_A}{R_a + R_A} a. \qquad (4.6.22)$$

In the above expression $\gamma \approx 0.9$ MeV/fm^2 is the surface tension, $a = 0.63$ fm the diffusivity of the potential, $R_i (=(1.233 A^{1/3} - 0.98 A^{-1/3})$ fm) being the radii of nuclei $i = a, A$. For two identical nuclei $R_a = R_A = R$ and $V_0 = 8\pi\gamma Ra$. In the case in which the interacting nuclei are two ^{120}Sn systems $U_{aA}^N(r_g + a) = U_{aA}^N(13.43$ fm$) \approx -9.1$ MeV.

Being leptodermous systems, nuclei can be described at profit concerning a number of properties, with the help of the liquid drop model. Because at the grazing distance the two leptodermous objects overlap, although weakly, two "unit" areas disappear. To reconstruct them one has to separate the two liquid drops until these areas are formed again. The energy needed to do so has to compensate for the value (4.6.22). Now, this quantity is a measure of the force that must be applied to surface molecules (nucleons) so that they experience the same situation as a molecule does when it is in the interior of the liquid (nucleus).

Similarly, the weak link (4.6.19) between the two superconductors 1 and 2 is associated with the situations in which each partner of the Cooper pair which tunnels is in a different superconductor, a kind of incipient (covalent) phenomenon, the Cooper pair being simultaneously shared by the two superconductors. The resulting tunneling interaction energy is proportional to the reduced pairing gap, that is $(\Delta_a \Delta_A/(\Delta_a + \Delta_A))$.

4.7 Rotation in Gauge Space

The occurrence of rotation as a feature of the nuclear spectrum, for example, of pairing rotational bands, originates in the phenomenon of spontaneous symmetry breaking of rotational invariance in the two-dimensional gauge space. In other words, violation of particle number conservation, which introduces a deformation that makes it possible to specify an orientation of the system in gauge space.

The condensate in the nuclear superfluid system involve a deformation of the field that creates the fermion pairs. The process of addition or removal of a Cooper pair constitutes *a rotational mode in gauge space in which particle number plays the role of angular momentum.* Pairing rotational bands represents the collective mode associated with restoration of particle number conservation. Let us elaborate on the above points within the framework of a simple model which contains the basic physical features one is interested in discussing.

We consider N nucleons moving in a single j-shell[71] of energy ϵ_j and total angular momentum $(ls)j$. The number of pairs moving in time-reversal states which can be accommodated in the shell is $\Omega = (2j + 1)/2$. Consequently, the values of the BCS occupation parameters are $V = (N/2\Omega)^{1/2}$ and $U = (1 - N/2\Omega)^{1/2}$. The solution of the BCS number and gap equations associated with a pairing force of constant matrix elements G are

$$\lambda = -\frac{G}{2}(\Omega - N), \tag{4.7.1}$$

and

$$\Delta = \frac{G}{2}\sqrt{N(2\Omega - N)}, \tag{4.7.2}$$

respectively.

The BCS ground state energy of the superfluid system is

$$U = 2\sum_{\nu>0}(\epsilon_\nu - \lambda)V_\nu^2 - \frac{\Delta^2}{G}. \tag{4.7.3}$$

The fact that the single-particle energies are measured with respect to λ implies that the nucleons feel the Coriolis force ($\hbar\dot{\phi} = \lambda$) associated with rotation in gauge

---

[71] For details, see, e.g., App. H of Brink and Broglia (2005), and refs. therein.

space (Eq. (4.7.11)). In other words, the BCS solution is carried out in the intrinsic system of reference $\mathcal{K}'$.

We now calculate the ground-state energy in the laboratory frame $\mathcal{K}$ (see Fig. 1.4.1),

$$E_0 = U + \lambda N = 2 \sum_{\nu > 0} \epsilon_\nu V_\nu^2 - \frac{\Delta^2}{G}. \tag{4.7.4}$$

Because all m-substates of the single j-shell are degenerate $\epsilon_\nu = \epsilon$. Setting, for simplicity, $\epsilon_\nu = 0$ the above equation can be written as

$$E_0 = -G\frac{\Omega}{2}N + \frac{G}{4}N^2. \tag{4.7.5}$$

Assuming $\Omega \gg N$ one obtains from Eq. (4.7.1) $\lambda \approx -G\Omega/2$, leading to

$$E_0 = \lambda N + \frac{\hbar^2}{2\mathcal{J}}N^2, \tag{4.7.6}$$

where

$$\frac{\hbar^2}{2\mathcal{J}} = \frac{G}{4}, \tag{4.7.7}$$

$\mathcal{J}$ being the moment of inertia of the associated pairing rotational band.[72]

4.7.1 Phase Coherence

The phase of a wavefunction and the number of nucleons (electrons in condensed matter) are conjugate variables. In other words, gauge angle and particle number operators $\hat{\phi}$ and $\hat{N}$ satisfy the commutation relation,

$$\left[\hat{\phi}, \hat{N}\right] = i. \tag{4.7.8}$$

In the particle number representation

$$\hat{N} = -i\partial/\partial\phi, \quad \hat{\phi} = \phi, \tag{4.7.9}$$

while

$$\hat{N} = N, \quad \hat{\phi} = i\partial/\partial N, \tag{4.7.10}$$

in the gauge angle representation. The time derivative of the gauge angle is given by the equation of motion[73],

---

[72] See Bès and Broglia (1966); Broglia et al. (2000), and refs. therein.
[73] See, e.g., Brink and Broglia (2005), App. I.

$$\dot{\phi} = \frac{i}{\hbar} [H, \phi] = \frac{1}{\hbar} \frac{\partial H}{\partial N} = \frac{1}{\hbar} \lambda, \qquad (4.7.11)$$

where λ is the chemical potential (Fermi energy).

Gauge invariance, that is, invariance under phase changes, implies number of particle conservation in a similar way that rotational invariance implies angular momentum conservation.

Example: let us introduce the invariant many-body state[74],

$$|\Psi_N\rangle = a_1^\dagger a_2^\dagger \cdots a_N^\dagger |0\rangle, \qquad (4.7.12)$$

and rotate it an angle ϕ with the help of the operator $\mathcal{G}(\phi) = e^{-i\hat{N}\phi}$. In other words, making use of

$$\mathcal{G}(\phi)\, a_\nu^\dagger\, \mathcal{G}^{-1}(\phi) = e^{-i\phi} a_\nu^\dagger = a_\nu'^\dagger, \qquad (4.7.13)$$

and as a result of the relation $a_\nu^\dagger = e^{i\phi} a_\nu'^\dagger$, one obtains

$$|\Psi_N\rangle = e^{iN\phi}|\Psi_N'\rangle, \qquad (4.7.14)$$

where

$$|\Psi_N'\rangle = a_1'^\dagger a_2'^\dagger \cdots a_N'^\dagger |0\rangle. \qquad (4.7.15)$$

Thus,

$$\hat{N}|\Psi_N\rangle = -i \frac{\partial}{\partial \phi}|\Psi_N\rangle = N|\Psi_N\rangle. \qquad (4.7.16)$$

A phase change for a gauge invariant function is just a trivial operation. Like to rotate a rotational invariant function. Quantum mechanically nothing happens rotating a spherical, symmetry conserving, system (in 3D, gauge, etc.) space.

The situation is different in the case of the wavefunction

$$\begin{aligned}
|BCS(\phi)\rangle_\mathcal{K} &= \prod_{\nu>0} \left(U_\nu + V_\nu a_\nu^\dagger a_{\bar{\nu}}^\dagger \right) |0\rangle, \\
&= \prod_{\nu>0} \left(U_\nu' + e^{-2i\phi} V_\nu' a_\nu^\dagger a_{\bar{\nu}}^\dagger \right) |0\rangle, \\
&= \prod_{\nu>0} \left(U_\nu' + V_\nu' a_\nu'^\dagger a_{\bar{\nu}}'^\dagger \right) |0\rangle, \\
&= |BCS(\phi = 0)\rangle_{\mathcal{K}'}, \qquad (4.7.17)
\end{aligned}$$

where

$$U_\nu = |U_\nu| = U_\nu'; \quad V_\nu = e^{-2i\phi} V_\nu' \ (V_\nu' = |V_\nu|). \qquad (4.7.18)$$

---

[74] Anderson (1964b).

In fact,

$$|BCS(\phi)\rangle_{\mathcal{K}} = \left(\prod_{\nu>0} U'_\nu\right) \sum_{N \text{ even}} \frac{e^{-iN\phi}}{(N/2)!} \left(\sum_{\nu>0} c'_\nu P^\dagger_\nu\right)^{N/2} |0\rangle, \qquad (4.7.19)$$

with

$$c'_\nu = \frac{V'_\nu}{U'_\nu}; \qquad P^\dagger_\nu = a^\dagger_\nu a^\dagger_{\bar\nu} \qquad (4.7.20)$$

is a coherent state (see Eq. (1.4.39)) in particle number which can be written as

$$|BCS(\phi)\rangle_{\mathcal{K}} = \sum_N f_N(\phi)|\Psi_N\rangle. \qquad (4.7.21)$$

The sum in Eq. (4.7.21) has nonvanishing amplitudes for all even number of fermions $N(= 0, 2, 4 \ldots)$ and, equivalently, within BCS condensation for all number of (Cooper) fermion pairs $n(= 0, 1, 2, \ldots)$; see Eq. (1.4.39). Adjusting the parameter λ (Eqs. (1.4.35)–(1.4.38)), the distribution of normalized $|f_N|^2 (\sum_N |f_N|^2 = 1)$ values is peaked at the average particle number value $\bar{N} = 2\sum_{\nu>0} V_\nu^2 (= \frac{V}{(2\pi)^3} \int d\mathbf{k}\, 2V_k^2)$. As in the grand canonical ensemble, the mean square of this distribution $\langle(N-\bar{N})^2\rangle = 4\sum_{\nu>0} U_\nu^2 V_\nu^2 = \frac{V}{(2\pi)^3} \int d\mathbf{k}\, 4U_k^2 V_k^2$, is proportional to the volume, and thus to the number of particles ($V \sim \bar{N}$). As a result, the full width at half maximum of the distribution of $|f_N|$ values, is of the order of $\sqrt{\bar{N}}$. That is, $\delta N_{rms} = \langle(N - \bar{N})^2\rangle^{1/2} = (\langle N^2\rangle - \bar{N}^2)^{1/2} \sim \sqrt{V} \sim \sqrt{\bar{N}}$. Consequently, in the thermodynamic limit ($N \to \infty$), particle number fluctuation become large, but the relative (fractional) fluctuation $\delta N/\bar{N} \sim (\bar{N})^{-1/2}$ approaches zero.[75]

Let us now apply the gauge angle operator to it,

$$\hat{\phi}|BCS(\phi)\rangle_{\mathcal{K}} = \hat{\phi} \sum_N f_N(\phi)|\Psi_N\rangle, \qquad (4.7.22)$$

that is,

$$i\frac{\partial}{\partial N} f_N(\phi) = \phi f_N(\phi). \qquad (4.7.23)$$

Thus,

$$f_N(\phi) \sim e^{-iN\phi}, \qquad (4.7.24)$$

and, making use of Eq. (4.7.14),

$$|BCS(\phi)\rangle_{\mathcal{K}} \sim \sum_N e^{-iN\phi}|\Psi_N\rangle = \sum_N |\Psi'_N\rangle \sim |BCS(\phi = 0)\rangle_{\mathcal{K}'}, \qquad (4.7.25)$$

[75] See Schrieffer (1964); Tinkham (1980); de Gennes (1966).

the state $|BCS(\phi = 0)\rangle_{\mathcal{K}'}$ being aligned in gauge space in which it defines a privileged orientation (z') with respect to the laboratory system.

An isolated nucleus will not remain long in this product type state. Due to the term[76] $(G/4) \left(\sum_{\nu>0} \left(U_\nu^2 + V_\nu^2 \right) \left(\Gamma_\nu^\dagger - \Gamma_\nu \right) \right)^2$ in the residual quasiparticle Hamiltonian it will fluctuate, and decay into a state

$$|N\rangle \sim \int d\phi e^{iN\phi}|BCS(\phi)\rangle_{\mathcal{K}} \sim \sum_{N'} \int_0^{2\pi} d\phi \, e^{-i(N'-N)\phi}|\Psi_{N'}\rangle \sim |\Psi_N\rangle \quad (4.7.26)$$

in keeping with the fact that

$$\int_0^{2\pi} d\phi \, e^{-i(N'-N)\phi} = \begin{cases} \frac{i}{N'-N} e^{-i(N'-N)\phi}\Big|_0^{2\pi} = 0 & (N \neq N'), \\ \\ 2\pi \, \delta(N, N') & (N = N'). \end{cases} \quad (4.7.27)$$

The state $|N\rangle$ is a member of a pairing rotational band.

In the example discussed in connection with Fig. 1.4.5, $|N\rangle$ is one of the ground states of the Sn-isotopes and $N_0 = 68$. Making use of the fact that $E_R = \left(\hbar^2/2\mathfrak{I} \right) (N - N_0)^2$ and that $\delta N \delta \phi \approx 1$, the coherent state (4.7.21) will decay into one of the states $|N\rangle$, likely the one corresponding to $N = \sum_{\nu>0} 2V_\nu^2$, in a time $\hbar/E_R$. The number of Cooper pairs participating in the nuclear condensate in the case of, for example, ^{120}Sn is $\alpha_0' \approx$ 5–6, the number of neutrons being $2\alpha_0' \approx 10$. Because $\delta N \sim \sqrt{N} \sim 3$ ($\delta\phi \sim 0.3$ rad, that is[77] $\delta\phi \sim 0.3/0.017 \approx 17°$), the energy of rotation in gauge space of a state which is defined with an uncertainty of $\delta\phi \approx 0.3$ rad is $E_R \approx 0.092$ MeV $\times (3)^2 \approx 1$ MeV ($\hbar^2/2\mathfrak{I} \approx G/4 \approx 25$ MeV$/(4N_0) \approx 0.092$ MeV, see Fig. 1.4.5). Consequently $\hbar/E_R \approx 10^{-21}$s.

This is also the case for metallic superconductors. In fact, the state (4.7.17) even if prepared in isolation will dissipate because there is a term in the energy of the superconductor depending on N, namely, the electrostatic energy $e^2(N - N_0)^2/2C$, where C is the electrostatic capacity.[78] Because the number of overlapping Cooper pairs contributing to superconductivity lying around the Fermi surface[79] is $\alpha_0' \approx$

---

[76] Within BCS theory of pairing, there are two parameters which determine spontaneous symmetry breaking in gauge space: the probability amplitude with which a pair state ($\nu\bar{\nu}$) is occupied, and that with which it is empty, namely, V_ν and U_ν, respectively. As a consequence, there are only two fields F which contribute, through terms of type $FF^\dagger$, to the residual interaction H_{res} acting among quasiparticles, which is neglected in the mean field solution of the pairing Hamiltonian: one which is antisymmetric with respect to the Fermi surface, namely, $(U_\nu^2 - V_\nu^2)$, and which leads to pairing vibrations of the gauge-deformed state $|BCS\rangle$. The other one, $(U_\nu^2 + V_\nu^2)$, is symmetric with respect to ϵ_F and leads to fluctuations which diverge in the long-wavelength limit ($W_1'' \to 0$) in precisely the right way to set $|BCS\rangle$ into rotation with a finite inertia and restore gauge symmetry (see Eq. 1.4.57).

[77] $\delta\phi \approx 0.3/(\pi/180) \approx 0.3/0.017 \approx 17°$.

[78] The capacity of a sphere is $C = R$, and we use $R = 1$cm in the example discussed (see Anderson (1964b)).

[79] In the case of Pb, $\xi \approx 0.3 \times 10^4$ Å, $r_s = 1.22$ Å, $\Delta \approx 1.4$ meV, and $\epsilon_F \approx 9.47$ eV are the correlations length, the Wigner–Seitz radius, the pairing gap, and the Fermi energy, respectively. Consequently, $(\xi/r_s)^3 \times (\Delta/\epsilon_F) \approx 10^6$.

10^6 and thus the associated number of electrons $2\alpha'_0 \approx 2 \times 10^6$, $\delta N \sim \sqrt{N} \sim 10^3$, one can write $E_{el} = \frac{e^2}{2C}(\delta N)^2 \approx \frac{14.4\,\text{eV} \times 10^{-8}\,\text{cm}}{2 \times 1\,\text{cm}}(10^3)^2 \approx 10^{-7}$ MeV, and $\hbar/E_{el} \approx 10^{-14}$ s. In this case $\delta\phi \sim 10^{-3}$ radians, that is $\delta\phi \sim 10^{-3}/0.017 \sim 0.1°$.

A different situation is found when one considers different parts of the same superconductor. In this case one can define relative variables $n = N_1 - N_2$ and $\phi = \phi_1 - \phi_2$ and again $n = -i\partial/\partial\phi$ and $\phi = i\partial/\partial n$. Thus, locally there is a superposition of different n states: ϕ is fixed so n is uncertain. There is a dividing line between these two behaviors, perfect phase coherence and negligible coherence, namely, the Josephson effect.

Again, the total phase of the assembly is not physical. However, the relative phases can be given a meaning when one observes, as one does in, for example, junctions between metallic superconductors, that pairs of electrons can pass back and forth through the barrier, leading to the possibility of coherence between states in which the total number of electrons is not fixed *locally*. Under such conditions there is, for instance, a coherence between the state with $N/2$ electrons in one of the weak linked superconductors and $N/2$ in the other, and that with $(N/2) + 2$ on one and $(N/2) - 2$ on the other.

4.8 Medium Polarization Effects and Pairing

In many-body systems, medium polarization effects play an important role in renormalizing single-particle motion and the four-point vertices, namely, Coulomb interaction in condensed matter and the bare NN-potential in nuclei. In what follows, special attention is paid to this mechanism in connection with the pairing interaction.

4.8.1 Nuclei

Elementary modes of excitation constitute a basis of states in which correlations, as found in observables, play an important role. As a consequence, it allows for an economic, low-dimensionality solution of the nuclear many-body problem of structure and reaction.

As a result of their interweaving, the variety of elementary modes of excitation may break in a number of states, eventually acquiring a lifetime and, within a coarse grain approximation, a damping width (imaginary component of the self-energy). Moving into the continuum, as for example in the case of direct reactions, one such component is the imaginary part of the optical potential[80] operating in the particular channel selected. It can, in principle, be calculated microscopically using similar techniques and elements as, for example, those used in the calculation

---

[80] See footnote 26 of Chapter 1.

of the damping width of giant resonances or of single-particle states. In this way, the consistency circle structure-reaction based on elementary modes and codified by NFT could be closed. The rich variety of emergent properties found along the way eventually acquiring a conspicuous level of physical validation. At that time it would be possible, arguably if there is one, to posit that the *ultima ratio* of structure and reactions, in any case that associated with pairing and Cooper pair transfer in nuclei, have been unveiled.[81]

Effective Moments

At the basis of the coupling between elementary modes of excitation, for example of single-particle motion and of collective vibrations, one finds the fact that, in describing the nuclear structure it is necessary to make reference to both of them simultaneously and in an unified way.

Within the harmonic approximation the above statement is economically embodied in, for example, the relation existing between the collective $(\hat{\alpha})$ and the single-particle $(\hat{F})$ representation of the operator creating, for example, a particle-hole excitation. That is[82],

$$
\begin{aligned}
\hat{F} &= \left\{ \langle k|F|\tilde{i}\rangle \Gamma_{ki}^{\dagger} + \langle \tilde{i}|F|k\rangle \Gamma_{ki} \right\} \\
&= \sum_{k,i,\alpha'} X_{ki}^{\alpha'} \Gamma_{\alpha'}^{\dagger} - Y_{ki}^{\alpha'} \Gamma_{\alpha'} \\
&= \sum_{\alpha'} \Lambda_{\alpha'} \sum_{ki} \frac{\left|\langle \tilde{i}|F|k\rangle\right|^2 2(\epsilon_i - \epsilon_k)}{(\epsilon_k - \epsilon_i)^2 - (\hbar\omega_{\alpha'})^2} \left(\Gamma_{\alpha'}^{\dagger} + \Gamma_{\alpha'}\right) \\
&= \sum_{\alpha'} \frac{\Lambda_{\alpha'}}{\kappa} \left(\Gamma_{\alpha'}^{\dagger} + \Gamma_{\alpha'}\right) = \sum_{\alpha'} \sqrt{\frac{\hbar\omega_{\alpha'}}{2C_{\alpha}'}} \left(\Gamma_{\alpha'}^{\dagger} + \Gamma_{\alpha'}\right) = \hat{\alpha}.
\end{aligned}
\tag{4.8.1}
$$

This is a consequence of the self-consistency relation

$$
\delta U(r) = \int d\mathbf{r}' \delta\rho(r) v(|\mathbf{r}-\mathbf{r}'|),
\tag{4.8.2}
$$

existing between density (collective) and potential (single-particle) distortion, typical of normal modes of many-body systems. It is of note that the assumption is

---

[81] In the above paragraph we allowed ourselves to paraphrase Jacques Monod writing in connection with biology and life: "L'*ultima ratio* de toutes les structures et performances téléonomiques des êtres vivants est donc enfermée dans les séquences des radicaux des fibres polypeptidiques "embryons" de ces démons de Maxwell biologiques que sont les protéines globulaires. En un sens, très réel, c'est à ce niveau d'organisation chimique qui gît, s'il y en a un, le secret de la vie. Et saurait – on non seullement décrire les séquences, mais énoncer la loi d'assemblage à laquelle obéissent, on pourrait dire que le secret est percé, l'ultima ratio découverte" (Monod (1970)).

[82] Cf. Bohr and Mottelson (1975); cf. also Brink and Broglia (2005), App. C.

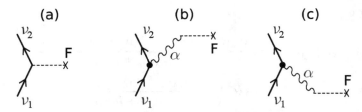

Figure 4.8.1 (a) F-moment of a single-particle and (b, c) renormalization effects induced by the collective vibration α.

made that F is dimensionless while κ (see Eq. (4.8.3)) has energy dimensions (see, e.g., Eqs. (1.2.7) and (1.2.13)).

Relation (4.8.1) implies that at the basis of these normal modes one finds the (attractive $\kappa < 0$) separable interaction

$$H = \frac{\kappa}{2}\hat{F}^{\dagger}\hat{F},\tag{4.8.3}$$

but where now (Fig. 4.8.1 (a))

$$\hat{F} = \sum_{\nu_1,\nu_2}\langle\nu_1|F|\nu_2\rangle a^{\dagger}_{\nu_1} a_{\nu_2},\tag{4.8.4}$$

is a general single-particle operator, while $\hat{F}$ in Eq. (4.8.1) is its harmonic representation acting in the particle (k)–hole (i) space, $\Gamma^{\dagger}_{ki}$ and Γ_{ki} being (quasi) bosons, that is, respecting the commutation relation

$$\left[\Gamma_{ki}, \Gamma^{\dagger}_{k'i'}\right] = \delta(k, k')\delta(i, i').\tag{4.8.5}$$

In other words, the representation (4.8.1), which is at the basis of the RPA (as well as QRPA), does not allow for scattering vertices, processes which become operative by rewriting (4.8.3) in terms of the particle–vibration coupling Hamiltonian

$$H_c = \kappa\hat{\alpha}\hat{F}.\tag{4.8.6}$$

It is of note that κ is negative for an attractive field. Let us now calculate the effective single-particle moments (cf. Fig. 4.8.1 (b)),

$$\langle\nu_2|\hat{F}|\nu_1\rangle_{(b)} = \frac{\langle\nu_2|\hat{F}|\nu_2, n_\alpha = 1\rangle\langle\nu_2, n_\alpha = 1|H_c|\nu_1\rangle}{(\epsilon_{\nu_1} - \epsilon_{\nu_2}) - \hbar\omega_\alpha},$$

$$= \frac{\langle 0|\hat{\alpha}|n_\alpha = 1\rangle\kappa\alpha\langle\nu_2|F|\nu_1\rangle}{(\epsilon_{\nu_1} - \epsilon_{\nu_2}) - \hbar\omega_\alpha},$$

$$= \kappa\alpha^2\frac{\langle\nu_2|F|\nu_1\rangle}{(\epsilon_{\nu_1} - \epsilon_{\nu_2}) - \hbar\omega_\alpha},\tag{4.8.7}$$

and (Fig. 4.8.1 (c))[83]

$$
\langle v_2|\hat{F}|v_1\rangle_{(c)} = \frac{\langle v_2|H_c|v_1, n_\alpha = 1\rangle\langle v_1, n_\alpha = 1|F|v_1\rangle}{\epsilon_{v_2} - (\epsilon_{v_1} + \hbar\omega_\alpha)},
$$

$$
= \kappa\alpha^2\left(-\frac{\langle v_2|F|v_1\rangle}{(\epsilon_{v_1} - \epsilon_{v_2}) + \hbar\omega_\alpha}\right), \tag{4.8.8}
$$

leading to

$$
\langle v_2|\hat{F}|v_1\rangle_{(b)} + \langle v_2|\hat{F}|v_1\rangle_{(c)} = \kappa\alpha^2\frac{2\hbar\omega_\alpha\langle v_2|F|v_1\rangle}{(\epsilon_{v_1} - \epsilon_{v_2})^2 - (\hbar\omega_\alpha)^2},
$$

$$
= \frac{\kappa}{C_\alpha}\frac{(\hbar\omega_\alpha)^2\langle v_2|F|v_1\rangle}{(\epsilon_{v_1} - \epsilon_{v_2})^2 - (\hbar\omega_\alpha)^2}. \tag{4.8.9}
$$

This is in keeping with the fact that the ZPF of the α-vibrational mode is

$$
\alpha = \sqrt{\frac{\hbar\omega_\alpha}{2C_\alpha}}. \tag{4.8.10}
$$

The particle–vibration coupling strength used below is defined as

$$
\Lambda_\alpha = \kappa\alpha. \tag{4.8.11}
$$

Together with $\langle v_2|\hat{F}|v_1\rangle_{(a)} = \langle v_2|F|v_1\rangle$ (see Fig. 4.8.1 (a)) one obtains

$$
\langle v_2|\hat{F}|v_1\rangle = (1 + \chi(\omega))\langle v_2|F|v_1\rangle, \tag{4.8.12}
$$

where

$$
\chi_\alpha(\omega) = \frac{\kappa}{C_\alpha}\frac{\omega_\alpha^2}{\omega^2 - \omega_\alpha^2} \tag{4.8.13}
$$

is the polarizability coefficient, while

$$
\omega = |\epsilon_{v_1} - \epsilon_{v_2}|/\hbar. \tag{4.8.14}
$$

In the static limit, for example, in the case in which α is a giant resonance and $\omega_\alpha \gg \omega$ one obtains

$$
\chi_\alpha(0) = -\frac{\kappa}{C_\alpha}. \tag{4.8.15}
$$

The sign of $\chi_\alpha(0)$ is opposite to that of κ, since the static polarization effect produced by an attractive coupling ($\kappa < 0$) is in phase with the single-particle moment,

---

[83] In calculating the energy denominators, one takes the difference between the energy of the initial and of the intermediate states. However, when an external field like $\hat{F}$ acts on the system, before the PVC or four-point vertices operate, being equivalent to an observation, the energy denominator is to be calculated as the energy difference between the final and the intermediate states.

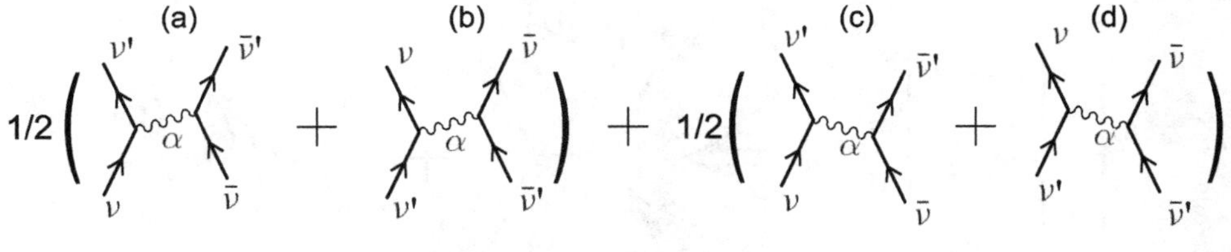

Figure 4.8.2 Diagrams associated with nuclear pairing–induced interaction.

while a repulsive coupling ($\kappa > 0$) implies opposite phases for the polarization effect and the one-particle moment.[84]

Let us now calculate the two-body pairing-induced interaction (Fig. 4.8.2) arising from the exchange of collective vibrations[85], summing over the two time orderings and symmetrizing between initial and final states[86]

$$
\begin{aligned}
v_{\nu\nu'}^{ind}(a) + v_{\nu\nu'}^{ind}(b) &= \kappa^2\alpha^2 \left|\langle \nu'|F|\nu\rangle\right|^2 \left(\frac{1}{\epsilon_\nu - \epsilon_{\nu'} - \hbar\omega_\alpha} + \frac{1}{\epsilon_{\nu'} - \epsilon_\nu - \hbar\omega_\alpha}\right), \\
&= \kappa^2\alpha^2 \left|\langle \nu'|F|\nu\rangle\right|^2 \left(\frac{1}{(\epsilon_\nu - \epsilon_{\nu'}) - \hbar\omega_\alpha} - \frac{1}{(\epsilon_\nu - \epsilon_{\nu'}) + \hbar\omega_\alpha}\right), \\
&= \Lambda_\alpha^2 \left|\langle \nu'|F|\nu\rangle\right|^2 \left(\frac{2\hbar\omega_\alpha}{(\epsilon_\nu - \epsilon_{\nu'})^2 - (\hbar\omega_\alpha)^2}\right), \\
&= v_{\nu\nu'}^{ind}(c) + v_{\nu\nu'}^{ind}(d). \tag{4.8.16}
\end{aligned}
$$

Thus

$$
\begin{aligned}
v_{\nu\nu'}^{ind} &= \frac{1}{2}\left(v_{\nu\nu'}^{ind}(a) + v_{\nu\nu'}^{ind}(b)\right) + \frac{1}{2}\left(v_{\nu\nu'}^{ind}(c) + v_{\nu\nu'}^{ind}(d)\right) \\
&= \Lambda_\alpha^2 \left|\langle \nu'|F|\nu\rangle\right|^2 \left(\frac{2\hbar\omega_\alpha}{(\hbar\omega)^2 - (\hbar\omega_\alpha)^2}\right). \tag{4.8.17}
\end{aligned}
$$

The diagonal matrix element,

$$
v_{\nu\nu}^{ind} \equiv -\frac{2\Lambda_\alpha^2 \left|\langle \nu|F|\nu\rangle\right|^2}{\hbar\omega_\alpha},
$$

testifies to the fact, for values of $\omega_\alpha \gtrsim \omega$, the induced pairing interaction is attractive. Summing to (4.8.17) the matrix element of the bare interaction (4.8.3), (Fig. 4.8.3 (b))[87]

$$
v_{\nu\nu'}^{bare} = \kappa \left|\langle \nu'|F|\nu\rangle\right|^2, \tag{4.8.18}
$$

---

[84] Bohr and Mottelson (1975); Mottelson (1962). Think in the first case about the effect the GQR ($\tau = 0$) plays in $(e(E2))_{eff}$, in the second that played by the GDR ($\tau = 1$) in $(e(E1))_{eff}$.

[85] In the present discussion, we do not consider spin modes. For details, see, e.g., Idini et al. (2015). See also Bortignon et al. (1983).

[86] Cf. Brink and Broglia (2005), p. 217.

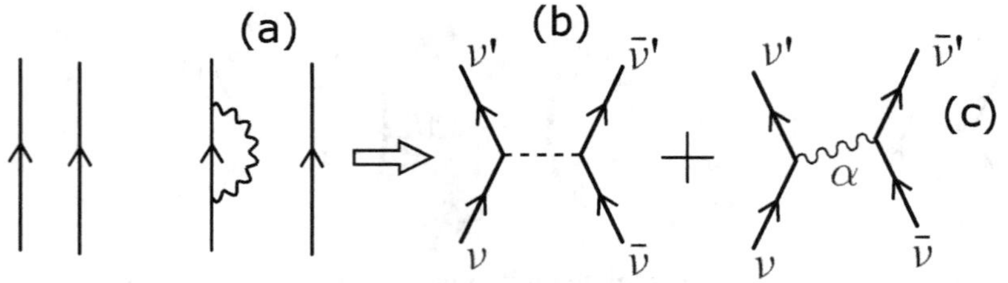

Figure 4.8.3 Starting with two bare nucleons moving around a closed-shell system N_0 in Hartree–Fock orbitals (arrowed lines far left), graphical (NFT) representations of (a) self-energy processes and of (b) bare and (c) induced pairing interactions are displayed.

one obtains for the total pairing matrix element[88]

$$v_{\nu\nu'} = \kappa \left| \langle \nu'|F|\nu \rangle \right|^2 + \frac{\kappa^2}{C_\alpha} \left| \langle \nu'|F|\nu \rangle \right|^2 \frac{\omega_\alpha^2}{\omega^2 - \omega_\alpha^2} = v_{\nu\nu'}^{bare} + \frac{\kappa}{C_\alpha} v_{\nu\nu'}^{bare} \frac{\omega_\alpha^2}{\omega^2 - \omega_\alpha^2}$$
$$= v_{\nu\nu'}^{bare} \left(1 + v_{\nu\nu'}^{bare} \Pi_{\nu\nu'}(\omega, \omega_\alpha) \right). \tag{4.8.19}$$

where

$$\Pi_{\nu,\nu'} = \begin{cases} \left(C_\alpha \left| \langle \nu'|F|\nu \rangle \right|^2 \right)^{-1} \frac{\omega_\alpha^2}{\omega^2 - \omega_\alpha^2}, \\ \left(D_\alpha \left| \langle \nu'|F|\nu \rangle \right|^2 \right)^{-1} \frac{1}{\omega^2 - \omega_\alpha^2}, \end{cases} \tag{4.8.20}$$

both expressions being equivalent in keeping with the fact that $\omega_\alpha = (C_\alpha/D_\alpha)^{1/2}$. In the second expression of $\Pi_{\nu\nu'}$ the inertia of the phonon appears in the denominator, similar to the factor (Z/AM) in (4.8.65). The units of both C_α and $(\hbar^2/D_\alpha)$ are MeV and, as a consequence, $\hbar\omega_\alpha = (\hbar^2 C_\alpha/D_\alpha)^{1/2}$ is also an energy. The units of $\Pi_{\nu\nu'}$ are MeV^{-1} in keeping with the units of $v_{\nu\nu'}^{bare}$.

It is of note that $v_{\nu\nu'}$ can display a resonant behavior leading to attraction regardless whether $v_{\nu\nu'}^{bare}$ is attractive or repulsive ($\lim_{\omega-\omega_\alpha\to 0^-} \Pi_{\nu\nu'} \to -\infty$), a situation of relevance in the case of dielectric polarization effects in metals (see Sects. 4.8.2 and 4.8.3). In that case, the resulting effective electron–electron interaction (phonon exchange) is attractive (see Eq. (4.8.64)), and at the basis of Cooper pair correlation and BCS superconductivity.

---

[87] Within the framework of (4.8.3) and of its role in (4.8.19), one finds, in the case of superconductivity in metals to be discussed below, that the bare unscreened Coulomb interaction can be written as

$$U_c(r) = \frac{1}{2} \sum_{i,j} \frac{q_i q_j}{|\mathbf{r}_i - \mathbf{r}_j|},$$

i, j running over all particles (nuclei and electrons) and $q_i = -e$ for electrons and Ze for nuclei.

[88] It is of note that $\Pi_{\nu\nu'}$ is closely related to Lindhard's function (Lindhard (1954)). See Eq. (4.8.65).

The importance of having available a bare interaction which allows for the calculation of HF and RPA solutions and thus of renormalized single-particle and of effective interactions on equal footing is apparent from Eq. (4.8.19) (see also the last paragraph of Sect. 4.8.3 in connection with pairing in metals).

Let us consider the nucleus ^{11}Li, in which case Cooper pair binding results mainly from the exchange of the low-lying dipole mode between the two halo nuclei. Making use of the bare screened interaction[89] $G_{scr} \approx 0.05G \approx 0.1$ MeV ($G \approx 28/A$ MeV) ($v_{\nu\nu'}^{bare} = -G_{scr} = -0.1$ MeV), and of the induced pairing interaction $v_{\nu\nu'}^{ind}(= M_{ind} \approx -0.6$ MeV), one can write for the last term in (4.8.19), $(0.1$ MeV$)^2 \Pi_{\nu\nu'}(\omega_\alpha, \omega) = -0.6$ MeV leading to $\Pi_{\nu\nu} = -60$ MeV^{-1}. This large effect is not a strictly resonant phenomenon, although the energies $\hbar\omega(= \tilde{\epsilon}_{p_{1/2}} - \tilde{\epsilon}_{s_{1/2}} \approx 0.5$ MeV$)$ and $\hbar\omega_\alpha(\approx 0.7$ MeV$)$ are rather similar, in keeping also with the fact that the pygmy dipole resonance[90] PDR has a width of the order of $\Gamma \approx 0.5$ MeV.

At the basis of the quite small value of the energy centroid of the ^{11}Li soft $E1$-mode, one finds the low-energy dipole particle-hole excitation between the parity inverted levels (0.45 MeV) and the screened symmetry potential ($V_1 = 25$ MeV) which arising from the poor overlap between the core (^{9}Li) and the halo neutron wavefunctions, as in the case of G_{scr} can be assumed, for order of magnitude estimates, to be equal to it (i.e., ≈ 0.05). Thus $(V_1)_{scr} \approx 1.25$ MeV.

One can compare the parameterization[91]

$$\hbar\omega_{GDR} = \hbar\omega_0 \left(1 + \frac{\kappa_1}{C^{(0)}}\right)^{1/2} = \frac{41 \text{ MeV}}{A^{1/3}} \left(1 + \frac{3.5V_1}{41 \text{ MeV}}\right)^{1/2}$$

$$\approx \frac{73}{A^{1/3}} \text{ MeV} \approx 36 \text{ MeV} \quad (A = 11) \tag{4.8.21}$$

used to estimate the value of the centroid of the giant dipole resonance with Eq. (3.6.23), rewritten as

$$\hbar\omega_{pygmy} = 0.45 \text{ MeV} \left(1 + \left(\frac{0.86}{0.45}\right)^2\right)^{1/2} = 0.45 \text{ MeV} \left(1 + \frac{0.6(V_1)_{scr}}{0.2 \text{ MeV}}\right)$$

$$= 1 \text{ MeV}. \tag{4.8.22}$$

Expressed it differently, the screened repulsive symmetry potential can hardly increase the energy of the dipole particle-hole excitation $\tilde{\epsilon}_{p_{1/2}} - \tilde{\epsilon}_{s_{1/2}} \approx 0.45$ MeV.

---

[89] See Broglia et al. (2019b).

[90] Broglia et al. (2019a).

[91] Concerning the first expression of $\hbar\omega_{GDR}$, we refer to Bohr and Mottelson (1975), Eq. (6-27a), while concerning $C^{(0)} = 41A^{-5/3}$ MeV fm^{-2} and $\kappa_1 = \frac{V_1'}{4A\langle x^2\rangle} = \frac{5V_1}{AR^2} = 3.5V_1 A^{-5/3}$ fm^{-2} ($V_1' = 130$ MeV; $V_1 = V_1'/4 = 33$ MeV), we refer to Eqs. (6-313) and (6-314) of this reference.

Let us now carry out a simple estimate of the contribution of the induced pairing interaction to the (empirical) nuclear pairing gap for nuclei lying along the stability valley. For this purpose we introduce the quantity

$$\lambda = N(0)v_{\nu\nu'}^{ind} \tag{4.8.23}$$

where $N(0)$ is the density of levels of a single spin orientation at the Fermi energy. The above quantity is known as the nuclear mass enhancement factor. This is because of the role it plays in the nucleon ω–mass (see Sects. 5.4 and 5.8)

$$m_\omega = (1 + \lambda)m. \tag{4.8.24}$$

Systematic studies of this quantity, and of the related discontinuity displayed by the single-particle occupation number at the Fermi energy, namely, $Z_\omega = (m/m_\omega)$ testifies to the fact that $\lambda \approx 0.4$, within an energy interval of the order of 10 MeV around the Fermi energy.

The BCS expressions of the pairing gap in terms of λ are

$$\Delta = \begin{cases} 2\hbar\omega_D e^{-1/\lambda}, & \text{(weak coupling } \lambda \ll 1) \\ \hbar\omega_D\lambda, & \text{(strong coupling } \lambda \geq 1) \end{cases} \tag{4.8.25}$$

where ω_D is the limiting frequency of the low-lying collective modes of nuclear excitation, typically of quadrupole and octupole vibrations.

In keeping with the fact that collective modes, in particular quadrupole vibrations are found at low energies, we choose $\hbar\omega_D \approx 5$ MeV which, together with $\lambda = 0.4$, intermediate between weak and strong coupling situations leads to,

$$\Delta \approx 0.8 \, \text{MeV} \tag{4.8.26}$$

if one uses the first expression (Eq. (4.8.25)), and to

$$\Delta \approx 2.0 \, \text{MeV}, \tag{4.8.27}$$

if one uses the second. Values to be compared with the empirical value

$$\Delta \approx 1.4 \, \text{MeV}, \tag{4.8.28}$$

of superfluid medium heavy mass nuclei like for example in the Sn-isotopes.

While the relations (4.8.25) can hardly be relied to provide a quantitative number, they testify to the fact that induced pairing is expected to play an important role in nuclei. These expectations have been confirmed by detailed confrontation of theory with experiment.[92]

---

[92] See footnote 55 of Chapter 3.

Hindsight

Static polarization effects can be important in dressing single-particle states. For example, effective charges and induced interactions associated with moments induced by giant resonances.[93] However, retarded ω-dependent self-energy effects and induced interactions associated with low-energy collective modes, are essential in describing structure and reactions of many-body systems. Examples are provided by the bootstrap binding of the halo neutrons (pair addition mode) to ^{9}Li, leading to the fragile $|^{11}Li(gs)\rangle$, displaying a $S_{2n} \approx 0.380$ MeV as compared to typical values of $S_{2n} \approx 16$ MeV as far as structure goes, and by the $^1H(^{11}Li, ^9Li(1/2^-; 2.69\,MeV))^3H$ population of the lowest member of the $(2^+ \times p_{3/2}(\pi))_{J^-}$ multiplet of ^{9}Li with a cross section $\sigma(1/2^-; 2.69\,MeV) \approx 1$ mb, as far as reaction goes. If there was need for support coming from other fields of research, one can mention just two: (1) the van der Waals force in, for example, the folding and the associated folding domain size of globular proteins[94], and (2) the electron–phonon interaction in superconductivity.

The overscreening effect which weakly binds Cooper pairs stems from a delicate ω-dependent interaction which in the case of low-temperature superconductors leads, eventually, to one of the first macroscopic manifestations of quantum mechanics, as, for example, "permanent" magnetic fields associated with persistent supercurrents.

The statement *"Life at the edge of chaos"* coined in connection with the study of emergent properties in biological molecules reflects the idea, expressed by de Gennes[95], that truly important new properties emerge in systems lying at the border between rigid order and randomness. We find physical examples of such systems, in, for example, the marginally stable, single halo Cooper pair nucleus ^{11}Li lying at the dripline, and the similarly marginally stable proteins of viral particles like the HIV-1- and HCV-proteases.[96]

Let us conclude by quoting again de Gennes but doing so with the hindsight of more than twenty years of nuclear research which have elapsed since "Les objets fragiles" was first published. The chapter entitled "Savoir s'arreter,

---

[93] See, e.g., Bohr and Mottelson (1975), Eqs. (6-217) and (6-228).

[94] It is of note that similar arguments (see Sect. 3.6) are at the basis of the estimate (3.6.13) (see also Sect. 3.7) concerning the size of the halo nucleus ^{11}Li, a quantity which is influenced to a large extent by the maximum distance (correlation length) over which partners of a Cooper pair are virtually but also solidly anchored to each other and have to be seen as an extended (quasi) bosonic entity. The fact that Cooper pair transfer proceeds mainly in terms of successive transfer controlled by the single-particle mean field reinforces the above physical picture of nuclear pairing. Even under the effect of extremely large, as compared to the pair correlation energy, external single-particle fields namely, that of target and projectile, the Cooper pair field extends over the two nuclei, permeating the whole summed nuclear volume also through small (normal) density overlap.

[95] De Gennes (1994).

[96] See, e.g., Broglia (2013).

savoir changer" starting at p. 180 opens with the statement[97] "En ce moment, la physique nucléaire (la science des noyaux atomiques) est une science qui, à mon avis, se trouve en fin de parcours.... C'est une physique qui demande des moyens coûteux, et qui s'est constituée par ailleurs en un puissant lobby. Mais elle me semble naturellement exténuée ...je suis tenté de dire: 'Arretons'... mais ce serait aussi absurde que de vouloir arreter un train a grande vitesse. Le mieux serait d'aiguiller ce train sur une autre voie, plus nouvelle et plus utile à la collectivité."

In a way, and even without knowing de Gennes remarks, part of the nuclear physics community have followed them. Capitalizing on the insight that novel embodiments which concepts like elementary modes of excitation, spontaneous symmetry breaking and phase transitions have had in the study of finite quantum many-body systems, of which the nucleus provides a paradigmatic example, one has applied them to other fields of research. In particular to the understanding of protein folding in an attempt to shed light on the possibility of designing leads to drugs (folding inhibitors) which are less prone to elicit resistance[98] finding unexpected connections, let alone the parallels that one has been able to establish and the associated progress resulting from them concerning atomic clusters and quantum dots physics, a particular example being the discovery of super shells.[99]

4.8.2 Metals

Plasmons and Phonons (Jellium Model)

The expression of the electron plasmon (**ep**) frequency of the antenna-like oscillations of the free, conduction electrons of mass m_e and charge $-e$, against the positive charged background (jellium model)[100] is

$$\omega_{ep}^2 = \frac{4\pi n_e e^2}{m_e} = \frac{3e^2}{m_e r_s^3}, \qquad (4.8.29)$$

where

$$n_e = \frac{3}{4\pi} \frac{1}{r_s^3}, \qquad (4.8.30)$$

[97] "In this moment, nuclear physics (the science of atomic nuclei) is a discipline which, in my opinion, has reached its final destination .... It is economically demanding, and has given rise to a powerful lobby. But I think it is naturally exhausted .... I am tempted to say: 'let's stop' ...but it would be as absurd as trying to stop a train traveling at full speed. It would be better to switch it to a different, newer, and more useful track."

[98] See, e.g., Broglia (2005); Rösner et al. (2017), and refs. therein. Within this connection see also Bergasa-Caceres and Rabitz (2020), and refs. therein, as well as Senior, et al. (2020); see also Alva et al. (2015), Service (2020), and Broglia and Tiana (2001).

[99] Pedersen et al. (1991); de Heer et al. (1987); Brack (1993); Pacheco et al. (1991); Lipparini (2003); Martin et al. (1994); Bjørnholm et al. (1994).

[100] Model in which the positive charge of the lattice ions is assumed to be uniformly distributed throughout the solid.

are the number of electrons per unit volume[101], r_s being

$$r_s = \left(\frac{3}{4\pi n_e}\right)^{1/3}, \tag{4.8.31}$$

that is, the radius of the Wigner–Seitz cell.

For[102] metallic Li

$$n_e = 4.70\frac{10^{22}}{\text{cm}^3} = \frac{4.7 \times 10^{-2}}{\text{Å}^3}, \tag{4.8.32}$$

while

$$r_s = \left(\frac{3\text{Å}^3}{4\pi \times 4.7 \times 10^{-2}}\right)^{1/3} = 1.72\text{Å}, \tag{4.8.33}$$

implying a value $(r_s/a_0) = 3.25$ in units of Bohr radius ($a_0 = 0.529\text{Å}$). Making use of

$$\alpha = 7.2973 \times 10^{-3} = \frac{e^2}{\hbar c} \tag{4.8.34}$$

and

$$e^2 = 14.4\,\text{eV Å}, \tag{4.8.35}$$

one obtains

$$\hbar c = \frac{14.4\,\text{eV Å}}{7.2973 \times 10^{-3}} = 1973.3\,\text{eV Å}. \tag{4.8.36}$$

With the help of the above values and of

$$m_e c^2 = 0.511\,\text{MeV}, \tag{4.8.37}$$

one can write

$$\hbar^2\omega_{ep}^2 = \frac{(\hbar c)^2}{m_e c^2}\frac{3e^2}{r_s^3} = \frac{(1973.3\,\text{eV Å})}{0.511 \times 10^6\,\text{eV}}\frac{3 \times 14.4\,\text{eV Å}}{(1.72\,\text{Å})^3} = 64.7\,\text{eV}^2 \tag{4.8.38}$$

leading to[103]

$$\hbar\omega_{ep} = 8.04\,\text{eV} \approx 1.94 \times 10^9\,\text{MHz} \tag{4.8.39}$$

For the case of metal clusters of Li, the Mie resonance frequency is[104]

$$\hbar\omega_M = \frac{\hbar\omega_{ep}}{\sqrt{3}} = 4.6\,\text{eV}. \tag{4.8.40}$$

[101] In the nuclear case, one can write $\rho_n = 1/(4\pi r_0^3/3) = 0.14\,\text{fm}^{-3}$ ($r_0 = 1.2$ fm), to be compared with the saturation density $\rho_0 = 0.17\,\text{fm}^{-3}$.

[102] Cf. page 5, Table 1.1 of Ashcroft and Mermin (1987).

[103] Kittel (1996), Table 2, p. 278.

[104] See Bertsch and Broglia (2005), Sect. 5.1, also Table 5.1.

4.8.3 Elementary Theory of Phonon Dispersion Relation

Within the framework of the jellium model, one can estimate the long-wavelength ionic plasma (**ip**) frequency introducing in (4.8.29) the substitution $e \rightarrow Ze$, $m_e \rightarrow AM$ ($A = N + Z$, mass number, M nucleon mass), $n_e \rightarrow n_i = n_e/Z$,

$$\omega_{ip}^2 = \frac{4\pi n_i (Ze)^2}{AM} = \frac{Zm_e}{AM} \omega_{ep}^2, \tag{4.8.41}$$

$AM(Ze)$ being the mass (charge) of the ions.[105] For metallic Li, one obtains

$$\hbar\omega_{ip} = \left(\frac{Zm_e}{AM}\right)^{1/2} \hbar\omega_{ep} = \left(\frac{3 \times 0.5}{9 \times 10^3}\right)^{1/2} \times 1.94 \times 10^{15} \, \text{sec}^{-1}$$
$$\approx 2.5 \times 10^{13} \, \text{sec}^{-1} \approx 10^{13} \, \text{sec}^{-1} \approx 1.04 \times 10^2 \, \text{meV}. \tag{4.8.42}$$

Now, both the relations (4.8.29) and (4.8.41), although being quite useful, are wrong from a many-body point of view: ω_{ep} because electrons appear as bare electrons not dressed by the phonons, neither by the plasmons; ω_{ip} because the static negative background does not allow for an exchange of electron plasmons between ions, exchange eventually leading to a screened, short-range ionic Coulomb repulsive field. Namely, ions interact, in the approximation used above, in terms of the "bare" ion–ion Coulomb interaction. Being it infinite range it does not allow for a dispersion relation linear in k at long wavelengths (sound waves) but forces a finite "mass" also to the lattice phonons. Allowing for electron screening of the "bare" ion–ion Coulomb interaction, as embodied in the Thomas–Fermi electron gas dielectric function $1/\epsilon_{TF} = q^2/(q^2 + k_s^2)$, one obtains the *dressed phonon frequency*

$$\omega_q^2 = \frac{\omega_{ip}^2}{\epsilon_{TF}} = \frac{Zm_e}{AM} \frac{\omega_{ep}^2}{q^2 + k_s^2} q^2 = \frac{\omega_{ip}^2}{q^2 + k_s^2} q^2. \tag{4.8.43}$$

The quantity

$$k_s = \left(\frac{6\pi n_e e^2}{\epsilon_F}\right)^{1/2} = \left(\frac{4k_F}{\pi a_0}\right)^{1/2} = 0.82 k_F \left(\frac{r_s}{a_0}\right)^{1/2}, \tag{4.8.44}$$

is the Thomas–Fermi screening wave vector, a quantity which is of the order of the Fermi momentum, the associated screening length being then of the order of the Wigner–Seitz radius. In writing the above relations use has been made of

$$k_F = \left(\frac{9\pi}{4}\right)^{1/3} \frac{1}{r_s} = \frac{1.92}{r_s}, \tag{4.8.45}$$

---

[105] Ketterson and Song (1999), p. 230.

and

$$\epsilon_F = \frac{e^2 a_0}{2} k_F^2 = \frac{50.1}{(r_s/a_0)^2}. \tag{4.8.46}$$

In the case of metallic Li ($k_F = 1.12 \text{ Å}^{-1}$, $r_s = 1.72 \text{ Å}$),

$$k_s = 1.64 \text{ Å}^{-1}. \tag{4.8.47}$$

and

$$\epsilon_F = 4.74 \text{ eV} \quad (r_s/a_0 = 3.25). \tag{4.8.48}$$

Let us return to (4.8.43), and take the long-wavelength limit of ω_q. One finds

$$\lim_{q \to 0} \omega_q = c_s q \tag{4.8.49}$$

the sound velocity (squared) being

$$c_s^2 = \frac{Z m_e}{AM} \frac{4\pi n_e e^2}{m_e} \frac{\epsilon_F}{6\pi n_e e^2} = \frac{2Z}{3AM} \epsilon_F = \frac{Z m_e}{3AM} v_F^2, \tag{4.8.50}$$

With the help of (4.8.45), of the velocity of light and of the proton mass,

$$c = 3 \times 10^{10} \text{ cm/sec}; \quad Mc^2 = 938.87 \text{ MeV}, \tag{4.8.51}$$

one obtains,

$$
\begin{aligned}
v_F = \left(\frac{\hbar}{m_e}\right) k_F &= \left(\frac{\hbar c}{m_e c^2}\right) \times 3 \times 10^{10} \frac{\text{cm}}{\text{sec}} \frac{1.92}{r_s} \\
&= \left(\frac{1973.3 \text{ ÅeV}}{0.511 \times 10^6 \text{ eV}}\right) \times 3 \times 10^{10} \frac{\text{cm}}{\text{sec}} \frac{1.92}{1.72 \text{ Å}} \approx 1.29 \times 10^8 \frac{\text{cm}}{\text{sec}},
\end{aligned}
\tag{4.8.52}
$$

and,

$$c_s^2 = \frac{1}{3} \frac{3 m_e}{9 M} v_F^2 \approx 6 \times 10^{-5} v_F^2, \tag{4.8.53}$$

that is,

$$c_s \approx 7.8 \times 10^{-3} v_F \approx 1.0 \times 10^6 \frac{\text{cm}}{\text{sec}}, \tag{4.8.54}$$

approximately equal to a hundredth of the Fermi velocity.[106]

Let us now discuss the effective electron–electron interaction. Within the jellium model used above, one can write it as

$$V(\mathbf{q}, \omega) = \frac{U_c(q)}{\epsilon(\mathbf{q}, \omega)}, \tag{4.8.55}$$

[106] Ashcroft and Mermin (1987), p. 51; Ketterson and Song (1999), p. 234.

where the dielectric function

$$\frac{1}{\epsilon(\mathbf{q}, \omega)} = \frac{\omega^2 q^2}{\omega^2(q^2 + k_s^2) - \omega_{ip}^2 q^2} \tag{4.8.56}$$

contains the effects due to both the ions and the background electrons, while

$$U_c(q) = \frac{4\pi e^2}{q^2} \tag{4.8.57}$$

is the Fourier transform of the bare Coulomb interaction[107]

$$U_c(r) = \frac{e^2}{r}. \tag{4.8.58}$$

For $\omega \gg \omega_{ip}$, $\epsilon(\mathbf{q}, \omega)^{-1} = q^2/(q^2 + k_s^2) = \epsilon_{TF}^{-1}$, and one obtains the screened Coulomb field,

$$U_c^{scr}(q) = \frac{U_c(q)}{\epsilon_{TF}(q)} = \frac{4\pi e^2}{q^2 + k_s^2}. \tag{4.8.59}$$

Its $\mathbf{r}$ space Fourier transform is

$$U_c^{scr}(r) = \frac{e^2}{r} e^{-k_s r}. \tag{4.8.60}$$

A quantity that for large values of r falls off exponentially. Thus, in the high-frequency limit, the electron–electron interaction, although strongly renormalized by the exchange of plasmons, as testified by the fact that (e.g., for Li),

$$U_c^{scr}(r = 5 \,\text{Å}) \approx U_c(r = 5 \,\text{Å}) e^{-1.6 \times 5} \approx 1 \text{meV}, \tag{4.8.61}$$

as compared to $U_c(r = 5 \,\text{Å}) \approx 2.9$ eV, is still repulsive.

Let us now consider frequencies $\omega \ll \omega_{ip}$ but for values of q of the order of a^{-1}, where a is the lattice constant ($a \approx 3 - 5\text{Å}$, $a^{-1} \approx 0.25\text{Å}^{-1}$) to be compared to $k_s \approx 1.6\text{Å}^{-1}$ and $k_F \approx 1.12\text{Å}^{-1}$ (metallic Li). In the case in which $\omega_{ip}^2/\omega^2 > (q^2 + k_s^2)/q^2$, V is attractive (see Eqs. (4.8.55) and (4.8.56)). This behavior explicitly involves the ions through ω_{ip} (electron–phonon coupling).

The dispersion relation of the associated frequency collective modes follows from

$$\epsilon(\mathbf{q}, \omega) = 0. \tag{4.8.62}$$

In other words, making use of Eq. (4.8.56) one obtains the relation (4.8.43). One can now rewrite the reciprocal of the dielectric functions in terms of ω_q, that is,

---

[107] The units being $[U_c(r)] = $ eV, $[U_c(q)] = $ eV$\times$ Å^3.

$$\frac{1}{\epsilon(\mathbf{q}, \omega)} = \frac{\omega^2 q^2}{\omega^2(q^2 + k_s^2) - \omega_{ip}^2 q^2} = \frac{q^2}{(q^2 + k_s^2)} \frac{\omega^2}{(\omega^2 - \omega_q^2)}$$

$$= \frac{q^2}{q^2 + k_s^2} \left[1 + \frac{\omega_q^2}{\omega^2 - \omega_q^2} \right]. \tag{4.8.63}$$

For $\omega \gg \omega_q$ one recovers the Thomas–Fermi dielectric function. For ω near, but smaller than ω_q the interaction (4.8.55) eventually becomes attractive.[108] The effective electron–electron interaction can be written as

$$V(\mathbf{q}, \omega) = \frac{4\pi e^2}{q^2 + k_s^2} + \frac{4\pi e^2}{q^2 + k_s^2} \frac{\omega_q^2}{\omega^2 - \omega_q^2}$$

$$= U_c^{scr}(q) + U_c^{scr}(q) \frac{\omega_q^2}{\omega^2 - \omega_q^2}$$

$$= U_c^{scr}(q) \left(1 + U_c^{scr}(q) \Pi(\mathbf{q}, \omega) \right), \tag{4.8.64}$$

where[109]

$$\Pi(\mathbf{q}, \omega) = \left(\frac{Z}{AM} \right) \frac{n_e q^2}{\omega^2 - \omega_q^2}. \tag{4.8.65}$$

The units of $\Pi(\mathbf{q}, \omega)$ are $\text{eV}^{-1}\text{Å}^{-3}$, in keeping with those of $U_c^{scr}(q)$.

In working out the last expression Eqs. (4.8.43) and (4.8.29) have been used. In other words $\omega_q^2 = (Zm_e/AM)(\omega_{ep}^2/(q^2+k_s^2))q^2 = (Z/AM)(4\pi n_e e^2/(q^2+k_s^2))q^2$. Of note is the close relation of (4.8.64) with the expression (4.8.19) of the nuclear renormalized pairing interaction. The first term of $V(\mathbf{q}, \omega)$ contains the screened Coulomb field arising ultimately from the exchange of plasmons between electrons (see Fig. 4.8.4).[110] The second term with the exchange of collective low-frequency phonons calculated making use of the same screened interaction.[111]

Concerning the parallels discussed above between pairing in nuclei and in metals, it is also important to point out the important differences. At least concerning the possibility of developing a unified theoretical working tool.

In metals, the bare interaction emerges from the exchange of photons, a process which can be described at profit in terms of an instantaneous Coulomb interaction.

---

[108] Schrieffer (1964), Fig. 6–11, p. 152.

[109] It is of note that $\omega_q^2 = (Zm_e/AM)\,\omega_{ep}^2 \times q^2/(q^2 + k_s^2) = (Z/AM)\,4\pi e^2 n_e q^2/(q^2 + k_s^2) = U_c^{scr}(q) \times (Z/AM) n_e q^2$, where the relations (4.8.43) and (4.8.29) have been used.

[110] Within this context, one can, apparently, note an important difference, namely that in the nuclear case, v^{bare} appears, while in the case of condensed matter, U_c^{scr} instead of U_c is present (see also footnote 87 of this chapter). This is because, in the nuclear case, it is not standard praxis to explicitly indicate the renormalization associated with the exchange (nor the emission and absorption by the same nucleon) of giant resonances (see Fig. 4.8.4). Nonetheless, the corresponding effects are taken into account, e.g., in the γ-decay of neutron states (effective charge).

[111] Lindhard (1954, no. 8).

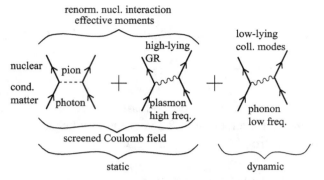

Figure 4.8.4   Schematic representation of the variety of contributions to the effective interaction in nuclei and in metals.

While major progress have been made concerning the bare nucleon interaction at large, and the NN-pairing interaction in particular, taking also into account three-body processes and carrying out ab initio calculations[112], we are not yet in possess of the equivalent to the Coulomb interaction. Let alone of a so-called low-k version of such interaction, which could allow to work out on equal footing Hartree-Fock-Boguliubov solutions, and QRPA microscopic calculations of low-lying collective modes in a variety of channels (density, spin, charge-exchange, etc.).

Concerning these phonons, they are built out of the same nucleon degrees of freedom which already exhaust the nuclear phase space. Double counting and Pauli principle violations have to be taken care of in using the collective modes as intermediate bosons. In metals, phonons are associated with lattice vibrations, that is degrees of freedom different from the electronic ones.

On the other hand, in nuclei one can describe in terms of individual quantal states and of single Cooper pairs, if not identical, rather similar phenomena as those leading to some of the most remarkable and technically transferable quantum phenomena. In particular, the Josephson effect. This fact makes the nuclear pairing paradigm a unique laboratory of low-temperature (finite) many-body physics.

We conclude this section with a technical remark. Making use of (4.8.64) we calculate the mass enhancement factor[113]

$$\lambda = N(0)V(\mathbf{q}, \omega) = N(0)U_c^{scr}\left(1 + U_c^{scr}\Pi\right). \tag{4.8.66}$$

In the weak coupling limit ($\lambda^2 \ll \lambda$)

$$\Delta = 2\omega_D e^{-1/\lambda}, \tag{4.8.67}$$

---

[112] See footnote 24 of Chapter 1.

[113] It is assumed that in this calculation, we have multiplied (4.8.64) by n_e, as well as shifted the corresponding factor appearing in $\Pi(\mathbf{q}, \omega)$ to $U_c^{scr}(q)$. Schematically, it is assumed

$$n_e V = (n_e U_c^{scr})\left(1 + (n_e U_c^{scr})\left(\frac{Z}{AM}\right)\frac{q^2}{\omega^2 - \omega_q^2}\right).$$

where ω_D is the Debye energy. Provided that one considers a situation in which ω is consistently different from ω_q,

$$\frac{1}{\lambda} = \frac{1}{N(0)U_c^{scr}\left(1 + U_c^{scr}\Pi\right)} \approx \frac{1}{N(0)U_c^{scr}}\left(1 - U_c^{scr}\Pi\right), \qquad (4.8.68)$$

Thus

$$\frac{1}{\lambda} = \frac{1}{N(0)U_c^{scr}} - \frac{\Pi}{N(0)}, \qquad (4.8.69)$$

and

$$\Delta = \left(2\omega_D e^{\frac{\Pi}{N(0)}}\right) e^{-\frac{1}{N(0)U_c^{scr}}}. \qquad (4.8.70)$$

Consequently, the renormalization effects of the pairing gap associated with phonon exchange are independent of the approximation used to calculate U_c^{scr} (Thomas–Fermi in the above discussion), provided one has used the same "bare" (screened) Coulomb interaction to calculate ω_q^2. Otherwise, the error introduced through a resonant renormalization process entering the expression of, for example, the pairing gap, could be consistent.

4.8.4 Pairing Condensation (Correlation) Energy

The condensation energy, namely, the energy difference $W_N - W_S$ between the normal N- and superfluid S-state is defined as (Eq. (2-35) of reference[114]):

$$W_{con} = W_N - W_S = \frac{1}{2}N(0)\Delta_0^2, \qquad (4.8.71)$$

where $N(0)$ is the density of single-electron states of one-spin orientation evaluated at the Fermi surface, and Δ_0 is the pairing gap at $T = 0$.

The correlation energy E_{corr} introduced in equation (6-618) of[115]

$$E_{corr} = -\frac{1}{2d}\Delta^2 \qquad (4.8.72)$$

to represent $W_S - W_N$ in the nuclear case, was calculated making use of a (single particle) spectrum of twofold degenerate (Kramer degeneracy) equally spaced (spacing d) single-particle levels. Consequently, $2/d$ corresponds to the total level density, and $1/d = N(0)$. In keeping with the fact that a nucleus in the ground state (or in any single quantal state), is at zero temperature, Eq. (4.8.72) coincides with (4.8.71), taking into account the difference in sign in the definitions.

Nuclei

The empirical value of the level density parameter for both states $(\nu, \bar{\nu})$ (Kramers degeneracy, both spin orientations) is $a \approx A/10\ \mathrm{MeV}^{-1}$, $A = N + Z$ being the

[114] Schrieffer (1964).
[115] Bohr and Mottelson (1975).

Table 4.8.1 *Summary of the quantities entering the calculation of the condensation energy of superconducting lead (Pb) and of the single open-shell superfluid nucleus ^{120}Sn.*

System	Δ_0		$N(0)$		W_{con}		E_{cohe}		BE/A	$\frac{W_{con}}{E_{cohe}}$	$\frac{W_{con}}{BE}$
	meV	MeV	$\frac{\text{meV}^{-1}}{\text{atom}}$	MeV^{-1}	$\frac{\text{meV}}{\text{atom}}$	MeV	$\frac{\text{meV}}{\text{atom}}$	MeV	$\frac{\text{MeV}}{A}$	10^{-7}	10^{-3}
Pb ^{120}Sn	1.4	1.46	0.276	4	3×10^{-4}	4.0	2030		8.5		

mass number. Thus, for neutrons one can write $a_N \approx N/10$ and $N_N(0) \approx N/20$ MeV^{-1}. For $^{120}_{50}$Sn$_{70}$, $N_N(0) \approx 4$ MeV^{-1}. Because[116] $\Delta = 1.46$ MeV, (Table 4.8.1)

$$W_{con} = \frac{1}{2} \times 4 \text{ MeV}^{-1} \times (1.46)^2 \text{ MeV}^2 \approx 4 \text{ MeV}. \tag{4.8.73}$$

The binding energy per nucleon is $BE/A = 8.504$ MeV. Thus $BE = 120 \times 8.504$ MeV $= 1.02 \times 10^3$ MeV, and

$$\frac{W_{con}}{BE} \approx 4 \times 10^{-3}. \tag{4.8.74}$$

Superconducting Lead

Making use of the value[117]

$$N(0) = \frac{0.276 \text{ eV}^{-1}}{\text{atom}}, \tag{4.8.75}$$

and of $\Delta_0 = 1.4$ meV, one obtains

$$W_{con} = 0.27 \times 10^{-6} \text{eV/atom}. \tag{4.8.76}$$

In keeping with the fact that the cohesive energy of lead, namely, the energy required to break all the bonds associated with one of its atoms is

$$E_{cohe} = 2.03 \frac{\text{eV}}{atom}, \tag{4.8.77}$$

[116] Idini et al. (2015).
[117] Beck and Claus (1970).

one obtains

$$\frac{W_{con}}{E_{cohe}} \approx 1.3 \times 10^{-7}. \tag{4.8.78}$$

The different quantities are summarized in Table 4.8.1.

4.9 Cooper Pair: Radial Dependence

The wavefunction of a Cooper pair can be written as

$$\Psi(\mathbf{r}_1\sigma_1; \mathbf{r}_2\sigma_1) = \phi_q(\mathbf{r})e^{i\mathbf{q}\cdot\mathbf{R}}\chi(\sigma_1, \sigma_2) \tag{4.9.1}$$

where $\mathbf{R} = (\mathbf{r}_1 + \mathbf{r}_2)/2$, $\mathbf{r} = \mathbf{r}_1 - \mathbf{r}_2$, and σ_1 and σ_2 denote the spins.[118]

Let us consider the state with zero center of mass momentum ($q = 0$) and with zero spin, so that the two electrons carry equal and opposite momenta, aside of being in the singlet spin state, with

$$\chi = \frac{1}{\sqrt{2}}\left[\begin{pmatrix} 1 \\ 0 \end{pmatrix}\begin{pmatrix} 0 \\ 1 \end{pmatrix} - \begin{pmatrix} 0 \\ 1 \end{pmatrix}\begin{pmatrix} 1 \\ 0 \end{pmatrix}\right]. \tag{4.9.2}$$

We have thus a pair of electrons moving in time-reversal states and can write[119]

$$\phi_0(\mathbf{r}) = \sum_{k>k_F} g(\mathbf{k})e^{i\mathbf{k}\cdot\mathbf{r}} = \sum_{k>k_F} g(\mathbf{k})e^{i\mathbf{k}\cdot\mathbf{r}_1}\, e^{-i\mathbf{k}\cdot\mathbf{r}_2}. \tag{4.9.3}$$

In the above wavefunction Pauli principle ($k > k_F$) and translational invariance (dependence on the relative coordinate $\mathbf{r}$) are apparent. Within this scenario, and in keeping with the fact that $|v_p^{eff}|/\epsilon_F \ll 1$, where $v_p^{eff} = v_p^{bare} + v_p^{ind}$, $\phi_0(\mathbf{r})$ con- *sists mainly of waves of wavenumber k_F*. Because the wavefunction of a Cooper pair represents a bound s-state, the motion it describes is a periodic back and forth movement of the two electrons in directions which are uniformly distributed, covering a relative distance $\approx \xi$, as schematically[120] shown in Fig. 4.9.1 (**i**). It is analogous to the motion of the two nucleons in a deuteron or the main ($L = 0$) component of the two neutrons in the triton. The hydrogen atom in s-state is also an example; in that case it is the electron that does most of the back and forth moving, whereas the proton only recoils slightly.

In keeping with the above arguments, $\phi_0(\mathbf{r})$ will look like $e^{i\mathbf{k}_F\cdot\mathbf{r}}$ for $r \ll \xi$, while for $r \gg \xi$ the waves $e^{i\mathbf{k}\cdot\mathbf{r}}$ weighted by $g(\mathbf{k})$ will destroy themselves by interference

---

[118] In the limit $q \to 0$ the relative coordinate problem is spherically symmetric so that $\phi_0(\mathbf{r})$ is an eigenfunction of the angular momentum operator (Schrieffer (1964)).

[119] In other words, one expands the $l = 0$ wavefunction ϕ_0 in terms of s-states of relative momentum k and total momentum zero.

[120] Weisskopf (1981).

(Fig. 4.9.2). In other words, $\phi_0(\mathbf{r})$ will look like $e^{i\mathbf{k}_F \cdot \mathbf{r}}$ for $r \ll \xi$ while for $r \gtrsim \xi$ one can approximate the weighting function as,

$$g(\mathbf{k}) \sim \delta(\mathbf{k}, \mathbf{k}_F + i\hat{\mathbf{k}}_F/\xi), \tag{4.9.4}$$

where $\hat{\mathbf{k}}_F$ is a unit vector. One then obtains (see also Eq. (4.3.2)),

$$\phi_0(\mathbf{r}) \sim e^{-r/\xi} e^{ik_F r}. \tag{4.9.5}$$

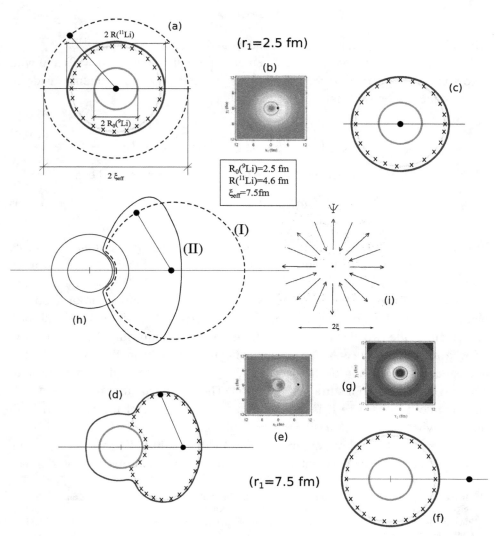

Figure 4.9.1 Aspects of the halo Cooper pair of the nucleus ^{11}Li. For the color version of this figure, refer to cambridge.org/nuclearcooperpair.

Because we are dealing with a singlet state, and the total wavefunction has to be antisymmetric,

$$\phi_0(\mathbf{r}) \sim e^{-r/\xi} \cos k_F r, \tag{4.9.6}$$

A more proper solution of the Cooper pair problem leads to[121]

$$\phi_0(\mathbf{r}) \sim K_0(r/\pi\xi) \cos k_F r, \tag{4.9.7}$$

where K_0 is the zeroth-order modified Bessel function. For $x \gg 0$, $K_0(x) \sim (\pi/2x)^{1/2} \exp(-x)$, where $x = r/\pi\xi$.

Figure 4.9.1 (Continued)
Synthesis of the spatial structure of ^{11}Li neutron halo Cooper pair calculated in NFT (Barranco et al. (2001)). We use, for illustrative purposes $\xi_{eff} = 7.5$ fm, as an *effective Cooper pair mean square radius*[122](dashed circle in (a) and (h)). Diagrams (a) and (d) are the schematic representations of the modulus square $|\Psi_0(\mathbf{r}_1, \mathbf{r}_2)|^2 = |\langle\mathbf{r}_1, \mathbf{r}_2|0\rangle|^2$ describing the motion of the two halo neutrons of ^{11}Li, moving around the ^{9}Li core as a function of the Cartesian coordinates of neutron 2, for fixed (small and large as compared to $R(^{11}\text{Li}) \approx 4.6$ fm) values of the position r_1 of neutron 1 (for more details see Caption to Fig. 3.6.2). Diagrams (b), (e), and (g) are the results of NFT (see also Fig. 3.6.2 (II) a) and b)). **(a)** The circles drawn with a continuous line and with crosses (x) correspond to the relative distance r at the radius of the ^{9}Li core and of that of ^{11}Li, respectively. The area between these two circles corresponds to the light gray area of (b). The Cooper pair "intrinsic coordinate" r_{12} is also shown. Particle 1 of the Cooper pair is assumed to occupy the center of the nucleus ($r_1 = 0$). **(b)** Result of NFT for a situation similar to the above but for $r_1 = 2.5$ fm. **(c)** Schematic representation of an uncorrelated pair in a potential weakly binding the pure configuration $p_{1/2}^2(0)(r_1 = 0)$. **(d)** Same as (a) but for $r_1 = 7.5$ fm (note however that the dashed curve associated with ξ_{eff} is not displayed here). The area within the closed curve drawn with crosses (x) corresponds to the light gray area in (e). **(e)** Result of the NFT calculation for a similar setup ($r_1 = 7.5$ fm). **(f)** Schematic representation of a pure configuration $p_{1/2}^2(0)(r_1 = 7.5$ fm). The area between the continuous curve and the curve drawn with crosses (x) corresponds to the light gray area of (g). **(g)** The result of the microscopic calculation for a pure $s_{1/2}^2(0)$ configuration ($r_1 = 6$ fm). **(h)** The variety of situations found in (a) and (d) in comparison to each other in a single cartoon. **(i)** Schematic picture of the dynamics of the partners fermions in the quantum state ϕ_0 of the Cooper pair. It is a linear combination of motions away and toward one another. The fermions stay within a distance of the order ξ, root mean square radius of the Cooper pair (see Weisskopf (1981), Kadin (2007) and van Witsen (2014)). For the color version of this figure, refer to cambridge.org/nuclearcooperpair.

---

[121] Kadin (2007); see also van Witsen (2014).

[122] In connection with this figure, we have estimated the Fermi momentum associated with the two halo neutrons of ^{11}Li as $k_F = (3\pi^2 \times 2/((4\pi/3)(4.6)^3 - (2.5)^3))^{1/3}$ fm$^{-1} \approx 0.56$ fm^{-1}, the denominator being the volume associated with the halo. Thus $(v_F/c) \approx 0.2(k_F)_{\text{fm}^{-1}} \approx 0.1$ and $\xi = \hbar v_F/(\pi \times 0.5$ MeV$) \approx 14$ fm. See also Sect. 3.6.1, App. 7.C and footnote 22 of the present chapter.

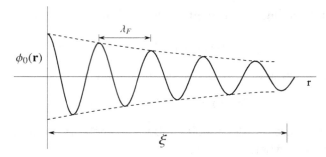

Figure 4.9.2 Schematic representation of the Cooper pair wavefunction. Indicated are the coherence length ξ and the Fermi wavelength $\lambda_F = h/p_F = 2\pi/k_F$. In the nuclear case, and for nuclei along the stability valley $\lambda_F \approx 4.6$ fm and $\xi \approx \hbar v_F/\pi\Delta \approx 14$ fm ($k_F \approx 1.36$ fm^{-1}, $v_F/c \approx 0.27$, $\Delta \approx 1.2$ MeV), while for ^{11}Li($k_F \approx 0.8$fm^{-1}), $\lambda_F \approx 8$ fm and $\xi \approx 20$ fm. Thus $\xi/\lambda_F \approx 3$.

A wavefunction which extends over distances much larger than the range of the binding potential is a well-known phenomenon when the binding energy is small. For example, in the case of the deuteron mentioned above. Be as it may, in the case of metallic superconductors the large size of the Cooper pair wavefunction also explains why the electrostatic repulsion between electron partners does not appreciably influence the binding.

Going back to Fig. 4.9.1, it is illustrative to elaborate on the two (NFT calculated) situations displayed in (**b**) and (**e**), concerning the relative distribution of the halo neutrons of ^{11}Li. Pairing correlations being, in particular in this case, mainly a surface phenomenon bring, for $r_1 = 7.5$ fm, the two nucleons close to each other as compared to the uncorrelated situation (diagram (**g**)[123]). The fact that this result, which is not under discussion, is more subtle than just expressed, emerges by looking at (**h**), where the situations displayed in (b) and (e) (see also (a) and (d)), are schematically drawn in a single plot corresponding to: **I**) the ("free from external fields") BCS-like, Cooper pair (dashed curve) which implies that the two neutrons recede from each other; **II**) the ("actual"), confined, situation in which the average potential $U(r) = \int d\mathbf{r}'\rho(\mathbf{r}')v(|\mathbf{r} - \mathbf{r}'|)$, acts as a strong external field. It determines where normal ($\rho(r)$), and thus abnormal densities can find themselves, thus distorting the halo Cooper pair. It leads to the situation represented with the continuous (solid) line. That is, a situation in which the two halo neutrons have come closer to each other as compared to the uncorrelated configuration.[124]

---

[123] Shown for $r_1 = 6$fm. No qualitative difference is expected for $r_1 = 7.5$fm in this uncorrelated configuration.

[124] And to do so, the finite quantal system under discussion uses the mechanism discussed in Sect. 4.11, namely, spatial quantization.

In principle, both situations can be experimentally observed. That labeled II) in terms of high-energy electron scattering or (p, pn) reactions[125], while that labeled I), making use of a two-particle pick up reaction, for example, $(^1H(^{11}Li,^9Li)^3H)$. In fact, during the period of time the proton is around the grazing distance, and one of the halo neutrons joins it to form a virtual deuteron, the other neutron can be at the antipodes, essentially one diameter apart. The situation will essentially evolve into the density distribution represented with the dashed curve in (h), the two members of the halo Cooper pair being at a distance of the order of the correlation length and carrying a much lower relative momentum than that typical of the situation (e) (continuous irregular curve). It is then natural that successive transfer dominates the absolute differential cross section.

While it is true that both I) and II) can be the outcome of experiments, the specific probe of dynamic or static distortion in gauge space, and thus of Cooper pair structure is Cooper pair transfer.

Specific Probe

From daily life experience one can state that, as a rule, macroscopic, condensed systems find themselves in the presence of external fields which fix the *order parameter* at some definite preferred value. Because of the long-range order, a very weak external field can pin down the *orientation* (violation of rotational invariance) and *position* (violation of translational invariance) of a crystal, or of a chair.[126] To determine the value of these order parameters, one needs to have instruments, probes, experimental setups, or whatever one likes to call them, which violate themselves rotational and translational invariance. Again, daily life objects (instruments) do so.

In the case of superconductivity, the order parameter is α_0, α_0' being its magnitude and ϕ its phase. It is gauge invariance which is spontaneously violated, the (BCS) superconductor displaying a rather perfect internal gauge phase order. There

---

[125] Kubota et al. (2020).

[126] Quoting from Weinberg (1996b), Chapter 19: "spontaneous symmetry breaking actually occurs only for idealized systems that are infinitely large. The appearance of broken symmetry for a chair arises because it has a macroscopic moment of inertia $\mathfrak{J}$, so that its ground state is part of a tower of rotationally excited states whose energies are separated by only tiny amounts, of the order of $\hbar^2/\mathfrak{J}$ ... even very weak ... rotationally asymmetric external fields will cause any ... state of the chair with definite angular momentum rapidly to develop components with other angular momentum quantum numbers. States of the chair that are relatively stable ... are ... those with a definite orientation, in which the rotational symmetry of the underlying theory is broken." In connection with superconductivity, distortion in gauge space is measured by the number of Cooper pairs $\alpha_0'(\approx 10^6)$ moving close to the Fermi energy, the associated number of electrons taking the place of angular momentum in 3D space. Now, is it 2×10^6 an infinitely large number of particles? And if yes, what about 10–12 nucleons ($\alpha_0' \approx$ 5–6 for Sn-isotopes)? An answer which can hardly be other than positive, in keeping with the observation of well defined pairing rotational bands, populated in two-nucleon transfer reactions. But if this is so, implying "angular momentum" fluctuations as large as $\approx$ 30 percent, why two nucleons ($\alpha_0' = 1$, ^{11}Li) with associated fluctuations of $\approx$ 70 percent cannot display, in an incipient fashion, basic features of BCS pairing?

are not many instruments which have such a property but another superconductor acting as a probe, weakly coupled[127] to the probed superconductor. Said it differently, a Josephson junction and associated special effects. In particular direct currents of carriers of charge $2e$ and mass $2m_e$.

Returning to the nuclear scenario, pair transfer, eventually back and forth[128] pair transfer, of nucleons correlated in time-reversal states over distances ξ, is the specific probe of pairing in atomic nuclei in general, and of halo nuclear Cooper pairs in particular.

To interact at profit through long-wavelength medium polarization pairing, pairs of nucleons have to have low momentum. To do so they have to reduce the effect of the strong external (mean) field by moving away from it, possible mechanisms being among others: halo (Fig. 4.9.1), transfer processes (see, e.g., Fig. 4.5.1; see also Fig. 7.3.1), exotic decay[129] (Fig. 4.9.3; see also Fig. 1.1.1) and, if nothing else, a consistent amount of spill out.

4.9.1 Number of Overlapping Pairs

The correlation length for low-temperature superconductors is of the order of 10^4 Å. In the case of, for example, bulk Pb, for which[130] $\Delta = 1.4$ meV and $v_F = 1.83 \times 10^8$ cm/s one has $\xi \approx 0.3 \times 10^{-4}$ cm.

Since electrons in metals typically occupy a volume of the order of $(1 \times \text{Å})^3$ (Wigner–Seitz cell radius), there would be of the order of[131] $\xi^3/(\text{Å})^3 \approx 3 \times 10^{10}$ other Cooper pairs within a "coherence volume." Eliminating the electrons deep within the Fermi sea as they behave essentially as if the metal was in the normal phase[132], one gets[133] $\approx 10^6$. *In other words, about a million of other Cooper pairs have their center of mass falling inside the coherence volume of a pair.* Thus, the isolated pair picture (Fig. 3.4.1) is not correct, but yes that displayed in Fig. 3.4.2.

---

[127] Linear response. See footnote 1 Chapter 2.

[128] See, e.g., Fig. 2.7.2 (a); see also Potel et al. (2021).

[129] In Fig. 4.9.3, a parallel is made between correlation lengths associated with (pairing) particle–particle or hole–hole modes and particle–hole vibrations. These last modes also display a consistent spatial correlation (see, e.g., Broglia et al. (1971c)).

[130] The experimental value of the critical temperature and of the pairing gap for Pb is $T_c = 7.193$ K ($k_B T_c = 0.62$ meV) and $\Delta_0(0) = 1.4$ meV, respectively (see Sect. 4.8.4). Then $\Delta_0(0)/k_B T_c = 2.26$, to be compared with the BCS prediction of 1.76. Within this context, Pb is considered a strong-coupling superconductor (see Tinkham (1980), Fig. 2-1).

[131] Ketterson and Song (1999) p. 198.

[132] The BCS ground state at $T = 0$ consists in two classes of electrons: those deep inside the Fermi sea, which behave essentially in the same way as those in the normal state, and those near the Fermi surface, which form the overlapping Cooper pairs. The latter electrons cannot scatter because they are in a coherent state, their intrinsic motion being frozen because of the pairing gap. The former electrons cannot either because they are blocked by the Pauli principle. At $T = 0$, all electrons of both classes contribute to the lossless supercurrent (see Sects. 2.2 and 9.4 of Waldram (1996)).

[133] Schrieffer (1964), p. 43. That is, $3 \times 10^{10} \times (\Delta/\epsilon_F) \approx 3 \times 10^{10} \times (1.4 \text{ meV}/9.47 \text{ eV}) \approx 4 \times 10^6$.

In the nuclear case, the number of Cooper pairs participating in the condensate, namely,

$$\alpha'_0 = \langle BCS|P'^{\dagger}|BCS \rangle = \sum_j \frac{2j+1}{2} U'_j V'_j, \tag{4.9.8}$$

can be estimated with the help of the single j-shell model, in which case $V_j = (N/2\Omega)^{1/2}$ and $U_j = (1 - N/2\Omega)^{1/2}$, where $\Omega = (2j+1)/2$. For a half-filled shell ($N = \Omega$), $\alpha'_0 = \Omega/2$. With the help of the approximate expression[134] $\Omega \approx (2/3)A^{2/3}$ one obtains, for ^{120}Sn ($N = 70$), $\alpha'_0 \approx 5$–6.

In keeping with the fact that $\xi > R_0$, in the nuclear case one has a complete overlap between all Cooper pairs participating in the condensate. This, together with the fact that the nuclear Cooper pairs press against the nuclear surface in an attempt to expand and are forced to bounce elastically off from it, receive circumstantial evidence from the following experimental results:

1. While the moment of inertia of rotational bands[135] is $\mathcal{J}_r/2$, it is also $5\mathcal{J}_{irrot}$. In other words, while pairing in nuclei is important, its role is only partially exhausted, and certainly strongly distorted by the mean field.

2. One- and two-nucleon transfer reactions in pairing correlated nuclei have the same order of magnitude. For example, $\sigma(^{120}\text{Sn}(p,d)^{119}\text{Sn}(5/2^+; 1.09 \text{ MeV})) = 5.35$mb ($2° < \theta_{cm} < 55°$), while $\sigma(^{120}\text{Sn}(p,t)^{118}\text{Sn}(gs)) = 2.25$mb ($7.6° < \theta_{cm} < 59.7°$) (see also footnote 28 of Chapter 3). Furthermore, in the Josephson-like junction, transiently established in the heavy ion collision ^{116}Sn+^{60}Ni at bombarding energies around, but still few MeV below the Coulomb barrier ($\tau_{coll} \approx 10^{-21}$ s), one finds for distances of closest approach $D_0 \lesssim \xi \approx 13.5$ fm, that σ_{2n} is of the order of σ_{1n}. However, while the Cooper pair channel ^{116}Sn+^{60}Ni$\rightarrow$^{114}Sn+^{62}Ni is dominated by the ground-ground state transition, the single-particle transfer one is inclusive. In fact, the theoretical calculations[136] of the differential cross section associated with the channel ^{116}Sn+^{60}Ni$\rightarrow$^{115}Sn+^{61}Ni indicates the incoherent contribution of a number of quasiparticle states of ^{61}Ni lying at energies $\lesssim 2.64$ MeV. A value which is consistent with twice the pairing gap of Ni. In other words, we are in presence of superconducting-quasiparticle (S-Q) tunneling across the junction, as discussed in connection with Eq. (4.6.20) (see also Sect. 7.3).

---

[134] Making use of the harmonic oscillator, one can write $\Omega = \frac{1}{2}(N + 1)(N + 2) \sim A^{2/3}$, where the proportionality constant has a value between 1/2 and 2/3.

[135] Bohr and Mottelson (1975), p. 75.

[136] Montanari et al. (2014); Potel et al. (2021).

3. The decay constant of the exotic decay $^{223}_{88}\text{Ra}_{135} \rightarrow ^{14}_{6}\text{C}_8 + ^{209}_{82}\text{Pb}_{127}$ has been measured to be $\lambda_{exp} = 4.3 \times 10^{-16}\text{sec}^{-1}$. For theoretical purposes it can be written as $\lambda = PfT$, product of the formation probability P of ^{14}C in ^{223}Ra (saddle configuration, see Fig. 1.1.1 (e)–(g) and bottom of Fig. 4.9.3), the knocking rate f and the tunneling probability T. These two last quantities hardly depend on pairing. On the other hand P changes[137] from $\approx 2 \times 10^{-76}$ to 2.3×10^{-10} by introducing pairing and, consequently, the associated lifetimes from 10^{75}y to the "observed" (empirical) value of 10^8y allowing Cooper pairs to be correlated over distances which can be as large as 20 fm.

Within the above context, exotic halo nuclei open new possibilities to understand the physics at the basis of pairing in nuclei. In fact, one may be able to include in a bare $^1S_0$ NN-potential the $\approx$50 percent contribution associated with the induced pairing interaction to the pairing gap in ^{120}Sn. Hardly, essentially all of the binding energy ($S_{2n} \approx 380$ keV) of the neutron halo Cooper pair of ^{11}Li to the core ^9Li. At the basis of the large magnitude of renormalization effects of elementary modes of excitation and of medium polarization contributions to the pairing interaction observed in light exotic halo nuclei like ^{11}Li,one finds a fundamental parameter of NFT, namely, the effective degeneracy $\Omega(\approx 2/3A^{2/3})$ of the single-particle phase space, $1/\Omega$ being the small expansion parameter. In the case of ^{11}Li, Ω is rather small (≈ 3) as compared with medium heavy nuclei like ^{120}Sn ($\Omega \approx 6$). Furthermore, in the case of ^{11}Li, the surface (S) to volume (V) ratio ($= aS/V$, a being the nuclear diffusivity) is much larger[138] (≈ 0.72) than in the case of medium heavy nuclei lying along the stability valley (≈ 0.33 in the case of ^{120}Sn).

4.9.2 Coherence Length and Quantality Parameter for (ph) Vibrations

As schematically indicated in Fig. 2.2.1, both $\beta = 0$ and $\beta = \pm 2$ modes contribute to give rise to a sigmoidal shape to the single-particle occupation probabilities within an energy range $\delta\epsilon = |E_{corr}|$ around the Fermi energy. Concerning $\beta = 0$ see also Fig. 1.2.5. Consequently, the arguments used in Sect. 4.5.1 and leading to

$$\xi = \frac{\hbar v_F}{\pi |E_{corr}|}, \qquad (4.9.9)$$

for pairing modes and resulting in $\xi = \hbar v_F / \pi \Delta$ for the case of superfluid nuclei, can be used equally well in connection with $\beta = 0$ modes. In other words, in a similar way in which pair addition and substraction modes (Cooper pairs) can

---

[137] See Brink and Broglia (2005), Chapter 7, and refs. therein.

[138] In carrying out this estimate, use was made of $a_{eff} = \left(R(^{11}\text{Li})/R_0(^{11}\text{Li})\right) a$ (within this context, see Fig. 4.3.2).

be viewed as correlated pp and hh modes over distances of the order ξ, $\beta = 0$ collective vibrations can be pictured as ph modes correlated again over distances inversely proportional to the correlation energy of, for example, the RPA collective roots. Summing up, ξ describes, both in the case of $\beta = \pm 2$ and $\beta = 0$ modes, a similar physical property of the vibration: the length over which pairs of fermions $((ph), (pp), (hh))$ are correlated in normal and in superfluid nuclei[139] (see Fig. 4.9.3).

Parallel between $\beta = \pm 2$ and $\beta = 0$ Modes

Vibrations can be classified by the transfer quantum number β (see Fig. 4.9.3). Collective modes with $\beta = 0$ correspond to correlated particle-hole (ph) excitations. For example low-lying quadrupole or octupole (surface) vibrations. Modes with $\beta = \pm 2$ correspond to correlated (pp) or (hh) modes, that is, pair addition and pair substraction modes. Thinking of these modes propagating in uniform nuclear matter, the associated correlation length can be calculated making use of Eq. (4.9.9), the correlation energy being, as a rule estimated from the energy shift of the lowest root of the RPA solution for both, ph and pairing modes, from the lowest unperturbed configuration (pole). The (generalized) quantality parameter, ratio of the quantal kinetic energy of localization and the correlation energy, gives a measure of the tendency to independent-particle ($q_\xi \approx 1$) or independent pp, hh, ph ($q_\xi \ll 1$) motion. A concrete example which testifies to the fact that (ph) excitations (large amplitude surface distortion) and phase-correlated independent-pair motion (superfluidity) are correlated over dimensions larger than typical nuclear dimensions, is provided by, for example, fission and exotic decay, in particular $^{223}\mathrm{Ra} \rightarrow {}^{14}\mathrm{C} + {}^{209}\mathrm{Pb}$. In the estimates shown in Fig. 4.9.3 use has been made of[140] $\Delta = 12/\sqrt{A}$ MeV; $C = 18.1$ MeV, $D/\hbar^2 = 29.1$ MeV^{-1} and[141] $\hbar\omega = \hbar(C/D)^{1/2} \approx 0.8$ MeV; $E = 2\Delta$ (uncorrelated 2-qp states), $E_{corr} = -(E - \hbar\omega)$. In keeping with the uncertainties affecting the above simple estimates (factor 2 or π in the denominator of ξ, $\langle r^2 \rangle_{Cooper}^{1/2}$, $\sqrt{\frac{5}{3}}\langle r^2 \rangle_{Cooper}^{1/2}$, etc.), it seems fair to conclude that 15 fm $\lesssim \xi \lesssim$ 20 fm. Thus, one is likely faced with a situation in which $2 \lesssim \xi/R \lesssim 3$.

The parallel which can be traced between Cooper pairs and correlated particle-hole excitations is further supported by the fact that two-nucleon transfer reaction do excite quite strongly also the $\beta = 0$ modes (see Tables 4.9.1 and 4.9.2). Such parallel emerged already in the analysis[142] of the $^{206}\mathrm{Pb}(t, p)^{208}\mathrm{Pb}$ which provided the experimental confirmation of the theoretical predictions concerning pairing vibrations. In particular concerning the population of the pair addition vibration

---

[139] See, e.g., Barranco et al. (2019b), and refs. therein.
[140] Bohr and Mottelson (1969).
[141] Brink and Broglia (2005), Sect. 7.1, and refs. therein.
[142] Broglia and Riedel (1967).

Vibrations

(ph) | (pp), (hh)

correlated excitations (E_{corr})
with transfer quantum number

$\beta = 0$ | $\beta = \pm2$

waves on

the nuclear | the Fermi

surface

correlation length

$$\xi = \frac{\hbar v_F}{\pi |E_{corr}|}$$

typical values (finite nuclei), E_{corr}=-1.2 MeV, (-0.5 MeV ^{11}Li), $v_F/c \approx 0.27$ (0.16, ^{11}Li)

$\xi = 14$ fm (20 fm, ^{11}Li)

generalized quantality parameter

$$q_\xi = \frac{\hbar^2}{2m\xi^2} \frac{1}{|E_{corr}|} \approx 0.085 \ (0.1, \ ^{11}\text{Li})$$

strongly correlated ($q_\xi \ll 1$), weakly "bound" ($|E_{corr}|/\epsilon_F \lesssim 0.03$)
very extended ($\xi/2d \gtrsim 6, d = \left(\frac{4\pi}{3}A\right)^{1/3}$) objects

subject to a strong external field

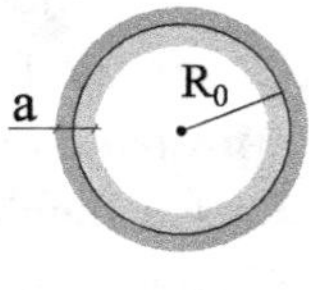

example

^{223}Ra$\rightarrow$^{14}C$+^{209}$Pb $(\lambda = PfT)$

$$P = \begin{cases} 10^{-76} & (\Delta = 0) \\ 10^{-10} & \Delta_{emp} \end{cases}$$

$$\langle r^2 \rangle_{def}^{1/2} = \xi = \frac{\hbar v_F}{\pi |E_{corr}|} \approx 21 \text{ fm} \qquad \langle r^2 \rangle_{Cooper}^{1/2} = \xi = \frac{\hbar v_F}{\pi \Delta} \approx 21 \text{ fm}$$

$$(E_{corr} \approx \text{-0.8 MeV}) \qquad\qquad (\Delta \approx 0.8 \text{ MeV})$$

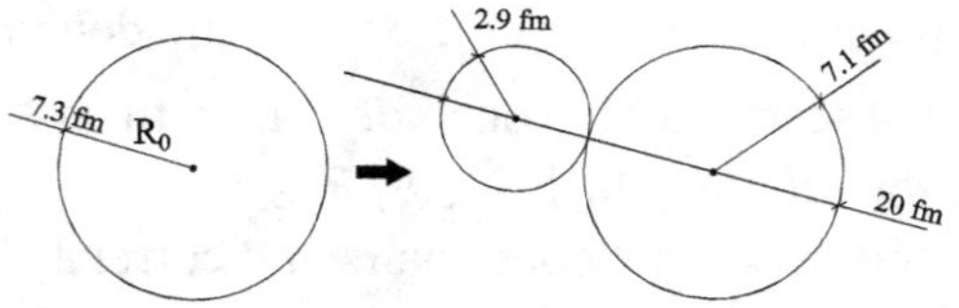

Figure 4.9.3 Parallel between pairing and particle-hole modes.

Table 4.9.1 *Relative two-nucleon transfer cross sections*
$\sigma\,(^{206}Pb\,(t,\,p)^{208}Pb(\,f\,)/\sigma\,(^{206}Pb\,(t,\,p)^{208}Pb(gs))$ *integrated in*
the range $5°–175°$ *of cm angles. After Broglia et al. (1973),*
Table A. VIII b.

	f	$\sigma(gs \to f)/\sigma(gs \to gs)$	Table A
J^π	E_x		
$0^+(gs)$	0	1	(hh) pair removal
3^-	2.62	0.21	(ph) collective mode
5^-	3.20	0.45	(ph) collective mode
0^+	4.87	0.45	(pp) pair addition

Table 4.9.2 *Absolute cross section associated with the*
reaction $^{120}Sn\,(p,\,t)^{118}Sn$ *to the ground state and first*
excited state integrated in the range $7.6° < \theta_{cm} < 69.7°$.
After Guazzoni et al. (2008).

J^π	$\sigma(gs \to f)$ (mb)	$\sigma(gs \to f)/\sigma(gs \to gs)$
$0^+(gs)$	2250±338	1
2^+	613± 92	0.27

$|a\rangle \equiv |gs(^{210}Pb)\rangle$ of the two-phonon pairing vibrational mode $|a\rangle \otimes |r\rangle$ (see Fig. 3.1.1 (b)). Predicted at an excitation energy[143] of 4.9 MeV (Figs. 3.5.1 and 1.3.2) in ^{208}Pb, it was observed[144] at 4.87 MeV. Because the lowest excited state populated with a sizable cross section (21 ±3 percent) of that associated with the ground state transition (and thus annihilating the pair removal mode ($|r\rangle \equiv |gs(^{206}Pb)\rangle$), see Fig. 3.1.1 (b) transition denoted r) was observed at $E_x = 2.619$ MeV, much effort and time was dedicated to experimentally disentangle an alleged $J^\pi = 0^+$ essentially degenerated with the octupole vibration at $E_x = 2.62 \pm 0.01$ MeV. This was in keeping with the idea that around close shell nuclei, only pairing vibrations were strongly populated, this not being the case for $\beta = 0$ collective surface states.

The analysis of the data dispelled such believe, as testified already by the title of the paper reporting the results. In fact, collective $\beta = 0$ and $\beta = \pm 2$ are similarly strongly populated in two-nucleon transfer reactions, the main difference being the fact that while the Y-components (ground state correlations, see Table 3.5.3) contribute in the case of the pairing modes constructively coherent to the cross

---

[143] See Fig. 5 of Bès and Broglia (1966).
[144] Bjerregaard et al. (1966).

section, the similar components of the $\beta = 0$ modes give rise to destructive inter-ference. Again an example of the competition for phase space between pairing and (dynamical) surface deformation (see, e.g., Fig. 2.6.2).[145] These results originally found in Pb, were confirmed in other mass regions and for different $\beta = 0$ modes.[146]

Within the above scenario one can look forward to the test of the $^9\mathrm{Li}(t, p)^{11}\mathrm{Li}$ predictions, concerning the population of both the ground state and the PDR ($E_x \lesssim 1$ MeV; see Fig. 4.12.1). Also to assess to which extent the 0^{*+} $^{12}\mathrm{Be}$ excited state and 1^- excitation on top of it, can be viewed as related elementary mode of excitation.

4.9.3 Tunneling Probabilities

The state $|BCS(\phi)\rangle_{\mathcal{K}} = \prod_{\nu>0}(U'_\nu + V'_\nu e^{-2i\phi}a^\dagger_\nu a^\dagger_{\bar\nu})|0\rangle$ displays off-diagonal long-range order (ODLRO) *because each pair is in a state* $(U'_\nu + V'_\nu e^{-2i\phi}a^\dagger_\nu a^\dagger_{\bar\nu})|0\rangle$ *with the same phase as all the others.* In fact, the above wavefunction leads to a two-particle density matrix with the property $\lim_{\mathbf{r}_1,\mathbf{r}_2;\mathbf{r}_3,\mathbf{r}_4\to\infty}\rho_2(\mathbf{r}_1,\mathbf{r}_2;\mathbf{r}_3,\mathbf{r}_4) \neq 0$ under the assumption that $r_{12}, r_{34} < \xi$, $(\mathbf{r}_1,\mathbf{r}_2)$ and $(\mathbf{r}_3,\mathbf{r}_4)$ being the coordinates of a Cooper pair, r_{ij} the modulus of the relative distance between the corre-sponding partner fermions, and ξ the coherence length.[147] Quoting from Anderson (1996): "ODLRO is the idea that the density matrix contains a factorizable part $\rho_2(x, x') = f^*(x)f(x')$+remainder, with the factorizable part playing the role of a condensate wavefunction."

Bringing the above argument into reaction implies, $P_2 \approx P_1$ (see Eq. (4.3.1)). The importance of this result concerning the mechanism at the basis of Cooper pair transfer is connected with the fact that the probability of one-electron tunneling across a typical dioxide layer giving rise to a weak $S - S$ coupling[148] is likely quite small, for example, of the order of 10^{-10}. Consequently, simultaneous pair transfer between two superconductors (S), with a probability $(10^{-10})^2$ would not be observed.[149] Thus, the observed Josephson current of carriers of charge $2e$ results from the tunneling of one Cooper pair partner at a time, equally pairing correlated than when they are both in the same superconductor (S) than when each of them is, within a correlation length, in a different (of the two weakly coupled) S.

---

[145] But more important, a further example of the fact that two-nucleon transfer reaction is the specific tool to probe pairing vibrational modes. In fact, setting $\hbar\omega(\beta = 0) \approx 0$ as well as $\hbar\omega(\beta = 2) \approx \hbar\omega(\beta = -2) \approx 0$ but finite (see footnote 145 of Chapter 2), in which case the (X, Y)-RPA amplitudes of the corresponding modes become about equal, the two-nucleon absolute cross section associated with the $\beta = 0$ mode vanish, while that of the $\beta = \pm 2$ essentially diverge.

[146] Broglia et al. (1971c).

[147] See footnote 47 of Chapter 1. See also Ambegaokar (1969) and Potel et al. (2017), and refs. therein.

[148] Pippard (2012).

[149] See, e.g., McDonald (2001).

There is experimental evidence which testifies to the fact that through the Josephson-like junctions, transiently established in heavy ion collisions between superfluid nuclei at energies below the Coulomb barrier, one- and two-nucleon transfer processes display absolute cross sections of the same order of magnitude, provided the distance of closest approach is[150] $D_0 \lesssim \xi$. *More correctly Cooper pair transfer cross section, dominated by ground–ground state transition (i.e., exclusive transitions between members of the corresponding pairing rotational bands), are approximately equal to the incoherent contribution of single quasiparticle states up to an energy of about the Cooper pair binding energy ($\approx 2\Delta$). In other words, the parallel of the equality, within a factor ($\pi/4$) of the maximum Josephson current of an unbiased junction, and the normal (quasiparticle) single electron carrier current across a Josephson junction biased with the equivalent potential ($2\Delta/e$) needed to break Cooper pairs (see Eq. (4.6.20) and Fig. 4.6.1).*

4.9.4 Correlation Energy Revisited

The pair-correlation energy is the difference between the energy with and without pairing. The energy including pair correlations is[151]

$$E_p = 2 \sum_{\nu>0} |V_\nu|^2 \epsilon_\nu - G |\alpha_0|^2, \qquad (4.9.10)$$

while the energy without correlation is

$$E_0 = \sum_{\nu>0} |V_\nu^0|^2 \epsilon_\nu. \qquad (4.9.11)$$

The occupation probabilities $|V_\nu^0|$ are unity below the Fermi energy level and zero above. In both Eqs. (4.9.10) and (4.9.11) the Fermi energy has to be chosen to give the correct number of particles. The pairing correlation energy is

$$E_{corr} = E_S - G |\alpha_0|^2, \qquad (4.9.12)$$

where

$$E_S = \sum_{\nu>0} 2(|V_\nu|^2 - |V_\nu^0|^2) \epsilon_\nu. \qquad (4.9.13)$$

The pairing energy $-G|\alpha_0|^2$ contribution is partially canceled by the first term describing the fact that, in the BCS ground state, particles moving in levels close to the Fermi energy are partially excited across the Fermi surface. That is, V_ν^2 changes smoothly from 1 to 0 around ϵ_F, being 1/2 at the Fermi energy.

[150] Potel et al. (2021).
[151] See Eq. (1.4.38).

In other words, the energy gain resulting from the potential energy term, where G is the pairing coupling constant while $|\alpha_0|$ measures the number of Cooper pairs, is partially compensated by a quantal, zero-point-fluctuation-like term. It can, in principle, be related to the Cooper pair kinetic energy of confinement $T_\xi = \frac{\hbar^2}{2m} \frac{1}{\xi^2}$ (already discussed in connection with the generalized quantality parameter (Sect. 3.4)), through the relation $2|\alpha_0|T_\xi$ (for one type of nucleons). Let us make a simple estimate which provides an embodiment of the above argument. Consider the nucleus ^{223}Ra and the associated parameters $G \approx (30/A)$ MeV, $|\alpha_0| \approx 5$ and $\xi \approx 17$ fm. Thus, $T_\xi \approx 0.1$ MeV and $2 \times (2 \times |\alpha_0| \times T_\xi) \approx 2$ MeV, while $2 \times (-G|\alpha_0|^2) \approx -6.5$ MeV is the pairing contribution[152] (factor of 2, both protons and neutrons). The resulting pairing correlation energy thus being $E_{corr} \approx -4.5$ MeV. This number can be compared with a "realistic" estimate provided by the relation[153]

$$E_{corr} = -\frac{g\Delta^2}{4}, \qquad (4.9.14)$$

where $g_n \approx N/10$ MeV^{-1} and $g_p \approx Z/10$ MeV^{-1} (see Bortignon et al. (1998), Eq. (5.2)). Taking into account both types of particles $g = g_n + g_p = A/10$ MeV^{-1} and making use of $\Delta = 12/\sqrt{A}$ MeV, one obtains $E_{corr} = -\frac{144}{40}$ MeV $\approx$ -3.6 MeV.

4.10 Absolute Cooper Pair Tunneling Cross Section: Quantitative Novel Physics at the Edge between Stability and Chaos

In the study of many-body systems, in particular of finite quantum many-body systems (FQMBS) like the atomic nucleus, it is the texture of the associated emergent properties, concrete embodiment of spontaneously symmetry breaking (potential energy) and of its restoration (zero-point fluctuations (collective modes), confinement kinetic energy), which provides insight into eventual new physics. Furthermore, the understanding of the many-body system under study in terms of the detailed motion of single particles (nucleons) and collective motion, taking properly into account their couplings, are instrumental to assess the transferability of these solutions to the study of other FQMBS like metal clusters, fullerenes[154], quantum dots[155], and eventually soft matter, for example, proteins[156], let alone the

---

[152] This quantity, but divided by 2, i.e., -3.3 MeV, can be compared with the effective pairing matrix element $v = \left(\frac{\Delta_\pi^2 + \Delta_\nu^2}{4G} \approx -2.9 \text{ MeV} \right)$, operative at level crossing in the calculation of the inertia of the exotic decay ^{223}Ra$\rightarrow$^{14}Ca$+$^{209}Pb; cf. Brink and Broglia (2005), p. 159, and refs. therein.

[153] See, e.g., Sect. 3.5 of Brink and Broglia (2005).

[154] See, e.g., Gunnarsson (2004); Broglia et al. (2004), and refs. therein.

[155] Lipparini (2003).

[156] Broglia (2013).

fact that with their help, one can make predictions. Predictions which, in connection with the study of halo nuclei, in particular of pairing[157], involve novel physics.[158]

Within this context one can quote from Leon Cooper's contribution to the volume[159] BCS: 50 years: "It has become fashionable ... to assert ... that once gauge symmetry is broken the properties of superconductors follow ... with no need to inquire into the mechanism by which the symmetry is broken.[160] This is not ... true, since broken gauge symmetry might led to molecule-like and a Bose–Einstein rather than BCS condensation ... in 1957 ... the major problem was to show ... how ... an order parameter or condensation in momentum space could come about ... to show how ... gauge-invariant symmetry of the Lagrangian could be spontaneously broken due to interactions which were themselves gauge invariant."

Nuclear physics has brought this quest a step further. This time in connection with the "extension" of the study of BCS condensation to its origin, a single Cooper pair in the rarified atmosphere resulting from the large radial (isotropic) deformation observed in light, exotic, halo nuclei in general, and in ^{11}Li in particular. During the last few years, the probing of this system in terms of absolute two-nucleon transfer (pickup) reactions, has helped at making this field a quantitative one, errors below the 10 percent limit being the rule. This achievement which has its basis on much experimental work, in particular the experiments of Tanihata et al. (2008), is also the result of the combined effort made in treating the structure and reaction aspects of the subject, two sides of the same physics, on equal footing.

4.10.1 Saturation Density, Spill-Out, and Halo

In the incipit to the section on bulk properties of nuclei of Bohr and Mottelson (Bohr and Mottelson, 1969), p. 139, one reads, "The almost constant density of nuclear matter is associated with the finite range of nuclear forces; the range of the forces is r_0 (where r_0 enters the nuclear radius in the expression $R = r_0 A^{1/3}$) thus small compared to nuclear size. This 'saturation' of nuclear matter is also reflected in the fact that the total binding energy of the nucleus is roughly proportional to

---

[157] Cf., e.g., Broglia and Zelevinsky (2013).

[158] Cf., e.g., Barranco et al. (2001); Tanihata et al. (2008); Potel et al. (2010), and refs. therein.

[159] Cooper (2011).

[160] Quoting (Weinberg (2011)): "In consequence of this spontaneous symmetry breaking, products of any even number of electron fields have non-vanishing expectation values in a superconductor, though a single electron field does not. All of the dramatic exact properties of superconductors – zero electric resistance, the expelling of the magnetic fields from superconductors known as the Meissner effect, the quantization of magnetic flux through a thick superconducting ring, and the Josephson formula for the frequency of the ac current at a junction between two superconductors with different voltages – follow from the assumption that electromagnetic gauge invariance is broken in this way, with no need to inquire into the mechanism by which the symmetry is broken." The above quotation is similar to saying that once the idea of a double DNA helix was thought, all about inheritance was solved and known (cf., e.g., Stent (1980), and refs. therein).

A. In a minor way, these features are modified by surface effects and long-range Coulomb forces acting between the protons."

Electron scattering experiments (see Fig. 2-1, p. 159, of the above reference) yield

$$\rho(0) = 0.17 \, \text{fm}^{-3}.$$ (4.10.1)

Thus, one can posit that

$$\frac{4\pi}{3} R_0^3 \rho(0) = A,$$ (4.10.2)

leading to

$$r_0 = \left(\frac{3}{4\pi} \frac{1}{\rho(0)} \right)^{1/3} \approx 1.12 \, \text{fm}.$$ (4.10.3)

Because the above relations imply a step function distribution, one should correct Eq. (4.10.3) by the nucleon spill out.[161] The relative correction being of the order of $(a_0/R_0) \ln 2 \approx (0.65/5.92) \ln 2 \approx 0.08 \, (A = 120))$ associated with the fact that a more realistic distribution is provided by a Fermi function of diffusivity $a_0 \approx 0.65$ fm. That is, an 8 percent correction leading to $r_0 = (1.12 + 0.08) \, \text{fm} \approx 1.2 \, \text{fm}$.

In the case of the nucleus ^{11}Li, observations[162] indicate a mean square (gyration radius[163]) $\langle r^2 \rangle^{1/2} = 3.55 \pm 0.1$ fm. Thus

$$R(^{11}\text{Li}) = \sqrt{\frac{5}{3}} \langle r^2 \rangle^{1/2} \approx 4.58 \pm 0.13 \, \text{fm}.$$ (4.10.4)

Making use of the relation $R_0 \approx 1.2 A^{1/3}$ fm, the quantity (4.10.4) leads to $(4.58/1.2)^3 \approx 56$, an effective mass number larger five times the actual value $A = 11$. To be noted that the actual mass number predicts a "systematic" value of the nuclear radius $R_0(^{11}\text{Li}) \approx 2.7$ fm.

The above results testify to a very large *"isotropic radial deformation,"* or halo region (skin), in keeping with the fact[164] that $R(^{11}\text{Li}) - R_0(^9\text{Li}) = R_0(^9\text{Li})(\frac{R(^{11}\text{Li})}{R_0(^9\text{Li})} - 1) = 0.83 R_0(^9\text{Li})$. In other words, ^{11}Li can be viewed as made out of a normal ^{9}Li core and of a skin made out of two neutrons extending over a shell of width of the order of that of the core radius. But even more important, it is the fashion in which the afore mentioned "deformation" affects nuclear matter. Matter which

---

[161] See, e.g., Bertsch and Broglia (2005).

[162] Kobayashi et al. (1989).

[163] The radius of gyration R_g is a measure of an object of arbitrary shape, R_g^2 being the second moment in 3D space. In the case of a sphere of radius R, $R_g^2 = 3R^2/5$.

[164] Let us parameterize the radius of ^{11}Li as (see Bohr and Mottelson (1975))
$R = R_0(1 + \alpha_{00}Y_{00}) = R_0(1 + \beta_0 \frac{1}{\sqrt{4\pi}})$. Thus $\beta_0 = \sqrt{4\pi}(\frac{R}{R_0} - 1) \approx 2.96$, which testifies to the extreme "exoticity" of the phenomenon.

is little compliant to undergo either compressions or, for that sake, "depressions," without resulting in nuclear instability. In one case, through a mini supernova. In the second, by obliterating the effect of the short-range strong force acting in the $^1S_0$ channel (screening effect of the bare pairing interaction).

In fact, in the case of the halo Cooper pair of ^{11}Li, that is of the last two weakly bound neutrons, one is dealing with a rarefied nuclear atmosphere of density

$$\rho \approx \frac{2}{\frac{4\pi}{3}(R^3(^{11}\text{Li}) - R_0^3(^9\text{Li}))} \approx 0.6 \times 10^{-2}\,\text{fm}^{-3} \qquad (4.10.5)$$

where the value $R_0(^9\text{Li}) \approx 2.5$ fm was used. That is, we are dealing with pairing in a nuclear system at a density which is only 4 percent of saturation density.

The quest for the long-range pairing mechanism which is at the basis of the binding of the halo Cooper pair of ^{11}Li to the ^{9}Li core ($S_{2n} \approx 0.380$ keV, to be compared to typical systematic values of $S_{2n} \approx 16$ MeV), has lead to the discovery of what can be considered a novel nuclear mode of elementary excitation. The symbiotic halo pair addition mode, which has to carry its own source of binding (glue). A novel embodiment of the Axel–Brink scenario in which not only the line shape, but the main structure of the resonance depends on the state on which it is built, and to which it is deeply interweaved as to guarantee its stability[165] and thus its own existence. It also provides a novel realization of the Bardeen–Frölich–Pines[166] microscopic mechanism to break gauge invariance: through the exchange of quite large dipole ZPF which ensures the same symmetries of the original Hamiltonian to a system displaying essentially a permanent dipole moment, as a consequence of the almost degeneracy of the pygmy dipole resonance (centroid ≈ 0.75 MeV) with the ground state.[167] An example of a van der Waals Cooper pair (Sect. 3.7).

The NFT diagram shown in Fig. 3.7.1 describing this binding seems quite involved and high order. Thus unlikely to be at the basis of a new elementary mode of nuclear excitation, if nothing else because of the apparent lack of "elementarity." This is not the case and, in fact, the physics at the basis of the process depicted by the oyster-like and eagle-like networks displayed in (a) and (b) is quite simple and recurrent throughout nuclear structure and reactions, let alone many-body theories and QED. In fact, it encompasses (see Fig. 3.7.1): (I,II) the changes in energy of single-particle levels as a function of quadrupole dynamical deformations leading to a Jahn–Teller like effect (III) the interaction between particles through the exchange of (quasi) bosons (vibrations), (IV,V) Pauli principle leading to a conspicuous Lamb shift–like effect, (VI,VII) the softening of collective modes

[165] Axel (1962); Brink (1955).
[166] Bardeen and Pines (1955); Fröhlich (1952).
[167] Aumann (2019).

due to ground state correlations ((ZPF)-components, QRPA),(VIII) the interaction between two nonpolar systems through virtual, ZPF associated dynamical dipoles.

Referring to general many-degrees-of-freedom systems, (I,II) and (III) are at the basis of the fact that, in QED, the coupling between one- and two- photons is zero (Furry's theorem). It is also related, through cancellation processes, to the small width displayed by (particle-hole) nuclear giant resonances, as well as of inhomogeneous damping of nuclear rotational motion at finite temperatures[168] –parallel to motional narrowing observed in NMR– as compared with the width of single particles at similar excitation energies.[169] Concerning (VIII), one can mention resonant interactions between fluctuating systems like two coupled harmonic oscillators.[170]

In the case of halo Cooper pair binding by PDR in ^{11}Li, the system although not being on resonance it is not far from it[171], in keeping with the fact $\widetilde{\epsilon}_{p_{1/2}} - \widetilde{\epsilon}_{s_{1/2}} \approx$ 0.45 MeV and $\hbar\omega_{PDR} \approx 0.75$ MeV with a width of ≈ 0.5 MeV. In other words the ^{10}Li inverted parity system is poised to acquire a permanent dipole moment or, almost equivalent, to display a large amplitude, dipole mode at very low energy as well as a collective $B(E1)$ to the halo ground state, of the order of a single-particle Weisskopf unit[172] $B_W(E1)$. The PDR (see Fig. 4.12.1, see also Fig. 2.7.1) is in ^{11}Li screened from the GDR through the poor overlap between core and halo single-particle wavefunctions. In this way, is able to retain essentially all of its $E1$-strength ($\approx 1 \times B_W$), and can likely be considered a new mode of excitation.[173] Said it differently, one is faced, already at the level of the single-particle spectrum (parity inversion), with the possibility of a plastic large amplitude dipole mode, as can be seen from

$$\frac{dn}{d\beta_L} = \frac{1}{4}\sqrt{\frac{2L+1}{3\pi}}A. \tag{4.10.6}$$

Relation which defines the number of level crossings n, in terms of the isotropic radial deformation[174] ($L = 0$ and $\beta_0 = 2.96$), leading to $n \approx 3$.

4.11 Pairing Spatial (Angular) Correlation: Simple Estimate

Let us assume two equal nucleons above the closed shell as the nuclear embodiment of Cooper's model. The two-particle wave function in configuration space can be written as,

---

[168] Broglia et al. (1987).
[169] See Sect. 5.5.1.
[170] See, e.g., Born (1969), App. XL, p. 471.
[171] See paragraphs before Eq. (4.8.21).
[172] See footnote 73 of Chapter 3.
[173] See Broglia et al. (2019a), and refs. therein.
[174] Brink and Broglia (2005), Eq. (7.35), and refs. therein. See also footnote 164 of this chapter.

$$\Psi(\mathbf{r}_1\sigma_1, \mathbf{r}_2\sigma_2) = \Psi_0(\mathbf{r}_1, \mathbf{r}_2)\chi_{S=0}(\sigma_1, \sigma_2) + [\Psi_1(\mathbf{r}_1, \mathbf{r}_2)\chi_{S=1}(\sigma_1\sigma_2)]_0, \quad (4.11.1)$$

where $\chi_{S=0}$ and $\chi_{S=1}$ are the singlet and triplet spin wavefunctions, respectively.

In what follows we shall consider a pairing interaction acting on pairs of particles moving in time-reversal states. Consequently we shall concentrate in the spin singlet radial component of (4.11.1). In the Tamm-Dancoff approximation (in keeping with Cooper ansatz)[175] one can write

$$\Psi_0(\mathbf{r}_1, \mathbf{r}_2) = \sum_{nn'lj} X_{nn'lj} R_{nl}(r_1) R_{n'l}(r_2) \sqrt{\frac{2j+1}{2(2l+1)}} \left[Y_l(\hat{r}_1) Y_l(\hat{r}_2) \right]_0. \quad (4.11.2)$$

This wave function can be rewritten as

$$\Psi_0(\mathbf{r}_1, \mathbf{r}_2) = \Psi_0(|\mathbf{r}_1|, |\mathbf{r}_2|, \theta)$$

$$= \sum_{n'nlj} X_{nn'lj} R_{nl}(r_1) R_{n'l}(r_2) \sqrt{\frac{2j+1}{2}} \frac{1}{4\pi} P_l(\cos\theta), \quad (4.11.3)$$

where $\theta = \widehat{\hat{r}_1 \hat{r}_2}$. A convenient way to display two-particle correlation is by plotting $|\Psi_0(|\mathbf{r}_1|, |\mathbf{r}_2|, \theta)|^2$ in the $x - z$ plane.

In the case of pure configurations $a \equiv nlj$,

$$\Psi_0(|\mathbf{r}_1|, |\mathbf{r}_2|, \theta) = R_{nl}(r_1) R_{nl}(r_2) \sqrt{\frac{2j+1}{2}} \frac{1}{4\pi} P_l(\cos\theta). \quad (4.11.4)$$

In keeping with the fact that the specific probe of pairing correlation is two–nucleon transfer, a phenomenon which takes place mainly, although not only, on the nuclear surface, we shall set $r_1 = r_2 = R_0$, and use the empirical relation[176]

$$R_{nl}(R_0) = \left(\frac{1.4}{R_0^3} \right)^{1/2}, \quad (4.11.5)$$

Thus

$$\Psi_0(R_0, R_0, \theta) = \left(\frac{1.4}{R_0^3} \right) \sqrt{\frac{2j+1}{2}} \frac{1}{4\pi} P_l(\cos\theta). \quad (4.11.6)$$

and

$$|\Psi_0(R_0, R_0, \theta)|^2 \sim |P_l(\cos\theta)|^2. \quad (4.11.7)$$

It is seen that the two particles have the same probability to be on top of each other ($\theta = 0°$; $P_l(1) = 1$), or on opposite sides of the nucleus ($\theta = 180°$; $P_l(-1) = (-1)^l$). Taking into account the actual radial dependence of $R_{nl}^2(r_1)$ for $r_1 = R_0$, the width of the two probability peaks is found to be ≈ 2 fm, that is, $d \approx \left(\frac{4\pi}{3} R^3 / A \right)^{1/3}$.

---

[175] See footnote 31 of Chapter 1.
[176] Bohr and Mottelson (1969).

Let us now consider the general expression (4.11.3), and assume, aside from $n = n'$, that the two nucleons are allowed to correlate in a phase space composed of N single-particle levels, and that all amplitudes are equal,

$$X \approx \frac{1}{\sqrt{N}}. \tag{4.11.8}$$

Thus

$$\Psi_0(R_0, R_0, \theta) = \left(\frac{1.4}{R_0^3}\right) \frac{1}{\sqrt{N}} \frac{1}{4\pi} \sqrt{\frac{2j+1}{2}} \sum_l P_l(\cos\theta), \tag{4.11.9}$$

where again (4.11.5) have been used. One can then write

$$|\Psi_0(R_0, R_0, \theta)|^2 \sim |\sum_l P_l(\cos\theta)|^2. \tag{4.11.10}$$

Assuming the closed-shell nucleus to be ^{208}Pb and the N single-particle levels the neutron valence orbitals $2g_{9/2}, 1i_{11/2}, 1j_{15/2}, 3d_{5/2}, 4s_{1/2}, 2g_{7/2}$, and $3d_{3/2}$, one obtains

$$\frac{|\Psi_0(R_0, R_0, \theta = 0°)|^2}{|\Psi_0(R_0, R_0, \theta = 180°)|^2} \approx \left(\frac{7}{5}\right)^2 \approx 2, \tag{4.11.11}$$

in keeping with the fact that there is only a single state of opposite parity (intruder $j_{15/2}$ state).

Making use of an extended basis, containing a similar amount of positive and negative natural parity states, that is taking into account a large number of major shells ($\pi = (-1)^N$, N principal quantum number), one can reduce in a consistent fashion the value of $|\Psi_0(R_0, R_0, \theta = 180°)|^2$. This of course materializes already within the basis of valence states in, for example, ^{11}Li, in keeping with the fact that in this case $s_{1/2}$ and $p_{1/2}$ play, in the ground state, a similar role (see Fig. 7.1.2).

Summing up, the above results have something to do with the Cooper pair problem, but much more with the peculiarities of spatial quantization associated with the nuclear self-bound many-body system at the level of mean field. That is, the nuclear Cooper pair phenomenon is to be expressed under the influence of a very strong external field which imposes not only confinement, but also spatial quantization with strong spin orbit effects resulting, among other things, in intruder states and thus parity mixing, let alone parity inversion due to quadrupole fluctuation of the mean field.

4.12 Hindsight

The formulation of superconductivity (BCS theory) described by Gor'kov[177] allows, among other things for a simple visualization of spatial dependences. In this formulation $F(\mathbf{x}, \mathbf{x}')$ is the amplitude for two Fermions (electrons) at $\mathbf{x}, \mathbf{x}'$, to belong to the Cooper pair (within the framework of nuclear physics, cf., e.g., Fig. 3.6.2 $\Psi_0(\mathbf{r}_1, \mathbf{r}_2)$; see also Sect. 4.9). The phase of F is closely related to the angular orientation of the spin variable in Anderson's quasispin formulation of BCS theory.[178] The gap function $\Delta(x)$ is given by $V(\mathbf{x})F(\mathbf{x}, \mathbf{x})$ where $V(\mathbf{x})$ is the local two–body interaction at the point $\mathbf{x}$. In the insulating barrier between the two superconductors of a Josephson junction, $V(\mathbf{x})$ is zero and thus $\Delta(x)$ is also zero.

The crucial point is that vanishing[179] $\Delta(x)$ does not imply vanishing F (abnormal density), provided, of course, that one has within the junction, a nonzero particle (electron) density, resulting from the overlap of densities from right (R) and left (L) superconductors. These junctions are such that they allow for one-electron-tunneling with a probability P_1 quite small, but finite and, consequently, the above requirement is fulfilled. Nonetheless, conventional simultaneous pair transfer, with a probability P_1^2 will not be observed.[180] *But because one electron at a time can tunnel profiting of the small, but finite electron density within the junction layer, $F(\mathbf{x}, \mathbf{x}')$ can have a sizable amplitude for Cooper pairs with partners electrons one on each side of the barrier (i.e., $\mathbf{x} \in L$ and $\mathbf{x}' \in R$), separated by distances $|\mathbf{x} - \mathbf{x}'|$ up to the coherence length.* Hence, for barriers thin compared with the coherence length, two electrons on opposite sides of the barrier can still be correlated and the superconducting, lossless, current be consistent. An evaluation of its value shows that, at zero temperature, the superconducting, lossless pair current is equal to $\pi/4$ times the normal, single (quasiparticle) electron carrier current, at an equivalent voltage[181] $2\Delta/e$. Evaluation which has been experimentally confirmed.[182]

The translation of the above parlance to the language of nuclear physics has to come to terms with the basic fact that nuclei are self-bound, finite many-body systems in which the surface, as well as the closely related property of spatial

[177] Gor'kov (1958, 1959).

[178] Anderson (1958); within the framework of nuclear physics, cf., e.g., Bohr and Ulfbeck (1988); Potel et al. (2013b), and refs. therein.

[179] This point was arguably misunderstood by Bardeen, who writes," In my view, virtual pair excitations do not extend across the layer"; see McDonald (2001); see also Bardeen (1961, 1962).

[180] Pippard (2012).

[181] In the case of Pb at low temperatures (≈ 7.19 K (0.62 meV)), this voltage is ≈ 1 meV/e = 1 mV, leading to ≈ 2 mA current for a barrier resistance of $R \sim 1\Omega$ (Ambegaokar and Baratoff (1963); McDonald (2001); Tinkham (1980)).

[182] Rogalla and Kes (2012).

quantization, play an important role both as a static element of confinement, as well as a dynamic source for renormalization effects.[183, 184] Under the influence of the average potential which can be viewed as very strong external field ($|V_0| \approx$ 50 MeV), Cooper pairs ($|E_{corr}| \approx$ 2 MeV) will become constrained within its boundaries with a consistent amount of spill out.

Let us now consider a two nucleon transfer reaction in the collision Sn+Sn assuming a distance of closest approach $D_0 \approx 14$ fm ($\lesssim \xi$), in which the two nuclear surfaces are separated by ≈ 2 fm (Fig 4.5.1). In keeping with the fact that this distance is about $3 \times a$, the heavy ion system will display a few percent overlap of (normal, saturation) density. Ever so small this overlap and so narrow the surface hole between the two leptodermic systems resulting from it, Cooper pairs can now extend over the two volumes, in a similar way as electron Cooper pairs could partially be found in the R and L superconductors in a Josephson junction. If this is the case, Cooper pair partners can, in principle, be at distances of the order of the correlation length (see Sects. 4.9.3 and 7.3).

An example of the fact that Cooper pairs will "expand" if the external mean field allows for it, is provided by ^{11}Li in which case, profiting of the weak binding (≈ 380 keV), the extension of the constrained Cooper pair (≈ 4.58 fm ± 0.13 fm) is similar to that expected in a nucleus of mass number $A \approx 60$, assuming a standard radial behavior, that is, $r_0 A^{1/3}$ fm. In keeping with this scenario, it could be expected that moving from one neutron pair addition 0^+ mode of the $N = 6$ isotones[185] to another one ($|^{11}$Li(gs)$\rangle$, $|^{12}$Be(gs)$\rangle$ and $|^{12}$Be(0^{+*}; 2.24 MeV)$\rangle$) one would see the system expanding, contracting and expanding again, respectively, in keeping with the fact that the external (mean) field is weak, strong, weak, respectively, as testified by S_{2n} (380 keV, 3672 keV, 1432 keV). Within this context, in Fig. 4.12.1 a schematic overall view of the pairing vibrational modes associated with $N = 6$ parity inverted closed-shell isotones, together with low-energy $E1$-strength modes is given. The possible candidates to the role of neutron halo pair addition modes and symbiotic state are explicitly indicated (boxed levels).

---

[183] Within this context it is of note that the liquid drop model is a very successful nuclear model, able to accurately describe not only large-amplitude motion (fission, exotic decay, low-lying collective surface vibrations, cf., e.g., Bohr and Wheeler (1939); Bertsch (1988); Barranco et al. (1990), and refs. therein), but also the masses of nuclides (see, e.g., Møller et al. (1995)), provided the superfluid inertia and shell corrections, respectively, are properly considered. Thus, it is likely that in the quest for more predictive theoretical tools of the global nuclear properties, one should develop on equal footing ever more "accurate" bare NN-potentials as well as effective forces and methods to deal with the long-wavelength renormalization effects, and induced interactions. Within this context, illuminating is the contribution of Anderson and Thouless (1962) concerning the nuclear surface diffuseness, at the time of the development of Brueckner theory (Brueckner et al. (1961)).

[184] Broglia (2002).

[185] See, e.g., Gori et al. (2004).

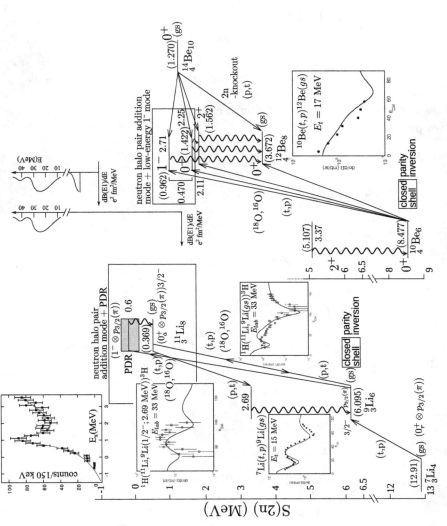

Figure 4.12.1 Monopole pairing vibrational modes associated with $N = 6$ parity inverted closed-shell isotones, together with low-energy $E1$-strength modes. In particular, in the upper left corner, the soft $E1$-mode (PDR) of ^{11}Li excited in a (p, p') reaction is shown (Tanaka et al. (2017); see also Aumann (2019)). The levels are displayed as a function of the two-neutron separation energies $S(2n)$. These quantities are shown in parentheses, and the excitation energies with respect to the ground state are quoted in MeV. Absolute differential cross sections for (t, p) and (p, t) reactions calculated as described in the text (cf. Potel et al. (2010, 2014)), in comparison with the experimental data, are also displayed (Young and Stokes (1971); Fortune et al. (1994). After Broglia et al. (2016).

4.A Coherent State

The BCS ground state can be written as,

$$|BCS(\phi)\rangle_{\mathcal{K}} = \prod_{\nu>0} \left(U_\nu + V_\nu a_\nu^\dagger a_{\bar\nu}^\dagger \right) |0\rangle = \prod_{\nu>0} U_\nu \left(1 + \frac{V_\nu}{U_\nu} a_\nu^\dagger a_{\bar\nu}^\dagger \right) |0\rangle$$

$$= \left(\prod_{\nu>0} U_\nu \right) \left(\prod_{\nu>0} (1 + c_\nu P_\nu^\dagger) \right) |0\rangle, \qquad (4.A.1)$$

where

$$c_\nu = \frac{V_\nu}{U_\nu} \quad \text{and} \quad P_\nu^\dagger = a_\nu^\dagger a_{\bar\nu}^\dagger. \qquad (4.A.2)$$

In what follows, we work out a couple of simple examples:
a) $\nu=1,2$ (**two pairs**),

$$\prod_{\nu>0} (1 + c_\nu P_\nu^\dagger) = \left(1 + c_1 P_1^\dagger \right) \left(1 + c_2 P_2^\dagger \right) = 1 + c_1 P_1^\dagger + c_2 P_2^\dagger$$

$$+ c_1 c_2 P_1^\dagger P_2^\dagger = 1 + \sum_{\nu>0} c_\nu P_\nu^\dagger + \frac{1}{2!} \left(\sum_{\nu>0} c_\nu P_\nu^\dagger \right)^2, \quad (4.A.3)$$

where use has been made of[186]

$$\left(c_1 P_1^\dagger + c_2 P_2^\dagger \right)^2 = 2 c_1 c_2 P_1^\dagger P_2^\dagger, \qquad (4.A.4)$$

in keeping with the fact that

$$\left(P_1^\dagger \right)^2 = \left(P_2^\dagger \right)^2 = 0. \qquad (4.A.5)$$

The above expression underscores the fact that Cooper pairs are not bosons.
b) $\nu=1,2,3$ (**three pairs**):

$$\prod_{\nu>0} (1 + c_\nu P_\nu^\dagger) = \left(1 + c_3 P_3^\dagger \right) \left(1 + c_2 P_2^\dagger \right) \left(1 + c_1 P_1^\dagger \right)$$

$$= \left(1 + c_3 P_3^\dagger \right) \left(1 + c_1 P_1^\dagger + c_2 P_2^\dagger + c_1 c_2 P_1^\dagger P_2^\dagger \right)$$

$$= 1 + \left(c_1 P_1^\dagger + c_2 P_2^\dagger + c_3 P_3^\dagger \right)$$

$$+ \left(c_1 c_2 P_1^\dagger P_2^\dagger + c_1 c_3 P_1^\dagger P_3^\dagger + c_2 c_3 P_2^\dagger P_3^\dagger \right)$$

$$+ c_1 c_2 c_3 P_1^\dagger P_2^\dagger P_3^\dagger = 1 + \sum_{\nu>0} c_\nu P_\nu^\dagger + \frac{1}{2!} \left(\sum_{\nu>0} c_\nu P_\nu^\dagger \right)^2$$

$$+ \frac{1}{3!} \left(\sum_{\nu>0} c_\nu P_\nu^\dagger \right)^3, \qquad (4.A.6)$$

---

[186] Since the pair creation operator $P_\nu^\dagger$ commutes for different values of ν, $|BCS(\phi)\rangle_{\mathcal{K}}$ represents uncorrelated occupancy of the various pair states.

where use has been made of the relations (4.A.5) and of,

$$[(a+b+c)(a+b+c)] = ab + ac + ba + bc + ca + cb = 2ab + 2ac + 2bc,$$
$$a^2 = b^2 = c^2 = 0,$$

together with

$$(a+b+c)[(a+b+c)(a+b+c)] = 2abc + 2bac + 2cab = 6abc. \quad (4.A.7)$$

Thus,

$$\left(\sum_{\nu>0} c_\nu P_\nu^\dagger\right)^3 = 6c_1 c_2 c_3 P_1^\dagger P_2^\dagger P_3^\dagger = 3! c_1 c_2 c_3 P_1^\dagger P_2^\dagger P_3^\dagger \qquad (4.A.8)$$

Making use of

$$e^x = 1 + x + \frac{x^2}{2!} + \frac{x^3}{3!} + \cdots, \qquad (4.A.9)$$

one can write

$$|BCS(\phi)\rangle_{\mathcal{K}} = \left(\prod_{\nu>0} U_\nu\right) \left\{ 1 + \frac{1}{1!}\left(\sum_{\nu>0} c_\nu P_\nu^\dagger\right) + \frac{1}{2!}\left(\sum_{\nu>0} c_\nu P_\nu^\dagger\right)^2 \right.$$

$$\left. + \frac{1}{3!}\left(\sum_{\nu>0} c_\nu P_\nu^\dagger\right)^3 + \cdots \right\} |0\rangle,$$

$$= \left(\prod_{\nu>0} U_\nu'\right) \left\{ 1 + \frac{e^{-2i\phi}}{1!}\left(\sum_{\nu>0} c_\nu' P_\nu^\dagger\right) + \frac{e^{-4i\phi}}{2!}\left(\sum_{\nu>0} c_\nu' P_\nu^\dagger\right)^2 \right.$$

$$\left. + \frac{e^{-6i\phi}}{3!}\left(\sum_{\nu>0} c_\nu' P_\nu^\dagger\right)^3 + \cdots \right\} |0\rangle, \qquad (4.A.10)$$

where

$$c_\nu = e^{-2i\phi} c_\nu', \quad c_\nu' = V_\nu'/U_\nu'. \qquad (4.A.11)$$

Thus,

$$|BCS(\phi)\rangle_{\mathcal{K}} = \left(\prod_{\nu>0} U_\nu\right) \exp\left(\sum_{\nu>0} c_\nu P_\nu^\dagger\right) |0\rangle = \left(\prod_{\nu>0} U_\nu\right) \sum_{n=0,1,2,\dots} \frac{\left(\sum_{\nu>0} c_\nu P_\nu^\dagger\right)^n}{n!} |0\rangle$$

$$= \left(\prod_{\nu>0} U_\nu'\right) \sum_{N\text{ even}} \frac{e^{-iN\phi}}{(N/2)!} \left(\sum_{\nu>0} c_\nu' P_\nu^\dagger\right)^{N/2} |0\rangle, \qquad (4.A.12)$$

(see Eq. (4.7.27)) and

$$|N_0\rangle = \int d\phi\, e^{iN_0\phi} |BCS(\phi)\rangle_{\mathcal{K}}$$

$$= \left(\prod_{\nu>0} U_\nu'\right) \sum_{N\text{ even}} \int d\phi\, e^{iN_0\phi} \frac{e^{-iN\phi}}{(N/2)!} \left(\sum_{\nu>0} c_\nu' P_\nu^\dagger\right)^{N/2} |0\rangle \sim \left(\sum_{\nu>0} c_\nu' P_\nu^\dagger\right)^{N_0/2} |0\rangle,$$

$$(4.A.13)$$

is the member with N_0 particles of the pairing rotational band, while

$$\left(\sum_{\nu>0} c_\nu' P_\nu^\dagger\right) |0\rangle, \qquad (4.A.14)$$

is the Cooper pair state. Because $U_\nu' \to 0$ for $\epsilon \ll \epsilon_F$, (4.A.14) is to be interpreted to be valid for values of ϵ_ν close to ϵ_F. In expression (4.A.12), $n = N/2$ is the number of Cooper pairs.

Making use of the single j-shell model

$$V' = \sqrt{\frac{N}{2\Omega}}, \quad U' = \sqrt{1 - \frac{N}{2\Omega}}, \qquad (4.A.15)$$

and

$$\frac{V'}{U'} = \sqrt{\frac{N}{2\Omega - N}} \approx U'V' + \mathcal{O}\left(\left(\frac{N}{2\Omega}\right)^{3/2}\right), \qquad (4.A.16)$$

for a number of particles considerably smaller than the full degeneracy of the single-particle subspace in which nucleons can correlate, that is for $N \ll 2\Omega$. Consequently, one can write

$$|\tilde{0}\rangle \approx \frac{1}{\sqrt{N_{norm}}} \sum_{\nu>0} (\alpha_0')_\nu P_\nu^\dagger |0\rangle, \qquad (4.A.17)$$

where

$$(\alpha_0')_\nu = \langle BCS | P_\nu^\dagger | BCS \rangle = U_\nu' V_\nu', \qquad (4.A.18)$$

and

$$N_{norm} = \sum_{\nu > 0} (\alpha_0')_\nu^2. \qquad (4.A.19)$$

5

One-Particle Transfer

Physics is an experimental science: it is concerned only with those statements which in some sense can be verified by an experiment. The purpose of the theory is to provide an unification, a codification, or however you want to say it, of those results which can be tested by means of some experiment.

J. Schwinger

In what follows we present a derivation of the one-particle transfer differential cross section within the framework of the distorted wave Born approximation (DWBA).[1]

The structure input in the calculations are, as a rule, single-particle states dressed, within the formalism of nuclear field theory, through the coupling with the variety of collective, (quasi-) bosonic vibrations, leading to renormalized energies, particle content and radial wavefunctions. With the help of the associated modified formfactors[2], and of global optical potentials, one can calculate the absolute differential cross sections, quantities which can be directly compared with the experimental findings.

In this way one avoids to introduce, let alone use spectroscopic factors, quantities which are rather elusive to calculate consistently.[3] This is in keeping with the fact that as a nucleon moves through the nucleus it feels the presence of the other nucleons whose configurations change as the motion progresses. It takes time for this information to be feed back on the nucleon. This renders the average potential nonlocal in time.[4] A time-dependent operator can always be transformed into an energy-dependent operator, implying an ω-dependence of the properties which are

---

[1] See Tobocman (1961); Austern (1963); Jackson (1970); Satchler (1980, 1983); Austern (1970); Glendenning (2004); Thompson and Nunes (2009), and refs. therein.

[2] See also Pinkston and Satchler (1982); Vaagen et al. (1979); Bang et al. (1980); Hamamoto (1970a), and refs. therein. See also Barranco et al.(2017).

[3] Duguet and Hagen (2012); Jenning (2011); Dickhoff and Barbieri (2004); Dickhoff and Van Neck (2005), and refs. therein. See also Feshbach (1992); Tamura (1974); Tamura et al. (1980).

[4] See Mahaux et al. (1985), and refs. therein; see also Sect. 5.8.

usually adscribed to particles like (effective) mass, charge, etc. Furthermore, due to Pauli principle, the average potential is also non local in space. Consequently, one is forced to deal with nucleons which carry around a cloud of (quasi) bosons, aside from exchanging its position with that of the other nucleons, and thus with renormalized energies, single-particle amplitude content and radial wavefunctions (formfactors) which eventually result in a *dynamical shell model*. It is of note that the afore mentioned phenomena are not only found in nuclear physics, but are universal within the framework of many-body systems as well as of field theories like quantum electrodynamics (QED). In fact, a basic result of such theories is that nothing is really free.[5] Within this context, in Sect. 5.2.1 we provide examples of one-particle transfer processes in nuclei lying along the stability valley (Sn-isotopes), where strongly renormalized quasiparticle states are populated, while in Sects. 5.2.2–5.2.3 we give examples of exotic, halo nuclei, namely, of the $N = 6$ isotones ^{11}Be and ^{10}Li, displaying also strong renormalization effects, as testified by the phenomenon of parity inversion.

5.1 General Derivation

We now proceed to derive the transition amplitude for the reaction (Fig. 5.1.1).

$$A + a(= b + 1) \longrightarrow B(= A + 1) + b. \tag{5.1.1}$$

For a simplified version – no recoil plus plane wave limit (see also Sect. 5.1.2) – we refer to Sect. 5.6, while for an alternative derivation within the framework of one-particle knockout reactions, we refer to Sect. 5.7.

Let us assume that the nucleon, bound initially to the core b is in a single-particle state with orbital and total angular momentum l_i and j_i respectively, and that the nucleon in the final state (bound to core A) is in the l_f, j_f state. The total spin and magnetic quantum numbers of nuclei A, a, B, b are $\{J_A, M_A\}, \{J_a, M_a\}, \{J_B, M_B\}, \{J_b, M_b\}$ respectively. Denoting ξ_A and ξ_b the intrinsic coordinates of the wavefunctions describing the structure of nuclei A and b, and $\mathbf{r}_{An}$ and $\mathbf{r}_{bn}$ the relative coordinates of the transferred nucleon with respect to the CM of nuclei A and b respectively, one can write the "intrinsic" wavefunctions of the colliding nuclei A, a as

$$\phi_{M_A}^{J_A}(\xi_A),$$
$$\Psi(\xi_b, \mathbf{r}_{b1}) = \sum_{m_i} \langle J_b\, j_i\, M_b\, m_i | J_a\, M_a \rangle \phi_{M_b}^{J_b}(\xi_b) \psi_{m_i}^{j_i}(\mathbf{r}_{bn}, \sigma), \tag{5.1.2}$$

---

[5] Feynman (1975).

one-particle transfer

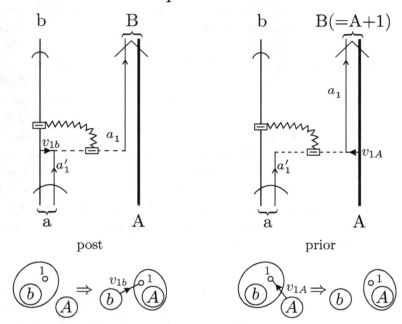

Figure 5.1.1 NFT graphical representation of the one-particle transfer reaction $a(= b + 1) + A \rightarrow b + B(= A + 1)$. The time arrow is assumed to point upward. The quantum numbers characterizing the states in which the transferred nucleon moves in projectile and target are denoted a_1' and a_1, respectively. The interaction inducing the nucleon to be transferred can act either in the entrance channel $((a, A); v_{1A}$, prior representation) or in the exit channel $((b, B); v_{1b}$, post representation), in keeping with energy conservation. In the transfer process, the nucleon changes orbital at the same time that a change in the mass partition takes place. The corresponding relative motion mismatch is known as the recoil process and is represented by a jagged curve (this is the recoil elementary mode which couples to the particle degrees of freedom through a Galilean transformation operator). The recoil mode provides information on the evolution of r_{1A} (r_{1b}), in other words, on the coupling between structure and reaction (relative motion) degrees of freedom.

while the "intrinsic" wavefunctions describing the structure of nuclei B and b are

$$\phi_{M_b}^{J_b}(\xi_b),$$

$$\Psi(\xi_A, \mathbf{r}_{A1}) = \sum_{m_f} \langle J_A \, j_f \, M_A \, m_f | J_B \, M_B \rangle \phi_{M_A}^{J_A}(\xi_A) \psi_{m_f}^{j_f}(\mathbf{r}_{An}, \sigma). \qquad (5.1.3)$$

For an unpolarized incident beam (sum over M_A, M_a and divide by $(2J_A + 1)$, $(2J_a + 1)$), and assuming that one does not detect the final polarization (sum over M_B, M_b), the differential cross section in the DWBA can be written as

$$\frac{d\sigma}{d\Omega} = \frac{k_f}{k_i}\frac{\mu_i\mu_f}{4\pi^2\hbar^4}\frac{1}{(2J_A+1)(2J_a+1)}$$

$$\times \sum_{\substack{M_A,M_a \\ M_B,M_b}} \left| \sum_{m_i,m_f} \langle J_b\, j_i\, M_b\, m_i | J_a\, M_a\rangle\langle J_A\, j_f\, M_A\, m_f | J_B\, M_B\rangle T_{m_i,m_f} \right|^2,$$

$$(5.1.4)$$

where k_i and k_f are the relative motion linear momentum in both initial and final channels (flux), while μ_i and μ_f are the corresponding relative masses. The two quantities within $\langle\ \rangle$ brackets are Clebsch–Gordan coefficients taking care of angular momentum conservation.[6]

The transition amplitude T_{m_i,m_f} is

$$T_{m_i,m_f} = \sum_\sigma \int d\mathbf{r}_f d\mathbf{r}_{bn}\chi^{(-)*}(\mathbf{r}_f)\psi_{m_f}^{j_f*}(\mathbf{r}_{An},\sigma)V(r_{bn})\psi_{m_i}^{j_i}(\mathbf{r}_{bn},\sigma)\chi^{(+)}(\mathbf{r}_i),$$

$$(5.1.5)$$

where

$$\psi_{m_i}^{j_i}(\mathbf{r}_{An},\sigma) = u_{j_i}(r_{bn})\left[Y^{l_i}(\hat{r}_i)\chi(\sigma)\right]_{j_i m_i}, \qquad (5.1.6)$$

is the single-particle wavefunction describing the motion of the nucleon to be transferred, when in the initial state, u, Y and χ being the radial, angular (spherical harmonics) and spin components. Similarly for $\psi_{m_f}^{j_f}$. The distorted waves describing the relative motion of the incoming projectile and of the target nucleus and of the outgoing system and the residual nucleus are,

$$\chi^{(+)}(\mathbf{k}_i,\mathbf{r}_i) = \frac{4\pi}{k_i r_i}\sum_{l'}i^{l'}e^{i\sigma_i^{l'}}g_{l'}(\hat{r}_i)\left[Y^{l'}(\hat{r}_i)Y^{l'}(\hat{k}_i)\right]_0^0, \qquad (5.1.7)$$

and

$$\chi^{(-)*}(\mathbf{k}_f,\mathbf{r}_f) = \frac{4\pi}{k_f r_f}\sum_l i^{-l}e^{i\sigma_f^l}f_l(\hat{r}_f)\left[Y^l(\hat{r}_f)Y^l(\hat{k}_f)\right]_0^0, \qquad (5.1.8)$$

respectively. In the above relations f and g are, respectively, the solutions of the radial Schrödinger equation describing the relative motion associated with the corresponding optical potential ("elastic" scattering) in entrance and exit channel. Let us now discuss the angular components involved in the reaction process, starting with the relation

---

[6] Brink and Satchler (1968); Edmonds (1960); also Bohr and Mottelson (1969).

$$\left[Y^l(\hat{r}_f)Y^l(\hat{k}_f)\right]_0^0 \left[Y^{l'}(\hat{r}_i)Y^{l'}(\hat{k}_i)\right]_0^0 = \sum_K ((ll)_0(l'l')_0|(ll')_K(ll')_K)_0$$

$$\times \left\{ \left[Y^l(\hat{r}_f)Y^{l'}(\hat{r}_i)\right]^K \left[Y^l(\hat{k}_f)Y^{l'}(\hat{k}_i)\right]^K \right\}_0^0. \tag{5.1.9}$$

The $9j$–symbol can be explicitly evaluated to give,

$$((ll)_0(l'l')_0|(ll')_K(ll')_K)_0 = \sqrt{\frac{2K+1}{(2l+1)(2l'+1)}}, \tag{5.1.10}$$

while the coupled expression can be written as

$$\left\{ \left[Y^l(\hat{r}_f)Y^{l'}(\hat{r}_i)\right]^K \left[Y^l(\hat{k}_f)Y^{l'}(\hat{k}_i)\right]^K \right\}_0^0$$

$$= \sum_M \langle K\,K\,M-M|0\,0\rangle \left[Y^l(\hat{r}_f)Y^{l'}(\hat{r}_i)\right]_M^K \times \left[Y^l(\hat{k}_f)Y^{l'}(\hat{k}_i)\right]_{-M}^K$$

$$= \sum_M \frac{(-1)^{K+M}}{\sqrt{2K+1}} \left[Y^l(\hat{r}_f)Y^{l'}(\hat{r}_i)\right]_M^K \left[Y^l(\hat{k}_f)Y^{l'}(\hat{k}_i)\right]_{-M}^K. \tag{5.1.11}$$

Thus,

$$\left[Y^l(\hat{r}_f)Y^l(\hat{k}_f)\right]_0^0 \left[Y^{l'}(\hat{r}_i)Y^{l'}(\hat{k}_i)\right]_0^0$$

$$= \sum_{K,M} \frac{(-1)^{K+M}}{\sqrt{(2l+1)(2l'+1)}} \left[Y^l(\hat{r}_f)Y^{l'}(\hat{r}_i)\right]_M^K \left[Y^l(\hat{k}_f)Y^{l'}(\hat{k}_i)\right]_{-M}^K. \tag{5.1.12}$$

For the angular integral to be different from zero, the integrand must be coupled to zero angular momentum (scalar). Noting that the only variables over which one integrates in the above expression are $\hat{r}_i$, $\hat{r}_f$, we have to couple the remaining functions of the angular variables, namely, the wavefunctions $\psi_{m_f}^{j_f*}(\mathbf{r}_{An}, \sigma) = (-1)^{j_f-m_f}\psi_{-m_f}^{j_f}(\mathbf{r}_{An}, -\sigma)$ and $\psi_{m_i}^{j_i}(\mathbf{r}_{bn}, \sigma)$ to angular momentum K, as well as to fulfill $M = m_f - m_i$. Let us then consider

$$(-1)^{j_f-m_f}\psi_{-m_f}^{j_f}(\mathbf{r}_{An}, -\sigma)\psi_{m_i}^{j_i}(\mathbf{r}_{bn}, \sigma)$$

$$= (-1)^{j_f-m_f}u_{j_f}(r_{An})u_{j_i}(r_{bn}) \times \sum_P \langle j_f\,j_i\,-m_f\,m_i|P\,m_i-m_f\rangle$$

$$\times \left\{ \left[Y^{l_f}(\hat{r}_{An})\chi^{1/2}(-\sigma)\right]^{j_f} \left[Y^{l_i}(\hat{r}_{bn})\chi^{1/2}(\sigma)\right]^{j_i} \right\}_{m_i-m_f}^P. \tag{5.1.13}$$

Recoupling the spherical harmonics to angular momentum K and the spinors to $S = 0$, only one term survives the angular integral in (5.1.5), namely,

$$(-1)^{j_f-m_f} u_{j_f}(r_{An}) u_{j_i}(r_{bn}) \left((l_f \tfrac{1}{2})_{j_f} (l_i \tfrac{1}{2})_{j_i} | (l_f l_i)_K (\tfrac{1}{2}\tfrac{1}{2})_0 \right)_K$$

$$\times \langle j_f \ j_i \ -m_f \ m_i | K \ m_i - m_f \rangle \left[Y^{l_f}(\hat{r}_{An}) Y^{l_i}(\hat{r}_{bn}) \right]^K_{m_i-m_f} [\chi(-\sigma)\chi(\sigma)]^0_0.$$

$$(5.1.14)$$

Making use of the fact that the sum over spins yields a factor $-\sqrt{2}$, and in keeping with the fact that $M = m_f - m_i$, one obtains,

$$T_{m_i,m_f} = (-1)^{j_f-m_f} \frac{-16\sqrt{2}\pi^2}{k_f k_i} \sum_{ll'} i^{l'-l} e^{\sigma^l_f + \sigma^{l'}_i} \sum_K \left((l_f \tfrac{1}{2})_{j_f} (l_i \tfrac{1}{2})_{j_i} | (l_f l_i)_K (\tfrac{1}{2}\tfrac{1}{2})_0 \right)_K$$

$$\times \langle j_f \ j_i \ -m_f \ m_i | K \ m_i - m_f \rangle \left[Y^l(\hat{k}_f) Y^{l'}(\hat{k}_i) \right]^K_{m_i-m_f} \int d\mathbf{r}_f d\mathbf{r}_{bn} \frac{f_l(r_f) g_{l'}(r_i)}{r_f r_i}$$

$$\times u_{j_f}(r_{An}) u_{j_i}(r_{bn}) V(r_{bn}) (-1)^{K+m_f-m_i}$$

$$\times \left[Y^l(\hat{r}_f) Y^{l'}(\hat{r}_i) \right]^K_{m_f-m_i} \left[Y^{l_f}(\hat{r}_{An}) Y^{l_i}(\hat{r}_{bn}) \right]^K_{m_i-m_f}.$$

$$(5.1.15)$$

Again, the only term of the expression

$$(-1)^{K+m_f-m_i} \left[Y^l(\hat{r}_f) Y^{l'}(\hat{r}_i) \right]^K_{m_f-m_i} \left[Y^{l_f}(\hat{r}_{An}) Y^{l_i}(\hat{r}_{bn}) \right]^K_{m_i-m_f}$$

$$= (-1)^{K+m_f-m_i} \sum_P \langle K \ K \ m_f - m_i \ m_i - m_f | P \ 0 \rangle$$

$$\times \left\{ \left[Y^l(\hat{r}_f) Y^{l'}(\hat{r}_i) \right]^K \left[Y^{l_f}(\hat{r}_{An}) Y^{l_i}(\hat{r}_{bn}) \right]^K \right\}^P_0$$

which survives after angular integration is the one with $P = 0$, that is,

$$\frac{1}{\sqrt{(2K+1)}} \left\{ \left[Y^l(\hat{r}_f) Y^{l'}(\hat{r}_i) \right]^K \left[Y^{l_f}(\hat{r}_{An}) Y^{l_i}(\hat{r}_{bn}) \right]^K \right\}^0_0$$

$$= \frac{1}{\sqrt{(2K+1)}} \sum_{M_K} \langle K \ K \ M_K \ -M_K | 0 \ 0 \rangle \left[Y^l(\hat{r}_f) Y^{l'}(\hat{r}_i) \right]^K_{M_K}$$

$$\times \left[Y^{l_f}(\hat{r}_{An}) Y^{l_i}(\hat{r}_{bn}) \right]^K_{-M_K} = \frac{1}{\sqrt{(2K+1)}} \sum_{M_K} \frac{(-1)^{K+M_K}}{\sqrt{(2K+1)}} \left[Y^l(\hat{r}_f) Y^{l'}(\hat{r}_i) \right]^K_{M_K}$$

$$\times \left[Y^{l_f}(\hat{r}_{An}) Y^{l_i}(\hat{r}_{bn}) \right]^K_{-M_K}$$

$$= \frac{1}{2K+1} \sum_{M_K} (-1)^{K+M_K} \left[Y^l(\hat{r}_f) Y^{l'}(\hat{r}_i) \right]^K_{M_K} \left[Y^{l_f}(\hat{r}_{An}) Y^{l_i}(\hat{r}_{bn}) \right]^K_{-M_K},$$

an expression which is spherically symmetric. One can evaluate it for a particular configuration, for example setting $\hat{r}_f = \hat{z}$ and the center of mass A, b, n in the $x-z$ plane (see Fig. 5.1.2). Once the orientation in space of this "standard" configuration

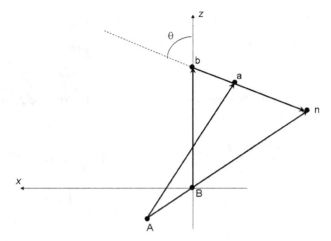

Figure 5.1.2 Coordinate system in the "standard" configuration. Note that $\mathbf{r}_f \equiv$ $\mathbf{r}_{Bb}$, and $\mathbf{r}_i \equiv \mathbf{r}_{Aa}$.

is specified (through, for example, a rotation $0 \le \alpha \le 2\pi$ around $\hat{z}$, a rotation $0 \le \beta \le \pi$ around the new x axis and a rotation $0 \le \gamma \le 2\pi$ around $\hat{r}_{bB}$), the only remaining angular coordinate is θ, while the integral over the other three angles yields $8\pi^2$. Setting $\hat{r}_f = \hat{z}$ one obtains

$$\left[Y^l(\hat{r}_f) Y^{l'}(\hat{r}_i) \right]^K_{M_K} = \langle l \; l' \; 0 \; M_K | K \; M_K \rangle \sqrt{\frac{2l+1}{4\pi}} Y^{l'}_{M_K}(\hat{r}_i). \tag{5.1.16}$$

Because of $M = m_i - m_f$, and $m = m_f$, $T_{m_i, m_f} \equiv T_{m,M}$ where

$$T_{m,M} = (-1)^{j_f - m} \frac{-64\sqrt{2}\pi^{7/2}}{k_f k_i} \sum_{ll'} i^{l'-l} e^{\sigma^l_f + \sigma^{l'}_i}$$

$$\times \sqrt{2l+1} \sum_K \frac{(-1)^K}{2K+1} \left((l_f \tfrac{1}{2})_{j_f} (l_i \tfrac{1}{2})_{j_i} | (l_f l_i)_K (\tfrac{1}{2}\tfrac{1}{2})_0 \right)_K$$

$$\times \langle j_f \; j_i \; -m \; M+m | K \; M \rangle \left[Y^l(\hat{k}_f) Y^{l'}(\hat{k}_i) \right]^K_M \int d\mathbf{r}_f d\mathbf{r}_{bn} \frac{f_l(r_f) g_{l'}(r_i)}{r_f r_i}$$

$$\times u_{j_f}(r_{An}) u_{j_i}(r_{bn}) V(r_{bn}) \sum_{M_K} (-1)^{M_K} \langle l \; l' \; 0 \; M_K | K \; M_K \rangle$$

$$\times \left[Y^{l_f}(\hat{r}_{An}) Y^{l_i}(\hat{r}_{bn}) \right]^K_{-M_K} Y^{l'}_{M_K}(\hat{r}_i). \tag{5.1.17}$$

We now turn our attention to the sum

$$\sum_{\substack{M_A, M_a \\ M_B, M_b}} \left| \sum_{m,M} \langle J_b \; j_i \; M_b \; m | J_a \; M_a \rangle \langle J_A \; j_f \; M_A \; M | J_B \; M_B \rangle T_{m,M} \right|^2, \tag{5.1.18}$$

appearing in the expression for the differential cross section (5.1.4). For any given value m', M' of m, M, the sum will be

$$\sum_{M_a,M_b} \left| \langle J_b\ j_i\ M_b\ m' | J_a\ M_a \rangle \right|^2 \sum_{M_A,M_B} \left| \langle J_A\ j_f\ M_A\ M' | J_B\ M_B \rangle \right|^2 \left| T_{m',M'} \right|^2$$

$$= \frac{(2J_a+1)(2J_B+1)}{(2j_i+1)(2j_f+1)} \sum_{M_a,M_b} \left| \langle J_b\ J_a\ M_b\ -M_a | j_i\ m' \rangle \right|^2$$

$$\times \sum_{M_A,M_B} \left| \langle J_A\ J_B\ M_A\ -M_B | j_f\ M' \rangle \right|^2 \left| T_{m',M'} \right|^2, \qquad (5.1.19)$$

by virtue of the symmetry property of Clebsch–Gordan coefficients

$$\langle J_b\ j_i\ M_b\ m | J_a\ M_a \rangle = (-1)^{J_b-M_b} \sqrt{\frac{(2J_a+1)}{(2j_i+1)}} \langle J_b\ J_a\ M_b\ -M_a | j_i\ m \rangle. \quad (5.1.20)$$

The sum over the Clebsch–Gordan coefficients in (5.1.19) is equal to 1, so (5.1.18) becomes

$$\frac{(2J_a+1)(2J_B+1)}{(2j_i+1)(2j_f+1)} \sum_{m,M} \left| T_{m,M} \right|^2, \qquad (5.1.21)$$

and the differential cross section can be written as

$$\frac{d\sigma}{d\Omega} = \frac{k_f}{k_i} \frac{\mu_i \mu_f}{4\pi^2\hbar^4} \frac{(2J_B+1)}{(2j_i+1)(2j_f+1)(2J_A+1)} \sum_{m,M} \left| T_{m,M} \right|^2. \qquad (5.1.22)$$

where

$$T_{m,M} = \sum_{Kll'} (-1)^{-m} \langle j_f\ j_i\ -m\ M+m | K\ M \rangle \left[Y^l(\hat{k}_f) Y^{l'}(\hat{k}_i) \right]_M^K t_{ll'}^K. \qquad (5.1.23)$$

Orienting $\hat{k}_i$ along the incident z-direction leads to

$$\left[Y^l(\hat{k}_f) Y^{l'}(\hat{k}_i) \right]_M^K = \langle l\ l'\ M\ 0 | K\ M \rangle \sqrt{\frac{2l'+1}{4\pi}} Y_M^l(\hat{k}_f), \qquad (5.1.24)$$

and

$$T_{m,M} = \sum_{Kll'} (-1)^{-m} \langle l\ l'\ M\ 0 | K\ M \rangle \langle j_f\ j_i\ -m\ M+m | K\ M \rangle Y_M^l(\hat{k}_f)\, t_{ll'}^K, \qquad (5.1.25)$$

with

$$t_{ll'}^K = (-1)^{K+j_f} \frac{-32\sqrt{2}\pi^3}{k_f k_i} i^{l'-l} e^{\sigma_f^l + \sigma_i^{l'}} \frac{\sqrt{(2l+1)(2l'+1)}}{2K+1}$$

$$\times \left((l_f \tfrac{1}{2})_{j_f} (l_i \tfrac{1}{2})_{j_i} | (l_f l_i)_K (\tfrac{1}{2}\tfrac{1}{2})_0 \right)_K$$

$$\times \int dr_f \, dr_{bn} d\theta r_{bn}^2 \sin\theta \, r_f \frac{f_l(r_f) g_{l'}(r_i)}{r_i} u_{j_f}(r_{An}) u_{j_i}(r_{bn}) V(r_{bn})$$

$$\times \sum_{M_K} (-1)^{M_K} \langle l \; l' \; 0 \; M_K | K \; M_K \rangle \left[Y^{l_f}(\hat{r}_{An}) Y^{l_i}(\hat{r}_{bn}) \right]_{-M_K}^K Y_{M_K}^{l'}(\hat{r}_i). \tag{5.1.26}$$

5.1.1 Coordinates

To perform the integral in (5.1.26), one needs the expression of $r_i, r_{An}, \hat{r}_{An}, \hat{r}_{bn}, \hat{r}_i$ in term of the integration variables r_f, r_{bn}, θ. Because one is interested in evaluating these quantities in the particular configuration depicted in Fig. 5.1.2, one has

$$\mathbf{r}_f = r_f \, \hat{z}, \tag{5.1.27}$$

$$\mathbf{r}_{bn} = -r_{bn}(\sin\theta \, \hat{x} + \cos\theta \, \hat{z}), \tag{5.1.28}$$

$$\mathbf{r}_{Bn} = \mathbf{r}_f + \mathbf{r}_{bn} = -r_{bn} \sin\theta \, \hat{x} + (r_f - r_{bn} \cos\theta) \, \hat{z}. \tag{5.1.29}$$

One can then write

$$\mathbf{r}_{An} = \frac{A+1}{A} \mathbf{r}_{Bn} = -\frac{A+1}{A} r_{bn} \sin\theta \, \hat{x} + \frac{A+1}{A}(r_f - r_{bn} \cos\theta) \, \hat{z}, \tag{5.1.30}$$

$$\mathbf{r}_{an} = \frac{b}{b+1} \mathbf{r}_{bn} = -\frac{b}{b+1} r_{bn}(\sin\theta \, \hat{x} + \cos\theta \, \hat{z}), \tag{5.1.31}$$

and

$$\mathbf{r}_i = \mathbf{r}_{An} - \mathbf{r}_{an} = -\frac{2A+1}{(A+1)A} r_{bn} \sin\theta \, \hat{x} + \left(\frac{A+1}{A} r_f - \frac{2A+1}{(A+1)A} r_{bn} \cos\theta \right) \hat{z}, \tag{5.1.32}$$

where A, b are the number of nucleons of nuclei A and b respectively.

5.1.2 Zero-Range Approximation

In the zero-range approximation,

$$\int dr_{bn} r_{bn}^2 u_{j_i}(r_{bn}) V(r_{bn}) = D_0; \quad u_{j_i}(r_{bn}) V(r_{bn}) = \delta(r_{bn})/r_{bn}^2. \tag{5.1.33}$$

It can be shown (see Fig. 5.1.2) that for $r_{bn} = 0$

$$\mathbf{r}_{An} = \frac{m_A + 1}{m_A} \mathbf{r}_f, \quad \mathbf{r}_i = \frac{m_A + 1}{m_A} \mathbf{r}_f. \tag{5.1.34}$$

One then obtains

$$t_{ll'}^K = \frac{-16\sqrt{2}\pi^2}{k_f k_i}(-1)^K \frac{D_0}{\alpha} i^{l'-l} e^{\sigma_f^l + \sigma_i^{l'}} \frac{\sqrt{(2l+1)(2l'+1)(2l_i+1)(2l_f+1)}}{2K+1}$$

$$\times \left((l_f\tfrac{1}{2})_{j_f}(l_i\tfrac{1}{2})_{j_i}|(l_f l_i)_K(\tfrac{1}{2}\tfrac{1}{2})_0\right)_K$$

$$\times \langle l\, l'\, 0\, 0|K\, 0\rangle\langle l_f\, l_i\, 0\, 0|K\, 0\rangle \int dr_f\, f_l(r_f)g_{l'}(\alpha r_f)u_{j_f}(\alpha r_f), \quad (5.1.35)$$

with

$$\alpha = \frac{A+1}{A}. \tag{5.1.36}$$

5.2 Examples and Applications

In the calculation of absolute reaction cross sections two elements melt together: reaction and structure. In the case of weakly coupled probes like, as a rule, direct one-particle transfer processes are, the first element can be further divided into two essentially separated components: elastic scattering (optical potentials), and transfer amplitudes connecting entrance and exit channels. In other words, the habitat of DWBA.

5.2.1 NFT of $^{120}Sn(p, d)^{119}Sn$ and $^{120}Sn(d, p)^{121}Sn$ Reactions[7]

The NFT calculation of the structure properties of the Sn-isotopes probed through (d, p) and (p, d) reactions, were carried out making use of an effective Skyrme interaction[8], SLy4, to determine the mean field (HFB) and the collective modes (QRPA), and a $v_{14}(^1S_0)(\equiv v_p^{bare})$ Argonne pairing interaction.[9] Hartree–Fock–Bogoliubov (HFB) results provided the bare quasiparticle spectrum, while QRPA a realization of density ($J^\pi = 2^+, 3^-, 4^+, 5^-$) and spin ($2^\pm, 3^\pm, 4^\pm, 5^\pm$) modes.

Taking into account renormalization processes (self-energy, vertex corrections) in terms of the PVC mechanism, the dressed particles and the induced pairing interaction v_p^{ind} were calculated. Adding it to v_p^{bare} the total pairing interaction v_p^{eff} was determined. It is found that v_p^{bare} and v_p^{ind} contribute about equally to the state-dependent pairing gap.

Making use of the above input, the lowest-order NFT renormalizing diagrams (self-energy and vertex correction) were propagated to infinite order making use of Nambu-Gorkov equation.[10] In Fig. 5.2.1, the absolute differential cross section associated with the population of the low-lying state $|^{119}Sn(7/2^+; 788\text{keV})\rangle$ in the

---

[7] Idini et al. (2015); see also Idini (2013).

[8] Chabanat et al. (1997).

[9] Wiringa et al. (1984).

[10] Idini et al. (2012).

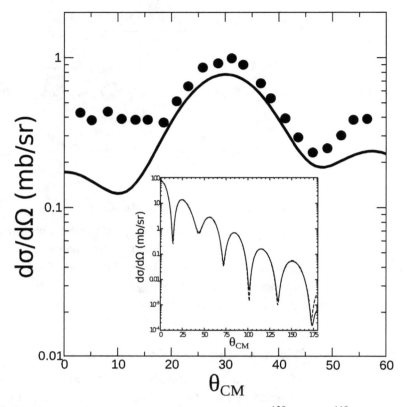

Figure 5.2.1 The absolute differential cross section ^{120}Sn$(p, d)^{119}$Sn(j^π) associated with the state $j^\pi = 7/2^+$. The theoretical prediction (continuous curve) is displayed in comparison with the experimental data (solid dots; Dickey et al. (1982)). The corresponding integrated cross sections are 5.0 and 5.2 ± 0.6 mb, respectively. In the inset, and for the sake of accuracy control, the absolute differential cross sections associated with the reaction ^{124}Sn$(p, d)^{123}$Sn(gs) calculated making use of the codes ONE (bold dashed) and FRESCO (thin continuous line (Thompson (1988))) are displayed.

one-particle pickup process ^{120}Sn$(p, d)^{119}$Sn and worked out with the help of the code ONE[11], of global optical parameters[12] and of NFT spectroscopic amplitudes, is compared with the experimental data.

Similar calculations have been carried for the reaction ^{120}Sn$(d, p)^{121}$Sn $(j^\pi; E_x)$ in connection with the population of the $|3/2^+; \text{gs}\rangle$ and $|11/2^-; E_x \approx 0\,\text{MeV}\rangle$ states. In the stripping experiment[13] the ground state and the $11/2^-$ state were not resolved in energy. This is the reason why theory and experiment are compared to

[11] Potel (2012a); see also App. 7.A.
[12] Dickey et al. (1982).
[13] Bechara and Dietzsch (1975).

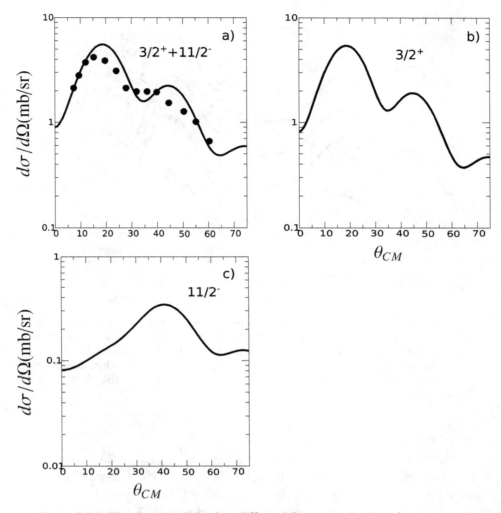

Figure 5.2.2 The theoretical absolute differential cross section (continuous curve) associated with the reaction $^{120}Sn(d, p)^{121}Sn$ populating the low-lying states $3/2^+$ and $11/2^-$ are shown in (b) and (c), while the incoherent summed differential cross section is displayed in (a) in comparison with the data (Bechara and Dietzsch (1975)).

the data for the summed $l = 2 + 5$ differential cross section (see Fig. 5.2.2 (a)), the separate theoretical predictions been displayed in Figs. 5.2.2 (b) and (c).

Let us now turn to the most fragmented low-lying quasiparticle state around ^{120}Sn, namely, that associated with the $d_{5/2}$ orbital As shown in the right panel of[14] Fig. 5.2.3 five low-lying $5/2^+$ states have been populated in the reaction

---

[14] See Idini et al. (2015).

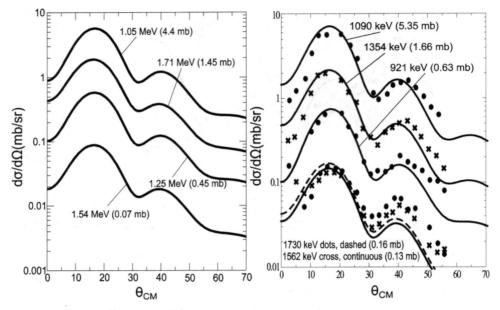

Figure 5.2.3 ^{120}Sn$(p,d)^{119}$Sn$(5/2^+)$ absolute experimental cross sections (solid dots; Dickey et al. (1982)), together with the DWBA fit carried out in the analysis of the data (right panel) in comparison with the DWBA calculations carried out with the same global optical potentials making use of NFT structure inputs as explained in the text and of the code ONE (Potel (2012a); see also App. 7.A). After Idini et al. (2015).

^{120}Sn$(p,d)^{119}$Sn with a summed cross section[15] $\sum_{i=1}^{5} \sigma(2° - 25°) \approx 8$ mb±2 mb while four are theoretically predicted with $\sum_{i=1}^{4} \sigma(2° - 25°) = 6.2$ mb. Within the present context, namely, that of probing the single-particle content of an elementary excitation (coupling to doorway states[16]), the study of the $5/2^+$ quasi-particle strength is a rather trying situation, providing a measure of the limitations encountered in such studies.

Analysis of the type presented above allows one to posit that structure and reactions are but just two aspects of the same physics. If one adds to this picture the fact that the optical potential – that is, the energy- and momentum-dependent nuclear dielectric function describing the medium where direct nuclear chemistry reactions take place – can be calculated microscopically[17] in terms of the same elements entering structure calculations (i.e., spectroscopic amplitudes, renormalized

---

[15] Dickey et al. (1982).

[16] Feshbach (1958); Rawitscher (1987); Bortignon and Broglia (1981); Bertsch et al. (1983).

[17] See footnote 26 of Chapter 1.

single-particle wavefunctions and transition densities and thus effective formfactors), the structure-reaction circle closes itself. Allowing halo nuclei to be part of the daily nuclear structure paradigm, the equivalence between structure and reactions becomes even stronger, in keeping with the central role the continuum plays in the structure of these nuclei.[18]

It seems then fair to state that the importance of the coupled channels approach to reactions[19] is not so much, or at least not only, that it is able to handle situations like for example one-particle transfer to members of a quadrupole rotational band, alas at the expenses of including effects to all orders which can be treated in the lowest one[20], but that it reminds us how intimately connected probed and probe are in nuclei.

On the other hand, for most of the situations dealt with in the present monograph, it is transparent the power, also to reflect the physics, of the approach based in perturbative DWBA (e.g., first order for one-nucleon transfer and second for Cooper pair tunneling), coupled with NFT elementary modes of nuclear excitation. To which extent a FRESCO like software built on a NFT basis will ever be attempted is an open question. Note in any case the important attempts made at incorporating so called core excitations within the FRESCO framework.[21]

We conclude this section by recalling the fact that the dressing of single particles with pairing vibrations plays also an important role in the structure properties of nuclei.[22]

5.2.2 NFT of ^{11}Be: One-Particle Transfer in Halo Nuclei

The nucleus $^{11}_{4}$Be$_7$ constitutes an example of one-neutron halo system, namely, a halo neutron outside the $N = 6$ closed shell resulting from the phenomenon of parity inversion[23]

---

[18] For example, in the case of ^{10}Li (see Sect. 5.2.3), in which one is confronted with continuum spectroscopy, the self-energy functionals describing the interweaving of the unbound neutron with vibrations of the core (^{9}Li) contribute both to the nuclear structure and nuclear reaction dielectric function.

[19] Thompson (1988, 2013); Tamura et al. (1970); Ascuitto and Glendenning (1969, 1970); Ascuitto et al. (1971); Ascuitto and Sørensen (1972); cf. also Fernández-García et al. (2010a, 2010b).

[20] Within this context, see, e.g., the couple channel Born approximation treatment of ^{208}Pb$(t, p)^{210}$Pb(3^-) carried out in Flynn et al. (1972b).

[21] Fernández-García et al. (2010a, 2010b).

[22] Barranco et al. (1987); Bès et al. (1988); Baroni et al. (2004).

[23] This nucleus has been extensively studied both experimentally (see Iwasaki et al. (2000); Fortier et al. (1999); Winfield et al. (2001); Auton (1970); Zwieglinski et al. (1979); Schmitt et al. (2013); Nörtershäuser et al. (2009); Kwan et al. (2014), and refs. therein) and theoretically (see Talmi and Unna (1960); Otsuka et al. (1993); Sagawa et al. (1993); Vinh Mau (1995); Gori et al. (2004); Nunes et al. (1996); Fossez et al. (2016); Hamamoto and Shimoura (2007); Kanada-En'yo and Horiuchi (2002); Calci et al. (2016); Krieger et al. (2012); Timofeyuk and Johnson (1999); Keeley et al. (2004); Deltuva (2009, 2013); Lay et al. (2014); de Diego et al. (2014), and refs. therein).

Outlook

In the core of ^{11}Be, namely, $^{10}_{4}$Be$_6$, six neutrons occupy the $1s_{1/2}$ and $1p_{3/2}$ levels (Fig. 5.2.4).[24] The dominant ZPF is of quadrupole type, the main neutron component being the $((p_{1/2}, p_{3/2}^{-1}) \otimes 2^+)_{0^+}$ one. In other words, because $\epsilon_{p1/2} - \epsilon_{p3/2} \approx$ 3.35 MeV and $\hbar\omega_{2^+} = 3.368$ MeV, the largest amplitude of the quadrupole mode is associated with the particle-hole excitation $(p_{1/2}, p_{3/2}^{-1})_{2^+}$.

The effect of dressing the $p_{1/2}$ state with the quadrupole vibration, amounts essentially to a Pauli principle correction. It results from the exchange of the $p_{1/2}$ nucleon considered explicitly and that participating in the vibration. The corresponding repulsion amounts to ≈ 2.8 MeV (Fig. 5.2.4 inset (**A**)). The clothing of the $2s_{1/2}$ bare level by the quadrupole mode (Fig. 5.2.4 inset (**B**)) makes it heavier, lowering its energy by almost one-half MeV (0.43 MeV). The result of the two processes is parity inversion and the appearance of the $N = 6$ new magic number together with the melting away of the $N = 8$ standard one. In a similar way in which the Lamb shift (Fig. 5.2.4, inset (**C**)) provides a measure of the fluctuations of the QED vacuum[25], parity inversion measures ZPF of the nuclear vacuum (ground) state.

Calculations

In the calculations one has simultaneously dealt with the $1p_{3/2}$, $1p_{1/2}$, $2s_{1/2}$, and $1d_{5/2}$ valence single-particle states, treating their interweaving with the low-lying quadrupole collective vibration of the ^{10}Be core and the mixing between bound and continuum states. The bare energies of the single-particle orbitals were determined by freely varying the depth, diffusivity, radius, and spin-orbit strength of a Woods–Saxon potential so that, making use of an effective radial-dependent effective mass $(m_k(r = 0) = 0.7m, m_k(r = \infty) = 0.91m)$, the fully dressed, renormalized energies best reproduce the experimental findings.[26]

The variety of self-energy diagrams, renormalizing self-consistently the motion of the odd neutron of ^{11}Be in both configurational – (Fig. 5.2.5) and conformational 3D-space (Fig. 5.2.6, formfactors), through the coupling to quadrupole vibrations, have been worked out.[27] The resulting states can be written as

$$|\widetilde{1/2^+}\rangle = \sqrt{0.80}|s_{1/2}\rangle + \sqrt{0.20}|(d_{5/2} \otimes 2^+)_{1/2^+}\rangle, \qquad (5.2.1)$$

[24] The present section is based on Barranco et al. (2017).

[25] Pais (1986), p. 451.

[26] Namely, parallel to mass renormalization in QED (see Sect. 5.3.3).

[27] The results, obtained taking also into account the coupling to the octupole vibration and the pair removal mode of the core ^{10}Be, are shown in Fig. 5.2.6. Their effect in the present case is rather small, and the final results are rather similar to the ones discussed in this section (for details, see Barranco et al. (2017), supplemental material).

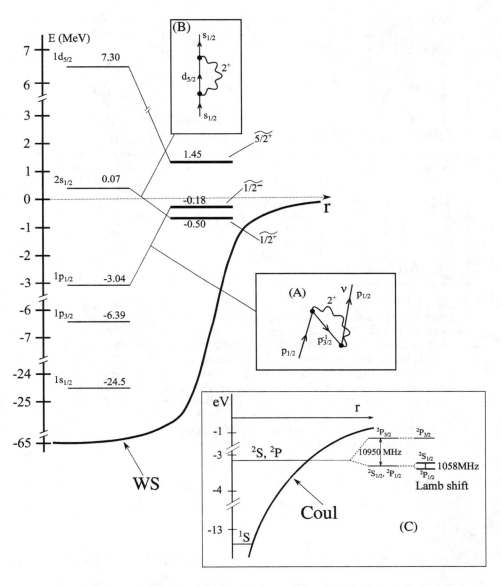

Figure 5.2.4 Bare ϵ_j (upper left thin horizontal lines) and dressed $\tilde{\epsilon}_j$ (boldface horizontal lines) single-particle levels of ^{11}Be. Due to the dressing of neutron motion with quadrupole vibrations of the core ^{10}Be (insets (A), (B)), inversion in sequence between the $2s_{1/2}$ and $1p_{1/2}$ levels (parity inversion) is observed. The numbers are energies in MeV. The Woods–Saxon (WS) bare mean field is indicated. In inset (C), the lowest energy levels of hydrogen are indicated; the Coulomb potential (Coul) is also schematically shown. The effects of the spin-orbit coupling and Lamb shift associated with the splitting of the $^2S_{1/2}$ and $^2P_{1/2}$ levels are displayed. After Barranco et al. (2019b).

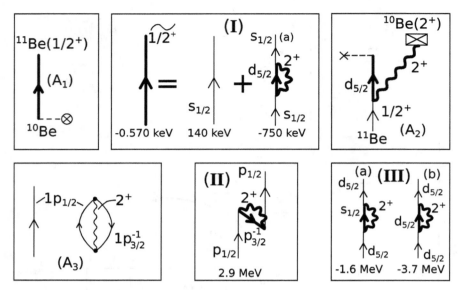

Figure 5.2.5 $(NFT)_{ren}$ diagrams describing the renormalization processes responsible for the different components of the clothed states (Eqs. (5.2.1)–(5.2.3) associated with (I)–(III)) and the stripping and pickup processes populating the ground state of ^{11}Be (A_1) and the first excited 2^+ state (A_2) of ^{10}Be, respectively. (A_3) Valence nucleon in presence of a virtual zero-point fluctuation of the core ^{10}Be. Bold (thin) arrowed lines pointing upward (downward) describe dressed (bare) particle (hole) states. The wavy line represents the quadrupole vibration. A cross (crossed circle) followed by a horizontal dashed line stands for an external one-neutron pickup (p, d) (stripping (d, p)) field. A crossed box indicates a γ-detector, eventually detecting the decay of the quadrupole vibrational state of ^{10}Be. After Barranco et al. (2017).

$$|\widetilde{1/2^-}\rangle = \sqrt{0.84}|(p_{1/2}\rangle + \sqrt{0.16}|(p_{1/2}, p_{3/2}^{-1})_{2^+} \otimes 2^+)_{0+}, p_{1/2}\rangle, \qquad (5.2.2)$$

and

$$|\widetilde{5/2^+}\rangle = \sqrt{0.49}|d_{5/2}\rangle + \sqrt{0.23}|(s_{1/2}\otimes 2^+)_{5/2^+}\rangle + \sqrt{0.28}|(d_{5/2}\otimes 2^+)_{5/2^+}\rangle. \quad (5.2.3)$$

The bare energies ϵ_j and the $((NFT)_{ren})$ values $\tilde{\epsilon}_j$ associated with the renormalized single-particle states are shown in Fig. 5.2.4. These last quantities reproduce quite accurately the experimental findings (Fig. 5.2.6, upper center). Also shown in this figure, are the corresponding bare and renormalized wavefunctions $\phi_j(r)$ and $\tilde{\phi}_j(r)$ respectively. The formfactors $\tilde{\phi}_j(r)$ were used, together with global optical potentials[28], to calculate the one-nucleon stripping and pickup absolute differential cross sections of the reactions ^{10}Be$(d, p)^{11}$Be$(1/2^+, 1/2^-, $ and $5/2^+)$ and

---

[28] Han et al. (2006); Koning and Delaroche (2003).

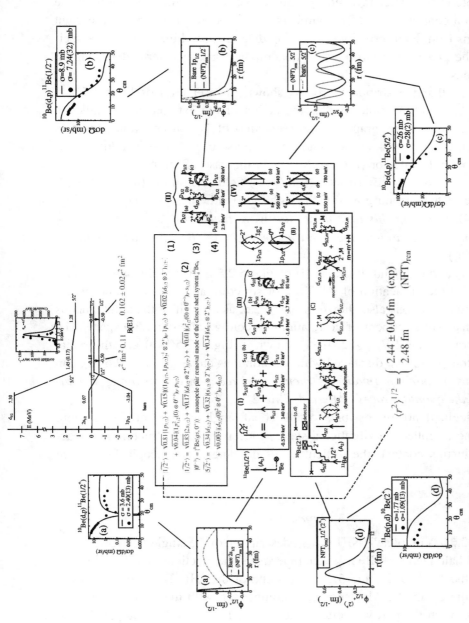

Figure 5.2.6 The clothing of the bare nucleons (single arrowed lines) with quadrupole and octupole particle-hole and monopole pair removal vibrations of the ^{10}Be core following the rules of renormalized nuclear field theory give rise to the (renormalized) energies $\tilde{\epsilon}_j$ and renormalized single-particle wavefunctions $\tilde{\phi}_j$ which were used as formfactors in connection with global optical potentials to calculate the one-nucleon stripping and pickup absolute differential cross sections (continuous curve (a), (b), (c), (d) (see also the corresponding formfactors), displayed in comparison with the experimental data (solid dots)). The same is true concerning the $B(E1)$ transition between the parity inverted $1/2^+$, $1/2^-$ states and the isotopic shift of the charge radius. After Barranco et al. (2017). Larger version of this figure available at cambridge.org/nuclearcooperpair.

^{11}Be$(p, d)^{10}$Be(2^+). The results provide an overall account of the experimental findings. It is of note that the $d_{5/2}$ plays, through its coupling to the 2^+ state, an essential role in the parity inversion phenomenon, by lowering the energy of the $s_{1/2}$ state in a consistent way. The numbers quoted below also contain important contributions from Pauli principle correcting diagrams see Fig. 5.2.6 (IV) associated with the presence of two-phonon quadrupole states in intermediate, virtual configurations.

We remark that the stripping process shown in inset (A$_1$) of Fig.5.2.5 and populating ^{11}Be ground state implies the action of the external (d, p) field on the left-hand side of the graphical representation of Dyson equation shown in Fig, 5.2.5(I), and involves, at the same time, the use of the corresponding radial wavefunction as formfactor (Fig. 5.2.6 (a)). In the case of the population of the first 2^+ excited state of ^{10}Be (inset A_2), a (p, d) field acts on the $(d_{5/2} \otimes 2^+)_{1/2^+}$ virtual state of the second graph of the right-hand side of Dyson equation (Fig. 5.2.5 (I) (a)), *involving this time the radial wave function* $\tilde{\phi}_{1/2^+}^{(2+)}(r)$, *namely, the odd neutron moving around the quadrupole excited* ^{10}Be *core, as formfactor* (Fig. 5.2.6 (d)).

Insets (A$_1$) and (A$_2$) and diagrams (I) of Fig. 5.2.5 testify to the subtle effects resulting from the unification of (NFT)$_{ren}$ of structure and reactions, and operative in the cross sections also shown in Fig. 5.2.6, as a result of the simultaneous and self-consistent treatment of correlations in configuration and 3D-space. The bold face drawn state $|(d_{5/2} \otimes 2^+)_{1/2^+}\rangle$ shown in Fig. 5.2.5(I)(a) and the radial wave function $(NFT)_{ren}$ displayed with a continuous curve in Fig. 5.2.6 (d), can be viewed as *on par renormalized structure and reaction intermediate (virtual) elements of the quantal process* $^{11}Be(p,d)\ ^{10}Be(2^+)$.

It is difficult, if not impossible, to talk about single-particle motion without also referring to collective vibrational states, concerning both structure and reactions. Within this context, the difficulties of defining spectroscopic factors are apparent.

Summary

In Figure 5.2.6 a "complete" (NFT)$_{ren\ (s+r)}$ description of the single-neutron outside closed-shell halo nucleus ^{11}Be is given, in terms of the reactions ^{2}H$(^{10}$Be,^{11}Be$)^1$H populating the $1/2^+$, $1/2^-$ and $5/2^+$ states and of the ^{1}H$(^{11}$Be,^{10}Be$)^2$H process populating the 2^+ mode. Also shown, in comparison with the data, are the $E1$-transition between the parity inverted states $1/2^+$, $1/2^-$ and the isotopic shift of the charge radius of ^{10}Be.

Repeating the last paragraph of Sect. 2.9 one can again state that *this, in a very real sense, is a nucleus.*

5.2.3 NFT of Continuum Structure and Reactions: ^{10}*Li*

As already stated, while the N-even isotopes ^{9}Li$_6$ and ^{11}Li$_8$ are bound, the N-odd system ^{10}Li$_7$ is not.[29] Based on the large corpus of experimental[30] and theoretical[31] evidence concerning the fact that ^{10}Li also displays, as its bound isotone ^{11}Be$_7$, parity inversion, the barely bound, very extended, ^{11}Li system can be viewed as two-weakly interacting fermions outside a closed shell ($N = 6$).

The consistency of the above picture was challenged by a ^{9}Li$(d, p)^{10}$Li experiment[32] which found no evidence for $\ell = 0$ strength at low energies. In what follows an analysis of the corresponding results within the framework of continuum (NFT)$_{\text{ren (s+r)}}$ is discussed.

The ^{9}Li$(d, p)^{10}$Li Reaction

The NFT structure calculations were carried out as already explained in connection with the single halo isotone ^{11}Be$_7$. In fact, in the determination of the bare potential parameters also the normal Mayer-Jensen level sequence, $N = 7$ isotones $^{12}_5$B and $^{13}_6$C, were included.

The self-energy matrix $\sum^{\mathbf{a}}_{\nu_1\nu_2}(E)$ ($\mathbf{a} \equiv \{lj\}$) for the single-particle levels ν_1, ν_2 of ^{10}Li fulfilling $50\,\text{MeV} > \epsilon_{\nu_1}, \epsilon_{\nu_2} > \epsilon_F$, was calculated in a spherical box of 60 fm (continuum discretization). The dressed single-particle states were determined by diagonalizing $\sum^{\mathbf{a}}_{\nu_1\nu_2}(E)$. Up to about 2 MeV the spectrum is essentially determined by the $\widetilde{1/2}^+$ and $\widetilde{1/2}^-$ waves.

The $1/2^+$ scattering length was found to be[33] $\alpha = -8$ fm, corresponding to an energy $\epsilon_{\widetilde{1/2}^+} = \hbar^2(1/\alpha)^2/2m \approx 0.3$ MeV. The eigenfunction of a state lying close to this energy and, thus, representative of this virtual state is:

$$|\widetilde{1/2}^+\rangle = \sqrt{0.98}|s_{1/2}\rangle + \sqrt{0.02}|(d_{5/2} \otimes 2^+)_{1/2^+}\rangle. \qquad (5.2.4)$$

---

[29] The present section is based on Barranco et al. (2020).

[30] Wilcox et al. (1975); Young et al. (1994); Bohlen et al. (1997); Caggiano et al. (1999); Santi et al. (2003); Jeppesen et al. (2006); Smith et al. (2015); Sanetullaev et al. (2016); Kryger et al. (1993); Zinser et al. (1995); Kobayashi et al. (1997); Thoennessen et al. (1999); Chartier et al. (2001); Simon et al. (2007); Aksyutina et al. (2013); Amelin et al. (1990); Gornov et al. (1998); Chernysev et al. (2015); Kanungo et al. (2015); Tanihata et al. (2013); Fortune (2018); Tanihata et al. (2008).

[31] Tanihata et al. (2008); Thompson and Zhukov (1994); Bertsch et al. (1998); Blanchon et al. (2007); Mau and Pacheco (1996); Barker and Hickey (1977); Poppelier et al. (1993); Kitagawa and Sagawa (1993); Descouvemont (1997); Wurzer and Hofmann (1996); Kato and Ikeda (1993); Garrido et al. (2002, 2003); Potel et al. (2010); Casal et al. (2017); Vinh Mau (1995); Nunes et al. (1996); Sagawa et al. (1993); T. Myo et al. (2008); Barranco et al. (2001); Orrigo and Lenske (2009); Moro et al. (2019); Barranco et al. (2020).

[32] Cavallaro et al. (2017).

[33] This value provides circumstantial evidence on how extended the system described by the virtual state $|virt.\rangle = |\widetilde{1/2}^+(\kappa = 1/\alpha)\rangle$ is, in particular, as compared to the closed-shell $N = 6$ core ^{9}Li of radius $R_0(^9\text{Li}) \approx 2.6$ fm. A rough estimate of the overlap between $|virt.\rangle$ and single-particle states of ^{9}Li can be attempted through the ratio $(R_0/|\alpha|)^3 \approx 3.4 \times 10^{-2}$. Within this scenario, one may better assess the importance the many-body mixing amplitude $\langle(d_{5/2} \otimes 2^+)_{1/2^+}|\widetilde{1/2}^+\rangle$ $\left(\approx \sqrt{0.02}\right)$ has to determine the properties of the renormalized $|\widetilde{1/2}^+\rangle$ state. Within this context see Eq. (5.2.1).

Similarly, the resonant $\widetilde{1/2}^-$ state can be written as

$$|\widetilde{1/2}^-\rangle = \sqrt{0.94}|p_{1/2}\rangle + \sqrt{0.07}\left|\left(\left(p_{1/2}, p_{3/2}^{-1}\right)_{2^+} \otimes 2^+\right)_{0^+} p_{1/2}; 1/2^-\right\rangle. \quad (5.2.5)$$

The centroid and width of the resonance are $\epsilon_{\widetilde{1/2}^-} = 0.5$ MeV, and $\Gamma_{\widetilde{1/2}^-} = 0.35$ MeV respectively.[34] Theory predicts also the existence of a $5/2^+$ resonance with a centroid at ≈ 3.5 MeV. This state has also a clear many-body character, coupling to the quadrupole vibrations of the core through itself and through the $s_{1/2}$ state, not only in terms of virtual states containing one phonon ($|(d_{5/2} \otimes 2^+)_{5/2^+}\rangle$, $|(s_{1/2} \otimes 2^+)_{5/2^+}\rangle$) but also many (non crossing) phonon states which can be summed to infinite order (rainbow series).[35] Anharmonicities associated with states of two or more quadrupole phonons, resulting essentially from Pauli principle violation (see inset (IV) Fig. 5.2.6), are found to play an important role. Larger in the present case than in the case of ^{11}Be, in keeping with the fact that in ^{10}Li the $\widetilde{5/2}^+$ resonance lies closer in energy to the $\widetilde{1/2}^+ \otimes 2^+$ state than in ^{11}Be.

The coupling of the dressed neutron to the odd $1p_{3/2}^{-1}(\pi)$ proton hole leads to the doublets $(\widetilde{1/2}^+ \otimes p_{3/2}^{-1}(\pi))_{1^-,2^-}$ and $(\widetilde{1/2}^- \otimes p_{3/2}^{-1}(\pi))_{1^+,2^+}$ and to the quadruplet $(\widetilde{5/2}^+ \otimes p_{3/2}^{-1}(\pi))_{1^-,2^-,3^-,4^-}$. The scattering length of the resulting 2^-, 1^- states lead to $\epsilon_{2^-} \approx 0.05$ MeV and $\epsilon_{1^-} = 0.8$ MeV. For the positive-parity doublet, one finds $\epsilon_{1^+} \approx 0.3$ MeV and $\epsilon_{2^+} \approx 0.6$ MeV, while the quadruplet spans the energy interval 2–6 MeV. Theory also predict a $3/2^-$ state which splits into four states $(\widetilde{3/2}^- \otimes p_{3/2}^{-1}(\pi))_{0^+,1^+,2^+,3^+}$ with energies within the range 3–6 MeV.

Making use of the nonlocal self-energy matrices $\Sigma^a(r, r'; E)$, whose configuration space representation correspond to $\Sigma^a_{\nu_1\nu_2}(E)$, together with global optical parameters[36] the absolute double differential cross sections $d\sigma/dEd\Omega$ containing the information the (d, p) reaction in question can provide on the structure of ^{10}Li, was calculated.

One can now put the following question: recording the outgoing particle in the range of angles $5.5° \leq \theta_{cm} \leq 16.5°$, what the probability of observing the transferred neutron in the $\widetilde{1/2}^+$ final state is? The answer which arises from the strength function $d\sigma_{\widetilde{1/2}^+}/dE\Big)_{5.5°-16.5°}$ calculated by integrating $d\sigma/dEd\Omega$ in the chosen range of angles is univocal and reads *negligible*, as it emerges from Fig. 5.2.7, where theory is confronted with the experimental data to satisfactory agreement.

---

[34] For details, see Barranco et al. (2020), also supplemental material.

[35] It is of note that those many-phonon diagrams containing crossings are associated with vertex corrections. Because such diagrams imply recoupling coefficients of varied complexity and somewhat random phases, their summed contributions are expected to be considerably smaller than that of the rainbow series which displays coherence regarding the different contributions (see Barranco et al. (2019a)).

[36] Schmitt et al. (2013).

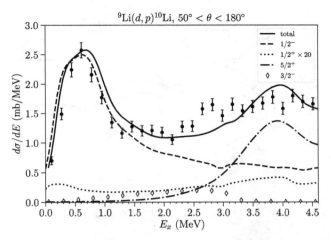

Figure 5.2.7 Theoretical predictions (continuous solid curve) of the absolute ^{10}Li strength function associated with the ^{9}Li$(d, p)^{10}$Li reaction at 100 MeV incident energy and integrated over the angular range $5.5° \leq \theta_{cm} \leq 16.5°$, in comparison with experimental data (solid dots with errors; Cavallaro et al. (2017)). The partial contributions are the incoherent sums of the strength functions of the multiplets $(\widetilde{j^{\pi}}, 1p_{3/2}(\pi))$, that is, (dashed line) $\widetilde{j^{\pi}} = \widetilde{1/2}^{-} (1^+, 2^+)$; (dash-dotted line) $\widetilde{5/2}^{+}(1^- - 4^-)$; (diamonds) $\widetilde{3/2}^{-} (0^+ - 3^+)$; (dotted line) $\widetilde{1/2}^{+}(1^-, 2^-)$ *multiplied by a factor of 20.*

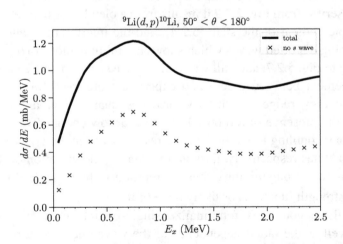

Figure 5.2.8 Predicted strength function obtained by integrating the absolute double differential cross section $d\sigma/dE d\Omega$ over the angular range $50° \leq \theta_{cm} \leq 180°$.

As seen from Fig. 5.2.8 the situation changes radically if *one puts the same question but changing the angular range from $5.5° \leq \theta_{cm} \leq 16.5°$ into $50° \leq \theta_{cm} \leq 180°$. Equally univocally the answer now reads large.*

$$0 < E_x < 0.2 \text{ MeV}$$

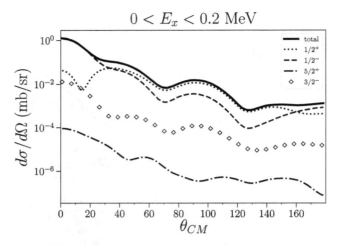

Figure 5.2.9 Predicted angular distributions ($d\sigma/d\Omega$) obtained by integrating the absolute double differential cross section $d\sigma/dEd\Omega$ in the energy interval 0–0.2 MeV. In the present case, and at variance with Fig. 5.2.7, no multiplicative factor was introduced in connection with $d\sigma_{\widetilde{1/2^+}}/d\Omega$.

The reason at the basis of the essential difference between the two questions, and thus answers, is provided by the absolute differential cross section[37] $d\sigma(\theta)_{\widetilde{1/2^+}}/d\Omega$ obtained integrating $d\sigma_{\widetilde{1/2^+}}/dEd\Omega$ over the energy interval 0–0.2 MeV. As observed from Fig. 5.2.9 it results from a blend of structure, related to the single-particle content of the state (5.2.4), and the interference pattern displayed by the outgoing distorted waves which shows a minimum close to $\theta_{cm} = 16.5°$.

Returning to Fig. 5.2.7, and still within the above scenario, one can posit that the overall agreement between theory and experiment observed for $d\sigma/dE)_{5°-16.5°}$ within the energy range 3–4.5 MeV provides quantitative, although indirect, evidence of the presence of a robust $1/2^+$ state at low energy. This is in keeping with the strong coupling found in the NFT results between the $s_{1/2}$ and $d_{5/2}$ virtual and resonant states respectively, through the quadrupole vibration of the core ^{9}Li. In fact, due to this strong mixing either one reproduces both the $d\sigma_{\widetilde{1/2^+}}/dE$ and the $d\sigma_{\widetilde{5/2^+}}/dE$ strength functions or likely none of them.[38]

Because these couplings renormalize single-particle content, energies, and widths, as well as the radial dependence of the wavefunctions (form factors), all these elements have to be calculated self-consistently. The role played by such requirements is likely emphasized in the case like the present one, of continuous spectroscopy, in which structure and reaction are, in a very real sense, just two aspects of the same physics.

[37] The consistency with the ^{9}Li$(d, p)^{10}$Li data of Jeppesen et al. (2006), recorded in the angular range $98° \leq \theta_{cm} \leq 134°$, is apparent. For details, see Barranco et al. (2020). See also Moro et al. (2019).

[38] Within this context, see Barranco et al. (2020). See also Moro et al. (2019).

Before concluding the present section, it may be useful to remind us what, within the framework of quantum mechanics, one can learn from a reaction experiment. It is not "what is the state after the collision" but "how probable is a given effect of the collision."[39]

If there is a lesson to be learned from the above discussion is the fact that, in dealing with a specific feature of a quantal many-body system, for example, single-particle motion in nuclei (structure) and one-particle transfer process (reaction), one can hardly avoid to talk about other elementary modes of excitation and reaction channels, respectively. Within the scenario of the chosen example, this is because a nucleon which, in first approximation is in a mean field stationary state, can actually be viewed as a fermion moving through a gas of quadrupole phonons (ZPF) to which it couples, becoming eventually heavier because dressed by them (case of $s_{1/2}$). But also due to Pauli principle, the nucleon is forced to exchange role with the nucleons of the composite (particle-hole) phonons undergoing repulsion (case of $p_{1/2}$).

5.3 Virtual States Forced to Become Real through Transfer Reactions

One of the main subjects which has been discussed in the present monograph concerns the melting of *structure and reactions* and of *bare, virtual and renormalized states* into a higher unity, which can be described in terms of Feynman diagrams where particles come in and the same or other particles come out and eventually interact with detectors whose clicks can be translated into absolute cross sections and lifetimes. Thus conveying to observations the properties of renormalized, dressed, physical elementary modes of excitation.

The initial and final asymptotic states of the $(NFT)_{ren(s+r)}$ Feynman diagrams are anchored to the laboratory (incoming beams and target-outgoing particles, including γ-rays and detectors). The intermediate states break open (virtually) the incoming systems, for example, a beam of the halo nucleus ^{11}Li, of half-life 8.75 ms, which impinges on a hydrogen target (proton) in an inverse kinematic experiment as displayed in, for example, Fig. 2.7.2 (a), in particular within the time interval $(t_1 - t_{11})$ – to allow the elementary modes of excitation which are probed by the nuclear reaction to become active, interact (melt) with other modes, and eventually pass the information concerning the properties of the physical modes to the outgoing particles.

Among these diagrams, one finds those associated with reaction processes in which a virtual state is acted upon and forced to become real, eventually reaching the detectors. A sort of Hawking radiation[40] to the extent that one concentrates on the (virtual)$\rightarrow$(real phenomenon) process, with the proviso of viewing the action

---

[39] Born (1926).
[40] Barranco et al. (2019b).

of the external field as the (no-return) event horizon of the black hole. An example of such diagrams which describes a possible event associated with the reaction ^{1}H(^{11}Li,^{9}Li(1/2$^-$))^{3}H is displayed in Fig. 2.7.3 (b): a virtual quadrupole vibration which, in the process of renormalizing the $s_{1/2}$ state, or of being exchanged between the partners of the neutron halo Cooper pair has been caught in the act by the pair transfer field produced by the ISAAC-2 facility at TRIUMF, and forced to become a real final state and to bring this information to the active detector MAYA (see also Fig. 2.7.3 (d)).

5.3.1 Empirical Renormalization

Most if not all theories, in particular those based on renormalization, able to provide a consistent description of structure and reactions of atomic nuclei can hardly avoid the separation between explicit and virtual configuration (phase) spaces.[41] Once this is recognized, a physical choice concerning the first type of degrees of freedom is elementary modes of excitation, that is, the response of the nuclear system, to the variety of experimental probes.

But here empiricism ends.[42] Because the *bare* elementary modes of excitation which potentially contains all of the physics that experiment can eventually provide, are not observables. To become so, they have to loose their elementarity and become mixed, dressed, renormalized, and melt together into effective fields. In fact, what we call a physical nucleon moving inside the nucleus is only partially to be associated with that nucleon field alone. It is also partially to be associated with the vibrational fields, because the two are in interaction. On the other hand, vibrational modes participating in virtual states which intervened upon with an appropriate external field can become on-shell, have to be dressed, fully renormalized modes, poised to be forced to become observable on short call.[43]

As shown in Fig. 5.3.2 (see also Fig. 5.3.1), the reaction ^{11}Be(p, d)^{10}Be(2^+) provides an embodiment of such processes. In fact, this reaction gives information

---

[41] See footnote 2 of Chapter 2. While configuration space traditionally refers to structure and phase space to reactions (see, e.g., Feshbach (1962)) these connotations tend to melt into a unity in the case of unstable, exotic nuclei like ^{11}Li ($\tau_{1/2} \approx 8$ ms), let alone ^{10}Li, transiently observed in terms of virtual and resonant states in a one-neutron transfer processes (see footnote 30 of the present chapter; see also Barranco et al. (2020), and refs. therein). As a consequence, both structure and reactions are to be worked out in the continuum, discretization being a connotation of methodological rather than of physical approach. Think only on the central role the tail of the neutron halo Cooper pair, bound by only 380 keV to the core (^{9}Li), play in stabilizing ^{11}Li, and in connection with two-nucleon transfer process (formfactors). Within this scenario, inverse kinematics in, e.g., the case of ^{2}H(^{9}Li,^{10}Li)^{1}H provides, in the laboratory set up, a large but finite box for both the probe (deuteron) and the probed (^{10}Li).

[42] To the extent of attributing the ancient Greek meaning of "find" and "discover" to the word *heuristic* ($\epsilon\upsilon\rho\iota\sigma\kappa\omega$) and of "serving to discover" of the Oxford dictionary, one can connotate the above empirical protocol as heuristic.

[43] As a rule, the collective modes participating in these virtual state are calculated making use of empirical input, thus, empirical renormalization.

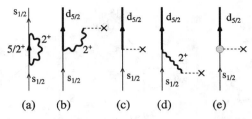

Figure 5.3.1 Inelastic population of the calculated $|^{11}\text{Be}(\widetilde{5/2}^+; 1.45 \text{ MeV})\rangle$ resonance. (a) Main process dressing the neutron moving in the $s_{1/2}$ orbital and leading to the ^{11}Be ground state. The virtual state is made out of the fully renormalized (empirical) quadrupole vibration of the core ^{10}Be and of the low-lying $5/2^+$ resonance of ^{11}Be and is indicated by using boldface symbols. (b) By intervening the process (a) with an external inelastic hadron field (e.g., (p, p')) of quadrupole character (dashed horizontal line starting at a cross), one can excite the $5/2^+$ resonance. This can also happen if the external field acts on the $1/2^+$ state as in (c), or if this field first excites the quadrupole vibration of ^{10}Be which eventually couples to the $1/2^+$ state as in (d). (e) If the vibration was a high-lying quadrupole giant resonance, the summed contribution of processes (b)–(d) could be replaced at profit by a single graph, namely, the equivalent of (c), but with an effective charge (hatched circle). For low-lying modes like the 3.368 MeV quadrupole vibration of ^{10}Be, retardation, ω-dependent effects are to be explicitly taken into account.

concerning the most important process clothing the $1/2^+$ parity inverted ground state of $^{11}_{4}\text{Be}_7$ through the coupling to the low-lying quadrupole vibration of the core $^{10}_{4}\text{Be}_6$ (see Fig. 5.3.2 **I** (**a**)). A schematic representation of the pickup of the neutron moving around a $N = 6$ closed shell and populating the low-lying quadrupole vibrational state of this core, in coincidence with the corresponding γ-decay is shown in **I(b)**. More detailed structure and reaction NFT diagrams, are shown in **I(d)** and **I(e)**, (the jagged line represents a graphic mnemonic of the recoil effect), together with a cartoon representation in (**f**). The predicted (continuous curve) and experimental (solid dots) absolute differential cross sections are displayed in **I(c)**. Protons and neutrons are labeled π and ν respectively, d stands for deuteron while γ-detectors are represented by a crossed rectangle. Curved arrows indicate motion in the continuum (reaction). Normal arrowed lines, motion inside target or projectile (structure).

At the basis of the γ-decay displayed in **I(b)** and **I(e)** we find the processes shown in the lower part of Fig. 5.3.2, as it is explained in what follows. (**II**) Interaction of protons in a nucleus with nuclear vibrations (solid dot, PVC vertex $\beta_L R_0 \partial U / \partial r \, Y^*_{LM}(\hat{r})$; β_L, dynamical distortion parameter; $U(r)$ central potential) and photons (normal vertex, electromagnetic interaction[44] $e \int d^4x \, J_\mu(x) A^\mu(x)$, the

---

[44] Holstein (1989).

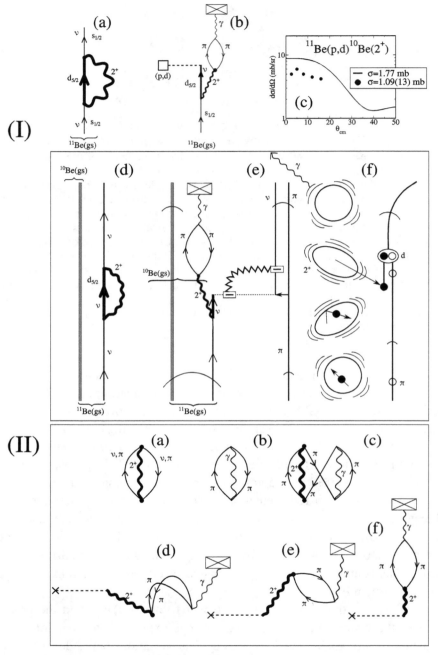

Figure 5.3.2 The reaction ^{11}Be$(p,d)^{10}$Be(2^+). A virtual process, namely, the self-energy contribution of the renormalization of the $s_{1/2}$ ground state of ^{11}Be through the coupling to the low-lying collective quadrupole vibration of ^{10}Be, becomes real through the action of a one-particle pickup external field. After Barranco et al. (2019b).

operator A^μ being the vector potential, and J_μ the current density ($\mu = 1, \ldots, 4$)). While the variety of diagrams shown have general validity, we have assumed one is dealing with the low-lying correlated particle-hole quadrupole vibration ($L = 2$) of $^{10}_{4}\text{Be}_6$ lying at 3.368 MeV, the corresponding quadrupole transition probability $B(E2; 0^+ \rightarrow 2^+) = 0.0052\ e^2b^2$ being associated with the dynamical deformation parameter $\beta_2 \approx 0.9$. An arrowed line pointing upward (downward) describes a proton (proton hole) moving in the $p_{1/2}$ ($1p_{3/2}$) orbital. Zero-point fluctuations of the nuclear ground state associated with: **(a)** the nuclear quadrupole vibration, **(b)** the electromagnetic field associated with the corresponding γ-decay. **(c)** Pauli principle correction to the simultaneous presence of the above two ZPF processes. **(d)** Intervening the virtual excitation of the nuclear vibrations (graph **(c)**) with an external (inelastic) field (cross followed by a dashed line), in coincidence with the γ−decay (γ−detector, crossed box), the virtual process **(c)** becomes real. **(e)**, **(f)** Time ordering of the above process correspond to the RPA contributions through backward-going and forward-going amplitudes and subsequent γ-decay.[45]

The above is an example of empirical renormalization (use of experimental elementary modes of excitation in both structure and reaction channels or, as in the present case, decay channels), namely, $((\text{NFT})_{\text{ren(s+r)}})$.[46]

5.3.2 On-Shell Energy

It was stated that intermediate states are, aside from their energy, fully dressed, renormalized physically observable elementary modes of excitation. Concerning the eventual value of the on-shell energy, it will naturally depend on the reaction process taking place. For example, in the reaction $^{11}\text{Be}(p, d)^{10}\text{Be}(2^+; 3.368\ \text{MeV})$ (Fig. 5.3.2 **(I)** (b) and (e)) populating the quadrupole vibration of ^{10}Be which, in the virtual $(\widetilde{5/2}^+ \otimes 2^+)_{1/2^+}$ state dress the $|\widetilde{1/2}^+\rangle$ ground state of ^{11}Be (Fig. 5.3.2 **(I)** (a)), the on-shell-energy coincides with that which is observed in the inelastic process $^{10}\text{Be}(\text{gs} \rightarrow 2^+)$. On the other hand, the $5/2^+$ pickup neutron, while displaying the structural properties of the $5/2^+$ resonance (Fig. 5.3.1) ($\tilde{\epsilon}_{5/2^+} = 1.45$ MeV), it has a binding energy equal to $\tilde{\epsilon}_{1/2^+} - \hbar\omega_{2^+} = -3.868\ \text{MeV}$ ($\tilde{\epsilon}_{1/2^+} = -0.5\ \text{MeV}$, $\hbar\omega_{2^+} = 3.368\ \text{MeV}$), unrelated to the experimental energy of $+1.28$ MeV (continuum, unbound), recorded in the reaction $^{10}\text{Be}(d, p)^{11}\text{Be}(5/2^+)$. This is a natural outcome of $(\text{NFT})_{\text{ren}}$ which, through the PVC and the Pauli mechanism, provides the proper clothing of the $d_{5/2}$ orbital so as to allow it to be

---

[45] It is of note that this is not a bubble diagram in the sense of rule (III) of NFT, as the initial and final vertices are associated with different processes (nuclear, electromagnetic).

[46] Broglia et al. (2016).

able "to exist" inside the $|\widetilde{s}_{1/2}\rangle$ state as a virtual, intermediate configuration. The asymptotic r–behavior results from the coherent superposition of many continuum states.

We now consider the process in which an inelastic field acting on $|^{11}\text{Be(gs)}\rangle$ populates the $5/2^+$ resonance (see Fig. 5.3.1). An important contribution to this excitation results from the action of the external field on the quadrupole vibration of the virtual state $(\widetilde{d}_{5/2} \otimes 2^+)_{1/2^+}$ which renormalizes the $1/2^+$ state (Fig. 5.3.1 (a)) leading to diagram (b) of the same figure. In this case, the external field has to provide an energy equal to $|\widetilde{\epsilon}_{1/2} - \widetilde{\epsilon}_{5/2^+}| = 1.95$ MeV, again unrelated to the on-shell energy of $|^{10}\text{Be}(2^+; 3.868 \text{ MeV})\rangle$ observed in the inelastic process $^{10}\text{Be(gs} \rightarrow 2^+)$, but needed to dress the bare single-particle valence orbital.

5.3.3 *Perturbation and Beyond*

In Q.E.D. at the one-loop diagrams level[47], the renormalized physical electron mass m_e, that is the observed mass (≈ 0.511 MeV), is related to the parameter m in the Dirac equation, the bare electron mass (not observable), according to $m_e = m\left(1 + \frac{3\alpha}{4\pi} \log\left(\frac{\Lambda_{cut}}{m}\right)^2\right)$, Λ_{cut} being the cutoff of the divergent integrals.[48] This quantity appears inside the log, and in front of it one has the fine structure constant α which is small ($\approx 1/137$). Therefore, even pushing the cutoff to the Planck scale, $\Lambda_{cut} \sim 10^{19}$ GeV, with $m \sim$ MeV one has $\delta m_e/m \approx \frac{3\alpha}{4\pi} \log\left(\frac{\Lambda_{cut}}{m}\right)^2 \approx 0.1$. So δm_e is really a small correction.

One can compare the value of the four parameters (plus the effective mass functional) which determine the mean field (Woods–Saxon) potential used to calculate the bare single-particle energies of ^{11}Be – imposing the condition that the dressed levels best reproduce the experimental findings – with the value of the corresponding parameters of the standard, global Woods–Saxon potential (and of effective mass equal to the observed mass m). The results displayed in Table 5.3.1 testify to the fact that renormalization in nuclear physics, in particular in the case of halo, exotic parity inverted nuclei, is less "perturbative" than in the case of Q.E.D. This fact becomes even clearer if one compares the overall centroid of the valence orbitals as well as the density of levels associated with the two potentials.

It is of note that the Lamb (-like) shift taking place between the $s_{1/2}$ and $p_{1/2}$ valence levels of ^{11}Be has a value of approximately 10 percent $((\Delta\epsilon_{1/2^+, 1/2^-})_{bare} - (\Delta\epsilon_{1/2^+, 1/2^-})_{ren} \approx 3.11 + 0.32\text{MeV})$ of that of the Fermi energy (≈ 36 MeV). This result can be compared with the ratio of the hydrogen $^2S_{1/2} - ^2P_{1/2}$ Lamb shift

---

[47] Diagram not allowed by rule III) of NFT in keeping with the fact that the nuclear (quasi) bosons are (RPA) "composite modes."

[48] See Bjorken and Drell (1998), pp. 162–166; Milloni (1994), p. 404.

Table 5.3.1 *Parameterization of the standard (Bohr and Mottelson (1969)) and of the bare mean field potential associated with ^{11}Be (Barranco et al. (2017); see also supplemental material). Changes of the order of 20–30 percent are observed. Within this context, in the case of Q.E.D., $\delta m_e/m = 0.1$:* [a]*$m^* = m$* [b]*$m^*(r = 0)$* *$= 0.7m$, $m^*(r = \infty) = (10/11) \times m$* [c] *$R_0 = r_0 A^{1/3}$, $r_0 = 1.2\,fm$* [d]*$r_0 \approx 1.03\,fm$.*

	V_0(MeV)	V_l(MeV)	R_0(fm)	a(fm)
Standard[a]	−50	17	2.7[c]	0.65
bare[b]	−68.9	14.47	2.15[d]	0.77

(1058 MHz$\approx 4.3 \times 10^{-9}$ eV) and the Rydberg constant ($R_H = 13.6$ eV), that is, $\approx 10^{-10}$, a result which underscores the strong coupled situation one is confronted with in trying to describe the structure of light halo exotic nuclei (Fig. 5.2.4).

5.3.4 One-Particle Transfer (s+r)-Dielectric Function

The fact that in spite of this nonperturbative situation, (NFT)$_{ren\,(r+s)}$ can provide an overall account of an essentially complete set of experimental data which characterizes ^{11}Be, within a 10 percent error, testifies to the power and flexibility Feynman version of Q.E.D. has. It can be used as paradigm to construct a field theory for both structure and reactions of a strongly interacting finite many-body system like the atomic nucleus. As already mentioned, examples of (NFT)$_{ren\,(r+s)}$ diagrams, aside from those discussed in detail in connection with Fig. 5.3.2, are displayed in Figs. 2.7.3 and 2.7.2. The first describes the process ^{1}H(^{11}Li,^{9}Li)^{3}H providing a quantitative account of the data and evidence of phonon induced pairing in nuclei. The second shows in (**a**), one of the most important channels contributing to the optical potential needed to describe the elastic scattering process ^{1}H(^{11}Li,^{11}Li)^{1}H. It is of note that in the case of ^{9}Li(d, p)^{10}Li (self-energy) and reaction (optical potentials) dielectric functions melt together. The calculation of the optical potential constitutes, within the scheme of (NFT)$_{ren\,(r+s)}$ a major challenge lying ahead.

5.3.5 (NFT)$_{ren(s+r)}$ Diagrams

Landau felt that Feynman diagrams, although usually derived from conventional field theory have an independent basic importance.[49] This work is closely connected with both Heisenberg statement that in quantum theory one should

---

[49] Landau (1959).

only introduce quantities[50] which are, in principle, observable and with the special role the scattering matrix plays in such a program. As the wavefunctions themselves cannot be observed and because the Hamiltonian formalism is intimately connected with wavefunctions, it may not be the most appropriate one to describe quantal systems. The quantities to be studied are the scattering amplitudes where particles go in and the same or different ones, including γ-rays, come out and determine the cross sections of the different processes. Within this context we refer to diagrams (b) and (e) of Fig. 5.3.2 (I) where the $(NFT)_{ren\,(r+s)}$ diagram describing the becoming of a concrete virtual process[51] real[52], is displayed.

5.4 Minimal Requirements for a Consistent Mean Field Theory

As seen from Figs. 1.2.2–1.2.4 the minimum requirements of self-consistency to be imposed upon single-particle motion requires both nonlocality in space (HF) and in time (TDHF)

$$i\hbar\frac{\partial\varphi_v}{\partial t} = -\frac{\hbar^2}{2m}\nabla^2\varphi_v(x,t) + \int dx'dt'U(x-x',t-t')\varphi_v(x',t'). \quad (5.4.1)$$

Assuming, for simplicity, infinite nuclear matter (confined by a constant potential of depth V_0), and thus plane wave solutions, the above time-dependent Schrödinger equation leads to the quasiparticle dispersion relation

$$\hbar\omega = \frac{\hbar^2k^2}{2m^*} + \frac{m}{m^*}V_0, \quad (5.4.2)$$

where the effective mass

$$m^* = \frac{m_k m_\omega}{m}, \quad (5.4.3)$$

is the product of the k-mass

$$m_k = m\left(1 + \frac{m}{\hbar^2k}\frac{\partial U}{\partial k}\right)^{-1}, \quad (5.4.4)$$

closely connected with the Pauli principle[53] $\frac{\partial U}{\partial k} \approx \frac{\partial U_x}{\partial k}$, while the ω-mass

$$m_\omega = m\left(1 - \frac{\partial U}{\partial\hbar\omega}\right), \quad (5.4.5)$$

---

[50] Heisenberg (1925); see English translation W. Heisenberg, Quantum-theoretical re-interpretation of kinematical and mechanical relations, in Van der Waerden (1967), p. 261.

[51] Dressing of the $s_{1/2}$ ground state of ^{11}Be, diagram (a), see also Table 5.3.1 for the parameters of the bare potential.

[52] Through the action of an external field, namely, ^{11}Be$(p,d)^{10}$Be(2^+).

[53] Within this context, see Sect. 2.3, in particular, footnote 47 of Chapter 2.

results from the dressing of the nucleon through the coupling with the (quasi) bosons.

The occupancy of levels around ε_F is related to Z_ω, a quantity which measures the discontinuity at the Fermi energy and which is equal to m/m_ω. This is in keeping with the fact that the time the nucleon is coupled to the vibrations it cannot behave as a single particle and can thus not contribute to, for example, the single-particle pickup cross section. *The particle–vibration coupling not only renormalizes energies and single-particle content but also the radial dependence of the wavefunction (formfactors).*

It is of note that the self-consistence requirements for the iterative solution of Eq. (5.4.1) reminds very much those associated with the solution of the Kohn-Sham equations in finite systems,

$$H^{KS}\varphi_\gamma(\mathbf{r}) = \lambda_\gamma \varphi_\gamma(\mathbf{r}), \tag{5.4.6}$$

where

$$H^{KS} = -\frac{\hbar^2}{2m_e}\nabla^2 + U_H(\mathbf{r}) + V_{ext}(\mathbf{r}) + U_{xc}(\mathbf{r}), \tag{5.4.7}$$

H^{KS} being known as the Kohn-Sham Hamiltonian, $V_{ext}(\mathbf{r})$ being the field created by the ions and acting on the electrons. Both the Hartree and the exchange–correlation potentials $U_H(\mathbf{r})$ and $U_{xc}(\mathbf{r})$ depend on the (local) density, hence on the whole set of wavefunctions $\varphi_\gamma(\mathbf{r})$. Thus, the set of KS–equations must be solved self-consistently.[54]

5.4.1 Density of Levels

Making use of Eq. (5.4.2) $(E = \hbar\omega)$, one can calculate dE/dk for a single nucleon and one spin orientation.[55] The inverse of this expression is

$$\frac{dk}{dE} = \frac{m^*}{\hbar^2 k}, \tag{5.4.8}$$

which testifies to the fact that the energy spacing between levels, that is, the density of levels (see below), changes as m^* does.

One can then calculate the average value over the Fermi distribution, obtaining

$$\left\langle \frac{dk}{dE} \right\rangle = \frac{m^*}{\hbar^2 (2/3)k_F}. \tag{5.4.9}$$

---

[54] See, e.g., Broglia et al. (2004), and refs. therein.
[55] Mahaux et al. (1985), p. 17.

Let us now take into account all nucleons, both spin orientations and eliminate the unit (inverse) length, i.e.,

$$\frac{2A}{k_F}\left\langle\frac{dk}{dE}\right\rangle = 3A\frac{m^*}{\hbar^2 k_F^2} = 3A\frac{m^*}{2m\epsilon_F}. \tag{5.4.10}$$

Assuming $m^* = m_\omega m_k/m \approx m$, where m is the experimental nucleon mass one obtains

$$\frac{3}{2}\frac{A}{\epsilon_F}, \tag{5.4.11}$$

a value which coincides with the Fermi gas model estimate for the one-particle level density g_0.[56] Taking properly into account the geometry of the system, one obtains

$$a = \frac{\pi^2}{6}\frac{3}{2}\frac{A}{\epsilon_F} \tag{5.4.12}$$

for the prefactor of the exponential of the Fermi expression of the total density of single-particle levels. Making use of $\epsilon_F = 36$ MeV leads to

$$a \approx \frac{A}{14}\text{MeV}^{-1}. \tag{5.4.13}$$

In keeping with the fact that one can interpret dE/dk as the rate of change in energy when the momentum changes or, equivalently, when the number of nodes per unit length changes, and this can be used to label the single-particle states, Eq. (5.4.13) can be confronted at profit with the average degeneracy per unit energy of valence orbitals (see Table 5.4.1).

Within this context it is of note that an estimate of the quantity a based on the harmonic oscillator, leads to $a \approx \frac{\pi^2}{6}\frac{(N_{max}+3/2)^2}{\hbar\omega_0}$, that is, an expression inversely proportional to the (constant) energy separation of levels.[57] In an attempt to bridge the gap between the nuclear matter expressions discussed above and finite nuclei, that is, potential wells of finite range, we consider

$$H = \frac{p^2}{2D} + \frac{C}{2}x^2, \tag{5.4.14}$$

($p = D\dot{x}$) which leads to a constant level spacing,

$$\hbar\omega_0 = \hbar\sqrt{\frac{C}{D}}, \tag{5.4.15}$$

and implies that the density of states is proportional to the square root of the particle inertia (mass). However, this result follows from the assumption that the potential

---

[56] See, e.g., Bohr and Mottelson (1969), Eq. (2-48).
[57] See Bohr and Mottelson (1969), p. 188, Eq. (2-125a).

Table 5.4.1 *Comparison of the factor (5.4.13)*
($a = n/14MeV^{-1}$) corresponding to ^{208}Pb for both
$n = N$ and $n = Z$ and for ^{120}Sn in the case of $n = N$, in
comparison with the empirical value associated with the
valence orbitals of these nuclei, that is, ($h_{9/2}$, $f_{7/2}$,
$i_{13/2}$, $p_{3/2}$, $f_{5/2}$, $p_{1/2}$, $g_{9/2}$, $i_{11/2}$, $d_{5/2}$, $j_{15/2}$, $s_{1/2}$, $g_{7/2}$, $d_{3/2}$)
and ($g_{7/2}$, $d_{5/2}$, $h_{11/2}$, $d_{3/2}$, $s_{1/2}$, $h_{9/2}$, $f_{7/2}$, $i_{13/2}$) for
^{208}Pb in the case of neutrons and protons, respectively,
leading to $\sum_j^N (2j + 1)/\Delta E_N \approx 102/10$ MeV$\approx$ 10
MeV^{-1} and $\sum_j^Z (2j + 1)/\Delta E_Z = 64/(9\ MeV) \approx 7$
MeV^{-1}. The quantity ΔE_N is the experimental energy
interval over which the valence orbitals are distributed
(see, e.g., Bohr and Mottelson (1969), p. 325, Fig. 3-3).
In the case of neutrons of ^{120}Sn, use is made of the
dressed valence orbitals $d_{5/2}$, $g_{7/2}$, $s_{1/2}$, $d_{3/2}$, $h_{11/2}$
resulting from the renormalization of HF-Sly4 levels
through the coupling of collective modes making use of
nuclear field theory plus Nambu–Gorkov techniques
((NFT)+(NG); for details, see Idini et al. (2015) and
Table I of Potel et al. (2017)). The result, taking into
account the breaking of the single-particle strength, in
particular, that of the $d_{5/2}$ orbital, is
$\sum_j^N (2j + 1)/\Delta E_N = 32/8$ MeV$\approx$ 4 MeV^{-1}.

	MeV^{-1}	
	empirical	a
$^{208}_{82}$Pb$_{126}$	$17(10^{a)} + 7^{b)})$	$15(9^{a)} + 6^{b)})$
$^{120}_{50}$Sn$_{70}$	$4^{a)}$	$5^{a)}$

a) neutrons
b) protons

remains unchanged if the bare mass (in which case D is, for example, set equal
to the HF k-mass, m_k) is replaced by an effective mass m^* (e.g., $m_k m_\omega/m$; Eq.
(5.4.3)). However, the ground state wavefunction

$$\varphi_0 \sim \exp\left(-\frac{x^2}{2b^2}\right), \qquad (5.4.16)$$

with,

$$b = \sqrt{\frac{\hbar}{m^*\omega_0}}, \qquad (5.4.17)$$

will, in the case of a dressed nucleon of effective mass $m^* > D$, shrink in space as compared with the one of mass D. Consequently, the mean square radius of the system

$$\langle r^2 \rangle = \frac{\hbar}{m^*\omega_0}\left(N + \frac{3}{2}\right) = b^2\left(N + \frac{3}{2}\right) \qquad (5.4.18)$$

will decrease. This is not correct, and one has to impose the condition b^2=constant. In other words, to use the relation (5.4.17) with the condition b =const., to relate $\hbar\omega_0$ and m^*. A condition which implies that the energy difference between levels is inversely proportional to the effective mass of the nucleon[58], that is,

$$\hbar\omega_0 = \frac{\hbar^2}{m^*b^2}. \qquad (5.4.19)$$

Let us conclude this section with a remark concerning the dimensions of the parameters D and C entering Eq. (5.4.14). Because the variable x has dimensions of length ($[x]$=fm), the dimensions of the inertia and restoring force parameters are

$$[C] = \text{MeV fm}^{-2} \quad \text{and} \quad [D] = \text{MeV fm}^{-2}\text{s}^2.$$

Consequently the associated zero-point fluctuations

$$\sqrt{\frac{\hbar\omega_0}{2C}} = \sqrt{\frac{\hbar^2}{2D}\frac{1}{\hbar\omega_0}} \qquad (5.4.20)$$

have dimensions of fm. It is of note that in the case of the harmonic oscillator Hamiltonian (1.1.9), the associated ZPF is a c-number, in keeping with the fact that the (dynamic) deformation parameter α is dimensionless and thus $[D]$=MeVs^2 and $[C]$=MeV.

5.5 Self-Energy and Vertex Corrections

In Fig. 5.5.1 a graphical example is displayed of the fact that in QED (and thus in NFT tailored after it) nothing is really free and that, for example, the bare mass of a fermion (electron or nucleon), is the parameter one adjusts so that the renormalized particle (graphs (b)–(d)) displays properties which agree with experimental findings. In this case, the observed mass m_e (single-particle energy ϵ_ν).

In Fig. 5.5.2, lowest-order diagrams associated with the renormalization of the fermion-boson interaction (vertex corrections) are shown. The sum of contributions (a) and (b) can, in principle, be represented by a renormalized vertex (see

[58] See Sect. 3.2, Eq. (3.2.3); also Fig. 2.6.1 ((a),(e)–(g)).

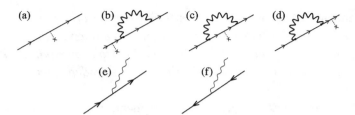

Figure 5.5.1 Self-energy (effective-mass-like) processes. The outcome of probing with an external field (dotted line started with a cross, observer) of the properties (mass, energy $\tilde{\epsilon}_\nu$, etc.) of a fermion (e.g., an electron or a nucleon, arrowed line) dressed through the coupling to bosons (photons or collective vibrations (in which case one talks about a quasi boson), wavy line), results from the sum of the contributions associated with diagrams (a)–(d) (cf. Feynman (1975)). In graphs (e) and (f), the amplitudes which eventually lead to the self-energy processes (b)–(d) are shown.

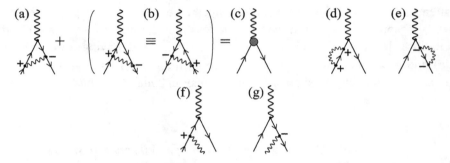

Figure 5.5.2 (a, b) Vertex corrections. These are triple-interaction diagrams (phonon, particle, and hole lines) in which none of the incoming lines can be detached from either of the other two by cutting one line. In connection with condensed matter, Migdal's theorem (Migdal (1958)) states that for phonons, (Bardeen and Pines (1955); Fröhlich (1952)), vertex corrections can be neglected (cf. also Anderson (1964a)). Vertex corrections are, as a rule, important in the nuclear case, where they lead to conspicuous cancellations of the self-energy contributions (d) and (e) (cf., e.g., Bortignon et al. (1983)). The solid gray circle in (c) represents the effective, renormalized vertex. In graphs (f) and (g) the amplitudes describing the coupling of a $p - h$ vibrational mode with the $2p - 2h$ doorway states which eventually lead to vertex and self-energy corrections ((a)–(e)) are schematically displayed.

diagram (c)). There is, as a rule, conspicuous interference (e.g., cancellation) in the nuclear case between vertex and self-energy contributions (see diagrams (a)+(b) and (d)+(e) of Fig. 5.5.2, a phenomenon closely related with conservation laws.[59] In particular, cancellation in the case in which the bosonic modes are of isoscalar,

[59] Schrieffer (1964); Bortignon and Broglia (1981); Bertsch et al. (1983); Bortignon et al. (1998), pp. 82–86.

spin-independent character.[60] *Consequently, one has to sum explicitly the appropriate amplitudes (diagrams (f)+(g)), with the corresponding phases before taking the modulus squared, to eventually obtain the quantities to be compared with the data, a fact that call attention to the use of an effective, ω-independent (renormalized) vertex (Fig. 5.5.2 (c)).*

Within the framework of QED the afore mentioned cancellations are exact implying, for example, that the interaction between one- and two-photon states vanishes (Furry theorem). The physics at the basis of the cancellation found in the nuclear case can be exemplified by looking at a spherical, closed-shell nucleus displaying a low-lying collective octupole vibration, like in ^{208}Pb where $|3^-; 2.615 \, \text{MeV}\rangle$ is the first excited state, its $B(E3)$ being equal to 32 B_{sp}. The associated zero-point fluctuations (ZPF) lead to time-dependent shapes with varied instantaneous values of the octupole moment. In other words, an important component of the ZPF associated with $|gs(^{208}\text{Pb})\rangle$ (vacuum) state can be written as $(|(j_p \otimes j_h^{-1})_{3^-} \otimes 3^-; 0^+\rangle)$. It can be viewed as a gas of octupole (quasi) bosons promoting a nucleon across the Fermi surface (particle-hole excitation), leading to fermionic states which behave as having a positive (particle) and a negative (hole) effective octupole moment. This is in keeping with the fact that the closed-shell system is spherical and thus has zero octupole moment.

5.5.1 *Phase Coherence and Damping of Vibrational Modes*

Paradigmatic examples of vibrational modes are provided by giant resonances. In particular by the giant dipole resonance (GDR). This collective mode is made out of correlated particle-hole excitations. An estimate of the associated damping width can be obtained in terms of the sum of the individual widths of the particle and of the hole. With he help of relations (2.3.20), (2.3.21) $(\Gamma_p(\frac{\hbar\omega}{2}) = 0.5 \times \frac{\hbar\omega}{2})$ valid for both particles and holes, one obtains

$$\Gamma_{GDR}(\hbar\omega_{GDR}) = \Gamma_p(\hbar\omega_{GDR}/2) + \Gamma_h(\hbar\omega_{GDR}/2) \approx 0.5 \, \hbar\omega_{GDR}, \qquad (5.5.1)$$

where the assumption was made that the excitation energy $\hbar\omega_{GDR}$ is divided equally between the particle and the hole. Making use of the empirical expression $\hbar\omega_{GDR} \approx 80/A^{1/2}$ MeV (Eq. (1.5.2) and following text), one can write

$$\Gamma_{GDR} \approx \frac{40}{A^{1/3}} \, \text{MeV}, \qquad (5.5.2)$$

[60] Bortignon et al. (1983).

which, in the case of ^{208}Pb leads to $\Gamma_{GDR} \approx 7$ MeV. A number to be compared with the experimental value[61] of ≈ 4 MeV. This overprediction of a factor of 2, is a consequence of taking the modulus squared of the amplitudes associated with graphs (e) and (f) of Fig. 5.5.1 separately, instead of doing so after having summed them with the proper phase (see Fig. 5.5.2 (f), (g)). That is

$$\Gamma_{GDR} \approx \left| \frac{\sqrt{\Gamma_p} + e^{-i\alpha}\sqrt{\Gamma_h}}{\sqrt{2}} \right|^2 \approx \Gamma_p \left| \frac{1 + e^{-i\alpha}}{\sqrt{2}} \right|^2 = \Gamma_p \left(1 + \cos\alpha\right), \quad (5.5.3)$$

where

$$\alpha = (1 - c)\,\pi/2, \quad (5.5.4)$$

c being defined by the relation $(\tau \hat{F} \tau^{-1})^\dagger = -c\hat{F}$, where $c(=\pm 1)$ is a c-number, $\hat{F}$ is the single-particle operator entering in the particle–vibration coupling Hamiltonian (see Eqs. (1.2.3)–(1.2.6)). In the case in which the (quasi-) bosonic mode is a low-lying surface vibration (see Eq. (1.2.7))[62], $c = -1$. As a result, a conspicuous cancellation between the particle and hole contributions[63] takes place (see diagrams (a) and (b) of Fig. 5.5.3, where typical amplitudes contributing to the damping width Γ_{GDR} of ^{208}Pb are displayed).

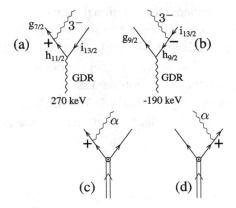

Figure 5.5.3 (a, b) Example of cancellation between the particle- and hole- doorway contributions to the damping width Γ_{GDR} (Bortignon et al. (1983)); (c, d) similar diagrams, this time associated with a pair addition mode in general, and with the GPV in particular (α: low-lying collective surface vibrations).

---

[61] Bertrand (1981).

[62] The precise formulation of this result is that in the case in which the field $\hat{F}$ is spin- and isospin-independent, $c = -1$ (see Bortignon et al. (1983); see also Bortignon and Broglia (1981)).

[63] The fact that expression (5.5.3) leads to complete cancellation is in keeping with the approximation that the single-particle and -hole subspaces are identical (i.e., $\Gamma_p = \Gamma_h$). In fact, the finite value of Γ_{GR} in general and of Γ_{GDR} in particular is associated with the asymmetry existing between particle and hole subspaces.

It is of note that for giant pairing vibrations, built out of two correlated particles (or holes), the corresponding contributions (see diagrams (c) and (d), Fig. 5.5.3) have the same sign, the expression $\Gamma_{GPV} \approx 0.5\,\hbar\omega_{GPV}$ (see also (5.5.1)) being, in this case, the correct one.

5.6 Single-Nucleon Transfer for Pedestrians

In what follows we discuss some aspects of the relations existing between nuclear structure and one-particle transfer cross sections. To do so, we repeat some of the steps carried out in Sect. 5.1, but this time in a simpler way, essentially ignoring the implications associated with the spin carried out by the particles, the spin-orbit dependence of the optical model potential, the recoil effect, etc.

We consider the case of $A(d, p)A+1$ reaction, namely, that of neutron stripping. The intrinsic wave functions ψ_α and ψ_β, where $\alpha = (A, d)$ and $\beta = ((A + 1), p)$,

$$\psi_\alpha = \psi_{M_A}^{I_A}(\xi_A)\phi_d(\mathbf{r}_{np}), \tag{5.6.1a}$$

$$\psi_\beta = \psi_{M_{A+1}}^{I_{A+1}}(\xi_{A+1})$$
$$= \sum_{l, I_A'}(I_A'; l|\}I_{A+1})[\psi^{I_A'}(\xi_A)\phi^l(\vec{r}_n)]_{M_{A+1}-M_A}^{I_{A+1}}, \tag{5.6.1b}$$

where $(I_A'; l|\}I_{A+1})$ is a generalized fractional parentage coefficient. It is of note that, as a rule, $(I_A'; l|\}I_{A+1})\,\phi^l(\vec{r}_n)$ should be able to describe a dressed quasiparticle state containing only a fraction of the single-particle strength. Although ignoring possible radial renormalization we assume, for simplicity, the expansion to be operative. To further simplify the derivation we assume we are dealing with spinless particles. This is the reason why no "intrinsic" proton wavefunction appears in Eq. (5.6.1b). The variable $\vec{r}_{np}$ is the relative coordinate of the proton and the neutron (see Fig. 5.6.1).

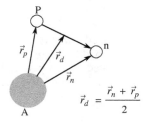

Figure 5.6.1 Coordinates used in the description of the $A(d, p)(A + 1)$ stripping process.

The transition matrix element can now be written as

$$T_{d,p} = \langle \psi_{M_{A+1}}^{I_{A+1}}(\xi_{A+1}) \chi_p^{(-)}(\mathbf{k}_p, \mathbf{r}_p), V_\beta' \psi_{M_A}^{I_A}(\xi_A) \chi_d^{(+)}(\mathbf{k}_d, \mathbf{r}_d) \rangle$$

$$= \sum_{\substack{l, I_A' \\ M_A'}} (I_A'; l|\}I_{A+1})(I_A' M_A' l M_{A+1} - M_A'|I_{A+1} M_{A+1})$$

$$\times \int d\mathbf{r}_n d\mathbf{r}_p \chi_p^{*(-)}(\mathbf{k}_p, \mathbf{r}_p) \phi_{M_{A+1}-M_A'}^{*l}(\mathbf{r}_n)(\psi_{M_A}^{I_A}(\xi_A), V_\beta' \psi_{M_A'}^{I_A'}(\xi_A))$$

$$\times \phi_d(\mathbf{r}_{np}) \chi_d^{(+)}(\mathbf{k}_d, \mathbf{r}_d) \, \delta_{I_A', I_A} \, \delta_{M_A', M_A}. \tag{5.6.2}$$

In the stripping approximation

$$V_\beta' = V_\beta(\xi, \mathbf{r}_\beta) - \bar{U}_\beta(r_\beta)$$

$$= V_\beta(\xi_A, \mathbf{r}_{pA}) + V_\beta(\mathbf{r}_{pn}) - \bar{U}_\beta(r_{pA}). \tag{5.6.3}$$

Then

$$(\psi_{M_A}^{I_A}(\xi_A), V_\beta' \psi_{M_A}^{I_A}(\xi_A)) = (\psi_{M_A}^{I_A}(\xi_A), V_\beta(\xi_A, \mathbf{r}_{pA}) \psi_{M_A}^{I_A}(\xi_A))$$

$$+ (\psi_{M_A}^{I_A}(\xi_A), V_\beta(\mathbf{r}_{pn}) \psi_{M_A}^{I_A}(\xi_A)) - \bar{U}_\beta(r_{pA}). \tag{5.6.4}$$

We assume

$$U_\beta(r_{pA}) = (\psi_{M_A}^{I_A}(\xi_A), V_\beta(\xi_A, \mathbf{r}_{pA}) \psi_{M_A}^{I_A}(\xi_A)). \tag{5.6.5}$$

Then

$$(\psi_{M_A}^{I_A}(\xi_A), V_\beta' \psi_{M_A}^{I_A}(\xi_A)) = V(\mathbf{r}_{pn}). \tag{5.6.6}$$

Inserting Eq. (5.6.6) into Eq. (5.6.2) we obtain

$$T_{d,p} = \sum_l (I_A; l|\}I_{A+1})(I_A M_A l M_{A+1} - M_A|I_{A+1} M_{A+1})$$

$$\times \int d\mathbf{r}_n d\mathbf{r}_p \chi_p^{*(-)}(\mathbf{k}_p, \mathbf{r}_p) \phi_{M_{A+1}-M_A}^{*l}(\mathbf{r}_n) V(\mathbf{r}_{pn}) \phi_d(\mathbf{r}_{np}) \chi_d^{(+)}(\mathbf{k}_d, \mathbf{r}_d). \tag{5.6.7}$$

The differential cross section is then equal to

$$\frac{d\sigma}{d\Omega} = \frac{2}{3} \frac{\mu_p \mu_d}{(2\pi\hbar^2)^2} \frac{(2I_{A+1}+1)}{(2I_A+1)} \frac{k_p}{k_d} \sum_{l, m_l} \frac{(I_A; l|\}I_{A+1})^2}{2l+1} |B_{m_l}^l|^2, \tag{5.6.8}$$

where

$$B_{m_l}^l(\theta) = \int d\mathbf{r}_n d\mathbf{r}_p \chi_p^{*(-)}(\mathbf{k}_p, \mathbf{r}_p) Y_m^{*l}(\hat{\mathbf{r}}_n) u_{nl}(r_n) V(\mathbf{r}_{pn}) \phi_d(\mathbf{r}_{np}) \chi_d^{(+)}(\mathbf{k}_d, \mathbf{r}_d) \tag{5.6.9}$$

and

$$\phi_m^l(\mathbf{r}_n) = u_{nl}(r_n)Y_m^l(\hat{\mathbf{r}}_n), \tag{5.6.10}$$

is the single-particle wave function of a neutron bound to the core A. For simplicity, the radial wave function $u_{nl}(r_n)$ can be assumed to be a solution of a Woods–Saxon potential of parameters $V_0 \approx 50$ MeV, $a = 0.65$ fm and $r_0 = 1.25$ fm.

The relation (5.6.8) gives the cross section for the stripping from the projectile of a neutron that would correspond to the nth valence neutron in the nucleus $(A+1)$. If we now want the cross section for stripping any of the valence neutrons of the final nucleus from the projectile, we must multiply Eq. (5.6.8) by n. A more careful treatment of the antisymmetry with respect to the neutrons shows this to be the correct answer.

Finally we get

$$\frac{d\sigma}{d\Omega} = \frac{(2I_{A+1} + 1)}{(2I_A + 1)} \sum_l S_l \sigma_l(\theta), \tag{5.6.11}$$

where

$$S_l = n(I_A; l|\}I_{A+1})^2, \tag{5.6.12}$$

and

$$\sigma_l(\theta) = \frac{2}{3} \frac{\mu_p \mu_d}{(2\pi \hbar^2)^2} \frac{k_p}{k_d} \frac{1}{2l + 1} \sum_m |B_m^l|^2 \tag{5.6.13}$$

Distorted wave codes evaluate numerically the quantity $B_{m_l}^l(\theta)$, using for the wave functions $\chi^{(-)}$ and $\chi^{(+)}$ the solution of the optical potentials that fit the elastic scattering, that is,

$$(-\nabla^2 + \bar{U} - k^2)\chi = 0, \tag{5.6.14}$$

Note that if the target nucleus is even-even, $I_A = 0$, $l = I_{A+1}$. That is, only one l value contributes in Eq. (5.6.8), and the angular distribution is uniquely given by $\sum_m |B_m^l|^2$. The l-dependence of the angular distributions helps to identify I_{A+1}. The factor S_l needed to normalize the calculated function to the data yields (assuming a good fit to the angular distribution), is known in the literature as the spectroscopic factor. In the early stages of studies of nuclear structure with one-particle transfer reactions it was assumed, not only that such quantity could be accurately defined, but also that it contained all the nuclear structure information (aside from that associated with the angular distribution) which could be extracted from single-particle transfer. In other words, that it was the bridge directly connecting theory with experiment. *Because nucleons are never bare, but are dressed by the coupling to collective modes, coupling which renormalizes not only the single-particle content but also the single-particle wavefunctions (formfactors), the spectroscopic factor approximation is not straightforward to justify.*

There is a fundamental problem which makes the handling of integrals like that of (5.6.9) difficult to handle from a structure point of view, if not numerically at least conceptually. This difficulty is connected with the so called recoil effect[64], namely, the fact that the center of mass of the two interacting particles in entrance ($\mathbf{r}_\alpha : \alpha = a + A$) and exit ($\mathbf{r}_\beta : \beta = b + B$) channels is different. This is at variance with what one is accustomed to deal with in nuclear structure calculations, in which the Hartree potential depends on a single coordinate, as well as in the case of elastic and inelastic reactions, situations in which $\mathbf{r}_\alpha = \mathbf{r}_\beta$. When $\mathbf{r}_\alpha \neq \mathbf{r}_\beta$ we enter a rather more complex many-body problem, in particular if continuum states are to be considered, than nuclear structure practitioners are accustomed to.

Returning to the subject of the present section, it is in general useful to be able to introduce approximations which can help the physics which is at the basis of the phenomenon under discussion (single-particle motion), to emerge in a natural way, if not to compare in detail with the experimental data. Within this context, to reduce the integral (5.6.9) one can assume that the proton–neutron interaction V_{np} has zero range, that is,

$$V(\mathbf{r}_{np})\phi_d(\mathbf{r}_{np}) = D_0\delta(\mathbf{r}_{np}) \qquad (5.6.15)$$

so that B_m^l becomes equal to

$$B_{m_l}^l(\theta) = D_0 \int d\mathbf{r}\, \chi_p^{*(-)}(\mathbf{k}_p, \mathbf{r}) Y_{m_l}^{*l}(\hat{\mathbf{r}}) u_l(r) \chi_d^{(+)}(\mathbf{k}_d, \mathbf{r}), \qquad (5.6.16)$$

which is a three dimensional integral, but in fact essentially a one–dimensional integral, as the integration over the angles can be worked out analytically.

5.6.1 Plane-Wave Limit

If in Eq. (5.6.14) one sets $\bar{U} = 0$, the distorted waves become plane waves, that is,

$$\chi_d^{(+)}(\mathbf{k}_d, \mathbf{r}) = e^{i\mathbf{k}_d\cdot\mathbf{r}}, \qquad (5.6.17a)$$

$$\chi_d^{*(-)}(\mathbf{k}_p, \mathbf{r}) = e^{-i\mathbf{k}_p\cdot\mathbf{r}}. \qquad (5.6.17b)$$

Equation (5.6.16) can now be written as

$$B_m^l = D_0 \int d\mathbf{r}\, e^{i(\mathbf{k}_d-\mathbf{k}_p)\cdot\mathbf{r}} Y_m^{*l}(\hat{\mathbf{r}}) u_l(r). \qquad (5.6.18)$$

---

[64] While this effect could be treated in a cavalier fashion in the case of light ion reactions ($m_a/m_A \ll 1$), this was not possible in the case of heavy ion reactions, as the change in momenta involved was always sizable (cf. Broglia and Winther (2004), and refs. therein).

The linear momentum transferred to the nucleus is $\mathbf{k}_d - \mathbf{r}_p = \mathbf{q}$. Let us expand $e^{i\mathbf{q}\cdot\mathbf{r}}$ in spherical harmonics, that is,

$$
\begin{aligned}
e^{i\mathbf{q}\cdot\mathbf{r}} &= \sum_l i^l j_l(qr)(2l+1) P_l(\hat{\mathbf{q}}\cdot\hat{\mathbf{r}}) \\
&= 4\pi \sum_l i^l j_l(qr) \sum_m Y_m^{*l}(\hat{\mathbf{q}}) Y_m^l(\hat{\mathbf{r}}),
\end{aligned}
\tag{5.6.19}
$$

so

$$
\int d\hat{\mathbf{r}}\, e^{i\mathbf{q}\cdot\mathbf{r}} Y_m^l(\hat{\mathbf{r}}) = 4\pi i^l j_l(qr) Y_m^{*l}(\hat{\mathbf{q}}).
\tag{5.6.20}
$$

Then

$$
\begin{aligned}
\sum_m |B_m^l|^2 &= \sum_m |Y_m^l(\hat{\mathbf{q}})|^2 D_0^2 16\pi^2 \\
&\quad \times \left| \int r^2 dr\, j_l(qr) u_l(r) \right|^2 \\
&= \frac{2l+1}{4\pi} D_0^2 16\pi^2 \left| \int r^2 dr\, j_l(qr) u_l(r) \right|^2.
\end{aligned}
\tag{5.6.21}
$$

Thus, the angular distribution is given by the integral $\left| \int r^2 dr\, j_l(qr) u_l(r) \right|^2$. If one assumes that the process takes place mostly on the surface, the angular distribution will be given by $|j_l(q R_0)|^2$, where R_0 is the nuclear radius.

We then have

$$
\begin{aligned}
q^2 &= k_d^2 + k_p^2 - 2k_d k_p \cos(\theta) \\
&= (k_d^2 + k_p^2 - 2k_d k_p) + 2k_d k_p (1 - \cos(\theta)) \\
&= (k_d - k_p)^2 + 4k_d k_p (\sin(\theta/2))^2 \\
&\approx 4k_d k_p (\sin(\theta/2))^2,
\end{aligned}
\tag{5.6.22}
$$

since $k_d \approx k_p$ for stripping reactions at typical bombarding energies. Thus the angular distribution has a diffraction-like structure given by

$$
|j_l(q R_0)|^2 = j_l^2(2R_0 \sqrt{k_d k_p} \sin(\theta/2)).
\tag{5.6.23}
$$

The function $j_l(x)$ has its first maximum at $x = l$, that is, where

$$
\sin(\theta/2) = \frac{l}{2R_0 k}, \qquad (k_p \approx k_d = k),
\tag{5.6.24}
$$

Examples of the above relation are provided in Fig. 5.6.2.

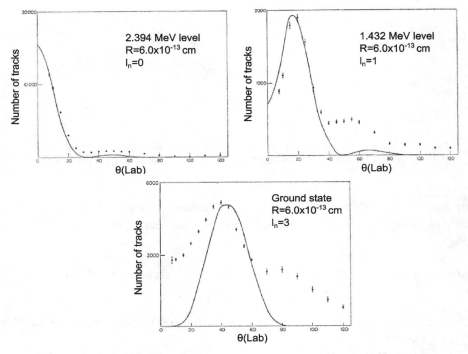

Figure 5.6.2 Plane wave approximation analysis of three ^{44}Ca(d,p)^{45}Ca differential cross sections leading to the ground state ($l_n = 3$) and to the 1.9 MeV ($l_n = 1$) and 2.4 MeV ($l_n = 0$) excited states, that is, $f_{9/2}$, $p_{1/2}$ and $s_{1/2}$ states (Cobb and Guthe (1957)).

5.7 One-Particle Knockout within DWBA

5.7.1 Spinless Particles

We consider the reaction $A + a \rightarrow a + b + c$, in which particle b (group of particles, cluster) is knocked out from the nucleus $A(= c + b)$. Cluster b is thus initially bound, while the final states of a, b and the initial state of a are all in the continuum, and can be described with distorted waves defined as scattering solutions of an optical potential. A schematic depiction of the situation is shown in Fig. 5.7.1. While the derivation presented below is quite general, special emphasis is set to one-particle knockout processes.

Transition Amplitude

A first derivation will be given in which, for simplicity, all the "particles" (nuclei) involved in the reaction process are spinless and inert. Use is made of central, complex optical potentials ($U(r_{aA})$, $U(r_{cb})$, $U(r_{ac})$) without a spin-orbit term. In addition, the interaction $v(r_{ab})$ between a and b is taken to be a function of

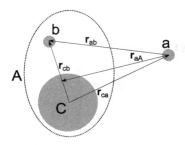

Figure 5.7.1 System of coordinates used to describe the reaction $A + a \rightarrow a + b + c$. The nucleus A is viewed as an inert cluster b bounded to an inert core c.

the distance r_{ab}. Within this scenario, the transition amplitude which is at the basis of the evaluation of the multidifferential cross section is the six-dimensional integral

$$T_{m_b} = \int d\mathbf{r}_{aA} d\mathbf{r}_{bc} \chi^{(-)*}(\mathbf{r}_{ac}) \chi^{(-)*}(\mathbf{r}_{bc}) v(r_{ab}) \chi^{(+)}(\mathbf{r}_{aA}) u_{l_b}(r_{bc}) Y_{m_b}^{l_b}(\hat{\mathbf{r}}_{bc}). \quad (5.7.1)$$

Coordinates

The vectors $\mathbf{r}_{ab}$, $\mathbf{r}_{ac}$ can easily be written in function of the integration variables $\mathbf{r}_{aA}$, $\mathbf{r}_{bc}$ (see Fig. 5.7.1), namely,

$$\begin{aligned}
\mathbf{r}_{ac} &= \mathbf{r}_{aA} + \frac{b}{A}\mathbf{r}_{bc}, \\
\mathbf{r}_{ab} &= \mathbf{r}_{aA} - \frac{c}{A}\mathbf{r}_{bc},
\end{aligned} \quad (5.7.2)$$

where $b, c,$ and A stand for the number of nucleons of the species $b, c,$ and A, respectively.

Distorted Waves in the Continuum

A standard way to reduce the dimensionality of the integral (5.7.1) consists in expanding the continuum wave functions $\chi^{(+)}(\mathbf{r}_{aA})$, $\chi^{(-)*}(\mathbf{r}_{ac})$, $\chi^{(-)*}(\mathbf{r}_{bc})$ in a basis of eigenstates of the angular momentum operator (partial waves). Then one can exploit the transformation properties of these eigenstates under rotations to conveniently carry out the angular integrations. Making use of time–reversed phasing, that is

$$Y_m^l(\theta, \phi) = i^l \sqrt{\frac{2l+1}{4\pi} \frac{(l-m)!}{(l+m)!}} P_l^m(\cos\theta) e^{im\phi}, \quad (5.7.3)$$

the general form of these expansions is

$$\chi^{(+)}(\mathbf{k}, \mathbf{r}) = \sum_l \frac{4\pi}{kr} i^l \sqrt{2l+1} e^{i\sigma^l} F_l(r) \left[Y^l(\hat{\mathbf{r}}) Y^l(\hat{\mathbf{k}}) \right]_0^0, \quad (5.7.4)$$

and

$$\chi^{(-)*}(\mathbf{k}, \mathbf{r}) = \sum_{l} \frac{4\pi}{kr} i^{-l} \sqrt{2l+1} e^{i\sigma^l} F_l(r) \left[Y^l(\hat{\mathbf{r}}) Y^l(\hat{\mathbf{k}}) \right]_0^0, \qquad (5.7.5)$$

σ_l being the Coulomb phase shift. The radial functions $F_l(r)$ are regular (finite at $r = 0$) solutions of the one–dimensional Schrödinger equation with an effective potential $U(r) + \frac{\hbar^2 l(l+1)}{2\mu r^2}$ and suitable asymptotic behavior at $r \to \infty$ as boundary conditions. Thus, the distorted waves appearing in (5.7.1) are

$$\chi^{(+)}(\mathbf{k_a}, \mathbf{r}_{aA}) = \sum_{l_a} \frac{4\pi}{k_a r_{aA}} i^{l_a} \sqrt{2l_a+1} e^{i\sigma^{l_a}} F_{l_a}(r_{aA}) \left[Y^{l_a}(\hat{\mathbf{r}}_{aA}) Y^{l_a}(\hat{\mathbf{k}}_a) \right]_0^0, \qquad (5.7.6)$$

describing the relative motion of A and a in the entrance channel as determined by the complex optical potential $U(r_{Aa})$,

$$\chi^{(-)*}(\mathbf{k_a'}, \mathbf{r}_{ac}) = \sum_{l_a'} \frac{4\pi}{k_a' r_{ac}} i^{-l_a'} \sqrt{2l_a'+1} e^{i\sigma^{l_a'}} F_{l_a'}(r_{ac}) \left[Y^{l_a'}(\hat{\mathbf{r}}_{ac}) Y^{l_a'}(\hat{\mathbf{k}}_a') \right]_0^0, \qquad (5.7.7)$$

which describes the relative motion of c and a, in the final channel controlled by the complex optical potential $U(r_{ac})$, and finally

$$\chi^{(-)*}(\mathbf{k_b'}, \mathbf{r}_{bc}) = \sum_{l_b'} \frac{4\pi}{k_b' r_{bc}} i^{-l_b'} \sqrt{2l_b'+1} e^{i\sigma^{l_b'}} F_{l_b'}(r_{bc}) \left[Y^{l_b'}(\hat{\mathbf{r}}_{bc}) Y^{l_b'}(\hat{\mathbf{k}}_b') \right]_0^0, \qquad (5.7.8)$$

final channel wavefunction describing the relative motion of b and c, as defined by the complex optical potential $U(r_{bc})$).

Recoupling of Angular Momenta

One now proceeds to the evaluation of the six-dimensional integral

$$\frac{64\pi^3}{k_a k_a' k_b'} \int d\mathbf{r}_{aA} d\mathbf{r}_{bc} u_{l_b}(r_{cb}) v(r_{ab}) \sum_{l_a, l_a', l_b'} \sqrt{(2l_a+1)(2l_a'+1)(2l_b'+1)}$$

$$\times e^{i(\sigma^{l_a}+\sigma^{l_a'}+\sigma^{l_b'})} \frac{F_{l_a}(r_{aA}) F_{l_a'}(r_{ac}) F_{l_b'}(r_{bc})}{r_{ac} r_{aA} r_{bc}}$$

$$\times \left[Y^{l_a}(\hat{\mathbf{r}}_{aA}) Y^{l_a}(\hat{\mathbf{k}}_a) \right]_0^0 \left[Y^{l_a'}(\hat{\mathbf{r}}_{ac}) Y^{l_a'}(\hat{\mathbf{k}}_a') \right]_0^0 \left[Y^{l_b'}(\hat{\mathbf{r}}_{bc}) Y^{l_b'}(\hat{\mathbf{k}}_b') \right]_0^0 Y_{m_b}^{l_b}(\hat{\mathbf{r}}_{bc}), \qquad (5.7.9)$$

an expression which explicitly depends on the asymptotic kinetic energies and scattering angles $(\hat{\mathbf{k}}_a, \hat{\mathbf{k}}_a', \hat{\mathbf{k}}_b')$ of a, b as determined by k_a, k_a', k_b' and $\hat{\mathbf{k}}_a, \hat{\mathbf{k}}_a', \hat{\mathbf{k}}_b'$ respectively. In what follows we will take advantage of the partial wave expansion to reduce the dimensionality of the integral from 6 to 3. A possible strategy to follow is that of recoupling together all the terms that depend on the integration variables to a global angular momentum and retain only the term coupled to 0 as the only one surviving the integration. Let us start to separately couple the terms corresponding to particles a and b. For particle a we write

$$\left[Y^{l_a}(\hat{\mathbf{r}}_{aA})Y^{l_a}(\hat{\mathbf{k}}_a)\right]_0^0 \left[Y^{l'_a}(\hat{\mathbf{r}}_{ac})Y^{l'_a}(\hat{\mathbf{k}}'_a)\right]_0^0 = \sum_K \left((l_a l_a)_0(l'_a l'_a)_0 | (l_a l'_a)_K (l_a l'_a)_K\right)_0$$

$$\times \left\{\left[Y^{l_a}(\hat{\mathbf{r}}_{aA})Y^{l'_a}(\hat{\mathbf{r}}_{ac})\right]^K \left[Y^{l_a}(\hat{\mathbf{k}}_a)Y^{l'_a}(\hat{\mathbf{k}}'_a)\right]^K\right\}_0^0.$$

$$(5.7.10)$$

We can now evaluate the $9j$-symbol,

$$\left((l_a l_a)_0(l'_a l'_a)_0 | (l_a l'_a)_K (l_a l'_a)_K\right)_0 = \sqrt{\frac{2K+1}{(2l'_a+1)(2l_a+1)}}, \qquad (5.7.11)$$

and expand the coupling,

$$\left\{\left[Y^{l_a}(\hat{\mathbf{r}}_{aA})Y^{l'_a}(\hat{\mathbf{r}}_{ac})\right]^K \left[Y^{l_a}(\hat{\mathbf{k}}_a)Y^{l'_a}(\hat{\mathbf{k}}'_a)\right]^K\right\}_0^0 = \sum_M \langle K\,K\,M\,-M|0\,0\rangle$$

$$\times \left[Y^{l_a}(\hat{\mathbf{r}}_{aA})Y^{l'_a}(\hat{\mathbf{r}}_{ac})\right]_M^K \left[Y^{l_a}(\hat{\mathbf{k}}_a)Y^{l'_a}(\hat{\mathbf{k}}'_a)\right]_{-M}^K = \sum_M \frac{(-1)^{K+M}}{\sqrt{2K+1}} \qquad (5.7.12)$$

$$\times \left[Y^{l_a}(\hat{\mathbf{r}}_{aA})Y^{l'_a}(\hat{\mathbf{r}}_{ac})\right]_M^K \left[Y^{l_a}(\hat{\mathbf{k}}_a)Y^{l'_a}(\hat{\mathbf{k}}'_a)\right]_{-M}^K.$$

Thus,

$$\left[Y^{l_a}(\hat{\mathbf{r}}_{aA})Y^{l_a}(\hat{\mathbf{k}}_a)\right]_0^0 \left[Y^{l'_a}(\hat{\mathbf{r}}_{ac})Y^{l'_a}(\hat{\mathbf{k}}'_a)\right]_0^0 = \sqrt{\frac{1}{(2l'_a+1)(2l_a+1)}}$$

$$\times \sum_{KM}(-1)^{K+M}\left[Y^{l_a}(\hat{\mathbf{r}}_{aA})Y^{l'_a}(\hat{\mathbf{r}}_{ac})\right]_M^K \left[Y^{l_a}(\hat{\mathbf{k}}_a)Y^{l'_a}(\hat{\mathbf{k}}'_a)\right]_{-M}^K.$$

$$(5.7.13)$$

One can further simplify the above expression by choosing the direction of the initial momentum to be parallel to the z axis, so that $Y_m^{l_a}(\hat{\mathbf{k}}_a) = Y_m^{l_a}(\hat{\mathbf{z}}) = \sqrt{\frac{2l_a+1}{4\pi}}\delta_{m,0}$. Then,

$$\left[Y^{l_a}(\hat{\mathbf{r}}_{aA})Y^{l_a}(\hat{\mathbf{k}}_a)\right]_0^0 \left[Y^{l'_a}(\hat{\mathbf{r}}_{ac})Y^{l'_a}(\hat{\mathbf{k}}'_a)\right]_0^0 = \sqrt{\frac{1}{4\pi(2l'_a+1)}}\sum_{KM}(-1)^{K+M}$$

$$\times \langle l_a\,0\,l'_a\,-M|K\,-M\rangle \left[Y^{l_a}(\hat{\mathbf{r}}_{aA})Y^{l'_a}(\hat{\mathbf{r}}_{ac})\right]_M^K Y_{-M}^{l'_a}(\hat{\mathbf{k}}'_a).$$

$$(5.7.14)$$

For particle b we have

$$Y_{m_b}^{l_b}(\hat{\mathbf{r}}_{bc}) \left[Y_b^{l_b'}(\hat{\mathbf{r}}_{bc}) Y_b^{l_b'}(\hat{\mathbf{k}}_b') \right]_0^0 = Y_{m_b}^{l_b}(\hat{\mathbf{r}}_{cb}) \sum_m \frac{(-1)^{l_b'+m}}{\sqrt{2l_b'+1}} Y_m^{l_b'}(\hat{\mathbf{r}}_{bc}) Y_{-m}^{l_b'}(\hat{\mathbf{k}}_b'), \quad (5.7.15)$$

and can write

$$Y_{m_b}^{l_b}(\hat{\mathbf{r}}_{bc}) Y_m^{l_b'}(\hat{\mathbf{r}}_{bc}) = \sum_{K'} \langle l_b \, m_b \, l_b' \, m | K' \, m_b + m \rangle \left[Y^{l_b}(\hat{\mathbf{r}}_{bc}) Y^{l_b'}(\hat{\mathbf{r}}_{bc}) \right]_{m_b+m}^{K'}.$$

$$(5.7.16)$$

In order to couple to 0 angular momentum with (5.7.14) we must only keep the term with $K' = K$, $m = -M - m_b$ so

$$Y_{m_b}^{l_b}(\hat{\mathbf{r}}_{bc}) \left[Y_b^{l_b'}(\hat{\mathbf{r}}_{bc}) Y_b^{l_b'}(\hat{\mathbf{k}}_b') \right]_0^0 = \frac{(-1)^{l_b'-M-m_b}}{\sqrt{2l_b'+1}} \langle l_b \, m_b \, l_b' \, - M - m_b | K \, - M \rangle$$

$$\times \left[Y^{l_b}(\hat{\mathbf{r}}_{bc}) Y^{l_b'}(\hat{\mathbf{r}}_{bc}) \right]_{-M}^{K} Y_{-M-m_b}^{l_b'}(\hat{\mathbf{k}}_b'),$$

$$(5.7.17)$$

and (5.7.9) becomes

$$\frac{32\pi^2}{k_a k_a' k_b'} \sum_{KM} (-1)^{K+l_b'-m_b} \langle l_a \, 0 \, l_a' \, - M | K \, - M \rangle \langle l_b \, m_b \, l_b' \, - M - m_b | K \, - M \rangle$$

$$\times \sum_{l_a, l_a', l_b'} \sqrt{(2l_a+1)} e^{i(\sigma^{l_a}+\sigma^{l_a'}+\sigma^{l_b'})} Y_{-M-m_b}^{l_b'}(\hat{\mathbf{k}}_b') Y_{-M}^{l_a'}(\hat{\mathbf{k}}_a') \int d\mathbf{r}_{aA} d\mathbf{r}_{bc} u_{l_b}(r_{bc}) v(r_{ab})$$

$$\times \frac{F_{l_a}(r_{aA}) F_{l_a'}(r_{ac}) F_{l_b'}(r_{bc})}{r_{ac} r_{aA} r_{bc}} \left[Y^{l_a}(\hat{\mathbf{r}}_{aA}) Y^{l_a'}(\hat{\mathbf{r}}_{ac}) \right]_M^{K} \left[Y^{l_b}(\hat{\mathbf{r}}_{bc}) Y^{l_b'}(\hat{\mathbf{r}}_{bc}) \right]_{-M}^{K}.$$

$$(5.7.18)$$

Note that

$$\left[Y^{l_a}(\hat{\mathbf{r}}_{aA}) Y^{l_a'}(\hat{\mathbf{r}}_{ac}) \right]_M^{K} \left[Y^{l_b}(\hat{\mathbf{r}}_{bc}) Y^{l_b'}(\hat{\mathbf{r}}_{bc}) \right]_{-M}^{K} = \sum_P \langle K \, M \, K \, - M | P \, 0 \rangle$$

$$\times \left\{ \left[Y^{l_a}(\hat{\mathbf{r}}_{aA}) Y^{l_a'}(\hat{\mathbf{r}}_{ac}) \right]^{K} \left[Y^{l_b}(\hat{\mathbf{r}}_{bc}) Y^{l_b'}(\hat{\mathbf{r}}_{bc}) \right]^{K} \right\}_0^{P},$$

$$(5.7.19)$$

and that to survive the integration the rotational tensors must be coupled to $P = 0$. Keeping only this term in the sum over P, we get

$$\left[Y^{l_a}(\hat{\mathbf{r}}_{aA}) Y^{l_a'}(\hat{\mathbf{r}}_{ac}) \right]_M^{K} \left[Y^{l_b}(\hat{\mathbf{r}}_{bc}) Y^{l_b'}(\hat{\mathbf{r}}_{bc}) \right]_{-M}^{K} = \frac{(-1)^{K+M}}{\sqrt{2K+1}}$$

$$\times \left\{ \left[Y^{l_a}(\hat{\mathbf{r}}_{aA}) Y^{l_a'}(\hat{\mathbf{r}}_{ac}) \right]^{K} \left[Y^{l_b}(\hat{\mathbf{r}}_{bc}) Y^{l_b'}(\hat{\mathbf{r}}_{bc}) \right]^{K} \right\}_0^0.$$

$$(5.7.20)$$

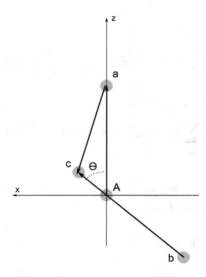

Figure 5.7.2 Coordinates in the "standard" configuration.

The coordinate-dependent part of the latter expression is a rotationally invariant scalar, so it can be evaluated in any conventional "standard" configuration such as the one depicted in Fig. 5.7.2. It must then be multiplied by a factor resulting of the integration of the remaining angular variables, which accounts for the rigid rotations needed to connect any arbitrary configuration to one of this type. This factor turns out to be $8\pi^2$ (a 4π factor for all possible orientations of, say, $\mathbf{r}_{aA}$ and a 2π factor for a complete rotation around its direction). According to Fig. 5.7.2,

$$\mathbf{r}_{bc} = r_{bc}\left(\sin\theta\,\hat{x} + \cos\theta\,\hat{z}\right),$$

$$\mathbf{r}_{aA} = -r_{aA}\,\hat{z},$$

$$\mathbf{r}_{ac} = \frac{b}{A}r_{bc}\sin\theta\,\hat{x} + \left(\frac{b}{A}r_{bc}\cos\theta - r_{aA}\right)\hat{z}. \tag{5.7.21}$$

As $\mathbf{r}_{aA}$ lies parallel to the z axis, $Y_{M_K}^{l_a}(\hat{\mathbf{r}}_{aA}) = \sqrt{\frac{2l_a+1}{4\pi}}\delta_{M_K,0}$ and

$$\left[Y^{l_a}(\hat{\mathbf{r}}_{aA})\,Y^{l'_a}(\hat{\mathbf{r}}_{ac})\right]_{M_K}^{K} = \sum_{m}\langle l_a\,m\,l'_a\,M_K - m|K\,M_K\rangle Y_m^{l_a}(\hat{\mathbf{r}}_{aA})Y_{M_K-m}^{l'_a}(\hat{\mathbf{r}}_{ac})$$

$$= \sqrt{\frac{2l_a+1}{4\pi}}\langle l_a\,0\,l'_a\,M_K|K\,M_K\rangle Y_{M_K}^{l'_a}(\hat{\mathbf{r}}_{ac}). \tag{5.7.22}$$

Then

$$
\left\{ \left[Y^{l_a}(\hat{\mathbf{r}}_{aA}) Y^{l'_a}(\hat{\mathbf{r}}_{ac}) \right]^K \left[Y^{l_b}(\hat{\mathbf{r}}_{bc}) Y^{l'_b}(\hat{\mathbf{r}}_{bc}) \right]^K \right\}_0^0
$$

$$
= \sum_{M_K} \langle K \; M_K \; K \; -M_K | 0 \; 0 \rangle \left[Y^{l_a}(\hat{\mathbf{r}}_{aA}) Y^{l'_a}(\hat{\mathbf{r}}_{ac}) \right]^K_{M_K} \left[Y^{l_b}(\hat{\mathbf{r}}_{bc}) Y^{l'_b}(\hat{\mathbf{r}}_{bc}) \right]^K_{-M_K}
$$

$$
= \sqrt{\frac{2l_a+1}{4\pi}} \sum_{M_K} \frac{(-1)^{K+M_K}}{\sqrt{2K+1}} \langle l_a \; 0 \; l'_a \; M_K | K \; M_K \rangle
$$

$$
\times \left[Y^{l_b}(\hat{\mathbf{r}}_{bc}) Y^{l'_b}(\hat{\mathbf{r}}_{bc}) \right]^K_{-M_K} Y^{l'_a}_{M_K}(\hat{\mathbf{r}}_{ac}). \tag{5.7.23}
$$

Remembering the $8\pi^2$ factor, the term arising from (5.7.20) to be considered in the integral is

$$
4\pi^{3/2} \frac{\sqrt{2l_a+1}}{2K+1} (-1)^K \sum_{M_K} (-1)^{M_K} \langle l_a \; 0 \; l'_a \; M_K | K \; M_K \rangle
$$

$$
\times \left[Y^{l_b}(\cos\theta, 0) Y^{l'_b}(\cos\theta, 0) \right]^K_{-M_K} Y^{l'_a}_{M_K}(\cos\theta_{ac}, 0), \tag{5.7.24}
$$

with

$$
\cos\theta_{ac} = \frac{\frac{b}{A} r_{bc} \cos\theta - r_{aA}}{\sqrt{\left(\frac{b}{A} r_{bc} \sin\theta\right)^2 + \left(\frac{b}{A} r_{bc} \cos\theta - r_{aA}\right)^2}}, \tag{5.7.25}
$$

(see (5.7.21)). The final expression of the transition amplitude is

$$
T_{m_b}(\mathbf{k}'_a, \mathbf{k}'_b) = \frac{128\pi^{7/2}}{k_a k'_a k'_b} \sum_{KM} \frac{(-1)^{l'_b+m_b}}{2K+1} \langle l_a \; 0 \; l'_a \; -M | K \; -M \rangle
$$

$$
\langle l_b \; m_b \; l'_b \; -M-m_b | K \; -M \rangle
$$

$$
\times \sum_{l_a, l'_a, l'_b} (2l_a+1) e^{i(\sigma^{l_a}+\sigma^{l'_a}+\sigma^{l'_b})} Y^{l'_b}_{-M-m_b} \tag{5.7.26}
$$

$$
(\hat{\mathbf{k}}'_b) Y^{l'_a}_{-M}(\hat{\mathbf{k}}'_a) \, \mathfrak{I}(l_a, l'_a, l'_b, K),
$$

where

$$
\mathfrak{I}(l_a, l'_a, l'_b, K) = \int dr_{aA} dr_{bc} d\theta \, r_{aA} r_{bc} \frac{\sin\theta}{r_{ac}} u_{l_b}(r_{bc}) v(r_{ab}) F_{l_a}(r_{aA}) F_{l'_a}(r_{ac}) F_{l'_b}(r_{bc})
$$

$$
\times \sum_{M_K} (-1)^{M_K} \langle l_a \; 0 \; l'_a \; M_K | K \; M_K \rangle \left[Y^{l_b}(\cos\theta, 0) Y^{l'_b}(\cos\theta, 0) \right]^K_{-M_K}
$$

$$
Y^{l'_a}_{M_K}(\cos\theta_{ac}, 0) \tag{5.7.27}
$$

is a three-dimensional integral that can be numerically evaluated with the help of, for example, Gaussian integration.

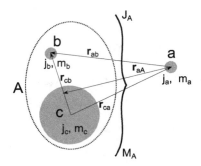

Figure 5.7.3 In the present case, all three clusters a, b, c have definite spins and projections. The nucleus A is coupled to total spin J_A, M_A.

5.7.2 Particles with Spin

We now treat the case in which the clusters have a definite spin (see Fig. 5.7.3), and the complex optical potentials $U(r_{aA}), U(r_{cb}), U(r_{ac})$ contain now a spin–orbit term proportional to the product $\mathbf{l} \cdot \mathbf{s} = 1/2(j(j+1) - l(l+1) - 3/4)$ for particles with spin 1/2. In addition, the interaction $V(r_{ab}, \sigma_a, \sigma_b)$ between a and b is taken to be a separable function of the distance r_{ab} and of the spin orientations, $V(r_{ab}, \sigma_a, \sigma_b) = v(r_{ab})v_\sigma(\sigma_a, \sigma_b)$. Note that this ansatz rules out spin orbit as well as tensor terms in the NN-interaction. For the time being we will assume that the spin-dependent interaction is rotationally invariant (scalar with respect to rotations), such as $v_\sigma(\sigma_a, \sigma_b) \propto \sigma_a \cdot \sigma_b$. Again, this assumption excludes tensor terms in the interaction. The transition amplitude is then

$$
\begin{aligned}
T_{m_a, m_b}^{m_a', m_b'} = \sum_{\sigma_a, \sigma_b} \int d\mathbf{r}_{aA} d\mathbf{r}_{bc} \chi_{m_a'}^{(-)*}(\mathbf{r}_{ac}, \sigma_a) \chi_{m_b'}^{(-)*}(\mathbf{r}_{bc}, \sigma_b) \\
\times v(r_{ab}) v_\sigma(\sigma_a, \sigma_b) \chi_{m_a}^{(+)}(\mathbf{r}_{aA}, \sigma_a) \psi_{m_b}^{l_b, j_b}(\mathbf{r}_{bc}, \sigma_b).
\end{aligned}
\tag{5.7.28}
$$

Distorted Waves

The distorted waves in (5.7.28) $\chi_m(\mathbf{r}, \sigma) = \chi(\mathbf{r})\phi_m^{1/2}(\sigma)$ have a spin dependence contained in the spinor $\phi_m^{1/2}(\sigma)$, where σ is the spin degree of freedom and m the projection of the spin along the quantization axis. The superscript $1/2$ reminds us that we are considering spin 1/2 particles, which have important consequences when dealing with the spin–orbit term of the optical potentials. As for the spin-dependent term $v_\sigma(\sigma_a, \sigma_b)$, the actual value of the spin of particles involved in the reaction process do not make much difference, *as long as this term is rotationally invariant*. Following (5.7.4),

$$\chi^{(+)}(\mathbf{k}, \mathbf{r})\phi_m(\sigma) = \sum_{l,j} \frac{4\pi}{kr} i^l \sqrt{2l+1} e^{i\sigma^l} F_{l,j}(r) \left[Y^l(\hat{\mathbf{r}}) Y^l(\hat{\mathbf{k}}) \right]_0^0 \phi_m^{1/2}(\sigma).$$

$$(5.7.29)$$

Note that now one also sums over the total angular momentum j, as the radial functions $F_{l,j}(r)$ depend both on j and on l, in keeping with the fact that they are solutions of an optical potential containing a spin-orbit term proportional to $1/2\,(j(j+1) - l(l+1) - 3/4)$. One must then couple the radial and spin functions to total angular momentum j, noting that

$$\left[Y^l(\hat{\mathbf{r}})\ Y^l(\hat{\mathbf{k}}) \right]_0^0 \phi_m^{1/2}(\sigma) = \sum_{m_l} \langle l\ m_l\ l\ -m_l|0\ 0\rangle Y_{m_l}^l(\hat{\mathbf{r}}) Y_{-m_l}^l(\hat{\mathbf{k}})\phi_m^{1/2}(\sigma)$$

$$= \sum_{m_l} \frac{(-1)^{l-m_l}}{\sqrt{2l+1}} Y_{m_l}^l(\hat{\mathbf{r}}) Y_{-m_l}^l(\hat{\mathbf{k}})\phi_m^{1/2}(\sigma),$$

$$(5.7.30)$$

and

$$Y_{m_l}^l(\hat{\mathbf{r}})\phi_m^{1/2}(\sigma) = \sum_j \langle l\ m_l\ 1/2\ m|j\ m_l + m\rangle \left[Y^l(\hat{\mathbf{r}})\phi^{1/2}(\sigma) \right]_{m_l+m}^j, \qquad (5.7.31)$$

we can write

$$\left[Y^l(\hat{\mathbf{r}})\ Y^l(\hat{\mathbf{k}}) \right]_0^0 \phi_m^{1/2}(\sigma) = \sum_{m_l,j} \frac{(-1)^{l+m_l}}{\sqrt{2l+1}} \langle l\ m_l\ 1/2\ m|j\ m_l + m\rangle$$

$$\times \left[Y^l(\hat{\mathbf{r}})\phi^{1/2}(\sigma) \right]_{m_l+m}^j Y_{-m_l}^l(\hat{\mathbf{k}}),$$

$$(5.7.32)$$

and the distorted waves in (5.7.28) are

$$\chi_{m_a}^{(+)}(\mathbf{r}_{aA}, \mathbf{k}_a, \sigma_a) = \sum_{l_a, m_{l_a}, j_a} \frac{4\pi}{k_a r_{aA}} i^{l_a} (-1)^{l_a + m_{l_a}} e^{i\sigma^{l_a}} F_{l_a, j_a}(r_{aA})$$

$$\times \langle l_a\ m_{l_a}\ 1/2\ m_a|j_a\ m_{l_a} + m_a\rangle$$

$$\left[Y^{l_a}(\hat{\mathbf{r}}_{aA})\phi^{1/2}(\sigma_a) \right]_{m_{l_a}+m_a}^{j_a} Y_{-m_{l_a}}^{l_a}(\hat{\mathbf{k}}_a),$$

$$(5.7.33)$$

$$\chi_{m_b'}^{(-)*}(\mathbf{r}_{bc}, \mathbf{k}_b', \sigma_b) = \sum_{l_b', m_{l_b'}, j_b'} \frac{4\pi}{k_b' r_{bc}} i^{-l_b'} (-1)^{l_b' + m_{l_b'}} e^{i\sigma^{l_b'}} F_{l_b', j_b'}(r_{bc})$$

$$\times \langle l_b'\ m_{l_b'}\ 1/2\ m_b'|j_b'\ m_{l_b'} + m_b'\rangle$$

$$\left[Y^{l_b'}(\hat{\mathbf{r}}_{bc})\phi^{1/2}(\sigma_b) \right]_{m_{l_b'}+m_b'}^{j_b'*} Y_{-m_{l_b'}}^{l_b'*}(\hat{\mathbf{k}}_b'),$$

$$(5.7.34)$$

$$\chi_{m_a'}^{(-)*}(\mathbf{r}_{ac}, \mathbf{k}_a', \sigma_a) = \sum_{l_a', m_{l_a'}, j_a'} \frac{4\pi}{k_a' r_{ac}} i^{-l_a'}(-1)^{l_a' + m_{l_a'}} e^{i\sigma_{l_a'}} F_{l_a', j_a'}(r_{ac})$$

$$\times \langle l_a' \, m_{l_a'} \, 1/2 \, m_a' | j_a' \, m_{l_a'} + m_a' \rangle \qquad (5.7.35)$$

$$\left[Y^{l_a'}(\hat{\mathbf{r}}_{ac}) \phi^{1/2}(\sigma_a) \right]_{m_{l_a'} + m_a'}^{j_a'*} Y_{-m_{l_a'}}^{l_a'*}(\hat{\mathbf{k}}_a').$$

The initial bound particle b wavefunction is

$$\psi_{m_b}^{l_b, j_b}(\mathbf{r}_{bc}, \sigma_b) = u_{l_b, j_b}(r_{bc}) \left[Y^{l_b}(\hat{\mathbf{r}}_{bc}) \phi^{1/2}(\sigma_b) \right]_{m_b}^{j_b}, \qquad (5.7.36)$$

Substituting in (5.7.28), one obtains,

$$T_{m_a, m_b}^{m_a', m_b'}(\mathbf{k}_a', \mathbf{k}_b') = \frac{64\pi^3}{k_a k_a' k_b'} \sum_{\sigma_a, \sigma_b} \sum_{l_a, m_{l_a}, j_a} \sum_{l_a', m_{l_a'}, j_a'} \sum_{l_b', m_{l_b'}, j_b'} e^{i(\sigma_{la} + \sigma_{l_a'} + \sigma_{l_b'})} i^{l_a - l_a' - l_b'}$$

$$(-1)^{l_a - m_{l_a} + l_a' - j_a' + l_b' - j_b'}$$

$$\times \langle l_a' \, m_{l_a'} \, 1/2 \, m_a' | j_a' \, m_{l_a'} + m_a' \rangle \langle l_a \, m_{l_a} \, 1/2 \, m_a | j_a \, m_{l_a} + m_a \rangle$$

$$\langle l_b' \, m_{l_b'} \, 1/2 \, m_b' | j_b' \, m_{l_b'} + m_b' \rangle$$

$$\times Y_{-m_{l_a}}^{l_a}(\hat{\mathbf{k}}_a) Y_{-m_{l_b'}}^{l_b'}(\hat{\mathbf{k}}_b') Y_{-m_{l_a'}}^{l_a'}(\hat{\mathbf{k}}_a') \int d\mathbf{r}_{aA} d\mathbf{r}_{bc}$$

$$\left[Y^{l_a'}(\hat{\mathbf{r}}_{ac}) \phi^{1/2}(\sigma_a) \right]_{-m_{l_a'} - m_a'}^{j_a'} \left[Y^{l_b'}(\hat{\mathbf{r}}_{bc}) \phi^{1/2}(\sigma_b) \right]_{-m_{l_b'} - m_b'}^{j_b'}$$

$$\times \frac{F_{l_a, j_a}(r_{aA}) F_{l_a', j_a'}(r_{ac}) F_{l_b', j_b'}(r_{bc})}{r_{ac} r_{aA} r_{bc}} u_{l_b, j_b}(r_{bc}) v(r_{ab}) v_\sigma(\sigma_a, \sigma_b)$$

$$\times \left[Y^{l_a}(\hat{\mathbf{r}}_{aA}) \phi^{1/2}(\sigma_a) \right]_{m_{l_a} + m_a}^{j_a} \left[Y^{l_b}(\hat{\mathbf{r}}_{bc}) \phi^{1/2}(\sigma_b) \right]_{m_b}^{j_b}, \qquad (5.7.37)$$

where use was made of the relation

$$\left[Y^l(\hat{\mathbf{r}}) \phi^{1/2}(\sigma) \right]_m^{j*} = (-1)^{j-m} \left[Y^l(\hat{\mathbf{r}}) \phi^{1/2}(\sigma) \right]_{-m}^j. \qquad (5.7.38)$$

Recoupling of Angular Momenta

Let us now separate spatial and spin coordinates, noting that the spin functions must be coupled to $S = 0$, a consequence of the fact that the interaction $v_\sigma(\sigma_a, \sigma_b)$ is rotationally invariant. Starting with particle a,

$$\left[Y^{l_a'}(\hat{\mathbf{r}}_{ac}) \phi^{1/2*}(\sigma_a) \right]_{-m_{l_a'} - m_a'}^{j_a'} \left[Y^{l_a}(\hat{\mathbf{r}}_{aA}) \phi^{1/2}(\sigma_a) \right]_{m_{l_a} + m_a}^{j_a}$$

$$= \sum_K \left((l_a' \tfrac{1}{2}) j_a' (l_a \tfrac{1}{2}) j_a | (l_a l_a') K (\tfrac{1}{2} \tfrac{1}{2}) 0 \right)_K$$

$$\times \left[Y^{l_a'}(\hat{\mathbf{r}}_{ac}) Y^{l_a}(\hat{\mathbf{r}}_{aA}) \right]_{-m_{l_a'} - m_a' + m_{l_a} + m_a}^{K} \left[\phi^{1/2*}(\sigma_a) \phi^{1/2}(\sigma_a) \right]_0^0. \qquad (5.7.39)$$

For particle b,

$$
\left[Y^{l'_b}(\hat{\mathbf{r}}_{bc})\phi^{1/2*}(\sigma_b)\right]^{j'_b}_{-m_{l'_b}-m'_b}\left[Y^{l_b}(\hat{\mathbf{r}}_{bc})\phi^{1/2}(\sigma_b)\right]^{j_b}_{m_b}
$$

$$
= \sum_{K'}\left((l'_b\tfrac{1}{2})_{j'_b}(l_b\tfrac{1}{2})_{j_b}|(l_bl'_b)_{K'}(\tfrac{1}{2}\tfrac{1}{2})_0\right)_{K'}
$$

$$
\times \left[Y^{l'_b}(\hat{\mathbf{r}}_{bc})Y^{l_b}(\hat{\mathbf{r}}_{bc})\right]^{K'}_{-m_{l'_b}-m'_b+m_b}\left[\phi^{1/2*}(\sigma_b)\phi^{1/2}(\sigma_b)\right]^0_0. \quad (5.7.40)
$$

The spin summation yields a constant factor,

$$
\sum_{\sigma_a,\sigma_b}\left[\phi^{1/2*}(\sigma_a)\phi^{1/2}(\sigma_a)\right]^0_0\left[\phi^{1/2*}(\sigma_b)\phi^{1/2}(\sigma_b)\right]^0_0 v_\sigma(\sigma_a,\sigma_b) \equiv T_\sigma, \quad (5.7.41)
$$

and what we have yet to do is very similar to what we have done in the case of spinless particles. First of all note that the constrain of coupling all angular momenta to 0, imposes $K' = K$ and $m_{l_a} + m_a - m_{l'_a} - m'_a = m_{l'_b} + m'_b - m_b$ (see (5.7.39) and (5.7.40)). If we set $M = m_{l_a} + m_a - m_{l'_a} - m'_a$ and take, as before, $\hat{\mathbf{k}}_a \equiv \hat{z}$

$$
T^{m'_a,m'_b}_{m_a,m_b}(\mathbf{k}'_a,\mathbf{k}'_b) = \frac{32\pi^{5/2}}{k_ak'_ak'_b}T_\sigma\sum_{l_a,j_al'_a,j'_al'_b,j'_b}\sum_{K,M}e^{i(\sigma^{l_a}+\sigma^{l'_a}+\sigma^{l'_b})}i^{l_a-l'_a-l'_b}(-1)^{l_a+l'_a+l'_b-j'_a-j'_b}
$$

$$
\times \sqrt{2l_a+1}\left((l'_a\tfrac{1}{2})_{j'_a}(l_a\tfrac{1}{2})_{j_a}|(l_al'_a)_K(\tfrac{1}{2}\tfrac{1}{2})_0\right)_K\left((l'_b\tfrac{1}{2})_{j'_b}(l_b\tfrac{1}{2})_{j_b}|(l_bl'_b)_K(\tfrac{1}{2}\tfrac{1}{2})_0\right)_K
$$

$$
\times \langle l'_a\ m_a - m'_a - M\ 1/2\ m'_a|j'_a\ m_a - M\rangle\langle l_a\ 0\ 1/2\ m_a|j_a\ m_a\rangle
$$

$$
\times \langle l'_b\ m_b - m'_b + M\ 1/2\ m'_b|j'_b\ M + m_b\rangle
$$

$$
\times Y^{l'_b}_{m'_b-m_b-M}(\hat{\mathbf{k}}'_b)Y^{l'_a}_{m'_a-m_a+M}(\hat{\mathbf{k}}'_a)\int d\mathbf{r}_{aA}d\mathbf{r}_{bc}\frac{F_{l_a,j_a}(r_{aA})F_{l'_a,j'_a}(r_{ac})F_{l'_b,j'_b}(r_{bc})}{r_{ac}r_{aA}r_{bc}}
$$

$$
\times u_{l_b,j_b}(r_{bc})v(r_{ab})\left[Y^{l_a}(\hat{\mathbf{r}}_{aA})Y^{l'_a}(\hat{\mathbf{r}}_{ac})\right]^K_M\left[Y^{l_b}(\hat{\mathbf{r}}_{bc})Y^{l'_b}(\hat{\mathbf{r}}_{bc})\right]^K_{-M}. \quad (5.7.42)
$$

The integral of the above expression is similar to the one in (5.7.18), so we obtain

$$
T^{m'_a,m'_b}_{m_a,m_b}(\mathbf{k}'_a,\mathbf{k}'_b) = \frac{128\pi^4}{k_ak'_ak'_b}T_\sigma\sum_{l_a,j_al'_a,j'_al'_b,j'_b}\sum_{K,M}e^{i(\sigma^{l_a}+\sigma^{l'_a}+\sigma^{l'_b})}i^{l_a-l'_a-l'_b}(-1)^{l_a+l'_a+l'_b-j'_a-j'_b}
$$

$$
\times \frac{2l_a+1}{2K+1}\left((l'_a\tfrac{1}{2})_{j'_a}(l_a\tfrac{1}{2})_{j_a}|(l_al'_a)_K(\tfrac{1}{2}\tfrac{1}{2})_0\right)_K\left((l'_b\tfrac{1}{2})_{j'_b}(l_b\tfrac{1}{2})_{j_b}|(l_bl'_b)_K(\tfrac{1}{2}\tfrac{1}{2})_0\right)_K
$$

$$
\times \langle l'_a\ m_a - m'_a - M\ 1/2\ m'_a|j'_a\ m_a - M\rangle\langle l'_b\ m_b - m'_b + M\ 1/2\ m'_b|j'_b\ M + m_b\rangle
$$

$$
\times \langle l_a\ 0\ 1/2\ m_a|j_a\ m_a\rangle Y^{l'_b}_{m'_b-m_b-M}(\hat{\mathbf{k}}'_b)Y^{l'_a}_{m_a-m'_a+M}(\hat{\mathbf{k}}'_a)\Im(l_a,l'_a,l'_b,j_a,j'_a,j'_b,K),
$$

$$
(5.7.43)
$$

with

$$
\mathfrak{I}(l_a, l'_a, l'_b, j_a, j'_a, j'_b, K) = \int dr_{aA} dr_{bc} d\theta r_{aA} r_{bc} \frac{\sin\theta}{r_{ac}} u_{l_b}(r_{bc}) v(r_{ab})
$$

$$
\times F_{l_a, j_a}(r_{aA}) F_{l'_a, j'_a}(r_{ac}) F_{l'_b, j'_b}(r_{bc})
$$

$$
\times \sum_{M_K} \langle l_a\, 0\, l'_a\, M_K | K\, M_K \rangle \left[Y^{l_b}(\cos\theta, 0) Y^{l'_b}(\cos\theta, 0) \right]_{-M_K}^{K} Y_{M_K}^{l'_a}(\cos\theta_{ac}, 0).
$$

$$(5.7.44)$$

Again, this is a three-dimensional integral that can be evaluated with the method of Gaussian quadratures. The transition amplitude $T_{m_a, m_b}^{m'_a, m'_b}(\mathbf{k}'_a, \mathbf{k}'_b)$ depends explicitly on the initial (m_a, m'_a) and final (m'_a, m'_b) polarizations of a, b. If the particle b is initially coupled to core c to total angular momentum J_A, M_A, the amplitude to be considered is rather

$$
T_{m_a}^{m'_a, m'_b}(\mathbf{k}'_a, \mathbf{k}'_b) = \sum_{m_b} \langle j_b\, m_b\, j_c\, M_A - m_b | J_A\, M_A \rangle T_{m_a, m_b}^{m'_a, m'_b}(\mathbf{k}'_a, \mathbf{k}'_b), \qquad (5.7.45)
$$

and the multidifferential cross section for detecting particle c (or a) is

$$
\left. \frac{d\sigma}{d\mathbf{k}'_a d\mathbf{k}'_b} \right]_{m_a}^{m'_a, m'_b} = \frac{k'_a}{k_a} \frac{\mu_{aA}\mu_{ac}}{4\pi^2\hbar^4} \left| \sum_{m_b} \langle j_b\, m_b\, j_c\, M_A - m_b | J_A\, M_A \rangle T_{m_a, m_b}^{m'_a, m'_b}(\mathbf{k}'_a, \mathbf{k}'_b) \right|^2.
$$

$$(5.7.46)$$

All spin–polarization observables (analyzing powers, etc.) can be derived from this expression. But let us now work out the expression of the cross section for an unpolarized beam (sum over initial spin orientations divided by the number of such orientations) and when we do not detect the final polarizations (sum over final spin orientations),

$$
\frac{d\sigma}{d\mathbf{k}'_a d\mathbf{k}'_b} = \frac{k'_a}{k_a} \frac{\mu_{aA}\mu_{ac}}{4\pi^2\hbar^4} \frac{1}{(2J_A + 1)(2j_a + 1)}
$$

$$
\times \sum_{\substack{m_a, m'_a \\ M_A, m'_b}} \left| \sum_{m_b} \langle j_b\, m_b\, j_c\, M_A - m_b | J_A\, M_A \rangle T_{m_a, m_b}^{m'_a, m'_b}(\mathbf{k}'_a, \mathbf{k}'_b) \right|^2. \qquad (5.7.47)
$$

The sum above can be simplified a bit. Let us consider a single particular value of m_b in the sum over m_b,

$$\sum_{m_a,m'_a,m'_b} \left| T^{m'_a,m'_b}_{m_a,m_b}(\mathbf{k}'_a, \mathbf{k}'_b) \right|^2 \sum_{M_A} \left| \langle j_b \ m_b \ j_c \ M_A - m_b | J_A \ M_A \rangle \right|^2$$

$$= \frac{2J_A + 1}{2j_b + 1} \sum_{m_a,m_a,m'_b} \left| T^{m'_a,m'_b}_{m_a,m_b}(\mathbf{k}'_a, \mathbf{k}'_b) \right|^2 \qquad (5.7.48)$$

$$\times \sum_{M_A} \left| \langle J_A \ - M_A \ j_c \ M_A - m_b | j_b \ m_b \rangle \right|^2,$$

where we have used

$$\langle j_b \ m_b \ j_c \ M_A - m_b | J_A \ M_A \rangle = (-1)^{j_c - M_A + m_b} \sqrt{\frac{2J_A + 1}{2j_b + 1}} \langle J_A - M_A \ j_c \ M_A - m_b | j_b \ m_b \rangle. \qquad (5.7.49)$$

As

$$\sum_{M_A} \left| \langle J_A \ - M_A \ j_c \ M_A - m_b | j_b \ m_b \rangle \right|^2 = 1, \qquad (5.7.50)$$

we finally have

$$\frac{d\sigma}{d\mathbf{k}'_a d\mathbf{k}'_b} = \frac{k'_a}{k_a} \frac{\mu_{aA}\mu_{ac}}{4\pi^2\hbar^4} \frac{1}{(2j_b + 1)(2j_a + 1)} \sum_{m_a,m'_a,m'_b} \left| \sum_{m_b} T^{m'_a,m'_b}_{m_a,m_b}(\mathbf{k}'_a, \mathbf{k}'_b) \right|^2. \qquad (5.7.51)$$

Zero-Range Approximation

The zero-range approximation (Eq. (5.6.15)) consists, in the present case to assume $v(r_{ab}) = D_0\delta(r_{ab})$. Then, (see (5.7.21))

$$\mathbf{r}_{aA} = \frac{c}{A}\mathbf{r}_{bc}, \qquad (5.7.52)$$

$$\mathbf{r}_{ac} = \mathbf{r}_{bc}.$$

The angular dependence of the integral can be readily evaluated. From (6.A.20), noting that $\hat{\mathbf{r}}_{aA} = \hat{\mathbf{r}}_{ac} = \hat{\mathbf{r}}_{bc} \equiv \hat{\mathbf{r}}$,

$$\left[Y^{l_a}(\hat{\mathbf{r}})Y^{l'_a}(\hat{\mathbf{r}}) \right]^K_M \left[Y^{l_b}(\hat{\mathbf{r}})Y^{l'_b}(\hat{\mathbf{r}}) \right]^K_{-M} = \frac{(-1)^{K-M}}{\sqrt{2K+1}} \left\{ \left[Y^{l_a}(\hat{\mathbf{r}})Y^{l'_a}(\hat{\mathbf{r}}) \right] \right. \qquad (5.7.53)$$
$$\left. \times {}^K \left[Y^{l_b}(\hat{\mathbf{r}})Y^{l'_b}(\hat{\mathbf{r}}) \right]^K \right\}^0_0.$$

We can as before evaluate this expression in the configuration shown in Fig. 5.7.2 ($\hat{\mathbf{r}} = \hat{z}$), but now the multiplicative factor is 4π. The corresponding contribution to the integral is

$$\frac{(-1)^K}{4\pi(2K+1)}\langle l_a\, 0\, l_a'\, 0|K\, 0\rangle\sqrt{(2l_a+1)(2l_a'+1)(2l_b+1)(2l_b'+1)}, \quad (5.7.54)$$

and

$$T^{m_a',m_b'}_{m_a,m_b}(\mathbf{k}_a',\mathbf{k}_b') = \frac{16\pi^2}{k_a k_a' k_b'}\frac{c}{A}D_0 T_\sigma \sum_{l_a,j_a}\sum_{l_a',j_a'}\sum_{l_b',j_b'}\sum_{K,M} e^{i(\sigma_{la}+\sigma_{la}'+\sigma_{lb}')}i^{l_a-l_a'-l_b'}$$

$$(-1)^{l_a+l_a'+l_b'-j_a'-j_b'}$$

$$\times\sqrt{(2l_a+1)(2l_a'+1)(2l_b+1)(2l_b'+1)}\,\langle l_a\, 0\, l_a'\, 0|K\, 0\rangle$$

$$\times\frac{2l_a+1}{2K+1}\left((l_a'\tfrac{1}{2})_{j_a'}(l_a\tfrac{1}{2})_{j_a}|(l_a l_a')_K(\tfrac{1}{2}\tfrac{1}{2})_0\right)_K\left((l_b'\tfrac{1}{2})_{j_b'}(l_b\tfrac{1}{2})_{j_b}|(l_b l_b')_K(\tfrac{1}{2}\tfrac{1}{2})_0\right)_K$$

$$\times\langle l_a'\, m_a-m_a'-M\, 1/2\, m_a'|j_a'\, m_a-M\rangle\langle l_b'\, m_b-m_b'+M\, 1/2\, m_b'|j_b'\, M+m_b\rangle$$

$$\times\langle l\, 0\, 1/2\, m_a|j\, m_a\rangle Y^{l_b'}_{M+m_b+m_b'}(\hat{\mathbf{k}}_b')Y^{l_a'}_{m_a+m_a'-M}(\hat{\mathbf{k}}_a')\mathfrak{I}_{ZR}(l_a,l_a',l_b',j_a,j_a',j_b'),$$

$$(5.7.55)$$

where now the one-dimensional integral to solve is

$$\mathfrak{I}_{ZR}(l_a,l_a',l_b',j_a,j_a',j_b') = \int dr\, u_{l_b}(r)F_{l_a,j_a}(\tfrac{c}{A}r)F_{l_a',j_a'}(r)F_{l_b',j_b'}(r)/r. \quad (5.7.56)$$

One-Particle Transfer

We now state the expression for the one-particle transfer reaction within the present context and using similar elements. In this way one can better compare the corresponding types of experiments. In particle transfer, the final state of b is a bound state of the $B(=a+b)$ nucleus (cf. Fig. 5.7.4), and we can carry on in a similar way as done previously just by substituting the distorted wave (continuum) wave function (5.7.34) with

$$\psi^{l_b',j_b'*}_{m_b'}(\mathbf{r}_{ab},\sigma_b) = u^*_{l_b',j_b'}(r_{ab})\left[Y^{l_b'}(\hat{\mathbf{r}}_{ab})\phi^{1/2}(\sigma_b)\right]^{j_b'*}_{m_b'}, \quad (5.7.57)$$

so the transition amplitude is now

$$T^{m_a',m_b'}_{m_a,m_b}(\mathbf{k}_a') = \frac{8\pi^{3/2}}{k_a k_a'}\sum_{\sigma_a,\sigma_b}\sum_{l_a,j_a}\sum_{l_a',m_{l_a'},j_a'} e^{i(\sigma_{la}+\sigma_{la}')}i^{l_a-l_a'}(-1)^{l_a+l_a'-j_a'-j_b'}$$

$$\times\sqrt{2l_a+1}\,\langle l_a'\, m_{l_a'}\, 1/2\, m_a'|j_a'\, m_{l_a'}+m_a'\rangle\langle l_a\, 0\, 1/2\, m_a|j_a\, m_a\rangle$$

$$\times Y^{l_a'}_{-m_{l_a'}}(\hat{\mathbf{k}}_a')\int d\mathbf{r}_{aA}d\mathbf{r}_{bc}\left[Y^{l_a'}(\hat{\mathbf{r}}_{Bc})\phi^{1/2}(\sigma_a)\right]^{j_a'}_{-m_{l_a'}-m_a'}\left[Y^{l_b'}(\hat{\mathbf{r}}_{ab})\phi^{1/2}(\sigma_b)\right]^{j_b'}_{-m_b'}$$

$$\times\frac{F_{l_a,j_a}(r_{aA})F_{l_a',j_a'}(r_{Bc})}{r_{Bc}r_{aA}}u^*_{l_b',j_b'}(r_{ab})u_{l_b,j_b}(r_{bc})v(r_{ab})v_\sigma(\sigma_a,\sigma_b)$$

$$\times\left[Y^{l_a}(\hat{\mathbf{r}}_{aA})\phi^{1/2}(\sigma_a)\right]^{j_a}_{m_a}\left[Y^{l_b}(\hat{\mathbf{r}}_{bc})\phi^{1/2}(\sigma_b)\right]^{j_b}_{m_b}. \quad (5.7.58)$$

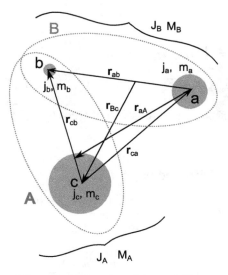

Figure 5.7.4 One-particle transfer reaction $A(= c + b) + a \rightarrow B(= a + b) + c$. For the color version of this figure, refer to cambridge.org/nuclearcooperpair.

Using (5.7.39), (5.7.40), (6.C.4), and setting $M = m_a - m'_a - m_{l'_a}$

$$
T_{m_a,m_b}^{m'_a,m'_b}(\mathbf{k}'_a) = \frac{8\pi^{3/2}}{k_a k'_a} T_\sigma \sum_{l_a,j_a} \sum_{l'_a,j'_a} \sum_{K,M} e^{i(\sigma^{l_a}+\sigma^{l'_a})} i^{l_a-l'_a} (-1)^{l_a+l'_a-j'_a-j'_b}
$$

$$
\times \left((l'_a\tfrac{1}{2})_{j'_a}(l_a\tfrac{1}{2})_{j_a}|(l_a l'_a)_K(\tfrac{1}{2}\tfrac{1}{2})_0\right)_K \left((l'_b\tfrac{1}{2})_{j'_b}(l_b\tfrac{1}{2})_{j_b}|(l_b l'_b)_K(\tfrac{1}{2}\tfrac{1}{2})_0\right)_K
$$

$$
\times \sqrt{2l_a+1} \langle l'_a\, m_a - m'_a - M\, 1/2\, m'_a|j'_a\, m_a - M\rangle \langle l_a\, 0\, 1/2\, m_a|j_a\, m_a\rangle
$$

$$
\times Y_{m_a-m'_a-M}^{l'_a}(\hat{\mathbf{k}}'_a) \int d\mathbf{r}_{aA} d\mathbf{r}_{bc} \frac{F_{l_a,j_a}(r_{aA}) F_{l'_a,j'_a}(r_{Bc})}{r_{Bc} r_{aA}} u_{l'_b,j'_b}^*(r_{ab}) u_{l_b,j_b}(r_{bc}) v(r_{ab})
$$

$$
\times \left[Y^{l_a}(\hat{\mathbf{r}}_{aA}) Y^{l'_a}(\hat{\mathbf{r}}_{Bc})\right]_M^K \left[Y^{l_b}(\hat{\mathbf{r}}_{bc}) Y^{l'_b}(\hat{\mathbf{r}}_{ab})\right]_{-M}^K . \quad (5.7.59)
$$

Aside from (5.7.21), we also need

$$
\mathbf{r}_{Bc} = \frac{a+B}{B}\mathbf{r}_{aA} + \frac{b}{A}\mathbf{r}_{bc}. \quad (5.7.60)
$$

From (5.7.20–5.7.25), one gets

$$
T_{m_a,m_b}^{m'_a,m'_b}(\mathbf{k}'_a) = \frac{32\pi^3}{k_a k'_a} T_\sigma \sum_{l_a,j_a} \sum_{l'_a,j'_a} \sum_{K,M} e^{i(\sigma^{l_a}+\sigma^{l'_a})} i^{l_a-l'_a} (-1)^{l_a+l'_a-j'_a-j'_b}
$$

$$
\times \left((l'_a\tfrac{1}{2})_{j'_a}(l_a\tfrac{1}{2})_{j_a}|(l_a l'_a)_K(\tfrac{1}{2}\tfrac{1}{2})_0\right)_K \left((l'_b\tfrac{1}{2})_{j'_b}(l_b\tfrac{1}{2})_{j_b}|(l_b l'_b)_K(\tfrac{1}{2}\tfrac{1}{2})_0\right)_K
$$

$$
\times \frac{2l_a+1}{2K+1} \langle l'_a\, m_a - m'_a - M\, 1/2\, m'_a|j'_a\, m_a - M\rangle
$$

$$
\times \langle l_a\, 0\, 1/2\, m_a|j_a\, m_a\rangle Y_{m_a-m'_a-M}^{l'_a}(\hat{\mathbf{k}}'_a) \mathfrak{I}(l_a, l'_a, j_a, j'_a, j'_b, K), \quad (5.7.61)
$$

with

$$
\begin{aligned}
I(l_a, l_a', j_a, j_a', K) = \int dr_{aA} dr_{bc} d\theta \, r_{aA}^2 r_{bc}^2 \frac{\sin\theta}{r_{Bc}} \\
\times F_{l_a, j_a}(r_{aA}) F_{l_a', j_a'}(r_{ac}) u_{l_b', j_b'}^*(r_{ab}) u_{l_b, j_b}(r_{bc}) v(r_{ab}) \\
\times \sum_{M_K} \langle l_a \, 0 \, l_a' \, M_K | K \, M_K \rangle \left[Y^{l_b}(\cos\theta, 0) Y^{l_b'}(\cos\theta_{ab}, 0) \right]_{-M_K}^K Y_{M_K}^{l_a'}(\cos\theta_{Bc}, 0),
\end{aligned}
$$

$$(5.7.62)$$

where (see (5.7.21), (5.7.60), and Fig. 5.7.2)

$$
\cos\theta_{ab} = \frac{-r_{aA} - \frac{c}{A} r_{bc} \cos\theta}{\sqrt{\left(\frac{c}{A} r_{bc} \sin\theta\right)^2 + \left(r_{aA} + \frac{c}{A} r_{bc} \cos\theta\right)^2}},
\tag{5.7.63}
$$

$$
\cos\theta_{Bc} = \frac{\frac{a+B}{B} r_{aA} + \frac{b}{A} r_{bc} \cos\theta}{\sqrt{\left(\frac{b}{A} r_{bc} \sin\theta\right)^2 + \left(\frac{a+B}{B} r_{aA} + \frac{b}{A} r_{bc} \cos\theta\right)^2}},
\tag{5.7.64}
$$

and

$$
r_{Bc} = \sqrt{\left(\frac{b}{A} r_{bc} \sin\theta\right)^2 + \left(\frac{a+B}{B} r_{aA} + \frac{b}{A} r_{bc} \cos\theta\right)^2}.
\tag{5.7.65}
$$

It is noted that Eq. (5.7.61) can also be used when particle b populates a resonant state of nucleus B which lies in the continuum.

5.8 Dynamical Shell Model

In the extreme shell model the nucleons move independently of each other in the average potential. The properly normalized probability for removing a nucleon from such orbitals is equal to 1 for ω equal to the (unperturbed) energy of the orbital in question and zero otherwise (see Fig. 5.8.1 (**a**)).

In the dynamical shell model[65], the nucleons can bounce inelastically off the nuclear surface setting the nucleus in a vibrational state, changing their state of motion as well as remaining in the same state. In this case the strength of the levels becomes, in general, distributed over a range of energies both below and above the Fermi energy. That is, the state k is found both in the system $(A - 1)$ populated in a pickup process and in the system $(A + 1)$ populated through a stripping reaction, as indicated in Fig 5.8.1 (**b**).

---

[65] Mahaux et al. (1985), and refs. therein; in connection with Eq. (5.8.1), see Sect. 4.7 (in particular, Eq. (4.7.20)) of this reference.

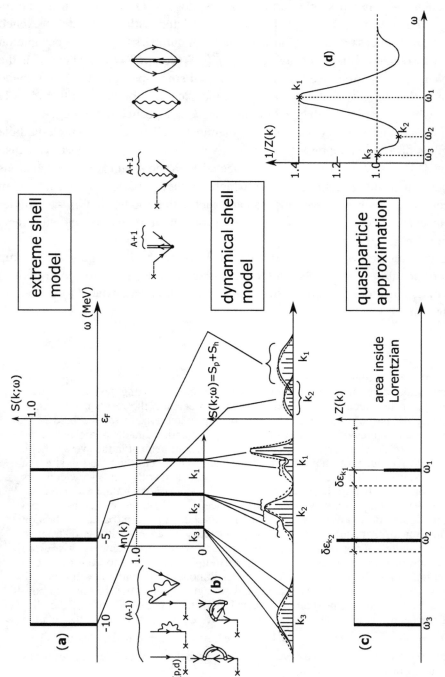

Figure 5.8.1

The next question is whether this distribution is concentrated in discrete states or displays a continuous behavior. Both situations can in principle be found depending of whether the original single-particle state is close or far away from the Fermi energy. The sum of the strength functions associated with all the states excited in the pickup process of a nucleon with quantum numbers k gives the occupation number associated with the orbital, that is, $\int_{-\infty}^{\epsilon_F} S_h(k; \omega)d\omega = n_k$, where S_h is the (hole) strength function (see Fig. 1.2.5). The full single-particle strength is found adding to this quantity that associated with the excitation of states in the $(A + 1)$ system where a particle with quantum numbers k is deposited in the target.

It is noted that the fact that in the dynamical shell model the single-particle strength is distributed not only over an energy range in the $(A - 1)$ system but also in the $(A + 1)$ system is intimately connected with the ground state correlations associated with the vibrational modes which produce particle-hole excitations in the ground state as well as pair addition and subtraction modes. In this way single-particle states which originally were filled become partially empty, and vice versa (see Fig. 1.8.3).

The transfer processes shown in Fig. 5.8.1 (b) for both stripping and pickup reactions are closely associated with the polarization and correlation contributions to the mass operator $\mathcal{M}(k; \omega) = \mathcal{V}(k; \omega) + i\mathcal{W}(k; \omega)$. The associated total strength distribution

---

Figure 5.8.1 (Continued)
Schematic representation of the main quantities characterizing the single-particle motion in the nuclear shell model taking into account the residual interaction among the particles at different levels of approximation. In (a) the bare NN-interaction is treated in the Hartree–Fock approximation and the particles feel the presence of the other particles through their own confinement in the average field. The strength function shows sharp peaks, each of them carrying the full strength of the states. In fact, the occupation number associated with each state k contains only one contribution. In (b) the particles still couple only with the average field. However, in this case, they can set the surface into vibrations as well as excite a pair addition or removal mode, by changing its state of motion (particle–vibration coupling mechanism). The strength associated with each orbital is distributed over a finite energy range. The corresponding occupation numbers arise from the sum of many contributions. Fitting a Breit–Wigner shape to each peak, one can regain the simplicity of the extreme shell model by defining new levels with energy equal to that of the centroid and strength equal to that of the area covered by the Lorentzian shape (cf. (c)). In (d) the energy variation of the shift of the centroids is contained into an effective ω-mass according to the relation $\frac{m_\omega}{m} = \left(1 - \frac{\partial \Delta E}{\partial \hbar \omega}\right)$ (see Eq. (5.4.5)). The resulting curve resembles the shape obtained by calculating the inverse of the area below the different Lorentzians (quasiparticle approximation).

$$S(k; \omega) = S_h(k; \omega) + S_p(k; \omega),$$

$$= \frac{1}{2\pi} \frac{\mathcal{W}(k; \omega) + \Delta}{(E_k + \mathcal{V}(k; \omega) - E)^2 + \frac{1}{4}(\mathcal{W}(k; \omega) + \Delta)^2}, \tag{5.8.1}$$

can display a variety of shapes, as $\mathcal{V}(k; \omega)$ and $\mathcal{W}(k; \omega)$ are energy dependent.

The intermediate states of the polarization and correlation diagrams which, in the present discussion are the doorway states to the renormalization of the single-particle motion, must be furthermore mixed with more complicated configurations. The extension to the complex plane of the associated expressions in terms of an imaginary parameter Δ is made to take these effects into account in some average way (see Sect. 2.3). In any case, most of the results are not dependent on the detailed value this quantity has. The picture shown in Fig. 5.8.1 (**b**), although accurate and controlled by only few parameters, is much too rich as compared to the originally, extreme shell model picture.

Within this context, a major simplification can be achieved recognizing that the essential difference between the situation depicted in Figs 5.8.1 (**a**) and (**b**) is the value of the mean free path. In the first case it is infinite while in the second it is finite. That is, due to the coupling to the surface modes (and eventually pair modes) single-particle motion acquires a lifetime. In other terms, the coupling to doorway states leads to a breaking of the single-particle strength (see, e.g., Fig. 5.2.3). The associated energy range (width Γ), over which this phenomenon takes place, provides insight into the time it takes for the different components to come out of phase ($\approx \hbar / \Gamma$).

Making the ansatz that the single-particle motion decays exponentially, a Breit–Wigner shape can be fitted to the different strength concentration resulting from the detailed calculation (Fig. 5.8.1 (**b**)) The centroid of the corresponding peaks can be viewed as the energy of the dressed single-particle state. On the other hand, caution should be exercised in interpreting the area under the fitted curve as single-particle content (spectroscopic factor), if only because this quantity can become larger than one (Fig. 5.8.1 (**c**)). It is only for the levels not too far away from the Fermi energy that such interpretation can be reasonably accurate.

Another quantity which can be used to characterize the dressed single-particle states is the rate of change of the energy shift as a function of the energy, that is, the effective mass. Again, for the orbitals close to the Fermi energy this quantity is inversely proportional to the area under the Breit–Wigner shapes. The energy range over which the effective mass deviates from one determine the region where the dynamical couplings change the density of single-particle levels (see Sect. 5.4.1, in particular Eq. (5.4.19)).

The trend of the effective masses shown in Fig. 5.8.1 (**d**) can be quantitatively understood as follows. For single-particle states close to the Fermi energy the intermediate states have all energies larger than the unperturbed single-particle energy, and the resulting shape is δ-like with a tail extending away from the Fermi energy. For single-particle orbitals 5–7 MeV away from ϵ_F there are a number of intermediate states which have the same energy of the initial state leading to vanishing energy denominators and thus to a marked structure in the strength function. Finally, for single-particle states far away from ϵ_F, the density of intermediate states is so large that the matrix elements of these couplings average out to a constant with a value much larger than typical distances between successive intermediate states. These are the conditions which lead to a Breit–Wigner shape.

It is then to be expected that the single-particle levels in the intermediate region will be associated with strength functions which deviate much from a smooth function and for which comparatively large errors can be made through a Breit–Wigner fitting. This is also seen from Fig. 5.8.1 (d) where the effective mass becomes smaller than one, indicating that the area of the fitted shapes are larger than that of the original strength function $S(k; \omega)$. This result emphasizes the difficulties found in trying to extract spectroscopic factors from the data in a general situation.

Starting from the extreme picture shown in Fig. 5.8.1 (a) where the relation $\omega = \hbar^2 k^2 / 2m$ holds, one arrives at the picture (c) where the relation $d\omega = \hbar^2 k \delta k / m^*$ again accounts for the main properties of the single-particle motion and where the effect of the couplings are contained in m^* (see Eq. (5.4.3)).

The question then arises of how to define an average quantity which depends only on the energy of the orbital and is state independent. Rather different prescriptions have been used to deal with the question. In any case the main result obtained was that the ω-dependent effective mass shows a well-defined peak at the Fermi surface, its maximum value being considerably larger than one (≈ 1.4). The associated full width at half maximum is of the order of 10 MeV (i.e., ± 5 MeV around ϵ_F). This quantity is much smaller than the Fermi energy (~ 36 MeV) and is essentially determined by the energy of the low-lying collective modes and of the single-particle gap around the closed-shell system, for example, ^{208}Pb.

6

Two-Particle Transfer

A unified picture of pairing and two-nucleon transfer, where the two subjects are blended together (which is what happens in nature).

B. F. Bayman

The present chapter is structured in the following way. In Sect. 6.1 we present a summary of two-nucleon transfer reaction theory. It provides, together with Sect. 4.1, the elements needed to calculate the absolute two-nucleon transfer differential cross sections in second-order DWBA, and thus to compare theory with experiment. For the practitioner in search of details and clarification we carry out, in Sect. 6.2 a derivation of the equations presented in Sect. 6.1. These equations are implemented and made operative in the software COOPER used in the applications (cf. App. 7.A).

A number of appendices are provided. Apps. 6.A–6.D contain relations used in Sect. 6.2 as well as in the derivation of some two-nucleon transfer spectroscopic amplitudes. Finally, App. 6.E provides a glimpse of original material due to Ben Bayman[1] which was instrumental to render quantitative the studies of two-nucleon transfer, studies which can now be carried out in terms of absolute differential cross sections and not only of relative ones.

6.1 Summary of Second-Order DWBA

Let us illustrate the theory of second-order DWBA two-nucleon transfer reactions with the $A + t \rightarrow B(\equiv A + 2) + p$ reaction, in which $A + 2$ and A are even nuclei in their 0^+ ground state. The extension of the expressions to the transfer of pairs coupled to arbitrary angular momentum is discussed in Sect. 6.2.10.

---

[1] Bayman (1970, 1971); Bayman and Chen (1982).

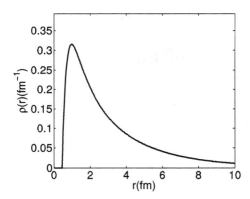

Figure 6.1.1 Radial function $\rho(r)$ (hardcore 0.45 fm) entering the tritium wave-function. After Potel et al. (2013b).

The wavefunction of the nucleus $A + 2$ can be written as

$$\Psi_{A+2}(\xi_A, \mathbf{r}_{A1}, \sigma_1, \mathbf{r}_{A2}, \sigma_2) = \psi_A(\xi_A) \sum_{l_i, j_i} [\phi_{l_i, j_i}^{A+2}(\mathbf{r}_{A1}, \sigma_1, \mathbf{r}_{A2}, \sigma_2)]_0^0, \qquad (6.1.1)$$

where

$$[\phi_{l_i, j_i}^{A+2}(\mathbf{r}_{A1}, \sigma_1, \mathbf{r}_{A2}, \sigma_2)]_0^0 = \sum_{nm} a_{nm} \left[\varphi_{n, l_i, j_i}^{A+2}(\mathbf{r}_{A1}, \sigma_1) \varphi_{m, l_i, j_i}^{A+2}(\mathbf{r}_{A2}, \sigma_2) \right]_0^0, \qquad (6.1.2)$$

while $\varphi_{n, l_i, j_i}^{A+2}(\mathbf{r})$ are single-particle wavefunctions. The radial dependence of the wavefunction of the two neutrons in the triton is written as $\phi_t(\mathbf{r}_{p1}, \mathbf{r}_{p2}) = \rho(r_{p1})\rho(r_{p2})\rho(r_{12})$, where r_{p1}, r_{p2}, r_{12} are the distances between neutron 1 and the proton, neutron 2 and the proton and between neutrons 1 and 2 respectively, while $\rho(r)$ is the hardcore ($r_{core} = 0.45$ fm) potential wavefunction depicted in[2] Fig 6.1.1.

The two-nucleon transfer differential cross section is written as

$$\frac{d\sigma}{d\Omega} = \frac{\mu_i \mu_f}{(4\pi\hbar^2)^2} \frac{k_f}{k_i} \left| T^{(1)}(\theta) + T_{succ}^{(2)}(\theta) - T_{NO}^{(2)}(\theta) \right|^2, \qquad (6.1.3)$$

where[3],

$$T^{(1)}(\theta) = 2 \sum_{l_i, j_i} \sum_{\sigma_1 \sigma_2} \int d\mathbf{r}_{tA} d\mathbf{r}_{p1} d\mathbf{r}_{A2} [\phi_{l_i, j_i}^{A+2}(\mathbf{r}_{A1}, \sigma_1, \mathbf{r}_{A2}, \sigma_2)]_0^{0*} \chi_{pB}^{(-)*}(\mathbf{r}_{pB})$$

$$\times v(r_{p1})\phi_t(\mathbf{r}_{p1}, \sigma_1, \mathbf{r}_{p2}, \sigma_2)\chi_{tA}^{(+)}(\mathbf{r}_{tA}), \qquad (6.1.4a)$$

[2] Tang and Herndon (1965).
[3] See Bayman and Chen (1982) and App. 6.E.

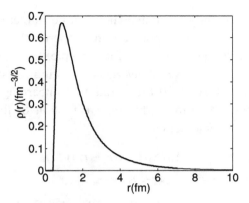

Figure 6.1.2 Radial wavefunction $\rho_d(r)$ (hardcore 0.45 fm) entering the deuteron wavefunction. After Potel et al. (2013b).

$$T^{(2)}_{succ}(\theta) = 2 \sum_{l_i,j_i} \sum_{l_f,j_f,m_f} \sum_{\substack{\sigma_1\sigma_2 \\ \sigma'_1\sigma'_2}} \int d\mathbf{r}_{dF} d\mathbf{r}_{p1} d\mathbf{r}_{A2} [\phi^{A+2}_{l_i,j_i}(\mathbf{r}_{A1}, \sigma_1, \mathbf{r}_{A2}, \sigma_2)]^{0*}_0$$

$$\times \chi^{(-)*}_{pB}(\mathbf{r}_{pB}) v(r_{p1}) \phi_d(\mathbf{r}_{p1}, \sigma_1) \varphi^{A+1}_{l_f,j_f,m_f}(\mathbf{r}_{A2}, \sigma_2) \int d\mathbf{r}'_{dF} d\mathbf{r}'_{p1} d\mathbf{r}'_{A2} G(\mathbf{r}_{dF}, \mathbf{r}'_{dF})$$

$$\times \phi_d(\mathbf{r}'_{p1}, \sigma'_1)^* \varphi^{A+1*}_{l_f,j_f,m_f}(\mathbf{r}'_{A2}, \sigma'_2) \frac{2\mu_{dF}}{\hbar^2} v(r'_{p2}) \phi_d(\mathbf{r}'_{p1}, \sigma'_1) \phi_d(\mathbf{r}'_{p2}, \sigma'_2) \chi^{(+)}_{tA}(\mathbf{r}'_{tA}),$$

(6.1.4b)

$$T^{(2)}_{NO}(\theta) = 2 \sum_{l_i,j_i} \sum_{l_f,j_f,m_f} \sum_{\substack{\sigma_1\sigma_2 \\ \sigma'_1\sigma'_2}} \int d\mathbf{r}_{dF} d\mathbf{r}_{p1} d\mathbf{r}_{A2} [\phi^{A+2}_{l_i,j_i}(\mathbf{r}_{A1}, \sigma_1, \mathbf{r}_{A2}, \sigma_2)]^{0*}_0$$

$$\times \chi^{(-)*}_{pB}(\mathbf{r}_{pB}) v(r_{p1}) \phi_d(\mathbf{r}_{p1}, \sigma_1) \varphi^{A+1}_{l_f,j_f,m_f}(\mathbf{r}_{A2}, \sigma_2) \int d\mathbf{r}'_{p1} d\mathbf{r}'_{A2} d\mathbf{r}'_{dF}$$

$$\times \phi_d(\mathbf{r}'_{p1}, \sigma'_1)^* \varphi^{A+1*}_{l_f,j_f,m_f}(\mathbf{r}'_{A2}, \sigma'_2) \phi_d(\mathbf{r}'_{p1}, \sigma'_1) \phi_d(\mathbf{r}'_{p2}, \sigma'_2) \chi^{(+)}_{tA}(\mathbf{r}'_{tA}).$$

(6.1.4c)

The quantities μ_i, $\mu_f(k_i, k_f)$ are the reduced masses (relative linear momenta) in both entrance (initial, i) and exit (final, f) channels, respectively. In the above expressions, $\varphi^{A+1}_{l_f,j_f,m_f}(\mathbf{r}_{A1})$ are the wavefunctions describing the intermediate states of the nucleus $F(\equiv (A+1))$, generated as solutions of a Woods–Saxon potential, $\phi_d(\mathbf{r}_{p2})$ being the deuteron bound wavefunction (see Fig. 6.1.2).[4] Note that some or all of the single-particle states described by the wavefunctions $\varphi^{A+1}_{l_f,j_f,m_f}(\mathbf{r}_{A1})$ may lie in the continuum (case in which the nucleus F is loosely bound or unbound). Although there are a number of ways to exactly treat such states, discretization

---

[4] Tang and Herndon (1965).

processes may be sufficiently accurate. They can be implemented by, for example, embedding the Woods–Saxon potential in a spherical box of sufficiently large radius. In actual calculations involving, for example, the halo nucleus ^{11}Li, and where $|F\rangle = |^{10}\text{Li}\rangle$, one achieved convergence making use of approximately 20 continuum states and a box of 30 fm of radius. Concerning the components of the triton wavefunction describing the relative motion of the dineutron, it was generated with the $p - n$ interaction[5]

$$v(r) = -v_0 \exp\left(-k(r - r_c)\right) \quad r > r_c \tag{6.1.5}$$

$$v(r) = \infty \quad r < r_c, \tag{6.1.6}$$

where $k = 2.5 \text{ fm}^{-1}$ and $r_c = 0.45 \text{ fm}$, the depth v_0 being adjusted to reproduce the experimental separation energies. The positive–energy wavefunctions $\chi_{tA}^{(+)}(\mathbf{r}_{tA})$ and $\chi_{pB}^{(-)}(\mathbf{r}_{pB})$ are the ingoing distorted wave in the initial channel and the outgoing distorted wave in the final channel respectively. They are continuum solutions of the Schrödinger equation associated with the corresponding optical potentials.

The transition potential responsible for the transfer of the pair is, in the *post* representation (cf. Fig. 6.5.1),

$$V_\beta = v_{pB} - U_\beta, \tag{6.1.7}$$

where v_{pB} is the interaction between the proton and nucleus B, and U_β is the optical potential in the final channel. We make the assumption that v_{pB} can be decomposed into a term containing the interaction between A and p and the potential describing the interaction between the proton and each of the transferred nucleons, namely,

$$v_{pB} = v_{pA} + v_{p1} + v_{p2}, \tag{6.1.8}$$

where v_{p1} and v_{p2} is the hardcore potential (6.1.5). The transition potential is

$$V_\beta = v_{pA} + v_{p1} + v_{p2} - U_\beta. \tag{6.1.9}$$

Assuming that $\langle\beta|v_{pA}|\alpha\rangle \approx \langle\beta|U_\beta|\alpha\rangle$ (i.e., assuming that the matrix element of the core–core interaction between the initial and final states is very similar to the matrix element of the real part of the optical potential), one obtains the final expression of the transfer potential in the *post-post* representation, namely,

$$V_\beta \simeq v_{p1} + v_{p2} = v(\mathbf{r}_{p1}) + v(\mathbf{r}_{p2}). \tag{6.1.10}$$

We make the further approximation of using the same interaction potential in all the (i.e., initial, intermediate, and final) channels.

The extension to a heavy-ion reaction $A + a(\equiv b + 2) \longrightarrow B(\equiv A + 2) + b$ imply no essential modifications in the formalism. The deuteron and triton wavefunctions

---

[5] Tang and Herndon (1965).

appearing in Eqs. (6.1.4a), (6.1.4b), and (6.1.4c) are to be substituted with the corresponding wavefunctions $\Psi_{b+2}(\xi_b, \mathbf{r}_{b1}, \sigma_1, \mathbf{r}_{b2}, \sigma_2)$, constructed in a similar way as those appearing in (6.1.1 and 6.1.2). The interaction potential used in Eqs. (6.1.4a), (6.1.4b) and (6.1.4c) will now be the Woods–Saxon used to define the initial (final) state in the post (prior) representation, instead of the proton–neutron interaction (6.1.5).

The Green's function $G(\mathbf{r}_{dF}, \mathbf{r}'_{dF})$ appearing in (6.1.4b) propagates the intermediate channel d, F. It can be expanded in partial waves as,

$$G(\mathbf{r}_{dF}, \mathbf{r}'_{dF}) = i \sum_l \sqrt{2l+1} \frac{f_l(k_{dF}, r_<) g_l(k_{dF}, r_>)}{k_{dF} r_{dF} r'_{dF}} \left[Y^l(\hat{r}_{dF}) Y^l(\hat{r}'_{dF}) \right]_0^0.$$

(6.1.11)

The $f_l(k_{dF}, r)$ and $g_l(k_{dF}, r)$ are the regular and the irregular solutions of a Schrödinger equation for a suitable optical potential and an energy equal to the kinetic energy of the intermediate state. In most cases of interest, the result is hardly altered if we use the same energy of relative motion for all the intermediate states. This representative energy is calculated when both intermediate nuclei are in their corresponding ground states. It is of note that the validity of this approximation can break down in some particular cases. If, for example, some relevant intermediate state become off shell, its contribution is significantly quenched. An subtle situation can arise when this happens to all possible intermediate states, so they can only be virtually populated.

6.2 Detailed Derivation of Second-Order DWBA

6.2.1 Simultaneous Transfer: Distorted Waves

For a (t, p) reaction, the triton is represented by an incoming distorted wave. We make the assumption that the two neutrons are in an $S = L = 0$ state, and that the relative motion of the proton with respect to the dineutron is also $l = 0$. Consequently, the total spin of the triton is entirely due to the spin of the proton. We will explicitly treat it, as we will consider a spin–orbit term in the optical potential acting between the triton and the target.[6]

Following (6.B.1), we can write the triton distorted wave as

$$\psi_{m_t}^{(+)}(\mathbf{R}, \mathbf{k}_i, \sigma_p) = \sum_{l_t} \exp\left(i\sigma_{l_t}^t\right) g_{l_t j_t} Y_0^{l_t}(\hat{\mathbf{R}}) \frac{\sqrt{4\pi(2l_t+1)}}{k_i R} \chi_{m_t}(\sigma_p), \qquad (6.2.1)$$

---

[6] In what follows, we will use the notation of Bayman (1971); see also App. 6.E.

where use was made of $Y_0^{l_t}(\hat{\mathbf{k}}_i) = i^{l_t}\sqrt{\frac{2l_t+1}{4\pi}}\delta_{m_t,0}$, in keeping with the fact that $\mathbf{k}_i$ is oriented along the z-axis. Note the phase difference with Eq. (7) of Bayman (1971), due to the use of time reversal rather than Condon–Shortley phase convention. Making use of the relation

$$Y_0^{l_t}(\hat{\mathbf{R}})\chi_{m_t}(\sigma_p) = \sum_{j_t}\langle l_t\ 0\ 1/2\ m_t|j_t\ m_t\rangle\left[Y^{l_t}(\hat{\mathbf{R}})\chi(\sigma_p)\right]_{m_t}^{j_t}, \qquad (6.2.2)$$

we have

$$\psi_{m_t}^{(+)}(\mathbf{R},\mathbf{k}_i,\sigma_p) = \sum_{l_t,j_t}\exp\left(i\sigma_{l_t}^t\right)\frac{\sqrt{4\pi(2l_t+1)}}{k_iR}g_{l_tj_t}(R)$$
$$\times\ \langle l_t\ 0\ 1/2\ m_t|j_t\ m_t\rangle\left[Y^{l_t}(\hat{\mathbf{R}})\chi(\sigma_p)\right]_{m_t}^{j_t}. \qquad (6.2.3)$$

We now turn our attention to the outgoing proton distorted wave, which, following (6.B.3), can be written as

$$\psi_{m_p}^{(-)}(\boldsymbol{\zeta},\mathbf{k}_f,\sigma_p) = \sum_{l_pj_p}\frac{4\pi}{k_f\zeta}i^{l_p}\exp\left(-i\sigma_{l_p}^p\right)f_{l_pj_p}^*(\zeta)\sum_m Y_m^{l_p}(\hat{\boldsymbol{\zeta}})Y_m^{l_p*}(\hat{\mathbf{k}}_f)\chi_{m_p}(\sigma_p).$$
$$(6.2.4)$$

Making use of the relation

$$\sum_m Y_m^{l_p}(\hat{\boldsymbol{\zeta}})Y_m^{l_p*}(\hat{\mathbf{k}}_f)\chi_{m_p}(\sigma_p)$$
$$= \sum_{m,j_p}Y_m^{l_p*}(\hat{\mathbf{k}}_f)\langle l_p\ m\ 1/2\ m_p|j_p\ m+m_p\rangle$$
$$\times\left[Y^{l_p}(\hat{\boldsymbol{\zeta}})\chi_{m_p}(\sigma_p)\right]_{m+m_p}^{j_p}$$
$$= \sum_{m,j_p}Y_{m-m_p}^{l_p*}(\hat{\mathbf{k}}_f)\langle l_p\ m-m_p\ 1/2\ m_p|j_p\ m\rangle\left[Y^{l_p}(\hat{\boldsymbol{\zeta}})\chi_{m_p}(\sigma_p)\right]_m^{j_p}, \quad (6.2.5)$$

one obtains

$$\psi_{m_p}^{(-)}(\boldsymbol{\zeta},\mathbf{k}_f,\sigma_p) = \frac{4\pi}{k_f\zeta}\sum_{l_pj_p,m}i^{l_p}\exp\left(-i\sigma_{l_p}^p\right)f_{l_pj_p}^*(\zeta)Y_{m-m_p}^{l_p*}(\hat{\mathbf{k}}_f)$$
$$\times\ \langle l_p\ m-m_p\ 1/2\ m_p|j_p\ m\rangle\left[Y^{l_p}(\hat{\boldsymbol{\zeta}})\chi(\sigma_p)\right]_m^{j_p}. \qquad (6.2.6)$$

6.2.2 Matrix Element for the Transition Amplitude

We now turn our attention to the evaluation of

$$\langle \Psi_f^{(-)}(\mathbf{k}_f)|V(r_{1p})|\Psi_i^{(+)}(k_i, \hat{\mathbf{z}})\rangle$$

$$= \frac{(4\pi)^{3/2}}{k_i k_f} \sum_{l_p l_t j_p j_t m} \left((\lambda \tfrac{1}{2})_k (\lambda \tfrac{1}{2})_k |(\lambda\lambda)_0 (\tfrac{1}{2}\tfrac{1}{2})_0\right)_0 \sqrt{2l_t + 1}$$

$$\times \langle l_p\ m - m_p\ 1/2\ m_p|j_p\ m\rangle \langle l_t\ 0\ 1/2\ m_t|j_t\ m_t\rangle\ i^{-l_p} \exp\left[i(\sigma_{l_p}^p + \sigma_{l_t}^t)\right]$$

$$\times 2Y_{m-m_p}^{l_p}(\hat{\mathbf{k}}_f) \sum_{\sigma_1\sigma_2\sigma_p} \int \frac{d\zeta\, dr\, d\eta}{\zeta R} u_{\lambda k}(r_1) u_{\lambda k}(r_2) \left[Y^\lambda(\hat{\mathbf{r}}_1)Y^\lambda(\hat{\mathbf{r}}_2)\right]_0^{0*} \qquad (6.2.7)$$

$$\times f_{l_p j_p}(\zeta) g_{l_t j_t}(R) \left[\chi(\sigma_1)\chi(\sigma_2)\right]_0^{0*} \left[Y^{l_p}(\hat{\boldsymbol{\zeta}})\chi(\sigma_p)\right]_m^{j_p*} V(r_{1p})$$

$$\times \theta_0^0(\mathbf{r}, \mathbf{s}) \left[\chi(\sigma_1)\chi(\sigma_2)\right]_0^0 \left[Y^{l_t}(\hat{\mathbf{R}})\chi(\sigma_p)\right]_{m_t}^{j_t},$$

where

$$\begin{aligned}
\mathbf{r} &= \mathbf{r}_2 - \mathbf{r}_1, \\
\mathbf{s} &= \frac{1}{2}(\mathbf{r}_1 + \mathbf{r}_2) - \mathbf{r}_p, \\
\eta &= \frac{1}{2}(\mathbf{r}_1 + \mathbf{r}_2), \\
\zeta &= \mathbf{r}_p - \frac{\mathbf{r}_1 + \mathbf{r}_2}{A + 2}.
\end{aligned} \qquad (6.2.8)$$

The sum over σ_1, σ_2 in (6.2.7) is found to be equal to 1. We will now simplify the term $\left[Y^{l_p}(\hat{\boldsymbol{\zeta}})\chi(\sigma_p)\right]_m^{j_p*} \left[Y^{l_t}(\hat{\mathbf{R}})\chi(\sigma_p)\right]_{m_t}^{j_t}$, noting that, (6.A.13)

$$\left[Y^{l_p}(\hat{\boldsymbol{\zeta}})\chi(\sigma_p)\right]_m^{j_p*} = (-1)^{1/2-\sigma_p+j_p-m} \left[Y^{l_p}(\hat{\boldsymbol{\zeta}})\chi(-\sigma_p)\right]_{-m}^{j_p}. \qquad (6.2.9)$$

and that

$$\left[Y^{l_p}(\hat{\boldsymbol{\zeta}})\ \chi(-\sigma_p)\right]_{-m}^{j_p} \left[Y^{l_t}(\hat{\mathbf{R}})\chi(\sigma_p)\right]_{m_t}^{j_t} = \sum_{JM} \langle j_p\ -m\ j_t\ m_t|J\ M\rangle$$

$$\times \left\{\left[Y^{l_p}(\hat{\boldsymbol{\zeta}})\chi(-\sigma_p)\right]^{j_p} \left[Y^{l_t}(\hat{\mathbf{R}})\chi(\sigma_p)\right]^{j_t}\right\}_M^J \qquad (6.2.10)$$

The only term which does not vanish after the integration is performed is the one in which the angular and spin functions are coupled to $L = 0$, $S = 0$, $J = 0$. Thus,

$$\langle j_p - m \ j_t \ m_t | 0 \ 0\rangle \left\{\left[Y^{l_p}(\hat{\boldsymbol{\zeta}})\chi(-\sigma_p)\right]^{j_p}\left[Y^{l_t}(\hat{\mathbf{R}})\chi(\sigma_p)\right]^{j_t}\right\}_0^0 \delta_{l_p l_t}\delta_{j_p j_t}\delta_{mm_t}$$

$$= \frac{(-1)^{j_p + m_t}}{\sqrt{2j_p + 1}}\left\{\left[Y^{l_p}(\hat{\boldsymbol{\zeta}})\chi(-\sigma_p)\right]^{j_p}\left[Y^{l_t}(\hat{\mathbf{R}})\chi(\sigma_p)\right]^{j_t}\right\}_0^0 \delta_{l_p l_t}\delta_{j_p j_t}\delta_{mm_t}. \quad (6.2.11)$$

Coupling separately the spin and angular functions, one obtains

$$\left\{\left[Y^l(\hat{\boldsymbol{\zeta}})\chi(-\sigma_p)\right]^j\left[Y^l(\hat{\mathbf{R}})\chi(\sigma_p)\right]^j\right\}_0^0$$

$$= \left((l\tfrac{1}{2})_j(l\tfrac{1}{2})_j|(ll)_0(\tfrac{1}{2}\tfrac{1}{2})_0\right)_0 \left[\chi(-\sigma_p)\chi(\sigma_p)\right]_0^0\left[Y^l(\hat{\boldsymbol{\zeta}})Y^l(\hat{\mathbf{R}})\right]_0^0. \quad (6.2.12)$$

We substitute (6.2.9), (6.2.30), (6.2.31) in (6.2.7) to obtain

$$\langle\Psi_f^{(-)}(\mathbf{k}_f)|V(r_{1p})|\Psi_i^{(+)}(k_i, \hat{\mathbf{z}})\rangle$$

$$= -\frac{(4\pi)^{3/2}}{k_i k_f}\sum_{lj}\left((\lambda\tfrac{1}{2})_k(\lambda\tfrac{1}{2})_k|(\lambda\lambda)_0(\tfrac{1}{2}\tfrac{1}{2})_0\right)_0\sqrt{\frac{2l+1}{2j+1}}$$

$$\times \langle l \ m_t - m_p \ 1/2 \ m_p|j \ m_t\rangle\langle l \ 0 \ 1/2 \ m_t|j \ m_t\rangle i^{-l}\exp\left[i(\sigma_l^p + \sigma_l^t)\right]$$

$$\times 2Y_{m_t - m_p}^l(\hat{\mathbf{k}}_f)\int\frac{d\boldsymbol{\zeta}\,d\mathbf{r}\,d\eta}{\zeta R}u_{\lambda k}(r_1)u_{\lambda k}(r_2)\left[Y^\lambda(\hat{\mathbf{r}}_1)Y^\lambda(\hat{\mathbf{r}}_2)\right]_0^{0*}$$

$$\times f_{lj}(\zeta)g_{lj}(R)\left[Y^l(\hat{\boldsymbol{\zeta}})Y^l(\hat{\mathbf{R}})\right]_0^0 V(r_{1p})\theta_0^0(\mathbf{r}, \mathbf{s})$$

$$\times \left((l\tfrac{1}{2})_j(l\tfrac{1}{2})_j|(ll)_0(\tfrac{1}{2}\tfrac{1}{2})_0\right)_0\sum_{\sigma_p}(-1)^{1/2-\sigma_p}\left[\chi(-\sigma_p)\chi(\sigma_p)\right]_0^0. \quad (6.2.13)$$

The last sum over σ_p leads to

$$\sum_{\sigma_p}(-1)^{1/2-\sigma_p}\left[\chi(-\sigma_p)\chi(\sigma_p)\right]_0^0 = \sum_{\sigma_p m}(-1)^{1/2-\sigma_p}\langle 1/2 \ m \ 1/2 \ -m|0 \ 0\rangle$$

$$\times \chi_m(-\sigma_p)\chi_{-m}(\sigma_p)$$

$$= \frac{1}{\sqrt{2}}\sum_{\sigma_p m}(-1)^{1/2-\sigma_p}(-1)^{1/2-m}\delta_{m,-\sigma_p}\delta_{-m,\sigma_p}$$

$$= -\sqrt{2}. \quad (6.2.14)$$

The $9j$-symbols can be evaluated to find

$$\left((\lambda\tfrac{1}{2})_k(\lambda\tfrac{1}{2})_k|(\lambda\lambda)_0(\tfrac{1}{2}\tfrac{1}{2})_0\right)_0 = \sqrt{\frac{2k+1}{2(2\lambda+1)}}$$

$$\left((l\tfrac{1}{2})_j(l\tfrac{1}{2})_j|(ll)_0(\tfrac{1}{2}\tfrac{1}{2})_0\right)_0 = \sqrt{\frac{2j+1}{2(2l+1)}}, \qquad (6.2.15)$$

and consequently,

$$\langle \Psi_f^{(-)}(\mathbf{k}_f) | V(r_{1p}) | \Psi_i^{(+)}(k_i, \hat{\mathbf{z}}) \rangle = \frac{(4\pi)^{3/2}}{k_i k_f} \sum_{lj} \sqrt{\frac{2k+1}{2\lambda+1}}$$

$$\times \langle l\ m_t - m_p\ 1/2\ m_p | j\ m_t \rangle \langle l\ 0\ 1/2\ m_t | j\ m_t \rangle\, i^{-l} \exp\left[i(\sigma_l^p + \sigma_l^t) \right]$$

$$\times \sqrt{2} Y_{m_t-m_p}^l(\hat{\mathbf{k}}_f) \int \frac{d\boldsymbol{\zeta}\, d\mathbf{r}\, d\boldsymbol{\eta}}{\zeta R} u_{\lambda k}(r_1) u_{\lambda k}(r_2) \left[Y^\lambda(\hat{\mathbf{r}}_1) Y^\lambda(\hat{\mathbf{r}}_2) \right]_0^{0*}$$

$$\times f_{lj}(\zeta) g_{lj}(R) \left[Y^l(\hat{\boldsymbol{\zeta}}) Y^l(\hat{\mathbf{R}}) \right]_0^0 V(r_{1p}) \theta_0^0(\mathbf{r}, \mathbf{s}). \tag{6.2.16}$$

The values of the Clebsh–Gordan coefficients are, for $j = l - 1/2$,

$$\langle l\ m_t - m_p\ 1/2\ m_p | l - 1/2\ m_t \rangle \langle l\ 0\ 1/2\ m_t | l - 1/2\ m_t \rangle$$

$$= \begin{cases} \dfrac{l}{2l+1} & \text{if } m_t = m_p \\[2ex] -\dfrac{\sqrt{l(l+1)}}{2l+1} & \text{if } m_t = -m_p \end{cases} \tag{6.2.17}$$

and, for $j = l + 1/2$:

$$\langle l\ m_t - m_p\ 1/2\ m_p | l + 1/2\ m_t \rangle \langle l\ 0\ 1/2\ m_t | l + 1/2\ m_t \rangle$$

$$= \begin{cases} \dfrac{l+1}{2l+1} & \text{if } m_t = m_p \\[2ex] \dfrac{\sqrt{l(l+1)}}{2l+1} & \text{if } m_t = -m_p \end{cases} \tag{6.2.18}$$

One thus can write,

$$\langle \Psi_f^{(-)}(\mathbf{k}_f) | V(r_{1p}) | \Psi_i^{(+)}(k_i, \hat{\mathbf{z}}) \rangle$$

$$= \frac{(4\pi)^{3/2}}{k_i k_f} \sum_l \frac{1}{(2l+1)} \sqrt{\frac{(2k+1)}{(2\lambda+1)}} \exp\left[i(\sigma_l^p + \sigma_l^t) \right] i^{-l}$$

$$\times \sqrt{2} Y_{m_t-m_p}^l(\hat{\mathbf{k}}_f) \int \frac{d\boldsymbol{\zeta}\, d\mathbf{r}\, d\boldsymbol{\eta}}{\zeta R} u_{\lambda k}(r_1) u_{\lambda k}(r_2) \left[Y^\lambda(\hat{\mathbf{r}}_1) Y^\lambda(\hat{\mathbf{r}}_2) \right]_0^{0*}$$

$$\times V(r_{1p}) \theta_0^0(\mathbf{r}, \mathbf{s}) \left[Y^l(\hat{\boldsymbol{\zeta}}) Y^l(\hat{\mathbf{R}}) \right]_0^0$$

$$\times \Big[\Big(f_{ll+1/2}(\zeta) g_{ll+1/2}(R)(l+1) + f_{ll-1/2}(\zeta) g_{ll-1/2}(R) l \Big) \delta_{m_p, m_t}$$

$$+ \Big(f_{ll+1/2}(\zeta) g_{ll+1/2}(R) \sqrt{l(l+1)} - f_{ll-1/2}(\zeta) g_{ll-1/2}(R) \sqrt{l(l+1)} \Big) \delta_{m_p, -m_t} \Big]. \tag{6.2.19}$$

We can further simplify this expression using

$$\left[Y^\lambda(\hat{\mathbf{r}}_1)Y^\lambda(\hat{\mathbf{r}}_2)\right]_0^{0*} = \left[Y^\lambda(\hat{\mathbf{r}}_1)Y^\lambda(\hat{\mathbf{r}}_2)\right]_0^0 = \sum_m \langle \lambda \ m \ \lambda \ -m|0 \ 0\rangle Y_m^\lambda(\hat{\mathbf{r}}_1)Y_{-m}^\lambda(\hat{\mathbf{r}}_2)$$

$$= \sum_m (-1)^{\lambda-m}\langle \lambda \ m \ \lambda \ -m|0 \ 0\rangle Y_m^\lambda(\hat{\mathbf{r}}_1)Y_m^{\lambda*}(\hat{\mathbf{r}}_2)$$

$$= \frac{1}{\sqrt{2\lambda+1}} \sum_m Y_m^\lambda(\hat{\mathbf{r}}_1)Y_m^{\lambda*}(\hat{\mathbf{r}}_2)$$

$$= \frac{\sqrt{(2\lambda+1)}}{4\pi} P_\lambda(\cos\theta_{12}). \tag{6.2.20}$$

Note that when using Condon–Shortley phases this last expression is to be multiplied by $(-1)^\lambda$, and that

$$\left[Y^l(\hat{\boldsymbol{\zeta}})Y^l(\hat{\mathbf{R}})\right]_0^0 = \sum_m \langle l \ m \ l \ -m|0 \ 0\rangle Y_m^l(\hat{\boldsymbol{\zeta}})Y_{-m}^l(\hat{\mathbf{R}})$$

$$= \frac{1}{\sqrt{(2l+1)}} \sum_m (-1)^{l+m} Y_m^l(\hat{\boldsymbol{\zeta}})Y_{-m}^l(\hat{\mathbf{R}}). \tag{6.2.21}$$

Because the integral of the above expression is independent of m, one can eliminate the m-sum and multiply by $2l+1$ the $m=0$ term, leading to

$$\left[Y^l(\hat{\boldsymbol{\zeta}})Y^l(\hat{\mathbf{R}})\right]_0^0 \Rightarrow (-1)^l \sqrt{(2l+1)} \ Y_0^l(\hat{\boldsymbol{\zeta}})_0 Y^l(\hat{\mathbf{R}})$$

$$= \sqrt{(2l+1)}Y_0^l(\hat{\boldsymbol{\zeta}})Y_0^{l*}(\hat{\mathbf{R}}). \tag{6.2.22}$$

We now change the integration variables from $(\boldsymbol{\zeta}, \mathbf{r}, \boldsymbol{\eta})$ to $(\mathbf{R}, \alpha, \beta, \gamma, r_{12}, r_{1p}, r_{2p})$, the quantity

$$\left|\frac{\partial(\mathbf{r}, \boldsymbol{\eta}, \boldsymbol{\zeta})}{\partial(\mathbf{R}, \alpha, \beta, \gamma, r_{12}, r_{1p}, r_{2p})}\right| = r_{12}r_{1p}r_{2p}\sin\beta, \tag{6.2.23}$$

being the Jacobian of the transformation. Finally,

$$\langle \Psi_f^{(-)}(\mathbf{k}_f)|V(r_{1p})|\Psi_i^{(+)}(k_i, \hat{\mathbf{z}})\rangle = \frac{\sqrt{8\pi}}{k_i k_f} \sum_l \sqrt{\frac{2k+1}{2l+1}} \exp\left[i(\sigma_l^p + \sigma_l^t)\right]i^{-l}$$

$$\times Y_{m_t-m_p}^l(\hat{\mathbf{k}}_f) \int d\mathbf{R} Y_0^{l*}(\hat{\mathbf{R}}) \int \frac{d\alpha \, d\beta \, d\gamma \, dr_{12} \, dr_{1p} \, dr_{2p} \, \sin\beta}{\zeta R} Y_0^l(\hat{\boldsymbol{\zeta}})$$

$$\times u_{\lambda k}(r_1)u_{\lambda k}(r_2)V(r_{1p})\theta_0^0(\mathbf{r}, \mathbf{s})P_\lambda(\cos\theta_{12})r_{12}r_{1p}r_{2p}$$

$$\times \left[\left(f_{ll+1/2}(\zeta)g_{ll+1/2}(R)(l+1) + f_{ll-1/2}(\zeta)g_{ll-1/2}(R)l\right)\delta_{m_p,m_t}\right.$$

$$\left.+\left(f_{ll+1/2}(\zeta)g_{ll+1/2}(R)\sqrt{l(l+1)} - f_{ll-1/2}(\zeta)g_{ll-1/2}(R)\sqrt{l(l+1)}\right)\delta_{m_p,-m_t}\right].$$

$$\tag{6.2.24}$$

It is noted that the second integral is a function of solely $\mathbf{R}$ transforming under rotations as $Y_0^l(\hat{\mathbf{R}})$, in keeping with the fact that the full dependence on the orientation of $\mathbf{R}$ is contained in the spherical harmonic $Y_0^l(\hat{\boldsymbol{\zeta}})$. The second integral can thus be cast into the form

$$A(R)Y_0^l(\hat{\mathbf{R}}) = \int d\alpha \, d\beta \, d\gamma \, dr_{12} \, dr_{1p} \, dr_{2p} \, \sin\beta$$
$$\times F(\alpha, \beta, \gamma, r_{12}, r_{1p}, r_{2p}, R_x, R_y, R_z). \tag{6.2.25}$$

To evaluate $A(R)$, we set $\mathbf{R}$ along the z-axis

$$A(R) = 2\pi i^{-l} \sqrt{\frac{4\pi}{2l+1}} \int d\beta \, d\gamma \, dr_{12} \, dr_{1p} \, dr_{2p} \, \sin\beta$$
$$\times F(\alpha, \beta, \gamma, r_{12}, r_{1p}, r_{2p}, 0, 0, R), \tag{6.2.26}$$

where a factor 2π results from the integration over α, the integrand not depending on α. Substituting (6.2.25) and (6.2.26) in (6.2.24) and, after integration over the angular variables of $\mathbf{R}$, we obtain

$$\langle \Psi_f^{(-)}(\mathbf{k}_f)| V(r_{1p})|\Psi_i^{(+)}(k_i, \hat{\mathbf{z}})\rangle = 2\frac{(2\pi)^{3/2}}{k_i k_f} \sum_l \sqrt{\frac{2k+1}{2l+1}} \exp[i(\sigma_l^p + \sigma_l^t)]i^{-l}$$

$$\times Y_{m_t - m_p}^l(\hat{\mathbf{k}}_f) \int dR \, d\beta \, d\gamma \, dr_{12} \, dr_{1p} \, dr_{2p} \, R \sin\beta \, r_{12} r_{1p} r_{2p}$$

$$\times u_{\lambda k}(r_1) u_{\lambda k}(r_2) V(r_{1p}) \theta_0^0(\mathbf{r}, \mathbf{s}) P_\lambda(\cos\theta_{12}) P_l(\cos\theta_\zeta)$$

$$\times \left[\left(f_{ll+1/2}(\zeta)g_{ll+1/2}(R)(l+1) + f_{ll-1/2}(\zeta)g_{ll-1/2}(R)l \right)\delta_{m_p,m_t} \right.$$

$$\left. + \left(f_{ll+1/2}(\zeta)g_{ll+1/2}(R)\sqrt{l(l+1)} - f_{ll-1/2}(\zeta)g_{ll-1/2}(R)\sqrt{l(l+1)} \right)\delta_{m_p,-m_t} \right]/\zeta, \tag{6.2.27}$$

where use was made of the relation

$$Y_0^l(\hat{\boldsymbol{\zeta}}) = i^l \sqrt{\frac{2l+1}{4\pi}} P_l(\cos\theta_\zeta). \tag{6.2.28}$$

The final expression of the differential cross section involves a sum over the spin orientations:

$$\frac{d\sigma}{d\Omega}(\hat{\mathbf{k}}_f) = \frac{k_f}{k_i} \frac{\mu_i \mu_f}{(2\pi\hbar^2)^2} \frac{1}{2} \sum_{m_t m_p} |\langle \Psi_f^{(-)}(\mathbf{k}_f)|V(r_{1p})|\Psi_i^{(+)}(k_i, \hat{\mathbf{z}})\rangle|^2. \tag{6.2.29}$$

When $m_p = 1/2$, $m_t = 1/2$ or $m_p = -1/2$, $m_t = -1/2$, the terms proportional to δ_{m_p,m_t} including the factor

$$|Y^l_{m_t - m_p}(\hat{\mathbf{k}}_f)\delta_{m_p, m_t}| = |Y^l_0(\hat{\mathbf{k}}_f)| = \left| i^l \sqrt{\frac{2l+1}{4\pi}} P^0_l(\cos\theta) \right|, \qquad (6.2.30)$$

in the case in which $m_p = -1/2, m_t = 1/2$

$$|Y^l_{m_t - m_p}(\hat{\mathbf{k}}_f)\delta_{m_p, -m_t}| = |Y^l_1(\hat{\mathbf{k}}_f)| = \left| i^l \sqrt{\frac{2l+1}{4\pi} \frac{1}{l(l+1)}} P^1_l(\cos\theta) \right|, \qquad (6.2.31)$$

and

$$|Y^l_{m_t - m_p}(\hat{\mathbf{k}}_f)\delta_{m_p, -m_t}| = |Y^l_{-1}(\hat{\mathbf{k}}_f)| = |Y^l_1(\hat{\mathbf{k}}_f)| = \left| i^l \sqrt{\frac{2l+1}{4\pi} \frac{1}{l(l+1)}} P^1_l(\cos\theta) \right|, \qquad (6.2.32)$$

when $m_p = 1/2, m_t = -1/2$ Taking the squared modulus of (6.2.27), the sum over m_t and m_p yields a factor 2 multiplying each one of the 2 different terms of the sum ($m_t = m_p$ and $m_t = -m_p$). This is equivalent to multiply each amplitude by $\sqrt{2}$, so the final constant that multiply the amplitudes is

$$\frac{8\pi^{3/2}}{k_i k_f}. \qquad (6.2.33)$$

Now, for the triton wavefunction we use

$$\theta^0_0(\mathbf{r}, \mathbf{s}) = \rho(r_{1p})\rho(r_{2p})\rho(r_{12}), \qquad (6.2.34)$$

$\rho(r)$ being a Tang–Herndon (1965) wave function also used by Bayman (1971). We obtain

$$\frac{d\sigma}{d\Omega}(\hat{\mathbf{k}}_f) = \frac{1}{2E_i^{3/2} E_f^{1/2}} \sqrt{\frac{\mu_f}{\mu_i}} \left(|I^{(0)}_{\lambda k}(\theta)|^2 + |I^{(1)}_{\lambda k}(\theta)|^2 \right), \qquad (6.2.35)$$

where

$$I^{(0)}_{\lambda k}(\theta) = \sum_l P^0_l(\cos\theta) \sqrt{2k+1} \exp[i(\sigma^p_l + \sigma^t_l)]$$

$$\times \int dR\, d\beta\, d\gamma\, dr_{12}\, dr_{1p}\, dr_{2p}\, R \sin\beta\, \rho(r_{1p})\rho(r_{2p})\rho(r_{12}) \qquad (6.2.36)$$

$$\times u_{\lambda k}(r_1) u_{\lambda k}(r_2) V(r_{1p}) P_\lambda(\cos\theta_{12}) P_l(\cos\theta_\zeta) r_{12} r_{1p} r_{2p}$$

$$\times \left(f_{ll+1/2}(\zeta) g_{ll+1/2}(R)\, (l+1) + f_{ll-1/2}(\zeta) g_{ll-1/2}(R)\, l \right)/\zeta,$$

and

$$I_{\lambda k}^{(1)}(\theta) = \sum_l P_l^1(\cos\theta)\sqrt{2k+1}\exp\left[i(\sigma_l^p + \sigma_l^t)\right]$$

$$\times \int dR\, d\beta\, d\gamma\, dr_{12}\, dr_{1p}\, dr_{2p}\, R\sin\beta\, \rho(r_{1p})\rho(r_{2p})\rho(r_{12}) \quad (6.2.37)$$

$$\times u_{\lambda k}(r_1)u_{\lambda k}(r_2)V(r_{1p})P_\lambda(\cos\theta_{12})P_l(\cos\theta_\zeta)r_{12}r_{1p}r_{2p}$$

$$\times \left(f_{ll+1/2}(\zeta)g_{ll+1/2}(R) - f_{ll-1/2}(\zeta)g_{ll-1/2}(R)\right)/\zeta.$$

Note that the absence of the $(-1)^\lambda$ factor with respect to what is found in the work of Bayman[7] is due to the use of time–reversed phases instead of Condon–Shortley phasing. This is compensated in the total result by a similar difference in the expression of the spectroscopic amplitudes. This ensures that, in either case, the contribution of all the single particle transitions tend to have the same phase for superfluid nuclei, adding coherently to enhance the transfer cross section.

Heavy-Ion Reactions

In dealing with a heavy ion reaction, $\theta_0^0(\mathbf{r}, \mathbf{s})$ is the spatial part of the wavefunction

$$\Psi(\mathbf{r}_{b1}, \mathbf{r}_{b2}, \sigma_1, \sigma_2) = \left[\psi^{j_i}(\mathbf{r}_{b1}, \sigma_1)\psi^{j_i}(\mathbf{r}_{b2}, \sigma_2)\right]_0^0$$

$$= \theta_0^0(\mathbf{r}, \mathbf{s})\left[\chi(\sigma_1)\chi(\sigma_2)\right]_0^0, \quad (6.2.38)$$

where $\mathbf{r}_{b1}, \mathbf{r}_{b2}$ are the positions of the two neutrons with respect to the b core. It can be shown to be

$$\theta_0^0(\mathbf{r}, \mathbf{s}) = \frac{u_{l_i j_i}(r_{b1})u_{l_i j_i}(r_{b2})}{4\pi}\sqrt{\frac{2j_i + 1}{2}}P_{l_i}(\cos\theta_i), \quad (6.2.39)$$

where θ_i is the angle between $\mathbf{r}_{b1}$ and $\mathbf{r}_{b2}$. Neglecting the spin–orbit term in the optical potential, as is usually done for heavy ion reactions, one obtains

$$\frac{d\sigma}{d\Omega}(\hat{\mathbf{k}}_f) = \frac{\mu_f\mu_i}{16\pi^2\hbar^4 k_i^3 k_f}|T^{(1)}(\theta)|^2, \quad (6.2.40)$$

where

$$T^{(1)}(\theta) = \sum_l(2l+1)P_l(\cos\theta)\sqrt{(2j_i+1)(2j_f+1)}\exp\left[i(\sigma_l^p + \sigma_l^t)\right]$$

$$\times \int dR\, d\beta\, d\gamma\, dr_{12}\, dr_{b1}\, dr_{b2}\, R\sin\beta\, u_{l_i j_i}(r_{b1})u_{l_i j_i}(r_{b2}) \quad (6.2.41)$$

$$\times u_{l_f j_f}(r_{A1})u_{l_f j_f}(r_{A2})V(r_{b1})P_\lambda(\cos\theta_{12})P_l(\cos\theta_\zeta)$$

$$\times r_{12}r_{b1}r_{b2}P_{l_i}(\cos\theta_i)\frac{f_l(\zeta)g_l(R)}{\zeta},$$

---

[7] Bayman (1971).

obtained by using Eq. (6.2.39) in Eq. (6.2.7) instead of (6.2.34), $\mathbf{r}_{A1}$, $\mathbf{r}_{A2}$ being the coordinates of the two transferred neutrons with respect to the A core.

For control, in what follows we work out the same transition amplitude but starting from the distorted waves for a reaction taking place between spinless nuclei, namely,

$$\psi^{(+)}(\mathbf{r}_{Aa}, \mathbf{k}_{Aa}) = \sum_l \exp\left(i\sigma_l^i\right) g_l Y_0^l(\hat{\mathbf{r}}_{aA}) \frac{\sqrt{4\pi(2l+1)}}{k_{aA}r_{aA}}, \tag{6.2.42}$$

and

$$\psi^{(-)}(\mathbf{r}_{bB}, \mathbf{k}_{bB}) = \frac{4\pi}{k_{bB}r_{bB}} \sum_{\tilde{l}} i^{\tilde{l}} \exp\left(-i\sigma_{\tilde{l}}^f\right) f_{\tilde{l}}^*(r_{bB}) \sum_m Y_m^{\tilde{l}*}(\hat{\mathbf{k}}_{bB}) Y_m^{\tilde{l}}(\hat{\mathbf{r}}_{bB}). \tag{6.2.43}$$

One can then write,

$$\begin{aligned}
T_{2N}^{1step} &= \langle \Psi_f^{(-)}(\mathbf{k}_{bB}) | V(r_{1p}) | \Psi_i^{(+)}(k_{aA}, \hat{\mathbf{z}}) \rangle \\
&= \frac{(4\pi)^{3/2}}{k_{aA}k_{bB}} \sum_{l\tilde{l}m} ((l_f \tfrac{1}{2})_{j_f} (l_f \tfrac{1}{2})_{j_f} | (l_f l_f)_0 (\tfrac{1}{2}\tfrac{1}{2})_0)_0 \\
&\quad \times ((l_i \tfrac{1}{2})_{j_i} (l_i \tfrac{1}{2})_{j_i} | (l_i l_i)_0 (\tfrac{1}{2}\tfrac{1}{2})_0)_0 \sqrt{2l+1} i^{-l_p} \exp[i(\sigma_{\tilde{l}}^f + \sigma_l^i)] \\
&\quad \times 2 Y_m^{\tilde{l}}(\hat{\mathbf{k}}_{bB}) \sum_{\sigma_1\sigma_2} \int \frac{d\mathbf{r}_{bB} d\mathbf{r} d\eta}{r_{bB}r_{aA}} u_{l_f j_f}(r_{A1}) u_{l_f j_f}(r_{A2}) u_{l_i j_i}(r_{b1}) u_{l_i j_i}(r_{b2}) \\
&\quad \times \left[Y^{l_f}(\hat{\mathbf{r}}_{A1}) Y^{l_f}(\hat{\mathbf{r}}_{A2})\right]_0^{0*} \left[Y^{l_i}(\hat{\mathbf{r}}_{b1}) Y^{l_i}(\hat{\mathbf{r}}_{b2})\right]_0^0 \\
&\quad \times f_{\tilde{l}}(r_{bB}) g_l(r_{aA}) \left[\chi(\sigma_1)\chi(\sigma_2)\right]_0^{0*} Y_m^{\tilde{l}*}(\hat{\mathbf{r}}_{bB}) V(r_{1p}) \\
&\quad \times \left[\chi(\sigma_1)\chi(\sigma_2)\right]_0^0 Y_0^l(\hat{\mathbf{r}}_{aA}), \tag{6.2.44}
\end{aligned}$$

which, after a number of simplifications becomes

$$\begin{aligned}
\langle \Psi_f^{(-)}(\mathbf{k}_{bB}) | V(r_{1p}) | \Psi_i^{(+)}(k_{aA}, \hat{\mathbf{z}}) \rangle &= \frac{(4\pi)^{3/2}}{k_{aA}k_{bB}} \sum_{l\tilde{l}m} \sqrt{\frac{(2j_f+1)(2j_i+1)}{(2l_f+1)(2l_i+1)}} \\
&\quad \times \sqrt{2l+1} i^{-\tilde{l}} \exp[i(\sigma_{\tilde{l}}^f + \sigma_l^i)] \\
&\quad \times Y_m^{\tilde{l}}(\hat{\mathbf{k}}_{bB}) \int \frac{d\mathbf{r}_{bB} d\mathbf{r} d\eta}{r_{bB}r_{aA}} u_{l_f j_f}(r_{A1}) u_{l_f j_f}(r_{A2}) u_{l_i j_i}(r_{b1}) u_{l_i j_i}(r_{b2}) \\
&\quad \times \left[Y^{l_f}(\hat{\mathbf{r}}_{A1}) Y^{l_f}(\hat{\mathbf{r}}_{A2})\right]_0^{0*} \left[Y^{l_i}(\hat{\mathbf{r}}_{b1}) Y^{l_i}(\hat{\mathbf{r}}_{b2})\right]_0^0 \\
&\quad \times f_{\tilde{l}}(r_{bB}) g_l(r_{aA}) Y_m^{\tilde{l}*}(\hat{\mathbf{r}}_{bB}) V(r_{1p}) Y_0^l(\hat{\mathbf{r}}_{aA}), \tag{6.2.45}
\end{aligned}$$

where $l = \tilde{l}$ and $m = 0$. Making use of Legendre polynomials leads to,

$$\langle \Psi_f^{(-)}(\mathbf{k}_{bB}) | V(r_{1p}) | \Psi_i^{(+)}(k_{aA}, \hat{\mathbf{z}}) \rangle = \frac{(4\pi)^{-1/2}}{k_{aA}k_{bB}} \sum_l \sqrt{(2j_f + 1)(2j_i + 1)}$$

$$\times \sqrt{2l + 1} i^{-l} \exp[i(\sigma_l^f + \sigma_l^i)] Y_0^l(\hat{\mathbf{k}}_{bB})$$

$$\times \int \frac{dr_{bB}drd\eta}{r_{bB}r_{aA}} u_{l_f j_f}(r_{A1}) u_{l_f j_f}(r_{A2}) u_{l_i j_i}(r_{b1}) u_{l_i j_i}(r_{b2})$$

$$\times P_{l_f}(\cos \theta_A) P_{l_i}(\cos \theta_b)$$

$$\times f_l(r_{bB}) g_l(r_{aA}) Y_0^{l*}(\hat{\mathbf{r}}_{bB}) V(r_{1p}) Y_0^l(\hat{\mathbf{r}}_{aA}). \qquad (6.2.46)$$

Changing the integration variables and proceeding as in last section (implying the multiplicative factor $2\pi \sqrt{\frac{4\pi}{2l+1}}$), the above expression becomes

$$\langle \Psi_f^{(-)}(\mathbf{k}_{bB}) | V(r_{1p}) | \Psi_i^{(+)}(k_{aA}, \hat{\mathbf{z}}) \rangle = \frac{2\pi}{k_{aA}k_{bB}} \sum_l \sqrt{(2j_f + 1)(2j_i + 1)}$$

$$\times i^{-l} \exp[i(\sigma_l^f + \sigma_l^i)] Y_0^l(\hat{\mathbf{k}}_{bB})$$

$$\times \int dr_{aA} d\beta d\gamma dr_{12} dr_{b1} dr_{b2} r_{aA} \sin \beta r_{12} r_{b1} r_{b2} \qquad (6.2.47)$$

$$\times P_{l_f}(\cos \theta_A) P_{l_i}(\cos \theta_b) u_{l_f j_f}(r_{A1}) u_{l_f j_f}(r_{A2}) u_{l_i j_i}(r_{b1}) u_{l_i j_i}(r_{b2})$$

$$\times f_l(r_{bB}) g_l(r_{aA}) Y_0^{l*}(\hat{\mathbf{r}}_{bB}) V(r_{1p})/r_{bB},$$

which eventually can be recast, through the use of Legendre polynomials, in the expression,

$$T_{2N}^{1step} = \langle \Psi_f^{(-)}(\mathbf{k}_{bB}) | V(r_{1p}) | \Psi_i^{(+)}(k_{aA}, \hat{\mathbf{z}}) \rangle = \frac{1}{2k_{aA}k_{bB}} \sum_l \sqrt{(2j_f + 1)(2j_i + 1)}$$

$$\times i^{-l} \exp[i(\sigma_l^f + \sigma_l^i)] P_l(\cos \theta)(2l + 1)$$

$$\times \int dr_{aA} d\beta d\gamma dr_{12} dr_{b1} dr_{b2} r_{aA} \sin \beta r_{12} r_{b1} r_{b2}$$

$$\times P_{l_f}(\cos \theta_A) P_{l_i}(\cos \theta_b) u_{l_f j_f}(r_{A1}) u_{l_f j_f}(r_{A2}) V(r_{1p})$$

$$\times u_{l_i j_i}(r_{b1}) u_{l_i j_i}(r_{b2}) f_l(r_{bB}) g_l(r_{aA}) P_l(\cos \theta_{if})/r_{bB}, \qquad (6.2.48)$$

expression which gives the same results as (6.2.41).

6.2.3 Coordinates for the Calculation of Simultaneous Transfer

In what follows we make explicit the coordinates used in the calculation of the above equations. Making use of the notation of Bayman (1971), we find the expression of the variables appearing in the integral as functions of the integration variables $r_{1p}, r_{2p}, r_{12}, R, \beta, \gamma$ (remember that $\mathbf{R} = R\hat{\mathbf{z}}$; see last section). $\mathbf{R}$ being the center of mass coordinate. Thus, one can write

$$\mathbf{R} = \frac{1}{3}\left(\mathbf{r}_1 + \mathbf{r}_2 + \mathbf{r}_p\right) = \frac{1}{3}\left(\mathbf{R} + \mathbf{d}_1 + \mathbf{R} + \mathbf{d}_2 + \mathbf{R} + \mathbf{d}_p\right), \qquad (6.2.49)$$

so

$$\mathbf{d}_1 + \mathbf{d}_2 + \mathbf{d}_p = 0. \qquad (6.2.50)$$

Together with

$$\mathbf{d}_1 + \mathbf{r}_{12} = \mathbf{d}_2 \qquad \mathbf{d}_2 + \mathbf{r}_{2p} = \mathbf{d}_p, \qquad (6.2.51)$$

we find

$$\mathbf{d}_1 = \frac{1}{3}\left(2\mathbf{r}_{12} + \mathbf{r}_{2p}\right), \qquad (6.2.52)$$

and

$$d_1^2 = \frac{1}{9}\left(4r_{12}^2 + r_{2p}^2 + 4\mathbf{r}_{12}\mathbf{r}_{2p}\right). \qquad (6.2.53)$$

Making use of

$$\begin{aligned}
\mathbf{r}_{12} + \mathbf{r}_{2p} &= \mathbf{r}_{1p} \\
r_{1p}^2 &= r_{12}^2 + r_{2p}^2 + 2\mathbf{r}_{12}\mathbf{r}_{2p} \\
2\mathbf{r}_{12}\mathbf{r}_{2p} &= r_{1p}^2 - r_{12}^2 - r_{2p}^2.
\end{aligned} \qquad (6.2.54)$$

one obtains

$$d_1 = \frac{1}{3}\sqrt{2r_{12}^2 + 2r_{1p}^2 - r_{2p}^2}. \qquad (6.2.55)$$

Similarly,

$$d_2 = \frac{1}{3}\sqrt{2r_{12}^2 + 2r_{2p}^2 - r_{1p}^2} \qquad d_p = \frac{1}{3}\sqrt{2r_{2p}^2 + 2r_{1p}^2 - r_{12}^2}. \qquad (6.2.56)$$

We now express the angle α between $\mathbf{d}_1$ and $\mathbf{r}_{12}$. We have

$$-\mathbf{d}_1\mathbf{r}_{12} = r_{12}d_1\cos(\alpha), \qquad (6.2.57)$$

and

$$\begin{aligned}
\mathbf{d}_1 + \mathbf{r}_{12} &= \mathbf{d}_2 \\
d_1^2 + r_{12}^2 + 2\mathbf{d}_1\mathbf{r}_{12} &= d_2^2.
\end{aligned} \qquad (6.2.58)$$

Consequently,

$$\cos(\alpha) = \frac{d_1^2 + r_{12}^2 - d_2^2}{2r_{12}d_1}. \qquad (6.2.59)$$

The complete determination of $\mathbf{r}_1, \mathbf{r}_2, \mathbf{r}_{12}$ can be made by writing their expression in a simple configuration, in which the triangle lies in the xz-plane with $\mathbf{d}_1$ pointing along the positive z-direction, and $\mathbf{R} = 0$. Then, a first rotation $\mathcal{R}_z(\gamma)$ of an angle γ around the z-axis, a second rotation $\mathcal{R}_y(\beta)$ of an angle β around the y-axis, and a translation along $\mathbf{R}$ will bring the vectors to the most general configuration. In other words,

$$\mathbf{r}_1 = \mathbf{R} + \mathcal{R}_y(\beta)\mathcal{R}_z(\gamma)\mathbf{r}_1',$$
$$\mathbf{r}_{12} = \mathcal{R}_y(\beta)\mathcal{R}_z(\gamma)\mathbf{r}_{12}', \tag{6.2.60}$$
$$\mathbf{r}_2 = \mathbf{r}_1 + \mathbf{r}_{12},$$

with

$$\mathbf{r}_1' = \begin{bmatrix} 0 \\ 0 \\ d_1 \end{bmatrix}, \tag{6.2.61}$$

$$\mathbf{r}_{12}' = r_{12} \begin{bmatrix} \sin(\alpha) \\ 0 \\ -\cos(\alpha) \end{bmatrix}, \tag{6.2.62}$$

and the rotation matrices are

$$\mathcal{R}_y(\beta) = \begin{bmatrix} \cos(\beta) & 0 & \sin(\beta) \\ 0 & 1 & 0 \\ -\sin(\beta) & 0 & \cos(\beta) \end{bmatrix}, \tag{6.2.63}$$

and

$$\mathcal{R}_z(\gamma) = \begin{bmatrix} \cos(\gamma) & -\sin(\gamma) & 0 \\ \sin(\gamma) & \cos(\gamma) & 0 \\ 0 & 0 & 1 \end{bmatrix}. \tag{6.2.64}$$

then

$$\mathbf{r}_1 = \begin{bmatrix} d_1 \sin(\beta) \\ 0 \\ R + d_1 \cos(\beta) \end{bmatrix}, \tag{6.2.65}$$

$$\mathbf{r}_{12} = \begin{bmatrix} r_{12} \cos(\beta) \cos(\gamma) \sin(\alpha) - r_{12} \sin(\beta) \cos(\alpha) \\ r_{12} \sin(\gamma) \sin(\alpha) \\ -r_{12} \sin(\beta) \cos(\gamma) \sin(\alpha) - r_{12} \cos(\alpha) \cos(\beta) \end{bmatrix}, \tag{6.2.66}$$

$$\mathbf{r}_2 = \begin{bmatrix} d_1 \sin(\beta) + r_{12} \cos(\beta) \cos(\gamma) \sin(\alpha) - r_{12} \sin(\beta) \cos(\alpha) \\ r_{12} \sin(\gamma) \sin(\alpha) \\ R + d_1 \cos(\beta) - r_{12} \sin(\beta) \cos(\gamma) \sin(\alpha) - r_{12} \cos(\alpha) \cos(\beta) \end{bmatrix}. \tag{6.2.67}$$

We also need $\cos(\theta_{12})$, ζ and $\cos(\theta_\zeta)$, θ_{12} being the angle between $\mathbf{r}_1$ and $\mathbf{r}_2$, $\zeta = \mathbf{r}_p - \frac{\mathbf{r}_1 + \mathbf{r}_2}{A+2}$ the position of the proton with respect to the final nucleus, and θ_ζ the angle between ζ and the z-axis:

$$\cos(\theta_{12}) = \frac{\mathbf{r}_1 \mathbf{r}_2}{r_1 r_2}, \tag{6.2.68}$$

and

$$\boldsymbol{\zeta} = 3\mathbf{R} - \frac{A+3}{A+2}(\mathbf{r}_1 + \mathbf{r}_2), \tag{6.2.69}$$

where we have used (6.2.49).

For heavy ions, we find instead

$$\mathbf{R} = \frac{1}{m_a}(\mathbf{r}_{A1} + \mathbf{r}_{A2} + m_b \mathbf{r}_{Ab}), \tag{6.2.70}$$

$$\mathbf{d}_1 = \frac{1}{m_a}(m_b \mathbf{r}_{b2} - (m_b + 1)\mathbf{r}_{12}), \tag{6.2.71}$$

$$d_1 = \frac{1}{m_a}\sqrt{(m_b + 1)r_{12}^2 + m_b(m_b + 1)r_{b1}^2 - m_b r_{b2}^2}, \tag{6.2.72}$$

$$d_2 = \frac{1}{m_a}\sqrt{(m_b + 1)r_{12}^2 + m_b(m_b + 1)r_{b2}^2 - m_b r_{b1}^2}, \tag{6.2.73}$$

and

$$\boldsymbol{\zeta} = \frac{m_a}{m_b}\mathbf{R} - \frac{m_B + m_b}{m_b m_B}(\mathbf{r}_{A1} + \mathbf{r}_{A2}). \tag{6.2.74}$$

The rest of the formulae are identical to the (t, p) ones. We list them for convenience,

$$\mathbf{r}_{A1} = \begin{bmatrix} d_1 \sin(\beta) \\ 0 \\ R + d_1 \cos(\beta) \end{bmatrix}, \tag{6.2.75}$$

$$\mathbf{r}_{A2} = \begin{bmatrix} d_1 \sin(\beta) + r_{12} \cos(\beta) \cos(\gamma) \sin(\alpha) - r_{12} \sin(\beta) \cos(\alpha) \\ r_{12} \sin(\gamma) \sin(\alpha) \\ R + d_1 \cos(\beta) - r_{12} \sin(\beta) \cos(\gamma) \sin(\alpha) - r_{12} \cos(\alpha) \cos(\beta) \end{bmatrix}. \tag{6.2.76}$$

We also find

$$\mathbf{r}_{b1} = \frac{1}{m_b}(\mathbf{r}_{A2} + (m_b + 1)\mathbf{r}_{A1} - m_a \mathbf{R}), \tag{6.2.77}$$

and

$$\mathbf{r}_{b2} = \frac{1}{m_b}(\mathbf{r}_{A1} + (m_b + 1)\mathbf{r}_{A2} - m_a \mathbf{R}). \tag{6.2.78}$$

One can readily obtain

$$\cos\theta_{12} = \frac{r_{A1}^2 + r_{A2}^2 - r_{12}^2}{2r_{A1}r_{A2}}, \tag{6.2.79}$$

and

$$\cos\theta_i = \frac{r_{b1}^2 + r_{b2}^2 - r_{12}^2}{2r_{b1}r_{b2}}. \tag{6.2.80}$$

6.2.4 *Alternative Derivation (Heavy Ions)*

In what follows we work out an alternative derivation of T_{2N}^{1step}, more closely related to heavy ion reactions[8], that is

$$T^{(1)}(\theta) = 2\frac{(4\pi)^{3/2}}{k_{Aa}k_{Bb}} \sum_{l_p j_p m l_t j_t} i^{-l_p} \exp\left[i(\sigma_{l_p}^p + \sigma_{l_t}^t)\right] \sqrt{2l_t + 1}$$

$$\times \langle l_p \, m - m_p \, 1/2 \, m_p | j_p \, m\rangle\langle l_t \, 0 \, 1/2 \, m_t | j_t \, m_t\rangle Y_{m-m_p}^{l_p}(\hat{\mathbf{k}}_{Bb})$$

$$\times \sum_{\sigma_1\sigma_2\sigma_p} \int d\mathbf{r}_{Cc}d\mathbf{r}_{b1}d\mathbf{r}_{A2} \left[\psi^{jf}(\mathbf{r}_{A1},\sigma_1)\psi^{jf}(\mathbf{r}_{A2},\sigma_2)\right]_0^{0*} \tag{6.2.81}$$

$$\times v(r_{b1}) \left[\psi^{ji}(\mathbf{r}_{b1},\sigma_1)\psi^{ji}(\mathbf{r}_{b2},\sigma_2)\right]_0^0 \frac{g_{l_t j_t}(r_{Aa})f_{l_p j_p}(r_{Bb})}{r_{Aa}r_{Bb}}$$

$$\times \left[Y^{l_t}(\hat{\mathbf{r}}_{Aa})\chi(\sigma_p)\right]_{m_t}^{j_t} \left[Y^{l_p}(\hat{\mathbf{r}}_{Bb})\chi(\sigma_p)\right]_m^{j_p*}.$$

As shown above one can write,

$$\sum_{\sigma_p}\langle l_p \, m - m_p \, 1/2 \, m_p | j_p \, m\rangle\langle l_t \, 0 \, 1/2 \, m_t | j_t \, m_t\rangle$$

$$\times \left[Y^{l_t}(\hat{\mathbf{r}}_{Aa})\chi(\sigma_p)\right]_{m_t}^{j_t} \left[Y^{l_p}(\hat{\mathbf{r}}_{Bb})\chi(\sigma_p)\right]_m^{j_p*}$$

$$= -\frac{\delta_{l_p,l_t}\delta_{j_p,j_t}\delta_{m,m_t}}{\sqrt{2l+1}} \left[Y^l(\hat{\mathbf{r}}_{Aa})Y^l(\hat{\mathbf{r}}_{Bb})\right]_0^0 \begin{cases} \dfrac{l}{2l+1} & \text{if } m_t = m_p \\[2mm] -\dfrac{\sqrt{l(l+1)}}{2l+1} & \text{if } m_t = -m_p \end{cases} \tag{6.2.82}$$

when $j = l - 1/2$ and

$$\sum_{\sigma_p}\langle l_p \, m - m_p \, 1/2 \, m_p | j_p \, m\rangle\langle l_t \, 0 \, 1/2 \, m_t | j_t \, m_t\rangle$$

$$\times \left[Y^{l_t}(\hat{\mathbf{r}}_{Aa})\chi(\sigma_p)\right]_{m_t}^{j_t} \left[Y^{l_p}(\hat{\mathbf{r}}_{Bb})\chi(\sigma_p)\right]_m^{j_p*}$$

$$= -\frac{\delta_{l_p,l_t}\delta_{j_p,j_t}\delta_{m,m_t}}{\sqrt{2l+1}} \left[Y^l(\hat{\mathbf{r}}_{Aa})Y^l(\hat{\mathbf{r}}_{Bb})\right]_0^0 \begin{cases} \dfrac{l+1}{2l+1} & \text{if } m_t = m_p \\[2mm] \dfrac{\sqrt{l(l+1)}}{2l+1} & \text{if } m_t = -m_p \end{cases} \tag{6.2.83}$$

---

[8] Bayman and Chen (1982).

if $j = l + 1/2$. One then gets

$$
T^{(1)}(\mu = 0; \theta) = 2 \frac{(4\pi)^{3/2}}{k_{Aa} k_{Bb}} \sum_l i^{-l} \frac{\exp[i(\sigma_l^p + \sigma_l^t)]}{2l+1} Y_{m_t - m_p}^l (\hat{\mathbf{k}}_{Bb})
$$

$$
\times \sum_{\sigma_1 \sigma_2} \int \frac{d\mathbf{r}_{Cc} d\mathbf{r}_{b1} d\mathbf{r}_{A2}}{r_{Aa} r_{Bb}} \left[\psi^{j_f}(\mathbf{r}_{A1}, \sigma_1) \psi^{j_f}(\mathbf{r}_{A2}, \sigma_2) \right]_0^{0*}
$$

$$
\times v(r_{b1}) \left[\psi^{j_i}(\mathbf{r}_{b1}, \sigma_1) \psi^{j_i}(\mathbf{r}_{b2}, \sigma_2) \right]_0^0 \left[Y^l(\hat{\mathbf{r}}_{Aa}) Y^l(\hat{\mathbf{r}}_{Bb}) \right]_0^0
$$

$$
\times \Big[\Big(f_{ll+1/2}(r_{Bb}) g_{ll+1/2}(r_{Aa})(l+1) + f_{ll-1/2}(r_{Bb}) g_{ll-1/2}(r_{Aa}) l \Big) \delta_{m_p, m_t}
$$

$$
+ \Big(f_{ll+1/2}(r_{Bb}) g_{ll+1/2}(r_{Aa}) \sqrt{l(l+1)}
$$

$$
- f_{ll-1/2}(r_{Bb}) g_{ll-1/2}(r_{Aa}) \sqrt{l(l+1)} \Big) \delta_{m_p, -m_t} \Big]. \tag{6.2.84}
$$

Making use of the relations,

$$
\left[\psi^{j_f}(\mathbf{r}_{A1}, \sigma_1) \psi^{j_f}(\mathbf{r}_{A2}, \sigma_2) \right]_0^{0*}
$$

$$
= \left((l_f \tfrac{1}{2})_{j_f} (l_f \tfrac{1}{2})_{j_f} \big| (l_f l_f)_0 (\tfrac{1}{2}\tfrac{1}{2})_0 \right)_0 u_{l_f}(r_{A1}) u_{l_f}(r_{A2})
$$

$$
\times \left[Y^{l_f}(\hat{\mathbf{r}}_{A1}) Y^{l_f}(\hat{\mathbf{r}}_{A2}) \right]_0^{0*} \left[\chi(\sigma_1) \chi(\sigma_2) \right]_0^{0*}
$$

$$
= \sqrt{\frac{2j_f + 1}{2(2l_f + 1)}} u_{l_f}(r_{A1}) u_{l_f}(r_{A2}) \tag{6.2.85}
$$

$$
\times \left[Y^{l_f}(\hat{\mathbf{r}}_{A1}) Y^{l_f}(\hat{\mathbf{r}}_{A2}) \right]_0^{0*} \left[\chi(\sigma_1) \chi(\sigma_2) \right]_0^{0*}
$$

$$
= \sqrt{\frac{2j_f + 1}{2}} \frac{u_{l_f}(r_{A1}) u_{l_f}(r_{A2})}{4\pi} P_{l_f}(\cos \omega_A) \left[\chi(\sigma_1) \chi(\sigma_2) \right]_0^{0*},
$$

and

$$
\left[\psi^{j_i}(\mathbf{r}_{b1}, \sigma_1) \psi^{j_i}(\mathbf{r}_{b2}, \sigma_2) \right]_0^0
$$

$$
= \left((l_i \tfrac{1}{2})_{j_i} (l_i \tfrac{1}{2})_{j_i} \big| (l_i l_i)_0 (\tfrac{1}{2}\tfrac{1}{2})_0 \right)_0 u_{l_i}(r_{b1}) u_{l_i}(r_{b2})
$$

$$
\times \left[Y^{l_i}(\hat{\mathbf{r}}_{b1}) Y^{l_i}(\hat{\mathbf{r}}_{b2}) \right]_0^0 \left[\chi(\sigma_1) \chi(\sigma_2) \right]_0^0
$$

$$
= \sqrt{\frac{2j_i + 1}{2(2l_i + 1)}} u_{l_i}(r_{b1}) u_{l_i}(r_{b2}) \tag{6.2.86}
$$

$$
\times \left[Y^{l_i}(\hat{\mathbf{r}}_{b1}) Y^{l_i}(\hat{\mathbf{r}}_{b2}) \right]_0^0 \left[\chi(\sigma_1) \chi(\sigma_2) \right]_0^0
$$

$$
= \sqrt{\frac{2j_i + 1}{2}} \frac{u_{l_i}(r_{b1}) u_{l_i}(r_{b2})}{4\pi} P_{l_i}(\cos \omega_b) \left[\chi(\sigma_1) \chi(\sigma_2) \right]_0^0,
$$

where ω_A is the angle between $\mathbf{r}_{A1}$ and $\mathbf{r}_{A2}$, and ω_b is the angle between $\mathbf{r}_{b1}$ and $\mathbf{r}_{b2}$. Consequently

$$
\begin{aligned}
T^{(1)}(\theta) = (4\pi)^{-3/2} &\frac{\sqrt{(2j_i + 1)(2j_f + 1)}}{k_{Aa}k_{Bb}} \sum_l i^{-l} \frac{\exp\left[i(\sigma_l^p + \sigma_l^t)\right]}{\sqrt{2l+1}} Y_{m_t - m_p}^l(\hat{\mathbf{k}}_{Bb}) \\
&\times \int \frac{d\mathbf{r}_{Cc} d\mathbf{r}_{b1} d\mathbf{r}_{A2}}{r_{Aa} r_{Bb}} P_{l_f}(\cos \omega_A) P_{l_i}(\cos \omega_b) P_l(\cos \omega_{if}) \\
&\times v(r_{b1}) u_{l_i}(r_{b1}) u_{l_i}(r_{b2}) u_{l_f}(r_{A1}) u_{l_f}(r_{A2}) \\
&\times \left[\left(f_{ll+1/2}(r_{Bb}) g_{ll+1/2}(r_{Aa})(l+1) + f_{ll-1/2}(r_{Bb}) g_{ll-1/2}(r_{Aa}) l \right) \delta_{m_p, m_t} \right. \\
&+ \left(f_{ll+1/2}(r_{Bb}) g_{ll+1/2}(r_{Aa}) \sqrt{l(l+1)} \right. \\
&\left. \left. - f_{ll-1/2}(r_{Bb}) g_{ll-1/2}(r_{Aa}) \sqrt{l(l+1)} \right) \delta_{m_p, -m_t} \right],
\end{aligned} \tag{6.2.87}
$$

where ω_{if} is the angle between $\mathbf{r}_{Aa}$ and $\mathbf{r}_{Bb}$. For heavy ions, we can consider that the optical potential does not have a spin–orbit term, and the distorted waves are independent of j. We thus have

$$
\begin{aligned}
T^{(1)}(\theta) = (4\pi)^{-3/2} &\frac{\sqrt{(2j_i + 1)(2j_f + 1)}}{k_{Aa}k_{Bb}} \sum_l i^{-l} \exp[i(\sigma_l^p + \sigma_l^t)] Y_0^l(\hat{\mathbf{k}}_{Bb}) \sqrt{2l+1} \\
&\times \int \frac{d\mathbf{r}_{Cc} d\mathbf{r}_{b1} d\mathbf{r}_{A2}}{r_{Aa} r_{Bb}} P_{l_f}(\cos \omega_A) P_{l_i}(\cos \omega_b) P_l(\cos \omega_{if}) \\
&\times v(r_{b1}) u_{l_i}(r_{b1}) u_{l_i}(r_{b2}) u_{l_f}(r_{A1}) u_{l_f}(r_{A2}) f_l(r_{Bb}) g_l(r_{Aa}).
\end{aligned} \tag{6.2.88}
$$

Changing variables one obtains,

$$
\begin{aligned}
T^{(1)}(\theta) = (4\pi)^{-1} &\frac{\sqrt{(2j_i + 1)(2j_f + 1)}}{k_{Aa}k_{Bb}} \sum_l \exp[i(\sigma_l^p + \sigma_l^t)] P_l(\cos \theta)(2l+1) \\
&\times \int dr_{1A} \, dr_{2A} \, dr_{Aa} \, d(\cos \beta) \, d(\cos \omega_A) \, d\gamma \, r_{1A}^2 r_{2A}^2 r_{Aa}^2 \\
&\times P_{l_f}(\cos \omega_A) P_{l_i}(\cos \omega_b) P_l(\cos \omega_{if}) v(r_{b1}) \\
&\times u_{l_i}(r_{b1}) u_{l_i}(r_{b2}) u_{l_f}(r_{A1}) u_{l_f}(r_{A2}) f_l(r_{Bb}) g_l(r_{Aa}).
\end{aligned} \tag{6.2.89}
$$

6.2.5 Coordinates Used to Derive the Transition Amplitude (Eq. (6.2.89))

We determine the relation between the integration variables in (6.2.87) and the coordinates needed to evaluate the quantities in the integrand. Noting that

$$
\mathbf{r}_{Aa} = \frac{\mathbf{r}_{A1} + \mathbf{r}_{A2} + m_b \mathbf{r}_{Ab}}{m_b + 2}, \tag{6.2.90}
$$

one has

$$\mathbf{r}_{b1} = \mathbf{r}_{bA} + \mathbf{r}_{A1} = \frac{(m_b + 1)\mathbf{r}_{A1} + \mathbf{r}_{A2} - (m_b + 2)\mathbf{r}_{Aa}}{m_b}, \tag{6.2.91}$$

$$\mathbf{r}_{b2} = \mathbf{r}_{bA} + \mathbf{r}_{A2} = \frac{(m_b + 1)\mathbf{r}_{A2} + \mathbf{r}_{A1} - (m_b + 2)\mathbf{r}_{Aa}}{m_b}, \tag{6.2.92}$$

and

$$\begin{aligned}
\mathbf{r}_{Cc} = \mathbf{r}_{CA} + \mathbf{r}_{A1} + \mathbf{r}_{1c} &= -\frac{1}{m_A + 1}\mathbf{r}_{A2} + \mathbf{r}_{A1} - \frac{m_b}{m_b + 1}\mathbf{r}_{b1} \\
&= \frac{m_b + 2}{m_b + 1}\mathbf{r}_{Aa} - \frac{m_b + 2 + m_A}{(m_b + 1)(m_A + 1)}\mathbf{r}_{A2}
\end{aligned} \tag{6.2.93}$$

Since

$$\mathbf{r}_{AB} = \frac{\mathbf{r}_{A1} + \mathbf{r}_{A2}}{m_A + 2}, \tag{6.2.94}$$

one obtains

$$\mathbf{r}_{Bb} = \mathbf{r}_{BA} + \mathbf{r}_{Ab} = \frac{m_b + 2}{m_b}\mathbf{r}_{Aa} - \frac{m_A + m_b + 2}{(m_A + 2)m_b}(\mathbf{r}_{A1} + \mathbf{r}_{A2}). \tag{6.2.95}$$

Using the same rotations as those used in Sect. 6.2.3, one gets

$$\mathbf{r}_{A1} = r_{A1}\begin{bmatrix} \sin\alpha \\ 0 \\ \cos\alpha \end{bmatrix} \tag{6.2.96}$$

and

$$\mathbf{r}_{A2} = r_{A2}\begin{bmatrix} -\cos\alpha\cos\gamma\sin\omega_A + \sin\alpha\cos\omega_A \\ -\sin\gamma\sin\omega_A \\ \sin\alpha\cos\gamma\sin\omega_A + \cos\alpha\cos\omega_A \end{bmatrix}, \tag{6.2.97}$$

with

$$\cos\alpha = \frac{r_{A1}^2 - d_1^2 + r_{Aa}^2}{2r_{A1}r_{Aa}}, \tag{6.2.98}$$

and

$$d_1 = \sqrt{r_{A1}^2 - r_{Aa}^2\sin^2\beta} - r_{Aa}\cos\beta. \tag{6.2.99}$$

Note that though β, r_{1A}, r_{Aa} are independent integration variables, they have to fulfill the condition

$$r_{Aa}\sin\beta \le r_{A1}, \quad \text{for } 0 \le \beta \le \pi. \tag{6.2.100}$$

The expression of the remaining quantities appearing in the integral are now straightforward,

$$r_{b1} = m_b^{-1} |(m_b + 1)\mathbf{r}_{A1} + \mathbf{r}_{A2} - (m_b + 2)\mathbf{r}_{Aa}|$$
$$= m_b^{-1} \Big((m_b + 2)^2 r_{Aa}^2 + (m_b + 1)^2 r_{A1}^2 + r_{A2}^2$$
$$- 2(m_b + 2)(m_b + 1)\mathbf{r}_{Aa}\,\mathbf{r}_{A1} - 2(m_b + 2)\mathbf{r}_{Aa}\,\mathbf{r}_{A2} + 2(m_b + 1)\mathbf{r}_{A1}\mathbf{r}_{A2} \Big)^{1/2},$$
$$(6.2.101)$$

$$r_{b2} = m_b^{-1} |(m_b + 1)\mathbf{r}_{A2} + \mathbf{r}_{A1} - (m_b + 2)\mathbf{r}_{Aa}|$$
$$= m_b^{-1} \Big((m_b + 2)^2 r_{Aa}^2 + (m_b + 1)^2 r_{A2}^2 + r_{A1}^2$$
$$- 2(m_b + 2)(m_b + 1)\mathbf{r}_{Aa}\,\mathbf{r}_{A2} - 2(m_b + 2)\mathbf{r}_{Aa}\,\mathbf{r}_{A1} + 2(m_b + 1)\mathbf{r}_{A2}\mathbf{r}_{A1} \Big)^{1/2},$$
$$(6.2.102)$$

$$r_{Bb} = \left| \frac{m_b + 2}{m_b}\mathbf{r}_{Aa} - \frac{m_A + m_b + 2}{(m_A + 2)m_b}(\mathbf{r}_{A1} + \mathbf{r}_{A2}) \right|$$
$$= \left[\left(\frac{m_b + 2}{m_b} \right)^2 r_{Aa}^2 + \left(\frac{m_A + m_b + 2}{(m_A + 2)m_b} \right)^2 (r_{A1}^2 + r_{A2}^2 + 2\mathbf{r}_{A1}\mathbf{r}_{A2}) \right. \quad (6.2.103)$$
$$\left. - 2\frac{(m_b + 2)(m_A + m_b + 2)}{(m_A + 2)m_b^2}\mathbf{r}_{Aa}(\mathbf{r}_{A1} + \mathbf{r}_{A2}) \right]^{1/2},$$

$$r_{Cc} = \left| \frac{m_b + 2}{m_b + 1}\mathbf{r}_{Aa} - \frac{m_b + 2 + m_A}{(m_b + 1)(m_A + 1)}\mathbf{r}_{A2} \right|$$
$$= \left[\left(\frac{m_a}{(m_a - 1)} \right)^2 r_{Aa}^2 + \left(\frac{m_A + m_a}{(m_A + 1)(m_a - 1)} \right)^2 r_{A2}^2 \right. \quad (6.2.104)$$
$$\left. - 2\frac{m_A m_a + m_a^2}{(m_A + 1)(m_a - 1)^2}\mathbf{r}_{Aa}\mathbf{r}_{A2} \right]^{1/2},$$

$$\cos \omega_b = \frac{\mathbf{r}_{b1}\mathbf{r}_{b2}}{r_{b1}r_{b2}}, \quad (6.2.105)$$

$$\cos \omega_{if} = \frac{\mathbf{r}_{Aa}\mathbf{r}_{Bb}}{r_{Aa}r_{Bb}}, \quad (6.2.106)$$

with

$$\mathbf{r}_{Aa}\mathbf{r}_{A1} = r_{Aa}r_{A1}\cos\alpha, \quad (6.2.107)$$

$$\mathbf{r}_{Aa}\mathbf{r}_{A2} = r_{Aa}r_{A2}(\sin\alpha\cos\gamma\sin\omega_A + \cos\alpha\cos\omega_A), \quad (6.2.108)$$

$$\mathbf{r}_{A1}\mathbf{r}_{A2} = r_{A1}r_{A2}\cos\omega_A. \quad (6.2.109)$$

6.2.6 Successive Transfer

The successive two–neutron transfer amplitudes can be written as (Bayman and Chen (1982)):

$$T_{succ}^{(2)}(\theta) = \frac{4\mu_{Cc}}{\hbar^2} \sum_{\substack{\sigma_1\sigma_2 \\ \sigma_1'\sigma_2' \\ KM}} \int d^3r_{Cc}d^3r_{b1}d^3r_{A2}d^3r_{Cc}'d^3r_{b1}'d^3r_{A2}' \, \chi^{(-)*}(\mathbf{k}_{Bb}, \mathbf{r}_{Bb})$$

$$\times \left[\psi^{jf}(\mathbf{r}_{A1}, \sigma_1)\psi^{jf}(\mathbf{r}_{A2}, \sigma_2) \right]_0^{0*} v(r_{b1}) \left[\psi^{jf}(\mathbf{r}_{A2}, \sigma_2)\psi^{ji}(\mathbf{r}_{b1}, \sigma_1) \right]_M^K$$

$$\times \, G(\mathbf{r}_{Cc}, \mathbf{r}_{Cc}') \left[\psi^{jf}(\mathbf{r}_{A2}', \sigma_2')\psi^{ji}(\mathbf{r}_{b1}', \sigma_1') \right]_M^{K*} v(r_{c2}')$$

$$\times \left[\psi^{ji}(\mathbf{r}_{b1}', \sigma_1')\psi^{ji}(\mathbf{r}_{b2}', \sigma_2') \right]_0^0 \chi^{(+)}(\mathbf{r}_{Aa}'). \tag{6.2.110}$$

It is of note that the time-reversal phase convention is used throughout. Expanding the Green function and the distorted waves in a basis of angular momentum eigenstate one can write,

$$\chi^{(-)*}(\mathbf{k}_{Bb}, \mathbf{r}_{Bb}) = \sum_{\tilde{l}} \frac{4\pi}{k_{Bb}r_{Bb}} i^{-\tilde{l}} e^{i\sigma_f^{\tilde{l}}} F_{\tilde{l}} \sum_m Y_m^{\tilde{l}}(\hat{r}_{Bb}) Y_m^{\tilde{l}*}(\hat{k}_{Bb}), \tag{6.2.111}$$

the sum over m being

$$\sum_m (-1)^{\tilde{l}-m} Y_m^{\tilde{l}}(\hat{r}_{Bb}) Y_{-m}^{\tilde{l}}(\hat{k}_{Bb}) = \sqrt{2\tilde{l}+1} \left[Y^{\tilde{l}}(\hat{r}_{Bb}) Y^{\tilde{l}}(\hat{k}_{Bb}) \right]_0^0, \tag{6.2.112}$$

where we have used (6.A.2) and (6.A.18), so

$$\chi^{(-)*}(\mathbf{k}_{Bb}, \mathbf{r}_{Bb}) = \sum_{\tilde{l}} \sqrt{2\tilde{l}+1} \frac{4\pi}{k_{Bb}r_{Bb}} i^{-\tilde{l}} e^{i\sigma_f^{\tilde{l}}} F_{\tilde{l}}(r_{Bb}) \left[Y^{\tilde{l}}(\hat{r}_{Bb}) Y^{\tilde{l}}(\hat{k}_{Bb}) \right]_0^0. \tag{6.2.113}$$

Similarly,

$$\chi^{(+)}(\mathbf{r}_{Aa}') = \sum_l i^l \sqrt{2l+1} \frac{4\pi}{k_{Aa}r_{Aa}'} e^{i\sigma_i^l} F_l(r_{Aa}') \left[Y^l(\hat{r}_{Aa}') Y^l(\hat{k}_{Aa}) \right]_0^0 \tag{6.2.114}$$

where we have taken into account the choice $\hat{k}_{Aa} \equiv \hat{z}$. The Green function can be written as

$$G(\mathbf{r}_{Cc}, \mathbf{r}_{Cc}') = i \sum_{l_c} \sqrt{2l_c+1} \frac{f_{l_c}(k_{Cc}, r_<) P_{l_c}(k_{Cc}, r_>)}{k_{Cc}r_{Cc}r_{Cc}'} \left[Y^{l_c}(\hat{r}_{Cc}) Y^{l_c}(\hat{r}_{Cc}') \right]_0^0. \tag{6.2.115}$$

Finally

$$T_{succ}^{(2)}(\theta) = \frac{4\mu_{Cc}(4\pi)^2 i}{\hbar^2 k_{Aa}k_{Bb}k_{Cc}} \sum_{l,l_c,\tilde{l}} e^{i(\sigma_i^l + \sigma_f^{\tilde{l}})} i^{l-\tilde{l}} \sqrt{(2l+1)(2l_c+1)(2\tilde{l}+1)}$$

$$\times \sum_{\substack{\sigma_1\sigma_2 \\ \sigma_1'\sigma_2'}} \int d^3r_{Cc}d^3r_{b1}d^3r_{A2}d^3r_{Cc}'d^3r_{b1}'d^3r_{A2}' v(r_{b1})v(r_{c2}') \left[Y^{\tilde{l}}(\hat{r}_{Bb})Y^{\tilde{l}}(\hat{k}_{Bb})\right]_0^0$$

$$\times \left[Y^l(\hat{r}_{Aa}')Y^l(\hat{k}_{Aa}')\right]_0^0 \left[Y^{l_c}(\hat{r}_{Cc})Y^{l_c}(\hat{r}_{Cc}')\right]_0^0 \frac{F_{\tilde{l}}(r_{Bb})}{r_{Bb}} \frac{F_l(r_{Aa}')}{r_{Aa}'}$$

$$\times \frac{f_{l_c}(k_{Cc},r_<)P_{l_c}(k_{Cc},r_>)}{r_{Cc}r_{Cc}'} \left[\psi^{j_f}(\mathbf{r}_{A1},\sigma_1)\psi^{j_f}(\mathbf{r}_{A2},\sigma_2)\right]_0^{0*}$$

$$\times \left[\psi^{j_i}(\mathbf{r}_{b1}',\sigma_1')\psi^{j_i}(\mathbf{r}_{b2}',\sigma_2')\right]_0^0 \sum_{KM} \left[\psi^{j_f}(\mathbf{r}_{A2},\sigma_2)\psi^{j_i}(\mathbf{r}_{b1},\sigma_1)\right]_M^K$$

$$\times \left[\psi^{j_f}(\mathbf{r}_{A2}',\sigma_2')\psi^{j_i}(\mathbf{r}_{b1}',\sigma_1')\right]_M^{K*}. \tag{6.2.116}$$

Let us now perform the integration over $\mathbf{r}_{A2}$,

$$\sum_{\sigma_1,\sigma_2} \int d\mathbf{r}_{A2} \left[\psi^{j_f}(\mathbf{r}_{A1},\sigma_1)\psi^{j_f}(\mathbf{r}_{A2},\sigma_2)\right]_0^{0*} \left[\psi^{j_f}(\mathbf{r}_{A2},\sigma_2)\psi^{j_i}(\mathbf{r}_{b1},\sigma_1)\right]_M^K$$

$$= \sum_{\sigma_1,\sigma_2} (-1)^{1/2-\sigma_1+1/2-\sigma_2} \int d\mathbf{r}_{A2}$$

$$\times \left[\psi^{j_f}(\mathbf{r}_{A1},-\sigma_1)\psi^{j_f}(\mathbf{r}_{A2},-\sigma_2)\right]_0^0 \left[\psi^{j_f}(\mathbf{r}_{A2},\sigma_2)\psi^{j_i}(\mathbf{r}_{b1},\sigma_1)\right]_M^K$$

$$= -\sum_{\sigma_1,\sigma_2} (-1)^{1/2-\sigma_1+1/2-\sigma_2} \int d\mathbf{r}_{A2}$$

$$\times \left[\psi^{j_f}(\mathbf{r}_{A2},-\sigma_2)\psi^{j_f}(\mathbf{r}_{A1},-\sigma_1)\right]_0^0 \left[\psi^{j_f}(\mathbf{r}_{A2},\sigma_2)\psi^{j_i}(\mathbf{r}_{b1},\sigma_1)\right]_M^K$$

$$= -\left((j_f j_f)_0(j_f j_i)_K|(j_f j_f)_0(j_f j_i)_K\right)_K \sum_{\sigma_1,\sigma_2} (-1)^{1/2-\sigma_1+1/2-\sigma_2}$$

$$\times \int d\mathbf{r}_{A2} \left[\psi^{j_f}(\mathbf{r}_{A2},-\sigma_2)\psi^{j_f}(\mathbf{r}_{A2},\sigma_2)\right]_0^0 \left[\psi^{j_f}(\mathbf{r}_{A1},-\sigma_1)\psi^{j_i}(\mathbf{r}_{b1},\sigma_1)\right]_M^K$$

$$= \frac{1}{2j_f+1}\sqrt{2j_f+1}\left((l_f\tfrac{1}{2})_{j_f}(l_i\tfrac{1}{2})_{j_i}|(l_f l_i)_K(\tfrac{1}{2}\tfrac{1}{2})_0\right)_K$$

$$\times u_{l_f}(r_{A1})u_{l_i}(r_{b1}) \left[Y^{l_f}(\hat{r}_{A1})Y^{l_i}(\hat{r}_{b1})\right]_M^K \sum_{\sigma_1} (-1)^{1/2-\sigma_1} \left[\chi^{1/2}(-\sigma_1)\chi^{1/2}(\sigma_1)\right]_0^0$$

$$= -\sqrt{\frac{2}{2j_f+1}}\left((l_f\tfrac{1}{2})_{j_f}(l_i\tfrac{1}{2})_{j_i}|(l_f l_i)_K(\tfrac{1}{2}\tfrac{1}{2})_0\right)_K$$

$$\times \left[Y^{l_f}(\hat{r}_{A1})Y^{l_i}(\hat{r}_{b1})\right]_M^K u_{l_f}(r_{A1})u_{l_i}(r_{b1}), \tag{6.2.117}$$

where we have evaluated the $9j$-symbol

$$((j_f j_f)_0 (j_f j_i)_K | (j_f j_f)_0 (j_f j_i)_K)_K = \frac{1}{2j_f + 1}, \tag{6.2.118}$$

as well as (6.A.19). We proceed in a similar way to evaluate the integral over $\mathbf{r}'_{b1}$,

$$\sum_{\sigma'_1, \sigma'_2} \int d\mathbf{r}'_{b1} \left[\psi^{j_i}(\mathbf{r}'_{b1}, \sigma'_1) \psi^{j_i}(\mathbf{r}'_{b2}, \sigma'_2) \right]_0^0 \left[\psi^{j_f}(\mathbf{r}'_{A2}, \sigma'_2) \psi^{j_i}(\mathbf{r}'_{b1}, \sigma'_1) \right]_M^{K*}$$

$$= -(-1)^{K-M} \sum_{\sigma'_1, \sigma'_2} \int d\mathbf{r}'_{b1} \left[\psi^{j_f}(\mathbf{r}'_{A2}, -\sigma'_2) \psi^{j_i}(\mathbf{r}'_{b1}, -\sigma'_1) \right]_{-M}^{K}$$

$$\times \left[\psi^{j_i}(\mathbf{r}'_{b2}, \sigma'_2) \psi^{j_i}(\mathbf{r}'_{b1}, \sigma'_1) \right]_0^0 (-1)^{1/2 - \sigma'_1 + 1/2 - \sigma'_2}$$

$$= -(-1)^{K-M} ((j_f j_i)_K (j_i j_i)_0 | (j_f j_i)_K (j_i j_i)_0)_K (-\sqrt{2j_i + 1})$$

$$\times ((l_f \tfrac{1}{2})_{j_f} (l_i \tfrac{1}{2})_{j_i} | (l_f l_i)_K (\tfrac{1}{2}\tfrac{1}{2})_0)_K (-\sqrt{2}) u_{l_f}(r'_{A2}) u_{l_i}(r'_{b2}) \left[Y^{l_f}(\hat{r}'_{A2}) Y^{l_i}(\hat{r}'_{b2}) \right]_{-M}^{K}$$

$$= -\sqrt{\frac{2}{2j_i + 1}} ((l_f \tfrac{1}{2})_{j_f} (l_i \tfrac{1}{2})_{j_i} | (l_f l_i)_K (\tfrac{1}{2}\tfrac{1}{2})_0)_K$$

$$\times \left[Y^{l_f}(\hat{r}'_{A2}) Y^{l_i}(\hat{r}'_{b2}) \right]_M^{K*} u_{l_f}(r'_{A2}) u_{l_i}(r'_{b2}). \tag{6.2.119}$$

Setting the different elements together one obtains

$$T_{succ}^{(2)}(\theta) = \frac{4\mu_{Cc}(4\pi)^2 i}{\hbar^2 k_{Aa} k_{Bb} k_{Cc}} \frac{2}{\sqrt{(2j_i + 1)(2j_f + 1)}} \sum_{K,M} ((l_f \tfrac{1}{2})_{j_f} (l_i \tfrac{1}{2})_{j_i} | (l_f l_i)_K (\tfrac{1}{2}\tfrac{1}{2})_0)_K^2$$

$$\times \sum_{l_c, l, \tilde{l}} e^{i(\sigma_l^l + \sigma_f^{\tilde{l}})} \sqrt{(2l_c + 1)(2l + 1)(2\tilde{l} + 1)} \, i^{l - \tilde{l}}$$

$$\times \int d^3 r_{Cc} d^3 r_{b1} d^3 r'_{Cc} d^3 r'_{A2} v(r_{b1}) v(r'_{c2}) u_{l_f}(r_{A1}) u_{l_i}(r_{b1}) u_{l_f}(r'_{A2}) u_{l_i}(r'_{b2})$$

$$\times \left[Y^{l_f}(\hat{r}'_{A2}) Y^{l_i}(\hat{r}'_{b2}) \right]_M^{K*} \left[Y^{l_f}(\hat{r}_{A1}) Y^{l_i}(\hat{r}_{b1}) \right]_M^{K}$$

$$\times \frac{F_l(r'_{Aa}) F_{\tilde{l}}(r'_{Bb}) f_{l_c}(k_{Cc}, r_<) P_{l_c}(k_{Cc}, r_>)}{r'_{Aa} r_{Bb} r_{Cc} r'_{Cc}}$$

$$\times \left[Y^{\tilde{l}}(\hat{r}_{Bb}) Y^{\tilde{l}}(\hat{k}_{Bb}) \right]_0^0 \left[Y^l(\hat{r}'_{Aa}) Y^l(\hat{k}_{Aa}) \right]_0^0 \left[Y^{l_c}(\hat{r}_{Cc}) Y^{l_c}(\hat{r}'_{Cc}) \right]_0^0. \tag{6.2.120}$$

We now proceed to write this expression in a more compact way. For this purpose one writes

$$\left[Y^{\tilde{l}}(\hat{r}_{Bb})Y^{\tilde{l}}(\hat{k}_{Bb})\right]_0^0 \left[Y^l(\hat{r}'_{Aa})Y^l(\hat{k}_{Aa})\right]_0^0$$

$$= \left((l\,l)_0(\tilde{l}\,\tilde{l})_0|(l\,\tilde{l})_0(l\,\tilde{l})_0\right)_0 \left[Y^{\tilde{l}}(\hat{r}_{Bb})Y^l(\hat{r}'_{Aa})\right]_0^0 \left[Y^{\tilde{l}}(\hat{k}_{Bb})Y^l(\hat{k}_{Aa})\right]_0^0$$

$$= \frac{\delta_{\tilde{l}l}}{2l+1} \left[Y^l(\hat{r}_{Bb})Y^l(\hat{r}'_{Aa})\right]_0^0 \left[Y^l(\hat{k}_{Bb})Y^l(\hat{k}_{Aa})\right]_0^0. \tag{6.2.121}$$

Taking into account the relations

$$\left[Y^l(\hat{k}_{Bb})Y^l(\hat{k}_{Aa})\right]_0^0 = \frac{(-1)^l}{\sqrt{4\pi}}Y_0^l(\hat{k}_{Bb})i^l, \tag{6.2.122}$$

and

$$\left[Y^l(\hat{r}_{Bb})Y^l(\hat{r}'_{Aa})\right]_0^0 \left[Y^{l_c}(\hat{r}_{Cc})Y^{l_c}(\hat{r}'_{Cc})\right]_0^0$$

$$= \left((l\,l)_0(l_c\,l_c)_0|(l\,l_c)_K(l\,l_c)_K\right)_0$$

$$\times \left\{\left[Y^l(\hat{r}_{Bb})Y^{l_c}(\hat{r}_{Cc})\right]^K \left[Y^l(\hat{r}'_{Aa})Y^{l_c}(\hat{r}'_{Cc})\right]^K\right\}_0^0$$

$$= \sqrt{\frac{2K+1}{(2l+1)(2l_c+1)}}$$

$$\times \sum_{M'} \frac{(-1)^{K+M'}}{\sqrt{2K+1}} \left[Y^l(\hat{r}_{Bb})Y^{l_c}(\hat{r}_{Cc})\right]^K_{-M'} \left[Y^l(\hat{r}'_{Aa})Y^{l_c}(\hat{r}'_{Cc})\right]^K_{M'}$$

$$= \sqrt{\frac{1}{(2l+1)(2l_c+1)}}$$

$$\times \sum_{M'} \left[Y^l(\hat{r}_{Bb})Y^{l_c}(\hat{r}_{Cc})\right]^{K*}_{M'} \left[Y^l(\hat{r}'_{Aa})Y^{l_c}(\hat{r}'_{Cc})\right]^K_{M'}. \tag{6.2.123}$$

It is of note that the integrals

$$\int d\hat{r}_{Cc}d\hat{r}_{b1} \left[Y^l(\hat{r}_{Bb})Y^{l_c}(\hat{r}_{Cc})\right]^{K*}_M \left[Y^{l_f}(\hat{r}_{A1})Y^{l_i}(\hat{r}_{b1})\right]^K_M, \tag{6.2.124}$$

and

$$\int d\hat{r}'_{Cc}d\hat{r}_{A2} \left[Y^l(\hat{r}'_{Aa})Y^{l_c}(\hat{r}'_{Cc})\right]^K_M \left[Y^{l_f}(\hat{r}'_{A2})Y^{l_i}(\hat{r}'_{b2})\right]^{K*}_M, \tag{6.2.125}$$

over the angular variables do not depend on M. Let us see why this is so with the help of (6.2.124),

$$\left[Y^l(\hat{r}_{Bb})Y^{l_c}(\hat{r}_{Cc})\right]_M^{K*}\left[Y^{l_f}(\hat{r}_{A1})Y^{l_i}(\hat{r}_{b1})\right]_M^K = (-1)^{K-M}\left[Y^l(\hat{r}_{Bb})Y^{l_c}(\hat{r}_{Cc})\right]_{-M}^K$$

$$\times\left[Y^{l_f}(\hat{r}_{A1})Y^{l_i}(\hat{r}_{b1})\right]_M^K = (-1)^{K-M}\sum_J\langle K\ K\ M\ -M|J\ 0\rangle$$

$$\times\left\{\left[Y^l(\hat{r}_{Bb})Y^{l_c}(\hat{r}_{Cc})\right]^K\left[Y^{l_f}(\hat{r}_{A1})Y^{l_i}(\hat{r}_{b1})\right]^K\right\}_0^J. \tag{6.2.126}$$

After integration, only the term

$$(-1)^{K-M}\langle K\ K\ M\ -M|0\ 0\rangle\left\{\left[Y^l(\hat{r}_{Bb})Y^{l_c}(\hat{r}_{Cc})\right]^K\left[Y^{l_f}(\hat{r}_{A1})Y^{l_i}(\hat{r}_{b1})\right]^K\right\}_0^0 = .$$

$$\frac{1}{\sqrt{2K+1}}\left\{\left[Y^l(\hat{r}_{Bb})Y^{l_c}(\hat{r}_{Cc})\right]^K\left[Y^{l_f}(\hat{r}_{A1})Y^{l_i}(\hat{r}_{b1})\right]^K\right\}_0^0$$

$$\tag{6.2.127}$$

corresponding to $J = 0$ survives, which is indeed independent of M. We can thus omit the sum over M in (6.2.120) and multiply by $(2K + 1)$, obtaining

$$T_{succ}^{(2)}(\theta) = \frac{64\mu_{Cc}(\pi)^{3/2}i}{\hbar^2 k_{Aa}k_{Bb}k_{Cc}}\frac{i^{-l}}{\sqrt{(2j_i+1)(2j_f+1)}}$$

$$\times\sum_K(2K+1)\left((l_f\tfrac{1}{2})_{j_f}(l_i\tfrac{1}{2})_{j_i}|(l_f l_i)_K(\tfrac{1}{2}\tfrac{1}{2})_0\right)_K^2$$

$$\times\sum_{l_c,l}\frac{e^{i(\sigma_i^l+\sigma_f^l)}}{\sqrt{(2l+1)}}Y_0^l(\hat{k}_{Bb})S_{K,l,l_c}, \tag{6.2.128}$$

where

$$S_{K,l,l_c} = \int d^3r_{Cc}d^3r_{b1}v(r_{b1})u_{l_f}(r_{A1})u_{l_i}(r_{b1})\frac{s_{K,l,l_c}(r_{Cc})}{r_{Cc}}\frac{F_l(r_{Bb})}{r_{Bb}}$$

$$\times\left[Y^{l_f}(\hat{r}_{A1})Y^{l_i}(\hat{r}_{b1})\right]_M^K\left[Y^{l_c}(\hat{r}_{Cc})Y^l(\hat{r}_{Bb})\right]_M^{K*}, \tag{6.2.129}$$

and

$$s_{K,l,l_c}(r_{Cc}) = \int_{r_{Cc}\,fixed}d^3r_{Cc}'d^3r_{A2}'v(r_{c2}')u_{l_f}(r_{A2}')u_{l_i}(r_{b2}')$$

$$\times\frac{F_l(r_{Aa}')}{r_{Aa}'}\frac{f_{l_c}(k_{Cc},r_<)P_{l_c}(k_{Cc},r_>)}{r_{Cc}'}$$

$$\times\left[Y^{l_f}(\hat{r}_{A2}')Y^{l_i}(\hat{r}_{b2}')\right]_M^{K*}\left[Y^{l_c}(\hat{r}_{Cc}')Y^l(\hat{r}_{Aa}')\right]_M^K. \tag{6.2.130}$$

It can be shown that the integrand in (6.2.129) is independent of M. Consequently, one can sum over M and divide by $(2K + 1)$, to get

$$\frac{1}{2K+1}v(r_{b1})u_{l_f}(r_{A1})u_{l_i}(r_{b1})\frac{s_{K,l,l_c}(r_{Cc})}{r_{Cc}}\frac{F_l(r_{Bb})}{r_{Bb}}$$

$$\times\sum_M\left[Y^{l_f}(\hat{r}_{A1})Y^{l_i}(\hat{r}_{b1})\right]_M^K\left[Y^{l_c}(\hat{r}_{Cc})Y^l(\hat{r}_{Bb})\right]_M^{K*}. \tag{6.2.131}$$

This integrand is rotationally invariant (it is proportional to a T_M^L spherical tensor with $L = 0$, $M = 0$), so one can evaluate it in the "standard" configuration in which $\mathbf{r}_{Cc}$ is directed along the z-axis and multiply by $8\pi^2$ (see Bayman and Chen (1982)), obtaining the final expression for S_{K,l,l_c}:

$$
\begin{aligned}
S_{K,l,l_c} =\ & \frac{4\pi^{3/2}\sqrt{2l_c+1}}{2K+1} i^{-l_c} \\
& \times \int r_{Cc}^2\, dr_{Cc}\, r_{b1}^2\, dr_{b1}\, \sin\theta\, d\theta\, v(r_{b1}) u_{l_f}(r_{A1}) u_{l_i}(r_{b1}) \\
& \times \frac{s_{K,l,l_c}(r_{Cc})}{r_{Cc}} \frac{F_l(r_{Bb})}{r_{Bb}} \\
& \times \sum_M \langle l_c\, 0\, l\, M | K\, M\rangle \left[Y^{l_f}(\hat{r}_{A1}) Y^{l_i}(\theta+\pi, 0) \right]_M^K Y_M^{l*}(\hat{r}_{Bb}).
\end{aligned}
\tag{6.2.132}
$$

Similarly, one has

$$
\begin{aligned}
s_{K,l,l_c}(r_{Cc}) =\ & \frac{4\pi^{3/2}\sqrt{2l_c+1}}{2K+1} i^{l_c} \\
& \times \int r_{Cc}'^2\, dr_{Cc}'\, r_{A2}'^2\, dr_{A2}'\, \sin\theta'\, d\theta'\, v(r_{c2}') u_{l_f}(r_{A2}') u_{l_i}(r_{b2}') \\
& \times \frac{F_l(r_{Aa}')}{r_{Aa}'} \frac{f_{l_c}(k_{Cc}, r_<) P_{l_c}(k_{Cc}, r_>)}{r_{Cc}'} \\
& \times \sum_M \langle l_c\, 0\, l\, M | K\, M\rangle \left[Y^{l_f}(\hat{r}_{A2}') Y^{l_i}(\hat{r}_{b2}') \right]_M^{K*} Y_M^l(\hat{r}_{Aa}').
\end{aligned}
\tag{6.2.133}
$$

Introducing the further approximations $\mathbf{r}_{A1} \approx \mathbf{r}_{C1}$ and $\mathbf{r}_{b2} \approx \mathbf{r}_{c2}$, one obtains the final expression

$$
\begin{aligned}
T_{succ}^{(2)}(\theta) =\ & \frac{1024\mu_{Cc}\pi^{9/2} i}{\hbar^2 k_{Aa} k_{Bb} k_{Cc}} \frac{1}{\sqrt{(2j_i+1)(2j_f+1)}} \\
& \times \sum_K \frac{1}{2K+1} \left((l_f \tfrac{1}{2})_{j_f} (l_i \tfrac{1}{2})_{j_i} | (l_f l_i)_K (\tfrac{1}{2}\tfrac{1}{2})_0 \right)_K^2 \\
& \times \sum_{l_c, l} e^{i(\sigma_i^l + \sigma_f^l)} \frac{(2l_c+1)}{\sqrt{2l+1}} Y_0^l(\hat{k}_{Bb}) S_{K,l,l_c},
\end{aligned}
\tag{6.2.134}
$$

with

$$
\begin{aligned}
S_{K,l,l_c} =\ & \int r_{Cc}^2\, dr_{Cc}\, r_{b1}^2\, dr_{b1}\, \sin\theta\, d\theta\, v(r_{b1}) u_{l_f}(r_{C1}) u_{l_i}(r_{b1}) \\
& \times \frac{s_{K,l,l_c}(r_{Cc})}{r_{Cc}} \frac{F_l(r_{Bb})}{r_{Bb}} \\
& \times \sum_M \langle l_c\, 0\, l\, M | K\, M\rangle \left[Y^{l_f}(\hat{r}_{C1}) Y^{l_i}(\theta+\pi, 0) \right]_M^K Y_M^{l*}(\hat{r}_{Bb}),
\end{aligned}
\tag{6.2.135}
$$

and

$$s_{K,l,l_c}(r_{Cc}) = \int r'^2_{Cc}\, dr'_{Cc}\, r'^2_{A2}\, dr'_{A2}\, \sin\theta'\, d\theta'\, v(r'_{c2}) u_{l_f}(r'_{A2}) u_{l_i}(r'_{c2})$$

$$\times\, \frac{F_l(r'_{Aa})}{r'_{Aa}}\, \frac{f_{l_c}(k_{Cc}, r_<)\, P_{l_c}(k_{Cc}, r_>)}{r'_{Cc}} \qquad (6.2.136)$$

$$\times\, \sum_M \langle l_c\, 0\, l\, M | K\, M\rangle \left[Y^{l_f}(\hat{r}'_{A2}) Y^{l_i}(\hat{r}'_{c2}) \right]^{K*}_M Y^l_M(\hat{r}'_{Aa}).$$

6.2.7 Coordinates for the Successive Transfer

In the standard configuration in which the integrals (6.2.135) and (6.2.136) are to be evaluated, we have

$$\mathbf{r}_{Cc} = r_{Cc}\,\hat{\mathbf{z}}, \qquad \mathbf{r}_{b1} = r_{b1}(-\cos\theta\,\hat{\mathbf{z}} - \sin\theta\,\hat{\mathbf{x}}). \qquad (6.2.137)$$

Now,

$$\mathbf{r}_{C1} = \mathbf{r}_{Cc} + \mathbf{r}_{c1} = \mathbf{r}_{Cc} + \frac{m_b}{m_b + 1}\mathbf{r}_{b1}$$

$$= \left(r_{Cc} - \frac{m_b}{m_b + 1}r_{b1}\cos\theta \right)\hat{\mathbf{z}} - \frac{m_b}{m_b + 1}r_{b1}\sin\theta\hat{\mathbf{x}}, \qquad (6.2.138)$$

and

$$\mathbf{r}_{Bb} = \mathbf{r}_{BC} + \mathbf{r}_{Cb} = -\frac{1}{m_B}\mathbf{r}_{C1} + \mathbf{r}_{Cb}. \qquad (6.2.139)$$

Substituting the relation

$$\mathbf{r}_{Cb} = \mathbf{r}_{Cc} + \mathbf{r}_{cb} = \mathbf{r}_{Cc} - \frac{1}{m_b + 1}\mathbf{r}_{b1}, \qquad (6.2.140)$$

in (6.2.139) one gets

$$\mathbf{r}_{Bb} = \left(\frac{m_B - 1}{m_B}r_{Cc} + \frac{m_b + m_B}{m_B(m_b + 1)}r_{b1}\cos\theta \right)\hat{\mathbf{z}} + \frac{m_b + m_B}{m_B(m_b + 1)}r_{b1}\sin\theta\hat{\mathbf{x}}.$$
$$(6.2.141)$$

The primed variables are arranged in a similar fashion,

$$\mathbf{r}'_{Cc} = r'_{Cc}\,\hat{\mathbf{z}}, \qquad \mathbf{r}'_{A2} = r'_{A2}(-\cos\theta'\,\hat{\mathbf{z}} - \sin\theta'\,\hat{\mathbf{x}}). \qquad (6.2.142)$$

Thus,

$$\mathbf{r}'_{c2} = \left(-r'_{Cc} - \frac{m_A}{m_A + 1}r'_{A2}\cos\theta' \right)\hat{\mathbf{z}} - \frac{m_A}{m_A + 1}r'_{A2}\sin\theta'\hat{\mathbf{x}}, \qquad (6.2.143)$$

and

$$\mathbf{r}'_{Aa} = \left(\frac{m_a - 1}{m_a} r'_{Cc} - \frac{m_A + m_a}{m_a(m_A + 1)} r'_{A2} \cos\theta' \right) \hat{\mathbf{z}} - \frac{m_A + m_a}{m_a(m_A + 1)} r'_{A2} \sin\theta' \hat{\mathbf{x}}. \tag{6.2.144}$$

6.2.8 Simplifying the Vector Coupling

We will now turn our attention to the vector-coupled quantities in (6.2.135) and (6.2.136),

$$\sum_M \langle l_c \, 0 \, l \, M | K \, M \rangle \left[Y^{l_f}(\hat{r}_{C1}) Y^{l_i}(\theta + \pi, 0) \right]_M^K Y_M^{l*}(\hat{r}_{Bb}), \tag{6.2.145}$$

and

$$\sum_M \langle l_c \, 0 \, l \, M | K \, M \rangle \left[Y^{l_f}(\hat{r}'_{A2}) Y^{l_i}(\hat{r}'_{c2}) \right]_M^{K*} Y_M^{l}(\hat{r}'_{Aa}). \tag{6.2.146}$$

We can express them both as

$$\sum_M f(M), \tag{6.2.147}$$

where, for example, in the case of (6.2.145), one has

$$f(M) = \langle l_c \, 0 \, l \, M | K \, M \rangle \left[Y^{l_f}(\hat{r}_{C1}) Y^{l_i}(\theta + \pi, 0) \right]_M^K Y_M^{l*}(\hat{r}_{Bb}). \tag{6.2.148}$$

Note that all the vectors that come into play in the above expressions are in the (x, z)-plane. Consequently, the azimuthal angle ϕ is always equal to zero. Under these circumstances and for time–reversed phases, $(Y_M^{L*}(\theta, 0) = (-1)^L Y_M^L(\theta, 0))$ one has

$$f(-M) = (-1)^{l_c + l_f + l_i + l} f(M). \tag{6.2.149}$$

Consequently,

$$\sum_M \langle l_c \, 0 \, l \, M | K \, M \rangle f(M) = \langle l_c \, 0 \, l \, 0 | K \, 0 \rangle f(0)$$
$$+ \sum_{M>0} \langle l_c \, 0 \, l \, M | K \, M \rangle f(M) \left(1 + (-1)^{l_c + l + l_i + l_f} \right). \tag{6.2.150}$$

Consequently, in the case in which $l_c + l + l_i + l_f$ is odd, we have only to evaluate the $M = 0$ contribution. This consideration is useful to restrict the number of numerical operations needed to calculate the transition amplitude.

6.2.9 Nonorthogonality Term

We write the nonorthogonality contribution to the transition amplitude (see Bayman and Chen (1982)):

$$
T_{NO}^{(2)}(\theta) = 2 \sum_{\substack{\sigma_1 \sigma_2 \\ \sigma_1' \sigma_2' \\ KM}} \int d^3 r_{Cc} d^3 r_{b1} d^3 r_{A2} d^3 r_{b1}' d^3 r_{A2}' \, \chi^{(-)*}(\mathbf{k}_{Bb}, \mathbf{r}_{Bb})
$$

$$
\times \left[\psi^{j_f}(\mathbf{r}_{A1}, \sigma_1) \psi^{j_f}(\mathbf{r}_{A2}, \sigma_2) \right]_0^{0*} v(r_{b1}) \left[\psi^{j_f}(\mathbf{r}_{A2}, \sigma_2) \psi^{j_i}(\mathbf{r}_{b1}, \sigma_1) \right]_M^{K}
$$

$$
\times \left[\psi^{j_f}(\mathbf{r}_{A2}', \sigma_2') \psi^{j_i}(\mathbf{r}_{b1}', \sigma_1') \right]_M^{K*} \left[\psi^{j_i}(\mathbf{r}_{b1}', \sigma_1') \psi^{j_i}(\mathbf{r}_{b2}', \sigma_2') \right]_0^{0} \chi^{(+)}(\mathbf{r}_{Aa}').
$$

$$
(6.2.151)
$$

This expression is equivalent to (6.2.110) if we make the replacement

$$
\frac{2\mu_{Cc}}{\hbar^2} G(\mathbf{r}_{Cc}, \mathbf{r}_{Cc}') v(r_{A2}') \rightarrow \delta(\mathbf{r}_{Cc} - \mathbf{r}_{Cc}').
$$

$$
(6.2.152)
$$

Looking at the partial–wave expansions of $G(\mathbf{r}_{Cc}, \mathbf{r}_{Cc}')$ and $\delta(\mathbf{r}_{Cc} - \mathbf{r}_{Cc}')$ (see App. 6.A), we find that we can use the above expressions for the successive transfer with the replacement

$$
i \frac{2\mu_{Cc}}{\hbar^2} \frac{f_{l_c}(k_{Cc}, r_<) P_{l_c}(k_{Cc}, r_>)}{k_{Cc}} \rightarrow \delta(r_{Cc} - r_{Cc}').
$$

$$
(6.2.153)
$$

We thus have

$$
T_{2NT}^{NO} = \frac{512\pi^{9/2}}{k_{Aa} k_{Bb}} \frac{1}{\sqrt{(2j_i + 1)(2j_f + 1)}}
$$

$$
\times \sum_K \left((l_f \tfrac{1}{2})_{j_f} (l_i \tfrac{1}{2})_{j_i} | (l_f l_i)_K (\tfrac{1}{2} \tfrac{1}{2})_0 \right)_K^2
$$

$$
\times \sum_{l_c, l} e^{i(\sigma_i^l + \sigma_f^l)} \frac{(2l_c + 1)}{\sqrt{2l + 1}} Y_0^l(\hat{k}_{Bb}) S_{K, l, l_c},
$$

$$
(6.2.154)
$$

with

$$
S_{K, l, l_c} = \int r_{Cc}^2 \, dr_{Cc} \, r_{b1}^2 \, dr_{b1} \, \sin\theta \, d\theta \, v(r_{b1}) u_{l_f}(r_{C1}) u_{l_i}(r_{b1})
$$

$$
\times \frac{s_{K, l, l_c}(r_{Cc})}{r_{Cc}} \frac{F_l(r_{Bb})}{r_{Bb}}
$$

$$
(6.2.155)
$$

$$
\times \sum_M \langle l_c \, 0 \, l \, M | K \, M \rangle \left[Y^{l_f}(\hat{r}_{C1}) Y^{l_i}(\theta + \pi, 0) \right]_M^K Y_M^{l*}(\hat{r}_{Bb}),
$$

and

$$
\begin{aligned}
s_{K,l,l_c}(r_{Cc}) = r_{Cc} \int dr'_{A2}\, r'^2_{A2} \, \sin\theta'\, d\theta'\, u_{l_f}(r'_{A2}) u_{l_i}(r'_{c2}) \frac{F_l(r'_{Aa})}{r'_{Aa}} \\
\times \sum_M \langle l_c\, 0\, l\, M | K\, M \rangle \left[Y^{l_f}(\hat{r}'_{A2}) Y^{l_i}(\hat{r}'_{c2}) \right]^{K*}_M Y^l_M(\hat{r}'_{Aa}).
\end{aligned}
\tag{6.2.156}
$$

6.2.10 General Orbital Momentum Transfer

We will now examine the case in which the two transferred nucleons carry an angular momentum Λ different from 0. Let us assume that two nucleons coupled to angular momentum Λ in the initial nucleus a are transferred into a final state of zero angular momentum in nucleus B. The transition amplitude is given by the integral

$$
\begin{aligned}
2 \sum_{\sigma_1\sigma_2} \int d\mathbf{r}_{cC} d\mathbf{r}_{A2} d\mathbf{r}_{b1} \chi^{(-)*}(\mathbf{r}_{bB}) \left[\psi^{j_f}(\mathbf{r}_{A1},\sigma_1)\psi^{j_f}(\mathbf{r}_{A2},\sigma_2) \right]^{0*}_0 \\
\times v(r_{b1})\Psi^{(+)}(\mathbf{r}_{aA},\mathbf{r}_{b1},\mathbf{r}_{b2},\sigma_1,\sigma_2).
\end{aligned}
\tag{6.2.157}
$$

If we neglect core excitations, the above expression is exact as long as $\Psi^{(+)}(\mathbf{r}_{aA},\mathbf{r}_{b1},\mathbf{r}_{b2},\sigma_1,\sigma_2)$ is the exact wavefunction. We can instead obtain an approximation for the transfer amplitude using

$$
\begin{aligned}
\Psi^{(+)}(\mathbf{r}_{aA},\mathbf{r}_{b1},\mathbf{r}_{b2},\sigma_1,\sigma_2) \approx \chi^{(+)}(\mathbf{r}_{aA}) \left[\psi^{j_{i1}}(\mathbf{r}_{b1},\sigma_1)\psi^{j_{i2}}(\mathbf{r}_{b2},\sigma_2) \right]^{\Lambda}_{\mu} \\
+ \sum_{K,M} \mathcal{U}_{K,M}(\mathbf{r}_{cC}) \left[\psi^{j_f}(\mathbf{r}_{A2},\sigma_2)\psi^{j_{i1}}(\mathbf{r}_{b1},\sigma_1) \right]^{K}_{M}
\end{aligned}
\tag{6.2.158}
$$

as an approximation for the incoming state. The first term of (6.2.158) gives rise to the simultaneous amplitude, while from second one leads to both the successive and the nonorthogonality contributions. To extract the amplitude $\mathcal{U}_{K,M}(\mathbf{r}_{cC})$, we define $f_{KM}(\mathbf{r}_{cC})$ as the scalar product

$$
f_{KM}(\mathbf{r}_{cC}) = \left\langle \left[\psi^{j_f}(\mathbf{r}_{A2},\sigma_2)\psi^{j_{i1}}(\mathbf{r}_{b1},\sigma_1) \right]^{K}_{M} \middle| \Psi^{(+)}(\mathbf{r}_{aA},\mathbf{r}_{b1},\mathbf{r}_{b2},\sigma_1,\sigma_2) \right\rangle
\tag{6.2.159}
$$

for fixed $\mathbf{r}_{cC}$, which can be seen to obey the equation

$$
\begin{aligned}
\left(\frac{\hbar^2}{2\mu_{cC}} k^2_{cC} + \frac{\hbar^2}{2\mu_{cC}} \nabla^2_{r_{cC}} - U(r_{cC}) \right) f_{KM}(\mathbf{r}_{cC}) \\
= \left\langle \left[\psi^{j_f}(\mathbf{r}_{A2},\sigma_2)\psi^{j_{i1}}(\mathbf{r}_{b1},\sigma_1) \right]^{K}_{M} \middle| v(r_{c2}) \middle| \Psi^{(+)}(\mathbf{r}_{aA},\mathbf{r}_{b1},\mathbf{r}_{b2},\sigma_1,\sigma_2) \right\rangle.
\end{aligned}
\tag{6.2.160}
$$

The solution can be written in terms of the Green function $G(\mathbf{r}_{cC}, \mathbf{r}'_{cC})$ defined by

$$\left(\frac{\hbar^2}{2\mu_{cC}} k_{cC}^2 + \frac{\hbar^2}{2\mu_{cC}} \nabla_{r_{cC}}^2 - U(r_{cC}) \right) G(\mathbf{r}_{cC}, \mathbf{r}'_{cC}) = \frac{\hbar^2}{2\mu_{cC}} \delta(\mathbf{r}_{cC} - \mathbf{r}'_{cC}). \quad (6.2.161)$$

Thus,

$$
\begin{aligned}
f_{KM}(\mathbf{r}_{cC}) &= \frac{2\mu_{cC}}{\hbar^2} \int d\mathbf{r}'_{cC} \, G(\mathbf{r}_{cC}, \mathbf{r}'_{cC}) \\
&\quad \times \left\langle \left[\psi^{j_f}(\mathbf{r}'_{A2}, \sigma'_2) \psi^{j_{i1}}(\mathbf{r}'_{b1}, \sigma'_1) \right]_M^K \middle| v(r_{C2}) \middle| \Psi^{(+)}(\mathbf{r}'_{aA}, \mathbf{r}'_{b1}, \mathbf{r}'_{b2}, \sigma'_1, \sigma'_2) \right\rangle \\
&\approx \frac{2\mu_{cC}}{\hbar^2} \sum_{\sigma'_1 \sigma'_2} \int d\mathbf{r}'_{cC} d\mathbf{r}'_{A2} d\mathbf{r}'_{b1} \, G(\mathbf{r}_{cC}, \mathbf{r}'_{cC}) \left[\psi^{j_f}(\mathbf{r}'_{A2}, \sigma'_2) \psi^{j_{i1}}(\mathbf{r}'_{b1}, \sigma'_1) \right]_M^{K*} \\
&\quad \times v(r'_{c2}) \chi^{(+)}(\mathbf{r}'_{aA}) \left[\psi^{j_{i1}}(\mathbf{r}'_{b1}, \sigma'_1) \psi^{j_{i2}}(\mathbf{r}'_{b2}, \sigma'_2) \right]_\mu^\Lambda = \mathcal{U}_{K,M}(\mathbf{r}_{cC}) \\
&\quad + \left\langle \left[\psi^{j_f}(\mathbf{r}'_{A2}, \sigma_2) \psi^{j_{i1}}(\mathbf{r}'_{b1}, \sigma_1) \right]_M^K \middle| \chi^{(+)}(\mathbf{r}'_{aA}) \left[\psi^{j_{i1}}(\mathbf{r}'_{b1}, \sigma'_1) \psi^{j_{i2}}(\mathbf{r}'_{b2}, \sigma'_2) \right]_\mu^\Lambda \right\rangle.
\end{aligned}
$$
$$(6.2.162)$$

Therefore

$$
\begin{aligned}
\mathcal{U}_{K,M}(\mathbf{r}_{cC}) &= \frac{2\mu_{cC}}{\hbar^2} \sum_{\sigma'_1 \sigma'_2} \int d\mathbf{r}'_{cC} d\mathbf{r}'_{A2} d\mathbf{r}'_{b1} \, G(\mathbf{r}_{cC}, \mathbf{r}'_{cC}) \left[\psi^{j_f}(\mathbf{r}'_{A2}, \sigma'_2) \psi^{j_{i1}}(\mathbf{r}'_{b1}, \sigma'_1) \right]_M^{K*} \\
&\quad \times v(r'_{c2}) \chi^{(+)}(\mathbf{r}'_{aA}) \left[\psi^{j_{i1}}(\mathbf{r}'_{b1}, \sigma'_1) \psi^{j_{i2}}(\mathbf{r}'_{b2}, \sigma'_2) \right]_\mu^\Lambda \\
&\quad - \left\langle \left[\psi^{j_f}(\mathbf{r}'_{A2}, \sigma_2) \psi^{j_{i1}}(\mathbf{r}'_{b1}, \sigma_1) \right]_M^K \middle| \chi^{(+)}(\mathbf{r}'_{aA}) \left[\psi^{j_{i1}}(\mathbf{r}'_{b1}, \sigma'_1) \psi^{j_{i2}}(\mathbf{r}'_{b2}, \sigma'_2) \right]_\mu^\Lambda \right\rangle.
\end{aligned}
$$
$$(6.2.163)$$

When we substitute $\mathcal{U}_{K,M}(\mathbf{r}_{cC})$ into (6.2.158) and (6.2.157), the first term gives rise to the successive amplitude for the two–particle transfer, while the second term is responsible for the nonorthogonal contribution.

Successive Transfer Contribution

We need to evaluate the integral

$$
\begin{aligned}
T_{succ}^{(2)}(\theta; \mu) &= \frac{4\mu_{cC}}{\hbar^2} \sum_{\sigma_1 \sigma_2} \sum_{KM} \int d\mathbf{r}_{cC} d\mathbf{r}_{A2} d\mathbf{r}_{b1} d\mathbf{r}'_{cC} d\mathbf{r}'_{A2} d\mathbf{r}'_{b1} \\
&\quad \times \left[\psi^{j_f}(\mathbf{r}_{A1}, \sigma_1) \psi^{j_f}(\mathbf{r}_{A2}, \sigma_2) \right]_0^{0*} \\
&\quad \times \chi^{(-)*}(\mathbf{r}_{bB}) G(\mathbf{r}_{cC}, \mathbf{r}'_{cC}) \left[\psi^{j_f}(\mathbf{r}'_{A2}, \sigma'_2) \psi^{j_{i1}}(\mathbf{r}'_{b1}, \sigma'_1) \right]_M^{K*} \chi^{(+)}(\mathbf{r}'_{aA}) v(r'_{c2}) v(r_{b1}) \\
&\quad \times \left[\psi^{j_{i1}}(\mathbf{r}'_{b1}, \sigma'_1) \psi^{j_{i2}}(\mathbf{r}'_{b2}, \sigma'_2) \right]_\mu^\Lambda \left[\psi^{j_f}(\mathbf{r}_{A2}, \sigma_2) \psi^{j_{i1}}(\mathbf{r}_{b1}, \sigma_1) \right]_M^K ,
\end{aligned}
$$
$$(6.2.164)$$

where we must substitute the Green function and the distorted waves by their partial wave expansions (see App. 6.B). The integral over $\mathbf{r}'_{b1}$ is:

$$\sum_{\sigma_1'} \int d\mathbf{r}_{b1}' \left[\psi^{j_f}(\mathbf{r}_{A2}', \sigma_2') \psi^{j_{i1}}(\mathbf{r}_{b1}', \sigma_1') \right]_M^{K*} \left[\psi^{j_{i1}}(\mathbf{r}_{b1}', \sigma_1') \psi^{j_{i2}}(\mathbf{r}_{b2}', \sigma_2') \right]_\mu^\Lambda$$

$$= \sum_{\sigma_1'} \int d\mathbf{r}_{b1}' (-1)^{-M+j_f+j_{i1}-\sigma_1-\sigma_2} \left[\psi^{j_{i1}}(\mathbf{r}_{b1}', -\sigma_1') \psi^{j_f}(\mathbf{r}_{A2}', -\sigma_2') \right]_{-M}^K$$

$$\times \left[\psi^{j_{i1}}(\mathbf{r}_{b1}', \sigma_1') \psi^{j_{i2}}(\mathbf{r}_{b2}', \sigma_2') \right]_\mu^\Lambda$$

$$= \sum_{\sigma_1'} \int d\mathbf{r}_{b1}' (-1)^{-M+j_f+j_{i1}-\sigma_1-\sigma_2} \sum_P \langle K \, \Lambda \, -M \, \mu | P \, \mu - M \rangle$$

$$\times \left((j_{i1} j_f)_K (j_{i1} j_{i2})_\Lambda | (j_{i1} j_{i1})_0 (j_f j_{i2})_P \right)_P$$

$$\times \left[\psi^{j_{i1}}(\mathbf{r}_{b1}', -\sigma_1') \psi^{j_{i1}}(\mathbf{r}_{b1}', \sigma_1') \right]_0^0 \left[\psi^{j_f}(\mathbf{r}_{A2}', -\sigma_2') \psi^{j_{i2}}(\mathbf{r}_{b2}', \sigma_2') \right]_{\mu-M}^P$$

$$= (-1)^{-M+j_f+j_{i1}} \sqrt{2 j_{i1} + 1} \, u_{l_f}(r_{A2}) u_{l_{i2}}(r_{b2}') \sum_P \langle K \, \Lambda \, -M \, \mu | P \, \mu - M \rangle$$

$$\times \left((j_{i1} j_f)_K (j_{i1} j_{i2})_\Lambda | (j_{i1} j_{i1})_0 (j_f j_{i2})_P \right)_P \left((l_f \tfrac{1}{2})_{j_f} (l_{i2} \tfrac{1}{2})_{j_{i2}} | (l_f l_{i2})_P (\tfrac{1}{2} \tfrac{1}{2})_0 \right)_P$$

$$\times \left[Y^{l_f}(\hat{\mathbf{r}}_{A2}') Y^{l_{i2}}(\hat{\mathbf{r}}_{b2}') \right]_{\mu-M}^P u_{l_f}(r_{A2}) u_{l_{i2}}(r_{b2}). \tag{6.2.165}$$

Integrating over $\mathbf{r}_{A2}$ (see (6.2.117)) leads to

$$\sum_{\sigma_2} \int d\mathbf{r}_{A2} \left[\psi^{j_f}(\mathbf{r}_{A1}, \sigma_1) \psi^{j_f}(\mathbf{r}_{A2}, \sigma_2) \right]_0^{0*} \left[\psi^{j_f}(\mathbf{r}_{A2}, \sigma_2) \psi^{j_{i1}}(\mathbf{r}_{b1}, \sigma_1) \right]_M^K$$

$$= -\sqrt{\frac{2}{2 j_f + 1}} \left((l_f \tfrac{1}{2})_{j_f} (l_{i1} \tfrac{1}{2})_{j_{i1}} | (l_f l_{i1})_K (\tfrac{1}{2} \tfrac{1}{2})_0 \right)_K$$

$$\times \left[Y^{l_f}(\hat{\mathbf{r}}_{A1}) Y^{l_{i1}}(\hat{\mathbf{r}}_{b1}) \right]_M^K u_{l_f}(r_{A1}) u_{l_{i1}}(r_{b1}). \tag{6.2.166}$$

Let us examine the term

$$\sum_M (-1)^M \langle K \, \Lambda - M \, \mu | P \, \mu - M \rangle \left[Y^{l_f}(\hat{\mathbf{r}}_{A1}) Y^{l_{i1}}(\hat{\mathbf{r}}_{b1}) \right]_M^K \left[Y^{l_f}(\hat{\mathbf{r}}_{A2}') Y^{l_{i2}}(\hat{\mathbf{r}}_{b2}') \right]_{\mu-M}^P \cdot$$

$$\tag{6.2.167}$$

Making use of the relation

$$\langle l_1 \, l_2 \, m_1 \, m_2 | L \, M_L \rangle = (-1)^{l_2 - m_2} \sqrt{\frac{2L+1}{2 l_1 + 1}} \langle L \, l_2 \, -M_L \, m_2 | l_1 \, -m_1 \rangle, \tag{6.2.168}$$

the expression (6.2.168) is equivalent to

$$(-1)^K \sqrt{\frac{2P+1}{2\Lambda+1}} \left\{ \left[Y^{l_f}(\hat{\mathbf{r}}_{A2}') Y^{l_{i2}}(\hat{\mathbf{r}}_{b2}') \right]^P \left[Y^{l_f}(\hat{\mathbf{r}}_{A1}) Y^{l_{i1}}(\hat{\mathbf{r}}_{b1}) \right]^K \right\}_\mu^\Lambda. \tag{6.2.169}$$

We now recouple the term

$$\left[Y^{l_a}(\hat{\mathbf{r}}'_{aA})Y^{l_a}(\hat{\mathbf{k}}_{aA})\right]_0^0 \left[Y^{l_b}(\hat{\mathbf{r}}_{bB})Y^{l_b}(\hat{\mathbf{k}}_{bB})\right]_0^0, \qquad (6.2.170)$$

arising from the partial wave expansion of the incoming and outgoing distorted waves, to have

$$((l_al_a)_0(l_bl_b)_0|(l_al_b)_\Lambda(l_al_b)_\Lambda)_0 \left\{\left[Y^{l_a}(\hat{\mathbf{r}}'_{aA})Y^{l_b}(\hat{\mathbf{r}}_{bB})\right]^\Lambda \left[Y^{l_a}(\hat{\mathbf{k}}_{aA})Y^{l_b}(\hat{\mathbf{k}}_{bB})\right]^\Lambda\right\}_0^0.$$
$$(6.2.171)$$

The only term which does not vanish upon integration is

$$\frac{(-1)^{\Lambda-\mu}}{\sqrt{(2l_a+1)(2l_b+1)}}\left[Y^{l_a}(\hat{\mathbf{r}}'_{aA})Y^{l_b}(\hat{\mathbf{r}}_{bB})\right]_{-\mu}^\Lambda \left[Y^{l_a}(\hat{\mathbf{k}}_{aA})Y^{l_b}(\hat{\mathbf{k}}_{bB})\right]_\mu^\Lambda. \qquad (6.2.172)$$

Again, the only term surviving

$$\left\{\left[Y^{l_f}(\hat{\mathbf{r}}'_{A2})Y^{l_{i2}}(\hat{\mathbf{r}}'_{b2})\right]^P \left[Y^{l_f}(\hat{\mathbf{r}}_{A1})Y^{l_{i1}}(\hat{\mathbf{r}}_{b1})\right]^K\right\}_\mu^\Lambda \left[Y^{l_a}(\hat{\mathbf{r}}'_{aA})Y^{l_b}(\hat{\mathbf{r}}_{bB})\right]_{-\mu}^\Lambda \quad (6.2.173)$$

is

$$\frac{(-1)^{\Lambda+\mu}}{\sqrt{2\Lambda+1}}\left[\left\{\left[Y^{l_f}(\hat{\mathbf{r}}'_{A2})Y^{l_{i2}}(\hat{\mathbf{r}}'_{b2})\right]^P\right.\right.$$
$$(6.2.174)$$
$$\left.\left.\left[Y^{l_f}(\hat{\mathbf{r}}_{A1})Y^{l_{i1}}(\hat{\mathbf{r}}_{b1})\right]^K\right\}^\Lambda \left[Y^{l_a}(\hat{\mathbf{r}}'_{aA})Y^{l_b}(\hat{\mathbf{r}}_{bB})\right]^\Lambda\right]_0^0.$$

We now couple this last term with the term $\left[Y^{l_c}(\hat{\mathbf{r}}'_{cC})Y^{l_c}(\hat{\mathbf{r}}_{cC})\right]_0^0$, arising from the partial wave expansion of the Green function. That is,

$$\left[\left\{\left[Y^{l_f}(\hat{\mathbf{r}}'_{A2})Y^{l_{i2}}(\hat{\mathbf{r}}'_{b2})\right]^P \left[Y^{l_f}(\hat{\mathbf{r}}_{A1})Y^{l_{i1}}(\hat{\mathbf{r}}_{b1})\right]^K\right\}^\Lambda \left[Y^{l_a}(\hat{\mathbf{r}}'_{aA})Y^{l_b}(\hat{\mathbf{r}}_{bB})\right]^\Lambda\right]_0^0$$

$$\times \left[Y^{l_c}(\hat{\mathbf{r}}'_{cC})Y^{l_c}(\hat{\mathbf{r}}_{cC})\right]_0^0$$

$$= ((l_al_b)_\Lambda(l_cl_c)_0|(l_al_c)_P(l_bl_c)_K)_\Lambda \left[\left\{\left[Y^{l_f}(\hat{\mathbf{r}}'_{A2})Y^{l_{i2}}(\hat{\mathbf{r}}'_{b2})\right]^P \left[Y^{l_f}(\hat{\mathbf{r}}_{A1})Y^{l_{i1}}(\hat{\mathbf{r}}_{b1})\right]^K\right\}^\Lambda\right.$$

$$\left.\left\{\left[Y^{l_a}(\hat{\mathbf{r}}'_{aA})Y^{l_c}(\hat{\mathbf{r}}'_{cC})\right]^P \left[Y^{l_b}(\hat{\mathbf{r}}_{bB})Y^{l_c}(\hat{\mathbf{r}}_{cC})\right]^K\right\}^\Lambda\right]_0^0$$

$$= ((l_al_b)_\Lambda(l_cl_c)_0|(l_al_c)_P(l_bl_c)_K)_\Lambda$$

$$\times ((PK)_\Lambda(PK)_\Lambda|(PP)_0(KK)_0)_0 \left[Y^{l_f}(\hat{\mathbf{r}}'_{A2})Y^{l_{i2}}(\hat{\mathbf{r}}'_{b2})\right]^P \left[Y^{l_a}(\hat{\mathbf{r}}'_{aA})Y^{l_c}(\hat{\mathbf{r}}'_{cC})\right]^P\right\}_0^0$$

$$\times \left\{\left[Y^{l_f}(\hat{\mathbf{r}}_{A1})Y^{l_{i1}}(\hat{\mathbf{r}}_{b1})\right]^K \left[Y^{l_b}(\hat{\mathbf{r}}_{bB})Y^{l_c}(\hat{\mathbf{r}}_{cC})\right]^K\right\}_0^0$$

$$= \left((l_a l_b)_\Lambda (l_c l_c)_0 | (l_a l_c)_P (l_b l_c)_K\right)_\Lambda$$

$$\times \sqrt{\frac{2\Lambda + 1}{(2K+1)(2P+1)}} \left\{\left[Y^{l_f}(\hat{\mathbf{r}}'_{A2}) Y^{l_{i2}}(\hat{\mathbf{r}}'_{b2})\right]^P \left[Y^{l_a}(\hat{\mathbf{r}}'_{aA}) Y^{l_c}(\hat{\mathbf{r}}'_{cC})\right]^P\right\}_0^0$$

$$\times \left\{\left[Y^{l_f}(\hat{\mathbf{r}}_{A1}) Y^{l_{i1}}(\hat{\mathbf{r}}_{b1})\right]^K \left[Y^{l_b}(\hat{\mathbf{r}}_{bB}) Y^{l_c}(\hat{\mathbf{r}}_{cC})\right]^K\right\}_0^0 . \tag{6.2.175}$$

Collecting all the contributions (including the constants and phases arising from the partial wave expansion of the distorted waves and the Green function), we get

$$T^{(2)}_{succ}(\theta; \mu) = (-1)^{j_f + j_{i1}} \frac{2048\pi^5 \mu_{Cc}}{\hbar^2 k_{Aa} k_{Bb} k_{Cc}} \sqrt{\frac{(2j_{i1}+1)}{(2\Lambda+1)(2j_f+1)}}$$

$$\times \sum_{K,P} \left((l_f \tfrac{1}{2})_{j_f} (l_{i2} \tfrac{1}{2})_{j_{i2}} | (l_f l_{i2})_P (\tfrac{1}{2} \tfrac{1}{2})_0\right)_P$$

$$\times \left((l_f \tfrac{1}{2})_{j_f} (l_{i1} \tfrac{1}{2})_{j_{i1}} | (l_f l_{i1})_K (\tfrac{1}{2} \tfrac{1}{2})_0\right)_K \left((j_{i1} j_f)_K (j_{i1} j_{i2})_\Lambda | (j_{i1} j_{i1})_0 (j_f j_{i2})_P\right)_P$$

$$\times \frac{(-1)^K}{(2K+1)\sqrt{2P+1}} \sum_{l_c, l_a, l_b} \left((l_a l_b)_\Lambda (l_c l_c)_0 | (l_a l_c)_P (l_b l_c)_K\right)_\Lambda e^{i(\sigma_i^{l_a} + \sigma_f^{l_b})} i^{l_a - l_b}$$

$$\times (2l_c+1)^{3/2} \left[Y^{l_a}(\hat{\mathbf{k}}_{aA}) Y^{l_b}(\hat{\mathbf{k}}_{bB})\right]_\mu^\Lambda S_{K,P,l_a,l_b,l_c}, \tag{6.2.176}$$

with (note that we have reduced the dimensionality of the integrals in the same fashion as for the $L=0$–angular momentum transfer calculation, see (6.2.132))

$$S_{K,P,l_a,l_b,l_c} = \int r_{Cc}^2 \, dr_{Cc} \, r_{b1}^2 \, dr_{b1} \, \sin\theta \, d\theta \, v(r_{b1}) u_{l_f}(r_{C1}) u_{l_i}(r_{b1})$$

$$\times \frac{s_{P,l_a,l_c}(r_{Cc})}{r_{Cc}} \frac{F_{l_b}(r_{Bb})}{r_{Bb}}$$

$$\times \sum_M \langle l_c \, 0 \, l_b \, M | K \, M \rangle \left[Y^{l_f}(\hat{r}_{C1}) Y^{l_{i1}}(\theta + \pi, 0)\right]_M^K Y^{l_b}_{-M}(\hat{r}_{Bb}), \tag{6.2.177}$$

and

$$s_{P,l_a,l_c}(r_{Cc}) = \int r_{Cc}'^2 \, dr_{Cc}' \, r_{A2}'^2 \, dr_{A2}' \, \sin\theta' \, d\theta' \, v(r_{c2}') u_{l_f}(r_{A2}') u_{l_i}(r_{c2}')$$

$$\times \frac{F_{l_a}(r_{Aa}')}{r_{Aa}'} \frac{f_{l_c}(k_{Cc}, r_<) P_{l_c}(k_{Cc}, r_>)}{r_{Cc}'}$$

$$\times \sum_M \langle l_c \, 0 \, l_a \, M | P \, M \rangle \left[Y^{l_f}(\hat{r}'_{A2}) Y^{l_{i2}}(\hat{r}'_{c2})\right]_M^P Y^{l_a}_{-M}(\hat{r}'_{Aa}). \tag{6.2.178}$$

We have evaluated the transition matrix element for a particular projection μ of the initial angular momentum of the two transferred nucleons. If they are coupled to a core of angular momentum J_f to total angular momentum J_i, M_i, the fraction of the initial wavefunction with projection μ is $\langle \Lambda \ \mu \ J_f \ M_i - \mu | J_i \ M_i \rangle$, and the cross section will be

$$\frac{d\sigma}{d\Omega}(\hat{\mathbf{k}}_{bB}) = \frac{k_{bB}}{k_{aA}} \frac{\mu_{aA}\mu_{bB}}{(2\pi\hbar^2)^2} \left| \sum_{\mu} \langle \Lambda \ \mu \ J_f \ M_i - \mu | J_i \ M_i \rangle T_{succ}^{(2)}(\theta; \mu) \right|^2 . \quad (6.2.179)$$

For a non polarized incident beam,

$$\frac{d\sigma}{d\Omega}(\hat{\mathbf{k}}_{bB}) = \frac{k_{bB}}{k_{aA}} \frac{\mu_{aA}\mu_{bB}}{(2\pi\hbar^2)^2} \frac{1}{2J_i + 1} \sum_{M_i} \left| \sum_{\mu} \langle \Lambda \ \mu \ J_f \ M_i - \mu | J_i \ M_i \rangle T_{succ}^{(2)}(\theta; \mu) \right|^2 . $$
$$(6.2.180)$$

This would be the differential cross section for a transition to a definite final state M_f. If we do not measure M_f, we have to sum for all M_f,

$$\frac{d\sigma}{d\Omega}(\hat{\mathbf{k}}_{bB}) = \frac{k_{bB}}{k_{aA}} \frac{\mu_{aA}\mu_{bB}}{(2\pi\hbar^2)^2} \frac{1}{2J_i + 1} \sum_{\mu} |T_{succ}^{(2)}(\theta; \mu)|^2 \sum_{M_i, M_f} \left| \langle \Lambda \ \mu \ J_f \ M_f | J_i \ M_i \rangle \right|^2 . $$
$$(6.2.181)$$

The sum over M_i, M_f of the Clebsh–Gordan coefficients gives $(2J_i + 1)/(2\Lambda + 1)$ (see Eq. (6.A.26)). One then gets,

$$\frac{d\sigma}{d\Omega}(\hat{\mathbf{k}}_{bB}) = \frac{k_{bB}}{k_{aA}} \frac{\mu_{aA}\mu_{bB}}{(2\pi\hbar^2)^2} \frac{1}{(2\Lambda + 1)} \sum_{\mu} |T_{succ}^{(2)}(\theta; \mu)|^2 , \quad (6.2.182)$$

where one can write

$$T_{succ}^{(2)}(\theta; \mu) = \sum_{l_a, l_b} C_{l_a, l_b} \left[Y^{l_a}(\hat{\mathbf{k}}_{aA}) Y^{l_b}(\hat{\mathbf{k}}_{bB}) \right]_{\mu}^{\Lambda}$$
$$= \sum_{l_a, l_b} C_{l_a, l_b} i^{l_a} \sqrt{\frac{2l_a + 1}{4\pi}} \langle l_a \ l_b \ 0 \ \mu | \Lambda \ \mu \rangle Y_{\mu}^{l_b}(\hat{\mathbf{k}}_{bB}). \quad (6.2.183)$$

Note that (6.2.182) takes into account only the spins of the heavy nucleus. In a (t, p) or (p, t) reaction, we have to sum over the spins of the proton and of the triton and divide by 2. If a spin-orbit term is present in the optical potential, the sum yields the combination of terms shown in Sect. 6.2.2:

$$\frac{d\sigma}{d\Omega}(\hat{\mathbf{k}}_{bB}) = \frac{k_{bB}}{k_{aA}} \frac{\mu_{aA}\mu_{bB}}{(2\pi\hbar^2)^2} \frac{1}{2(2\Lambda + 1)} \sum_{\mu} |A_{\mu}|^2 + |B_{\mu}|^2 . \quad (6.2.184)$$

6.3 ZPF, Exclusion Principle, and Medium Polarization Effects: Self-Energy, Vertex Corrections, Induced Interaction

Of all quantal phenomena, zero-point fluctuations (ZPF), closely connected with virtual states, are likely the most representative of the essential difference existing between quantum and classical mechanics. In fact, ZPF are intimately connected with the complementary principle[9], and thus with indeterminacy[10] and noncommutative[11] relations, and with the probabilistic interpretation[12] of the (modulus squared) of the wavefunctions, solution of Schrödinger's or Dirac's equations.[13]

Pauli principle[14] brings about essential modifications to the virtual fluctuations of the many-body system, modifications which are instrumental in the dressing and interweaving of the elementary modes of excitation.[15]

In Fig. 6.3.1, NFT diagrams are given which correspond to the lowest-order medium polarization effects renormalizing the properties of a particle-hole collective mode (wavy line), correlated particle-hole excitation ((upgoing)–(downgoing) arrowed lines) calculated within the random phase approximation (RPA,QRPA), and leading to the particle–vibration coupling vertex (formfactor and strength, that is, transition density (solid dot), see inset (I), bottom). The action of an external field on the zero-point fluctuations (ZPF) of the vacuum (inset (II)), forces a virtual process to become real, leading to a collective vibration by annihilating a (virtual, spontaneous) particle-hole excitation (backwards RPA Y-amplitude) or, in the time-ordered process, by creating a particle-hole excitation which eventually, through the particle–vibration coupling vertex, materializes into the collective (coherent) state (forwardsgoing X-amplitudes). Now, oyster-like diagrams associated with the vacuum ZPF can occur at any time (see inset (III) of Fig. 6.3.1). It is of note that while virtual states can violate energy conservation, they have to respect conservation rules, as well as Pauli principle. For example, a virtual state cannot violate angular momentum, and there should be a diagram correcting for the presence of two fermions in the same quantal state. The process shown in the inset III (α) leads, through Pauli principle correcting diagrams (exchange of fermionic arrowed lines) to self-energy (inset III (β), (δ)) and vertex corrections (induced $p - h$ interaction; inset III (γ), (ε), see also graphs (h) and (i)) processes. Similar processes are found in the interplay between ($p - h$-like) ZPF and pair addition modes as shown in Fig. 6.3.2. Note the parallel between diagrams of Figs. 6.3.1

---

[9] Bohr (1928).
[10] Heisenberg (1927).
[11] Born and Jordan (1925); Born et al. (1926).
[12] Born (1926).
[13] Schrödinger (1926a); Dirac (1930).
[14] Pauli (1925).
[15] Within the present context, see also Schrieffer (1964).

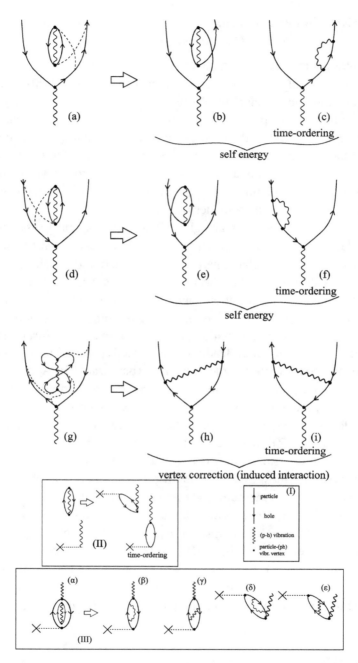

Figure 6.3.1 Nuclear field theory (NFT) diagrams describing renormalization processes associated with ZPF. For details, see the caption to Fig. 6.3.2.

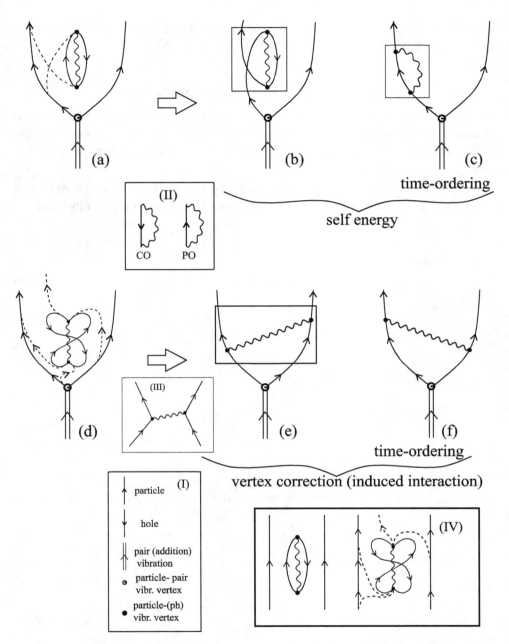

Figure 6.3.2 Pauli effects–associated (p-h) ZPF dressing a pairing vibrational (pair addition) mode (see inset I) in terms of self-energy (graphs (a)–(c); correlation (CO) and polarization (PO) diagrams, inset II) and vertex correction (graphs (d)–(f); induced particle–particle (pairing) interaction) processes (inset (III)) associated with phonon exchange between nucleons (inset (IV)).

(g)–(i) and of Figs. 6.3.2 (d)–(f). The collective vibrational modes can be viewed as coherent states[16] exhausting a consistent fraction of the EWSR.

6.4 Coherence and Effective Formfactors

In what follows we shall work out a simplified derivation of the simultaneous two-nucleon transfer amplitude, within the framework of first-order DWBA specially suited to discuss correlation aspects of pair transfer in general, and of the associated effective formfactors in particular.[17]

We will concentrate on (t, p) reactions, namely, reactions of the type $A(\alpha, \beta)B$ where $\alpha = \beta + 2$ and $B = A + 2$. The intrinsic wave functions are in this case

$$\psi_\alpha = \psi_{M_i}^{J_i}(\xi_A) \sum_{ss_f'} \left[\chi^s(\sigma_\alpha) \chi^{s_f'}(\sigma_\beta) \right]_{M_{s_i}}^{s_i} \phi_t^{L=0}\left(r_{12}, r_{1p}, r_{2p}\right)$$

$$= \psi_{M_i}^{J_i}(\xi_A) \sum_{M_s M_{s_f}'} (s M_s' s_f' M_{s_f}' | s_i M_{s_i}) \chi_{M_s'}^{s}(\sigma_\alpha) \chi_{M_{s_f}'}^{s_f'}(\sigma_\beta) \qquad (6.4.1)$$

$$\times \phi_t^{L=0}\left(r_{12}, r_{1p}, r_{2p}\right)$$

while

$$\psi_\beta = \psi_{M_f}^{J_f}(\xi_{A+2}) \chi_{M_{s_f}}^{s_f}(\sigma_\beta)$$

$$= \sum_{\substack{n_1 l_1 j_1 \\ n_2 l_2 j_2}} B(n_1 l_1 j_1, n_2 l_2 j_2; J J_i' J_f) \left[\phi^J(j_1 j_2) \phi^{J_i'}(\xi_A) \right]_{M_f}^{J_f} \qquad (6.4.2)$$

$$\times \chi_{M_{s_f}}^{s_f}(\sigma_\beta).$$

Making use of the above equation one can define the spectroscopic amplitude B as

$$B(n_1 l_1 j_1, n_2 l_2 j_2; J J_i' J_f) = \left\langle \psi^{J_f}(\xi_{A+2}) \left| \left[\phi^J(j_1 j_2) \phi^{J_i}(\xi_A) \right]^{J_f} \right\rangle, \qquad (6.4.3)$$

where

$$\phi^J(j_1 j_2) = \frac{\left[\phi_{j_1}(\mathbf{r}_1) \phi_{j_2}(\mathbf{r}_2) \right]^J - \left[\phi_{j_1}(\mathbf{r}_2) \phi_{j_2}(\mathbf{r}_1) \right]^J}{\sqrt{1 + \delta(j_1, j_2)}}, \qquad (6.4.4)$$

is an antisymetrized, normalized wave function of the two transferred particles. The function $\chi_{M_s}^{s}(\sigma_\beta)$ appearing both in Eqs. (6.4.1) and (6.4.2) is the spin wave function of the proton while

[16] See, e.g., Glauber (1969, 2007).
[17] Glendenning (1965); Bayman and Kallio (1967).

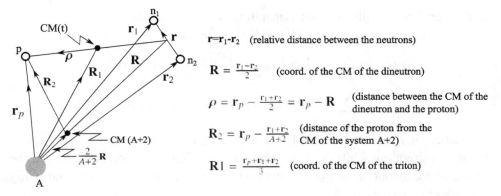

$r=r_1-r_2$ (relative distance between the neutrons)

$R = \frac{r_1-r_2}{2}$ (coord. of the CM of the dineutron)

$\rho = r_p - \frac{r_1+r_2}{2} = r_p - R$ (distance between the CM of the dineutron and the proton)

$R_2 = r_p - \frac{r_1+r_2}{A+2}$ (distance of the proton from the CM of the system A+2)

$R1 = \frac{r_p+r_1+r_2}{3}$ (coord. of the CM of the triton)

Figure 6.4.1 Coordinate system used in the calculation of the two-nucleon transfer amplitude.

$$\chi^s(\sigma_\alpha) = \left[\chi^{s_1}(\sigma_{n_1})\chi^{s_2}(\sigma_{n_2})\right]^s, \tag{6.4.5}$$

is the spin function of the two-neutron system.

A simple description of the intrinsic degrees of freedom of the triton is obtained by using a wavefunction symmetric in the coordinates of all particles, that is,

$$\begin{aligned}\phi_t^{L=0}\left(r_{12}, r_{1p}, r_{2p}\right) &= N_t\, e^{[(r_1-r_2)^2+(r_1-r_p)^2+(r_2-r_p)^2]}\\ &= \phi_{000}(\mathbf{r})\phi_{000}(\rho),\end{aligned} \tag{6.4.6}$$

where

$$\phi_{000}(\mathbf{r}) = R_{nl}(\nu^{1/2}r)Y_{lm}(\hat{\mathbf{r}}). \tag{6.4.7}$$

The coordinate ρ is the radius vector which measures the distance between the center of mass of the dineutron and the proton, while the vector $\mathbf{r}$ is the dineutron relative coordinate (cf. Fig. 6.4.1). To obtain the DWBA cross section we have to calculate the integral

$$T(\theta) = \int d\xi_A\, d\mathbf{r}_1\, d\mathbf{r}_2\, d\mathbf{r}_p \chi_p^{(-)}(\mathbf{R}_2)\psi_\beta^*(\xi_{A+2}, \sigma_\beta)V_\beta\psi_\alpha(\xi_A, \sigma_\alpha, \sigma_\beta)\psi_t^{(+)}(\mathbf{R}_1), \tag{6.4.8}$$

where the final state effective interaction $V_\beta(\rho)$ is assumed to depend only on the distance ρ between the center of mass of the di-neutron and of the proton.

To carry out the integral (6.4.8) we transform the wave function (6.4.4) into center of mass and relative coordinates. If we assume that both $\phi_{j_1}(\mathbf{r}_1)$ and $\phi_{j_2}(\mathbf{r}_2)$ are harmonic oscillator wave functions (used as a basis to expand the Woods–Saxon single-particle wavefunctions), this transformation can be carried with the aid of the Moshinsky brackets. If $|n_1l_1, n_2l_2; \lambda\mu\rangle$ is a complete system of wave functions in the harmonic oscillator basis, depending on $\mathbf{r}_1$ and $\mathbf{r}_2$ and $|nl, NL; \lambda\mu\rangle$ is the corresponding one depending on $\mathbf{r}$ and $\mathbf{R}$, we can write

$$
\begin{aligned}
|n_1l_1, n_2l_2; \lambda\mu\rangle &= \sum_{nlNL} (|nl, NL; \lambda\mu|) \langle nl, NL; \lambda\mu|) |n_1l_1, n_2l_2; \lambda\mu\rangle \\
&= \sum_{nlNL} |nl, NL; \lambda\mu\rangle \langle nl, NL; \lambda\mu|n_1l_1, n_2l_2; \lambda\rangle.
\end{aligned}
\tag{6.4.9}
$$

The labels n, l are the principal and angular momentum quantum numbers of the relative motion, while N, L are the corresponding ones corresponding to the center of mass motion of the two-neutron system. Because of energy and parity conservation we have

$$
\begin{aligned}
2n_1 + l_1 + 2n_2 + l_2 &= 2n + l + 2N + L, \\
(-1)^{l_1+l_2} &= (-1)^{l+L}.
\end{aligned}
\tag{6.4.10}
$$

The coefficients $\langle nl, NL, L|n_1l_1, n_2l_2, L\rangle$ were first discussed by Moshinsky.[18]
With the help of Eq. (6.4.9) we can write the wave function $\psi_{M_f}^{J_f}(\xi_{A+2})$ as

$$
\begin{aligned}
\psi_{M_f}^{J_f}(\xi_{A+2}) &= \sum_{\substack{n_1l_1j_1 \\ n_2l_2j_2 \\ JJ_i}} B(n_1l_1j_1, n_2l_2j_2; JJ_i'J_f) \left[\phi^J(j_1j_2)\phi^{J_i'}(\xi_A)\right]_{M_f}^{J_f} \\
&= \sum_{\substack{n_1l_1j_1 \\ n_2l_2j_2}} \sum_{JJ_i} B(n_1l_1j_1, n_2l_2j_2; JJ_i'J_f) \\
&\quad \times \sum_{M_JM_{J_i}'} \langle JM_J J_i'M_{J_i}|J_fM_{J_f}\rangle \psi_{M_{J_i}'}^{J_i'}(\xi_A) \\
&\quad \times \sum_{LS'} \langle S'LJ|j_1j_2J\rangle \sum_{M_LM_S'} \langle LM_L S'M_S'|JM_J\rangle \chi_{M_S'}^{S'}(\sigma_\alpha) \\
&\quad \times \sum_{nlN\Lambda} \langle nl, N\Lambda, L|n_1l_1, n_2l_2, L\rangle \\
&\quad \times \sum_{m_lM_\Lambda} \langle lm_l \Lambda M_\Lambda|LM_L\rangle \phi_{nlm_l}(\mathbf{r})\phi_{N\Lambda M_\Lambda}(\mathbf{R}),
\end{aligned}
\tag{6.4.11}
$$

Integration over $\vec{r}$ gives

$$
\langle \phi_{nlm_l}(\mathbf{r})|\phi_{000}(\mathbf{r})\rangle = \delta(l, 0)\delta(m_l, 0)\Omega_n,
\tag{6.4.12}
$$

where

$$
\Omega_n = \int R_{nl}(\nu_1^{1/2}r) R_{00}(\nu_2^{1/2}r)r^2 \, dr.
\tag{6.4.13}
$$

---

[18] Moshinsky (1959).

Note that there is no selection rule in the principal quantum number n, as the potential in which the two neutrons move in the triton has a frequency v_2 which is different from the one that the two neutrons are subjected to, when moving in the system A (nonorthogonality effect).

Integration over ξ_A and multiplication of the spin functions gives

$$\left(\psi_{M_{J_i}}^{J_i}, V'_\beta(\rho)\psi_{M'_{J_i}}^{J'_i}\right) = \delta(J_i, J'_i)\delta(M_{J_i}, M_{J'_i})V(\rho),$$

$$\left(\chi_{M_S}^S(\sigma_\alpha), \chi_{M_{S'}}^{S'}(\sigma_\alpha)\right) = \delta(S, S')\delta(M_S, M_{S'}), \tag{6.4.14}$$

$$\left(\chi_{M_{S_f}}^{S_f}(\sigma_\beta), \chi_{M_{S'_f}}^{S'_f}(\sigma_\beta)\right) = \delta(S_f, S'_f)\delta(M_{S_f}, M_{S'_f}).$$

The integral (6.4.8) can then be written as

$$\begin{aligned}
T(\theta) = \sum_{\substack{n_1 l_1 j_1 \\ n_2 l_2 j_2}} \sum_{JM_J} \sum_{nN} \sum_S &B(n_1 l_1 j_1, n_2 l_2 j_2; JJ_iJ_f) \\
&\times \langle JM_J J_i M_{J_i}|J_f M_{J_f}\rangle\langle SLJ|j_1 j_2 J\rangle \\
&\times \langle LM_L SM_S|JM_J\rangle\langle n0, NL, L|n_1 l_1, n_2 l_2, L\rangle \\
&\times \langle SM_S S_f M_{S_f}|S_i M_{S_i}\rangle\Omega_n \\
&\times \int d\mathbf{R}\, d\mathbf{r}_p\, \chi_t^{(+)*}(\mathbf{R}_1)\phi_{NLM_L}^*(\mathbf{R})V(\rho)\phi_{000}(\rho)\chi_t^{(+)}(\mathbf{R}_1),
\end{aligned} \tag{6.4.15}$$

where we have approximated V'_β by an effective interaction depending on $\rho = |\rho|$. We now define the effective two-nucleon transfer formfactor as

$$\begin{aligned}
u_{LSJ}^{J_i J_f}(R) = \sum_{\substack{n_1 l_1 j_1 \\ n_2 l_2 j_2}} \sum_{nN} &B(n_1 l_1 j_1, n_2 l_2 j_2; JJ_iJ_f)\langle SLJ|j_1 j_2 J\rangle \\
&\langle n0, NL, L|n_1 l_1, n_2 l_2; L\rangle\Omega_n R_{NL}(R).
\end{aligned} \tag{6.4.16}$$

One can then rewrite Eq. (6.4.15) as

$$\begin{aligned}
T(\theta) = \sum_J \sum_L \sum_S &\left(JM_J J_i M_{J_i}|J_f M_{J_f}\right)\left(SM_S S_f M_{S_f}|S_i M_{S_i}\right)\left(LM_L SM_S|JM_J\right) \\
&\times \int d\mathbf{R}\, d\mathbf{r}_p\, \chi_p^{*(-)}(\mathbf{R}_2)u_{LSJ}^{J_i J_f}(R)Y_{LM_L}^* V(\rho)\phi_{000}(\rho)\chi_t^{(+)}(\mathbf{R}_1).
\end{aligned} \tag{6.4.17}$$

Because the di-neutron has $S = 0$, we have that

$$\left(LM_L 00|JM_J\right) = \delta(J, L)\delta(M_L, M_J), \tag{6.4.18}$$

and the summations over S and L disappear from Eq. (6.4.17). Let us now make also here, as done in Sect. 5.6, Eq. (5.6.15) for one-particle transfer reactions, the zero-range approximation, that is,

$$V(\rho)\phi_{000}(\rho) = D_0\delta(\rho), \tag{6.4.19}$$

where D_0 is an empirical parameter ($D_0^2 = (31.6 \pm 9.3) \times 10^4 \text{ MeV}^2\text{fm}^2$) determined to reproduce, in average, the observed absolute cross sections[19] with only the simultaneous transfer. This means that the proton interacts with the center of mass of the di-neutron, only when they are at the same point in space. Within this approximation (cf. Fig. 6.4.1)

$$\mathbf{R} = \mathbf{R}_1 = \mathbf{r},$$
$$\mathbf{R}_2 = \frac{A}{A+2}\mathbf{R}. \tag{6.4.20}$$

Then Eq. (6.4.15) can be written as

$$T = D_0 \sum_L \left(LM_L J_i M_{J_i} | J_f M_{J_f}\right)$$
$$\times \int d\mathbf{R}\, \chi_p^{*(-)}\left(\frac{A}{A+2}\mathbf{R}\right) u_L^{J_i J_f}(R) Y_{LM_L}^*(\hat{\mathbf{R}}) \chi_t^{(+)}(\mathbf{R}) \tag{6.4.21}$$

From Eq. (6.4.21) it is seen that the change in parity implied by the reaction is given by $\Delta\pi = (-1)^L$. Consequently, the selection rules for (t, p) and (p, t) reactions in zero-range approximation are

$$\Delta S = 0,$$
$$\Delta J = \Delta L = L,$$
$$\Delta\pi = (-1)^L; \tag{6.4.22}$$

that is, only normal parity states are excited. The integral appearing in Eq. (6.4.21) has the same structure as the DWBA integral appearing in Eq. (5.6.16) which was derived for the case of one-nucleon transfer reactions.

The difference between the two processes manifests itself through the different structure of the two formfactors. While $u_l(r)$ is a single-particle bound state wave function (cf. Eq. (5.6.1a)), $u_L^{J_i J_f}$ is a coherent summation *over the center of mass states of motion of the two transferred neutrons* (see Eq. (6.4.16)). In other words, an effective quantity (function). It is of note that this difference between single-particle transfer formfactors and simultaneous two-nucleon transfer effective formfactors, becomes less pronounced when one considers dressed particles resulting from the coupling to collective vibrations and leading to renormalized energies, single-particle content, and renormalized radial wavefunction. In view of the relation (4.3.1), closely connected with Fig. 4.3.1 (b) found at the basis of the fact that successive transfer of entangled nucleons over distances of the order of the correlation length is the dominant contribution to pair transfer, the above results[20] are certainly dated.

---

[19] Broglia et al. (1973).
[20] See Potel et al. (2013a), in particular App. C and Fig. 10 of this reference.

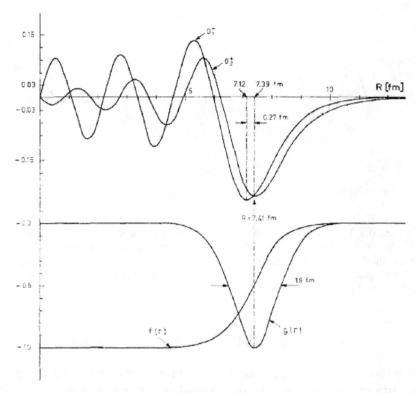

Figure 6.4.2 The upper part of the figure shows the modified formfactor for the ^{206}Pb(t,p)^{208}Pb transition to the ground state (0_1^+) and the pairing vibrational state (0_2^+) at 4.87 MeV. Both curves are matched with appropriate Hankel functions. In the lower part, the formfactors of the real ($f(r)$) and the imaginary ($g(r)$) part of the optical potential used to calculate the differential cross sections (see Fig. 4.5.4) are given. After Broglia and Riedel (1967).

The reason to bring the subject here is to set in evidence subtle aspects of the correlations existing between partner nucleons of a Cooper pair to be also found in the formfactors associated with successive transfer although, arguably, in a less evident fashion.

Examples of two-nucleon transfer formfactors are given in[21] Figs 6.4.2, 6.4.3, and 6.4.4.

6.5 Relative Importance of Successive and Simultaneous Transfer and Nonorthogonality Corrections (Semiclassical)

In what follows we discuss the relative importance of successive and simultaneous two-neutron transfer and of nonorthogonality corrections associated with the reaction

---

[21] Broglia and Riedel (1967).

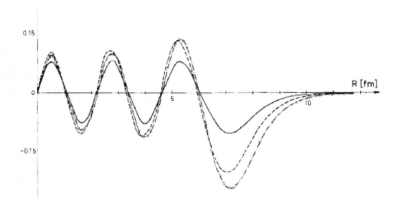

Figure 6.4.3 Modified formfactor for the transition to the ground state (^{206}Pb(t,p)^{208}Pb(gs); see Fig. 4.5.4) calculated in different spectroscopic models (pure shell-model configuration (solid), shell model plus pairing residual interaction (dashed), including ground state correlations (dashed with circles). After Broglia and Riedel (1967).

$$\alpha \equiv a(= b + 2) + A \rightarrow b + B(= A + 2) \equiv \beta, \qquad (6.5.1)$$

in the limits of independent particles and of strongly correlated Cooper pairs, making use of the semiclassical approximation[22], in which case the two-particle transfer differential cross section can be written as

$$\frac{d\sigma_{\alpha\rightarrow\beta}}{d\Omega} = P_{\alpha\rightarrow\beta}(t = +\infty) \sqrt{\left(\frac{d\sigma_\alpha}{d\Omega}\right)_{el}} \sqrt{\left(\frac{d\sigma_\beta}{d\Omega}\right)_{el}}, \qquad (6.5.2)$$

where P is the absolute value squared of a quantum mechanical transition amplitude. It gives the probability that the system at $t = +\infty$ is found in the final channel. The quantities $(d\sigma/d\Omega)_{el}$ are the classical elastic cross sections in the center of mass system, calculated in terms of the deflection function, namely, the functional relating the impact parameter and the scattering angle.

The transfer amplitude can be written as

$$a(t = +\infty) = a^{(1)}(\infty) - a^{(NO)}(\infty) + \tilde{a}^{(2)}(\infty), \qquad (6.5.3)$$

where $\tilde{a}^{(2)}(\infty)$ at $t = +\infty$ labels the successive transfer amplitude expressed in the post-prior representation (see below). The simultaneous transfer amplitude is given by (see Fig. 6.5.1 (I))

---

[22] For details, see Broglia and Winther (2004).

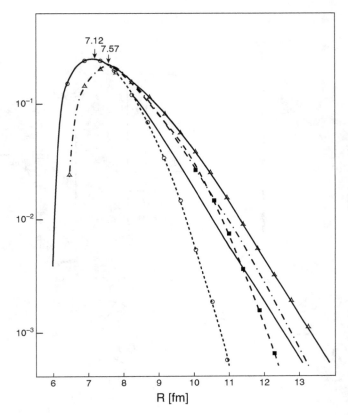

Figure 6.4.4 Asymptotic behavior of the modified formfactor for the ^{206}(t,p) ^{208}Pb(gs) ground state transition for oscillator plus Hankel wave functions (continuous solid curve), oscillator wave functions alone (dashed-point-dashed curve), and Woods–Saxon wave functions with a variety of asymptotic matchings. After Broglia and Riedel (1967).

$$a^{(1)}(\infty) = \frac{1}{i\hbar} \int_{-\infty}^{\infty} dt \, (\psi^b \psi^B, (V_{bB} - < V_{bB} >) \psi^a \psi^A) \times \exp[\frac{i}{\hbar}(E^{bB} - E^{aA})t]$$

$$\approx \frac{2}{i\hbar} \int_{-\infty}^{\infty} dt \, \left(\phi^{B(A)}(S_{(2n)}^B; \mathbf{r}_{1A}, \mathbf{r}_{2A}), U(r_{1b}) e^{i(\sigma_1 + \sigma_2)} \phi^{a(b)}(S_{(2n)}^a; \mathbf{r}_{1b}, \mathbf{r}_{2b}) \right)$$

$$\times \exp[\frac{i}{\hbar}(E^{bB} - E^{aA})t + \gamma(t)] \tag{6.5.4}$$

where[23],

$$\sigma_1 + \sigma_2 = \frac{1}{\hbar} \frac{m_n}{m_A} (m_{aA} \mathbf{v}_{aA}(t) + m_{bB} \mathbf{v}_{bB}(t)) \cdot (\mathbf{r}_{1\alpha} + \mathbf{r}_{2\alpha}), \tag{6.5.5}$$

[23] See Broglia (1975), Eq. (W12), p. 361.

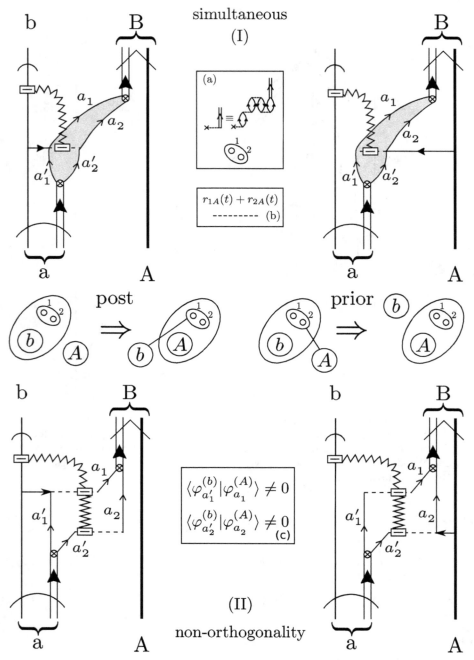

Figure 6.5.1 Graphical representation of simultaneous (I) and nonorthogonality (II) transfer processes. For details, see text and the caption to Fig. 6.5.2. In the present figure, as well as in Fig. 6.5.2, and at variance with, for example, Fig. 2.7.2, the particle-pair vibration coupling vertex is represented by a crossed circle. Concerning the jagged line representing recoil, see App. 2.A. After Potel et al. (2013a).

the "α-point" being defined by the relation[24]

$$\mathbf{r}_{\alpha A} = \frac{m_B}{m_a + m_B} \mathbf{r}_{bA}, \tag{6.5.6}$$

and, as a result,

$$\mathbf{r}_{iA} = \mathbf{r}_{i\alpha} + \frac{m_B}{m_a + m_B} \mathbf{r}_{bA}, \quad (i = 1, 2). \tag{6.5.7}$$

The $\exp(i(\sigma_1 + \sigma_2))$ takes care of recoil effects (Galilean transformation associated with the mismatch between entrance and exit channels). The phase $\gamma(t)$ is related with the effective $Q-$value of the reaction. In the above expression, ϕ indicates an antisymmetrized, correlated two-particle (Cooper pair) wavefunction, $S(2n)$ being the two-neutron separation energy (see Fig. 6.5.3), $U(r_{1b})$ being the single particle potential generated by nucleus b. The contribution arising from nonorthogonality effects can be written as (see Fig. 6.5.1 (II))

$$a^{(NO)}(\infty) = \frac{1}{i\hbar} \sum_{f,F} \int_{-\infty}^{\infty} dt \, (\psi^b \psi^B, (V_{bB} - \langle V_{bB} \rangle) \psi^f \psi^F)(\psi^f \psi^F, \psi^a \psi^A)$$

$$\times \exp[\frac{i}{\hbar}(E^{bB} - E^{aA})t]$$

$$\approx \frac{2}{i\hbar} \sum_{f,F} \int_{-\infty}^{\infty} dt \, \phi^{B(F)}(S_{(n)}^B, \mathbf{r}_{1A}), U(r_{1b})e^{i\sigma_1}(\phi^{f(b)}(S^f(n), \mathbf{r}_{1b})$$

$$\times \phi^{F(A)}(S^F(n), \mathbf{r}_{2A})e^{i\sigma_2}\phi^{a(f)}(S^a(n), \mathbf{r}_{2b}))\exp[\frac{i}{\hbar}(E^{bB} - E^{aA})t + \gamma(t)], \tag{6.5.8}$$

the reaction channel $f = (b+1) + F(= A + 1)$ having been introduced, the quantity $S(n)$ being the one-neutron separation energy (see Fig. 6.5.3). The summation over $f (\equiv a_1', a_2')$ and $F (\equiv a_1, a_2)$ involves a restricted number of states, namely, the valence shells in nuclei B and a. It is of note that the NFT$_{(s+r)}$ diagrams appearing in Fig. 6.5.1 (II) are the only ones we have encountered in which to a particle-recoil coupling vertex (dashed open rectangle) and thus the starting or/and ending of a recoil mode (jagged line) does not correspond the action of the nuclear interaction or mean field (horizontal short arrow) inducing a transfer process.

---

[24] See Broglia and Winther (1972), Eq. (2.26); see also Götz et al. (1975), Eq. (5.2 c).

successive
(III)

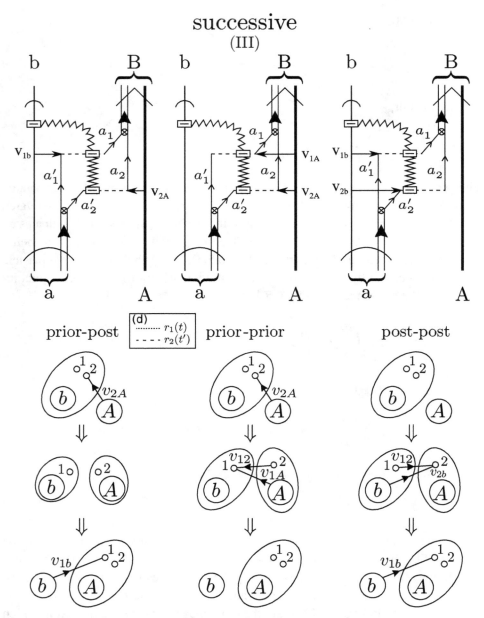

Figure 6.5.2 Graphical representation of the successive transfer of two nucleons. For details, see text. Because of energy conservation, the different choices concerning the interactions inducing transfer lead to identical results. After Potel et al. (2013a).

The successive transfer amplitude $\tilde{a}_\infty^{(2)}$ written making use of the post-prior representation is equal to (see Fig. 6.5.2)

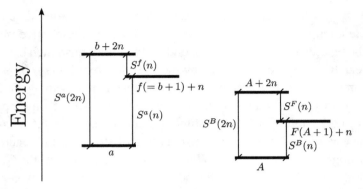

Figure 6.5.3 One- and two-neutron separation energies $S(n)$ and $S(2n)$ associated with the channels $\alpha \equiv a(= b + 2) + A \rightarrow \gamma \equiv f(= b + 1) + F(= A + 1) \rightarrow \beta \equiv b + B(= A + 2)$. After Potel et al. (2013a).

$$\tilde{a}^{(2)}(\infty) = \frac{1}{i\hbar} \sum_{f,F} \int_{-\infty}^{\infty} dt\,(\psi^b \psi^B, (V_{bB} - <V_{bB}>)e^{i\sigma_1} \psi^f \psi^F)$$

$$\times \exp[\frac{i}{\hbar}(E^{bB} - E^{fF})t + \gamma_1(t)]$$

$$\times \frac{1}{i\hbar} \int_{-\infty}^{t} dt'(\psi^f \psi^F, (V_{fF} - <V_{fF}>)e^{i\sigma_2} \psi^a \psi^A >$$

$$\times \exp[\frac{i}{\hbar}(E^{fF} - E^{aA})t' + \gamma_2(t')]. \tag{6.5.9}$$

To gain insight into the relative importance of the three terms contributing to Eq. (6.5.3) we discuss two situations, namely, the independent-particle model and the strong-correlation limits.

Before doing so, let us comment on the graphical description of the transfer amplitudes (6.5.4), (6.5.8), and (6.5.9) displayed in Figs. 6.5.1 and 6.5.2. The time arrow is assumed to point upwards: (**I**) Simultaneous transfer, in which one particle is transferred by the nucleon–nucleon interaction acting either in the entrance $\alpha \equiv a + A$ channel (prior) or in the final $\beta \equiv b + B$ channel (post), while the other particle follows suit making use of the particle-particle correlation (gray area) which binds the Cooper pair (see upper inset labeled (a)), represented by a solid arrow on a double line, to the projectile (curved arrowed lines) or to the target (opened arrowed lines). The above argument provides the explanation why when, for example, v_{1b} acts on one nucleon, the other nucleon also reacts instantaneously. In fact a Cooper pair displays generalized rigidity (emergent property in gauge space). A crossed open circle represents the particle-pair vibration coupling. The associated strength, together with an energy denominator, determines the amplitude $X_{a'_1 a'_2}$ with which the pair mode (Cooper pair) is in the (time reversed)[25] two particle configuration $a'_1 a'_2$. In the transfer process, the relative motion orbit changes, the readjustment of the corresponding trajectory mismatch being operated by a Galilean transformation induced by the operator $(\exp\{\mathbf{k} \cdot (\mathbf{r}_{1A}(t) + \mathbf{r}_{2A}(t))\})$. This phenomenon, known as recoil process, is represented by a jagged line which provides information on the two transferred nucleons (single time appearing as argument of both single-particle coordinates r_1 and r_2; see Fig. 6.5.1 inset labeled (b)). In other words, information on the coupling of structure and reaction modes. (**II**) Nonorthogonality contribution. While one of the nucleons of the Cooper pairs is transferred under the action of v, the other goes, uncorrelatedly over, profiting of the nonorthogonality of the associated single-particle wavefunctions (see inset (c)). (**III**) Successive transfer. In this case, there are two time dependences associated with the acting of the nucleon–nucleon interaction twice (see Fig. 6.5.2 inset (d)).

6.5.1 Independent Particle Limit

In the independent particle limit, the two transferred particles do not interact among themselves but for antisymmetrization. Thus, the separation energies fulfill the relations (see Fig. 6.5.3)

$$S^B(2n) = 2S^B(n) = 2S^F(n), \tag{6.5.10}$$

and

$$S^a(2n) = 2S^a(n) = 2S^f(n). \tag{6.5.11}$$

In this case

$$\phi^{B(A)}(S^B(2n), \mathbf{r}_{1A}, \mathbf{r}_{2A}) = \sum_{a_1 a_2} \phi_{a_1}^{B(F)}(S^B(n), \mathbf{r}_{1A})\phi_{a_2}^{F(A)}(S^F(n), \mathbf{r}_{2a}), \tag{6.5.12}$$

and

$$\phi^{a(b)}(S^a(2n), \mathbf{r}_{1b}, \mathbf{r}_{2b}) = \sum_{a'_1 a'_2} \phi_{a'_1}^{a(f)}(S^a(n), \mathbf{r}_{2b})\phi_{a'_2}^{f(b)}(S^f(n), \mathbf{r}_{1b}), \tag{6.5.13}$$

where $(a_1, a_2) \equiv F$ and $(a'_1, a'_2) \equiv f$ span, as mentioned above, shells in nuclei B and a respectively.

Inserting Eqs. (6.5.10–6.5.13) in Eq. (6.5.4) one can show that

$$a^{(1)}(\infty) = a^{(NO)}(\infty). \tag{6.5.14}$$

---

[25] Generalized to include also different number of nodes.

It can be further demonstrated that within the present approximation, $Im \, \tilde{a}^{(2)} = 0$, and that

$$
\begin{aligned}
\tilde{a}^{(2)}(\infty) &= \frac{1}{i\hbar} \sum_{f,F} \int_{-\infty}^{\infty} dt \, (\psi^b \psi^B, (V_{bB} - < V_{bB} >) e^{i\sigma_1} \psi^f \psi^F > \\
&\quad \times \exp[\frac{i}{\hbar}(E^{bB} - E^{fF})t + \gamma_1(t)] \\
&\quad \times \frac{1}{i\hbar} \int_{-\infty}^{\infty} dt' (\psi^f \psi^F, (V_{fF} - < V_{fF} >) e^{i\sigma_2} \psi^a \psi^A) \\
&\quad \times \exp[\frac{i}{\hbar}(E^{fF} - E^{aA})t' + \gamma_2(t)].
\end{aligned}
\tag{6.5.15}
$$

The total absolute differential cross section (6.5.2), where $P = |a(\infty)|^2 = |\tilde{a}^{(2)}|^2$, is then equal to the product of two one-particle transfer cross sections (see Fig. 5.1.1), associated with the (virtual) reaction channels

$$
\alpha \equiv a + A \rightarrow f + F \equiv \gamma,
\tag{6.5.16}
$$

and

$$
\gamma \equiv f + F \rightarrow b + B \equiv \beta.
\tag{6.5.17}
$$

In fact, Eq.(6.5.15) involves no time ordering and consequently the two processes above are completely independent of each other. This result was expected because being $v_{12} = 0$, the transfer of one nucleon cannot influence, aside form selecting the initial state for the second step, the behavior of the other nucleon.

6.5.2 Strong Correlation (Cluster) Limit

The second limit to be considered is the one in which the correlation between the two nucleons is so strong that (see Fig. 6.5.3)

$$
S^a(2n) \approx S^a(n) \gg S^f(n),
\tag{6.5.18}
$$

and

$$
S^B(2n) \approx S^B(n) \gg S^F(n).
\tag{6.5.19}
$$

That is, the magnitude of the one-nucleon separation energy is strongly modified by the pair breaking.

There is a different, although equivalent way to express (6.5.3) which is more convenient to discuss the strong coupling limit. In fact, making use of the post-prior representation one can write

$$a^{(2)}(t) = \tilde{a}^{(2)}(t) - a^{(NO)}(t)$$

$$= \frac{1}{i\hbar} \sum_{f,F} \int_{-\infty}^{\infty} dt (\psi^b \psi^B, (V_{bB} - <V_{bB}>)e^{i\sigma_1} \psi^f \psi^F)$$

$$\times \exp[\frac{i}{\hbar}(E^{bB} - E^{fF})t + \gamma_1(t)]$$

$$\times \frac{1}{i\hbar} \int_{-\infty}^{t} dt' (\psi^f \psi^F, (V_{aA} - <V_{aA}>)\psi^a \psi^A)$$

$$\times \exp[\frac{i}{\hbar}(E^{fF} - E^{aA})t' + \gamma_2(t')]. \tag{6.5.20}$$

The relations (6.5.18), (6.5.19) imply

$$E^{fF} - E^{aA} = S^a(n) - S^F(n) >> \frac{\hbar}{\tau}, \tag{6.5.21}$$

where τ is the collision time. Consequently the real part of $a^{(2)}(\infty)$ vanishes exponentially with the $Q-$value of the intermediate transition, while the imaginary part vanishes inversely proportional to this energy. One can thus write,

$$Re\ a^{(2)}(\infty) \approx 0, \tag{6.5.22}$$

and

$$a^{(2)}(\infty) \approx \frac{1}{i\hbar} \frac{\tau}{<E^{fF}> - E^{bB}} \sum_{fF} (\psi^b \psi^B, (V_{bB} - <V_{bB}>)\psi^f \psi^F)_{t=0}$$

$$\times (\psi^f \psi^F, (V_{aA} - <V_{aA}>)\psi^a \psi^A)_{t=0}, \tag{6.5.23}$$

where one has utilized the fact that $E^{bB} \approx E^{aA}$. For $v_{12} \to \infty$, $(<E^{fF}> - E^{bB}) \to \infty$) and, consequently,

$$\lim_{v_{12} \to \infty} a^{(2)}(\infty) = 0. \tag{6.5.24}$$

Thus the total two-nucleon transfer amplitude is equal, in the strong coupling limit, to the amplitude $a^{(1)}(\infty)$.

Summing up, only when successive transfer and non-orthogonal corrections are included in the description of the two-nucleon transfer process, does one obtain a consistent description of the process, which correctly converges to the weak and strong correlation limiting values.

Parallel with Josephson and Giaever Tunneling

As already stated (Sect. 4.5), actual nuclei are, as a rule, closer to the independent particle limit than to the cluster one. Thus, it is not surprising that successive constitutes the major contribution to two-nucleon transfer processes. Nonetheless, the slight deviation from independent particle motion arising from pairing correlations, entangles the fermions (nucleons) partners of a Cooper pair, in a subtle fashion.

This can be seen by expressing the pair transfer operator in terms of quasiparticles. Making use of the relation

$$a_\nu^\dagger = U_\nu \alpha_\nu^\dagger + V_\nu \alpha_{\tilde{\nu}}, \tag{6.5.25}$$

one obtains

$$P_\nu^\dagger = a_\nu^\dagger a_{\tilde{\nu}}^\dagger = U_\nu^2 \alpha_\nu^\dagger \alpha_{\tilde{\nu}}^\dagger - U_\nu V_\nu (\alpha_\nu^\dagger \alpha_\nu + \alpha_{\tilde{\nu}}^\dagger \alpha_{\tilde{\nu}}) - V_\nu^2 \alpha_{\tilde{\nu}} \alpha_\nu + U_\nu V_\nu. \tag{6.5.26}$$

Because one-quasiparticle transfer amplitude is $\sim U_\nu$ and the associated cross section is $\sim U_\nu^2$ (i.e., $\sim T^2$, T being the associated transition amplitude), it is natural that the two-quasiparticle transfer amplitude is $\sim U_\nu^2$ (within this context see Fig. 4.6.1, S-Q current) and that the associated cross section is $\sim U_\nu^4$ (i.e., $\sim T^4$). As it is apparent from (6.5.26), there exists a different possibility to transfer a pair of nucleons, leaving this time the quasiparticle distribution unchanged,[26] namely, through the last term of (6.5.26). Thus, a transfer amplitude $\sim U_\nu V_\nu$ (Fig. 4.6.1, S-S current) and associated two-nucleon transfer cross section $\sim (U_\nu V_\nu)^2$ ($\sim T^2$). A result which one finds at the basis of (4.3.1).

6.A Spherical Harmonics and Angular Momenta

With Condon–Shortley phases

$$Y_m^l(\hat{z}) = \delta_{m,0} \sqrt{\frac{2l+1}{4\pi}}, \qquad Y_m^{l*}(\hat{r}) = (-1)^m Y_{-m}^l(\hat{r}). \tag{6.A.1}$$

Time–reversed phases consist in multiplying Condon–Shortley phases with a factor i^l, so

$$Y_m^l(\hat{z}) = \delta_{m,0} i^l \sqrt{\frac{2l+1}{4\pi}}, \qquad Y_m^{l*}(\hat{r}) = (-1)^{l-m} Y_{-m}^l(\hat{r}). \tag{6.A.2}$$

With this phase convention, the relation with the associated Legendre polynomials includes an extra i^l factor with respect to the Condon–Shortley phase,

$$Y_m^l(\theta, \phi) = i^l \sqrt{\frac{2l+1}{4\pi} \frac{(l-m)!}{(l+m)!}} P_l^m(\cos\theta) e^{im\phi}. \tag{6.A.3}$$

---

[26] This was the main message of Josephson (1962) and was the possibility which was missed by Giaever (1973), who interpreted supercurrents with zero voltage drop through the oxide barrier separating two superconductors as a metallic short. As Giaever states, "to make an experimental discovery is not enough to observe something, one must also realize the significance of the observation." It is also not found in Cohen et al. (1962), who developed the many-body tunneling Hamiltonian (eventually extended by Josephson) and predicted four contributions to the current which correspond to normal (carriers of charge e) current of type S-Q, that is, implying quasiparticle excitation.

6.A.1 Addition Theorem

The addition theorem for the spherical harmonics states that

$$P_l(\cos\theta_{12}) = \frac{4\pi}{2l+1} \sum_m Y_m^l(\mathbf{r}_1) Y_m^{l*}(\mathbf{r}_2), \qquad (6.A.4)$$

where θ_{12} is the angle between the vectors $\mathbf{r}_1$ and $\mathbf{r}_2$. This result is independent of the phase convention. With *time–reversed phases*,

$$P_l(\cos\theta_{12}) = \frac{4\pi}{\sqrt{2l+1}} \left[Y^l(\hat{\mathbf{r}}_1) Y^l(\hat{\mathbf{r}}_2) \right]_0^0. \qquad (6.A.5)$$

With *Condon–Shortley phases*,

$$P_l(\cos\theta_{12}) = (-1)^l \frac{4\pi}{\sqrt{2l+1}} \left[Y^l(\hat{\mathbf{r}}_1) Y^l(\hat{\mathbf{r}}_2) \right]_0^0. \qquad (6.A.6)$$

6.A.2 Expansion of the Delta Function

The Dirac delta function can be expanded in multipoles, yielding

$$\delta(\mathbf{r}_2 - \mathbf{r}_1) = \sum_l \delta(r_1 - r_2) \frac{2l+1}{4\pi r_1^2} P_l(\cos\theta_{12})$$

$$= \sum_l \delta(r_1 - r_2) \frac{1}{r_1^2} \sum_m Y_m^l(\mathbf{r}_1) Y_m^{l*}(\mathbf{r}_2). \qquad (6.A.7)$$

This result is independent of the phase convention. With *time-reversed phases*,

$$\delta(\mathbf{r}_2 - \mathbf{r}_1) = \sum_l \delta(r_1 - r_2) \frac{\sqrt{2l+1}}{r_1^2} \left[Y^l(\hat{\mathbf{r}}_1) Y^l(\hat{\mathbf{r}}_2) \right]_0^0. \qquad (6.A.8)$$

6.A.3 Coupling and Complex Conjugation

If $\Psi_{M_1}^{I_1*} = (-1)^{I_1-M_1} \Psi_{-M_1}^{I_1}$ and $\Phi_{M_2}^{I_2*} = (-1)^{I_2-M_2} \Phi_{-M_2}^{I_2}$, as it happens to be the case for spherical harmonics with time-reversed phases, then

$$\left[\Psi^{I_1} \Phi^{I_2} \right]_M^{I*} = \sum_{\substack{M_1 M_2 \\ (M_1+M_2=M)}} \langle I_1\, I_2\, M_1\, M_2 | I M \rangle \Psi_{M_1}^{I_1*} \Phi_{M_2}^{I_2*}$$

$$= \sum_{\substack{M_1 M_2 \\ (M_1+M_2=M)}} (-1)^{I-M_1-M_2} \langle I_1\, I_2\, -M_1\, -M_2 | I -M \rangle \Psi_{-M_1}^{I_1} \Phi_{-M_2}^{I_2}$$

$$= (-1)^{I-M} \sum_{\substack{M_1 M_2 \\ (M_1+M_2=M)}} \langle I_1 \ I_2 \ -M_1 \ -M_2 | I - M \rangle \Psi^{I_1}_{-M_1} \Phi^{I_2}_{-M_2}$$

$$= (-1)^{I-M} \left[\Psi^{I_1} \Phi^{I_2} \right]^I_{-M} , \tag{6.A.9}$$

where we have used (6.A.23).

Let us care now about the spinor functions $\chi^{1/2}_m(\sigma)$, which have the form

$$\chi^{1/2}(\sigma = 1/2) = \begin{bmatrix} 1 \\ 0 \end{bmatrix} \quad \chi^{1/2}(\sigma = -1/2) = \begin{bmatrix} 0 \\ 1 \end{bmatrix}, \tag{6.A.10}$$

or

$$\chi^{1/2}_m(\sigma) = \delta_{m,\sigma}. \tag{6.A.11}$$

Thus, $\chi^{1/2*}_m(\sigma) = \chi^{1/2}_m(\sigma) = \delta_{m,\sigma}$, but we can also write

$$\chi^{1/2*}_m(\sigma) = (-1)^{1/2-m+1/2-\sigma} \chi^{1/2}_{-m}(-\sigma). \tag{6.A.12}$$

This trick enable us to write

$$\left[Y^l(\hat{r}) \chi^{1/2}(\sigma) \right]^{J*}_M = (-1)^{1/2-\sigma+J-M} \left[Y^l(\hat{r}) \chi^{1/2}(-\sigma) \right]^J_{-M}, \tag{6.A.13}$$

which can be derived in a similar way as (6.A.9).

6.A.4 Angular Momenta Coupling

Relation between Clebsh–Gordan and $3j$ coefficients:

$$\langle j_1 \ j_2 \ m_1 \ m_2 | JM \rangle = (-1)^{j_1-j_2+M} \sqrt{2J+1} \begin{pmatrix} j_1 & j_2 & J \\ m_1 & m_2 & -M \end{pmatrix}. \tag{6.A.14}$$

Relation between Wigner and $9j$ coefficients:

$$\left((j_1 j_2)_{j_{12}} (j_3 j_4)_{j_{34}} | (j_1 j_3)_{j_{13}} (j_2 j_4)_{j_{24}} \right)_j$$

$$= \sqrt{(2j_{12}+1)(2j_{13}+1)(2j_{24}+1)(2j_{34}+1)} \begin{Bmatrix} j_1 & j_2 & j_{12} \\ j_3 & j_4 & j_{34} \\ j_{13} & j_{24} & j \end{Bmatrix}. \tag{6.A.15}$$

6.A.5 Integrals

Let us now prove

$$\int d\Omega \left[Y^l(\hat{r}) Y^l(\hat{r}) \right]^I_M = \delta_{M,0} \delta_{I,0} \sqrt{2l+1}. \tag{6.A.16}$$

$$\int d\Omega \left[Y^l(\hat{r}) Y^l(\hat{r}) \right]_M^I = \sum_{\substack{m_1, m_2 \\ (m_1 + m_2 = M)}} \langle l\ l\ m_1\ m_2 | I M \rangle \int d\Omega Y_{m_1}^l(\hat{r}) Y_{m_2}^l(\hat{r})$$

$$= \sum_{\substack{m_1, m_2 \\ (m_1 + m_2 = M)}} (-1)^{l+m_1} \langle l\ l - m_1\ m_2 | I M \rangle \int d\Omega Y_{m_1}^{l*}(\hat{r}) Y_{m_2}^l(\hat{r})$$

$$= \delta_{M,0} \sum_m (-1)^{l+m} \langle l\ l\ -m\ m | I 0 \rangle$$

$$= \delta_{M,0} \sqrt{2l+1} \sum_m \langle l\ l\ -m\ m | I 0 \rangle \langle l\ l\ -m\ m | 0 0 \rangle$$

$$= \delta_{M,0} \delta_{I,0} \sqrt{2l+1},$$

(6.A.17)

where we have used

$$\langle l\ l\ -m\ m | 0\ 0 \rangle = \frac{(-1)^{l+m}}{\sqrt{2l+1}}$$

(6.A.18)

Let us now prove

$$\sum_\sigma \int d\Omega (-1)^{1/2-\sigma} \left[\Psi^j(\hat{r}, -\sigma) \Psi^j(\hat{r}, \sigma) \right]_M^I = -\delta_{M,0} \delta_{I,0} \sqrt{2j+1}. \quad (6.A.19)$$

$$\sum_\sigma \int d\Omega (-1)^{1/2-\sigma} \left[\Psi^j(\hat{r}, -\sigma) \Psi^j(\hat{r}, \sigma) \right]_M^I$$

$$= \sum_{\substack{m_1, m_2 \\ (m_1 + m_2 = M)}} \langle j\ j\ m_1\ m_2 | I M \rangle \sum_\sigma \int d\Omega \Psi_{m_1}^j(\hat{r}, -\sigma) \Psi_{m_2}^j(\hat{r}, \sigma)$$

$$= \sum_{\substack{m_1, m_2 \\ (m_1 + m_2 = M)}} \langle j\ j\ m_1\ m_2 | I M \rangle \sum_\sigma (-1)^{j+m_1} \int d\Omega \Psi_{-m_1}^{j*}(\hat{r}, \sigma) \Psi_{m_2}^j(\hat{r}, \sigma)$$

$$= \sum_{\substack{m_1, m_2 \\ (m_1 + m_2 = M)}} \langle j\ j\ m_1\ m_2 | I M \rangle (-1)^{j+m_1} \delta_{-m_1, m_2}$$

$$= \delta_{M,0} \sum_m (-1)^{j+m} \langle j\ j\ m\ -m | I 0 \rangle$$

$$= -\delta_{M,0} \sqrt{2j+1} \sum_m (-1)^{j+m} \langle j\ j\ m\ -m | I 0 \rangle \langle j\ j\ m\ -m | 0 0 \rangle$$

$$= -\delta_{M,0} \delta_{I,0} \sqrt{2j+1}.$$

(6.A.20)

6.A.6 Symmetry Properties

Note also another useful property

$$\left[\Psi^{I_1}\Psi^{I_2}\right]_M^I = (-1)^{I_1+I_2-I}\left[\Psi^{I_2}\Psi^{I_1}\right]_M^I, \tag{6.A.21}$$

by virtue of the symmetry property of the Clebsh–Gordan coefficients

$$\langle I_1\ I_2\ m_1\ m_2|I\ M\rangle = (-1)^{I_1+I_2-I}\langle I_2\ I_1\ m_2\ m_1|I\ M\rangle. \tag{6.A.22}$$

Here's another symmetry property of the Clebsh–Gordan coefficients

$$\langle I_1\ I_2\ m_1\ m_2|I\ M\rangle = (-1)^{I_1+I_2-I}\langle I_1\ I_2\ -m_2\ -m_1|I\ -M\rangle. \tag{6.A.23}$$

Another one, which can be derived from the simpler properties of $3j$-symbols

$$\langle I_1\ I_2\ m_1\ m_2|I\ M\rangle = (-1)^{I_1-m_1}\sqrt{\frac{2I+1}{2I_2+1}}\langle I_1\ I\ m_1\ -M|I_2m_2\rangle. \tag{6.A.24}$$

Let us use this last property to calculate sums of the type

$$\sum_{m_1,m_3}|\langle I_1\ I_2\ m_1\ m_2|I_3m_3\rangle|^2. \tag{6.A.25}$$

Using (6.A.24), we have

$$\sum_{m_1,m_3}|\langle I_1\ I_2\ m_1\ m_2|I_3m_3\rangle|^2 =$$

$$\frac{2I_3+1}{2I_2+1}\sum_{m_1,m_3}|\langle I_1\ I_3\ m_1\ -m_3|I_2m_2\rangle|^2 = \frac{2I_3+1}{2I_2+1}, \tag{6.A.26}$$

since

$$\sum_{m_1,m_3}|\langle I_1\ I_3\ m_1\ -m_3|I_2m_2\rangle|^2 = \sum_{m_1,m_3}|\langle I_1\ I_3\ m_1\ m_3|I_2m_2\rangle|^2 = 1. \tag{6.A.27}$$

6.B Distorted Waves

Let us have a closer look at the partial wave expansion of the distorted waves

$$\chi^{(+)}(\mathbf{k},\mathbf{r}) = \sum_l \frac{4\pi}{kr}i^l e^{i\sigma^l}F_l\sum_m Y_m^l(\hat{r})Y_m^{l*}(\hat{k}). \tag{6.B.1}$$

Of note is the very important fact that *this definition is independent of the phase convention*, since the l-dependent phase is multiplied by its complex conjugate.

$$\chi^{(-)}(\mathbf{k},\mathbf{r}) = \chi^{(+)*}(-\mathbf{k},\mathbf{r}) = \sum_l \frac{4\pi}{kr}i^{-l}e^{-i\sigma^l}F_l^*\sum_m Y_m^{l*}(\hat{r})Y_m^l(-\hat{k}). \tag{6.B.2}$$

As $Y_m^l(-\hat{k}) = (-1)^l Y_m^l(\hat{k})$, we have

$$\chi^{(-)}(\mathbf{k}, \mathbf{r}) = \sum_l \frac{4\pi}{kr} i^l e^{-i\sigma^l} F_l^* \sum_m Y_m^{l*}(\hat{r}) Y_m^l(\hat{k}), \qquad (6.B.3)$$

which is also independent of the phase convention. With time-reversed phase convention

$$\chi^{(+)}(\mathbf{k}, \mathbf{r}) = \sum_l \frac{4\pi}{kr} i^l \sqrt{2l+1} e^{i\sigma^l} F_l \left[Y^l(\hat{r}) Y^l(\hat{k}) \right]_0^0, \qquad (6.B.4)$$

while with Condon–Shortley phase convention we get an extra $(-1)^l$ factor:

$$\chi^{(+)}(\mathbf{k}, \mathbf{r}) = \sum_l \frac{4\pi}{kr} i^{-l} \sqrt{2l+1} e^{i\sigma^l} F_l \left[Y^l(\hat{r}) Y^l(\hat{k}) \right]_0^0. \qquad (6.B.5)$$

The partial-wave expansion of the Green function $G(\mathbf{r}, \mathbf{r}')$ is

$$G(\mathbf{r}, \mathbf{r}') = i \sum_l \frac{f_l(k, r_<) P_l(k, r_>)}{krr'} \sum_m Y_m^l(\hat{r}) Y_m^{l*}(\hat{r}'), \qquad (6.B.6)$$

where $f_l(k, r_<)$ and $P_l(k, r_>)$ are the regular and the irregular solutions of the homogeneous problem respectively. With *time-reversed* phase convention

$$G(\mathbf{r}, \mathbf{r}') = i \sum_l \sqrt{2l+1} \frac{f_l(k, r_<) P_l(k, r_>)}{krr'} \left[Y^l(\hat{r}) Y^l(\hat{r}') \right]_0^0. \qquad (6.B.7)$$

6.C Hole States and Time Reversal

Let us consider the state $|(jm)^{-1}\rangle$ obtained by removing a ψ_{jm} single-particle state from a $J = 0$ closed shell $|0\rangle$. The antisymmetrized product state

$$\sum_m \mathcal{A}\{\psi_{jm} | (jm)^{-1}\rangle\} \propto |0\rangle \qquad (6.C.1)$$

is clearly proportional to $|0\rangle$. This gives us the transformation rules of $|(jm)^{-1}\rangle$ under rotations, which must be such that, when multiplied by a j, m spherical tensor and summed over m, yields a $j = 0$ tensor. It can be seen that these properties imply that $|(jm)^{-1}\rangle$ transforms like $(-1)^{j-m} T_{j-m}$, T_{j-m} being a spherical tensor. It also follows that the hole state $|(j\bar{m})^{-1}\rangle$ transforms like a j, m spherical tensor if $\psi_{j\bar{m}}$ is defined as the $\mathcal{R}$–conjugate to ψ_{jm} by the relation

$$\psi_{j\bar{m}} \equiv (-1)^{j+m} \psi_{j-m}. \qquad (6.C.2)$$

In other words, with the latter definition a *hole state* transforms under rotations with the right phase. We will now show that $\mathcal{R}$-conjugation is equivalent to a rotation of spin and spatial coordinates through an angle $-\pi$ about the y-axis:

$$e^{i\pi J_y}\psi_{jm} = (-1)^{j+m}\psi_{j-m} \equiv \psi_{j\bar{m}}. \tag{6.C.3}$$

Let us begin by calculating $e^{i\pi L_y}Y_l^m$. The rotation matrix about the y-axis is

$$R_y(\theta) = \begin{pmatrix} \cos(\theta) & 0 & \sin(\theta) \\ 0 & 1 & 0 \\ -\sin(\theta) & 0 & \cos(\theta) \end{pmatrix}, \tag{6.C.4}$$

so for $R_y(-\pi)$ we get

$$R_y(-\pi) = \begin{pmatrix} -1 & 0 & 0 \\ 0 & 1 & 0 \\ 0 & 0 & -1 \end{pmatrix}. \tag{6.C.5}$$

When applied to the generic direction $(\sin(\theta)\cos(\phi), \sin(\theta)\sin(\phi), \cos(\theta))$, we obtain $(-\sin(\theta)\cos(\phi), \sin(\theta)\sin(\phi), -\cos(\theta))$, which corresponds to making the substitutions

$$\theta \to \pi - \theta, \quad \phi \to \pi - \phi. \tag{6.C.6}$$

When we substitute these angular transformations in the spherical harmonic $Y_l^m(\theta, \phi)$, we obtain the rotated $Y_l^m(\theta, \phi)$:

$$e^{i\pi L_y}Y_l^m = (-1)^{l+m}Y_l^{-m}. \tag{6.C.7}$$

Let us now turn our attention to the spin coordinates rotation $e^{i\pi S_y}\chi_m$. The rotation matrix in spin space is

$$\begin{pmatrix} \cos(\theta/2) & -\sin(\theta/2) \\ \sin(\theta/2) & \cos(\theta/2) \end{pmatrix}, \tag{6.C.8}$$

which, for $\theta = -\pi$ is

$$\begin{pmatrix} 0 & 1 \\ -1 & 0 \end{pmatrix}. \tag{6.C.9}$$

Applying it to the spinors, we find the rule

$$e^{i\pi S_y}\chi_m = (-1)^{1/2+m}\chi_{-m}, \tag{6.C.10}$$

so

$$e^{i\pi J_y}\psi_{jm} = \sum_{m_l m_s}\langle l\ m_l\ 1/2\ m_s | j\ m\rangle\ e^{i\pi L_y}Y_l^{m_l}\ e^{i\pi s_y}\chi_{m_s}$$

$$= \sum_{m_l m_s}(-1)^{1/2+m_s+l+m_l}\langle l\ m_l\ 1/2\ m_s|j\ m\rangle\ Y_l^{-m_l}\ \chi_{-m_s}$$

$$\qquad\qquad\qquad\qquad\qquad\qquad\qquad\qquad (6.C.11)$$

$$= \sum_{m_l m_s}(-1)^{1+m-j+2l}\langle l\ -m_l\ 1/2\ -m_s|j\ -m\rangle\ Y_l^{-m_l}\ \chi_{-m_s}$$

$$= (-1)^{m+j}\psi_{j-m} \equiv \psi_{j\bar{m}},$$

where we have used $(-1)^{1+m-j+2l} = -(-1)^{m-j} = (-1)^{m+j}$, as j, m are always half–integers and l is always an integer.

We now turn our attention to the time-reversal operation, which amounts to the transformations

$$\mathbf{r} \to \mathbf{r}, \quad \mathbf{p} \to -\mathbf{p}. \qquad\qquad (6.C.12)$$

This is enough to define the operator of time reversal of a spinless particle (see Messiah). In the position representation, in which $\mathbf{r}$ is real and $\mathbf{p}$ pure imaginary, this (unitary antilinear) operator is the complex conjugation operator.

As angular momentum $\mathbf{l} = \mathbf{r} \times \mathbf{p}$ changes sign under time reversal, so does spin:

$$\mathbf{s} \to -\mathbf{s}, \qquad\qquad (6.C.13)$$

which, along with (6.C.12), completes the set of rules that define the time-reversal operation on a particle with spin. In the representation of eigenstates of $\mathbf{s}^2$ and s_z, complex conjugation alone changes only the sign of s_y, so an additional rotation of $-\pi$ around the y-axis is necessary to change the sign of s_x, s_z and implement the transformation (6.C.13). If we call K the time-reversal operator, we have

$$K\psi_{jm} = e^{i\pi s_y}\psi_{jm}^*. \qquad\qquad (6.C.14)$$

This is completely general and independent of the phase convention. It only depends on the fact that we have used the $\mathbf{r}$ representation for the spatial wave function and the representation of the eigenstates of $\mathbf{s}^2$ and s_z for the spin part. *If we use time-reversal phases for the spherical harmonics* (see(6.A.2)),

$$Y_m^{l*} = (-1)^{l+m}Y_{-m}^l = e^{i\pi L_y}Y_m^l. \qquad\qquad (6.C.15)$$

So we can write

$$K\psi_{jm} = e^{i\pi J_y}\psi_{jm} = \psi_{j\bar{m}}. \qquad\qquad (6.C.16)$$

Note again that this last expression is valid only if we use time-reversal phases for the spherical harmonics. Only in this case, time reversal coincides with $\mathcal{R}$–conjugation and hole states.

In BCS theory, the quasiparticles are defined in terms of linear combinations of particles and holes. With time-reversal phases, holes are equivalent to time–reversed states, and we get the definitions

$$\begin{aligned}
\alpha_\nu^\dagger &= u_\nu a_\nu^\dagger - v_\nu a_{\bar\nu} & a_\nu^\dagger &= u_\nu \alpha_\nu^\dagger + v_\nu \alpha_{\bar\nu} \\
\alpha_{\bar\nu}^\dagger &= u_\nu a_{\bar\nu}^\dagger + v_\nu a_\nu & a_{\bar\nu}^\dagger &= u_\nu \alpha_{\bar\nu}^\dagger - v_\nu \alpha_\nu \\
\alpha_\nu &= u_\nu a_\nu - v_\nu a_{\bar\nu}^\dagger & a_\nu &= u_\nu \alpha_\nu + v_\nu \alpha_{\bar\nu}^\dagger \\
\alpha_{\bar\nu} &= u_\nu a_{\bar\nu} + v_\nu a_\nu^\dagger & a_{\bar\nu} &= u_\nu \alpha_{\bar\nu} - v_\nu \alpha_\nu^\dagger
\end{aligned}$$
(6.C.17)

6.D Spectroscopic Amplitudes in the BCS Approximation

The creation operator of a pair of fermions coupled to J, M can be expressed in second quantization as

$$T^\dagger(j_1, j_2, JM) = N \sum_m \langle j_1\, m\, j_2\, M-m | J\, M \rangle\, a_{j_1 m}^\dagger a_{j_2 M-m}^\dagger,$$
(6.D.1)

where N is a normalization constant. To determine it, we write the wave function resulting from the action of (6.D.1) on the vacuum

$$\Psi = T^\dagger(j_1, j_2, JM)|0\rangle = \frac{N}{\sqrt{2}} \sum_m \langle j_1\, m\, j_2\, M-m | J\, M \rangle$$
$$\times \left(\phi_{j_1 m}(\mathbf{r_1})\phi_{j_2 M-m}(\mathbf{r_2}) - \phi_{j_2 M-m}(\mathbf{r_1})\phi_{j_1, m}(\mathbf{r_2}) \right).$$
(6.D.2)

The norm is

$$|\Psi|^2 = \frac{N^2}{2} \sum_{mm'} \langle j_1\, m\, j_2\, M-m | J\, M \rangle \langle j_1\, m'\, j_2\, M-m' | J\, M \rangle$$
$$\times \left(\phi_{j_1 m}(\mathbf{r_1})\phi_{j_2 M-m}(\mathbf{r_2}) - \phi_{j_2 M-m}(\mathbf{r_1})\phi_{j_1, m}(\mathbf{r_2}) \right)$$
$$\times \left(\phi_{j_1 m'}(\mathbf{r_1})\phi_{j_2 M-m'}(\mathbf{r_2}) - \phi_{j_2 M-m'}(\mathbf{r_1})\phi_{j_1, m'}(\mathbf{r_2}) \right).$$
(6.D.3)

Integrating we get

$$1 = \frac{N^2}{2} \sum_{mm'} \langle j_1\, m\, j_2\, M-m | J\, M \rangle \langle j_1\, m'\, j_2\, M-m' | J\, M \rangle$$
$$\times \left(2\delta_{m,m'} - 2\delta_{j_1,j_2}\delta_{m,M-m'} \right)$$
$$= N^2 \left(\sum_m \langle j_1\, m\, j_2\, M-m | J\, M \rangle^2 \right.$$
(6.D.4)

$$\left. -\delta_{j_1,j_2} \sum_m \langle j_1\, m\, j_2\, M-m | J\, M \rangle \langle j_1\, M-m\, j_2\, m | J\, M \rangle \right)$$
$$= N^2 \left(1 - \delta_{j_1,j_2}(-1)^{2j-J} \right),$$

where we have used the closure condition for Clebsh–Gordan coefficients and (6.A.22), and δ_{j_1, j_2} must be interpreted as a δ function regarding all the quantum numbers but the magnetic one. We see that two fermions with identical quantum numbers (but the magnetic one) *cannot couple to J odd*. If J is even, the normalization constant is

$$N = \frac{1}{\sqrt{1 + \delta_{j_1, j_2}}}. \tag{6.D.5}$$

To sum up,

$$T^{\dagger}(j_1, j_2, JM) = \frac{1}{\sqrt{1 + \delta_{j_1, j_2}}} \sum_m \langle j_1 \; m \; j_2 \; M - m | J \; M \rangle \; a^{\dagger}_{j_1 m} a^{\dagger}_{j_2 M - m}. \tag{6.D.6}$$

The spectroscopic amplitude for finding in a $A + 2$, J_f, M_f nucleus a couple of nucleons with quantum numbers j_1, j_2 coupled to J on top of a A, J_i nucleus is

$$B(j_1, j_2(J)) = \sum_{M, M_i} \langle J_i \; M_i \; JM | J_f \; M_f \rangle \langle \Psi_{J_f M_f} | T^{\dagger}(j_1, j_2, JM) | \Psi_{J_i M_i} \rangle. \tag{6.D.7}$$

This is completely general. It depends on the structure model only through the way the $A + 2$ and A nuclei are treated. We now want to turn our attention to the expression of $B(J, j_1, j_2)$ in the BCS approximation when both the $A + 2$ and the A are 0^+, zero–quasiparticle ground states. In order to do this, we write (6.D.6) in terms of quasiparticle operators using (6.C.17)[27]:

$$
\begin{aligned}
T^{\dagger}(j_1, j_2, JM) = \frac{1}{\sqrt{1 + \delta_{j_1, j_2}}} \sum_{m_1, m_2} \langle j_1 \; m_1 \; j_2 \; m_2 | J \; M \rangle & \left(U_{j_1} U_{j_2} \alpha^{\dagger}_{j_1 m_1} \alpha^{\dagger}_{j_2 m_2} \right. \\
+ (-1)^{j_1 + j_2 - M} V_{j_1} V_{j_2} \alpha_{j_1 - m_1} \alpha_{j_2 - m_2} & \\
+ (-1)^{j_2 - m_2} U_{j_1} V_{j_2} \alpha^{\dagger}_{j_1 m_1} \alpha_{j_2 - m_2} & \\
- (-1)^{j_1 - m_1} V_{j_1} U_{j_2} \alpha^{\dagger}_{j_2 m_2} \alpha_{j_1 - m_1} & \\
+ (-1)^{j_1 - m_1} V_{j_1} U_{j_2} \delta_{j_1 j_2} \delta_{-m_1 m_2} & \left. \right).
\end{aligned}
\tag{6.D.8}
$$

If both nuclei are in the $|BCS\rangle$ ground state (zero-quasiparticles state), the only term that survives is the last one in the above expression, and (6.D.7) becomes

---

[27] In what follows, we use the phase convention $\alpha_{j\bar{m}} = (-1)^{j-m} \alpha_{j-m}$ instead of $\alpha_{j\bar{m}} = (-1)^{j+m} \alpha_{j-m}$, consistent with (6.C.2). Had we stuck to the definition (6.C.2), the amplitude $B(0, j, j)$ calculated below would have had a minus sign, which would not have had any physical consequence.

$$B_j = B(j^2(0)) = \frac{1}{\sqrt{2}} \sum_m \langle j\, m\, j\, -m | 0\, 0 \rangle (-1)^{j-m} V_j U_j$$

$$= \frac{1}{\sqrt{2}} \sum_m \frac{(-1)^{j-m}}{\sqrt{(2j+1)}} (-1)^{j-m} V_j U_j \qquad (6.D.9)$$

$$= \frac{1}{\sqrt{2}} \sum_m \frac{1}{\sqrt{(2j+1)}} V_j U_j.$$

After carrying out the summation one finds,

$$B_j = B(j^2(0)) = \sqrt{j + 1/2}\; V_j U_j. \qquad (6.D.10)$$

Note that in this final expression V_j refers to the $A + 2$ nucleus, while U_j is related to the A nucleus. In practice, it does not make a big difference to calculate both for the same nucleus.

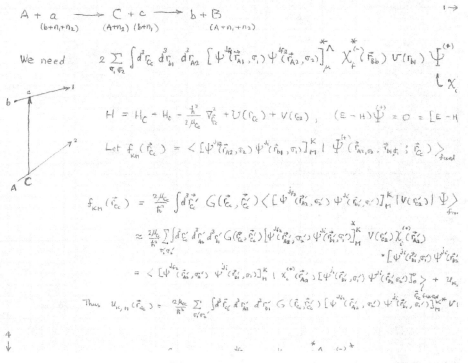

Figure 6.E.1 First manuscript page of Ben Bayman's derivation of the two-nucleon transfer reaction amplitude, in second-order DWBA approximation.

6.E Derivation of Two-Nucleon Transfer Transition Amplitudes, Including Recoil, Nonorthogonality, and Successive Transfer

In the present appendix we reproduce with the permission of the author the first (manuscript) page (Fig. 6.E.1) of what, arguably, was the first complete derivation[28] of the different contributions needed to calculate absolute two-nucleon transfer cross sections in a systematic way.[29]

---

[28] Bayman, unpublished.

[29] Bayman (1971) and Bayman and Chen (1982). Within this context, we refer to Broglia et al. (1973) and Potel et al. (2013a), in particular Fig. 10 of this reference.

7

Probing Cooper Pairs with Pair Transfer

> I advocated the application of the word *phenomenon* exclusively to refer to the observation obtained under specified circumstances, including the account of the whole experimental arrangement.
>
> *Niels Bohr*

Pairing density, both static and dynamic,[1] can exist only to the extent that normal density exists. At the basis of *normal density*, one finds independent nucleon (fermion) motion in a nonlocal mean field, created by all nucleons. At the basis of pair (*abnormal*) density, one finds the conversion of this motion into long-range correlated motion of extended, strongly overlapping, weakly bound pairs of nucleons (quasi-bosons), moving in time-reversal states close to the Fermi energy displaying *off-diagonal long-range order*. Because the mean square radius of a Cooper pair is considerably larger than nuclear dimensions, the average potential acts on them as a strong external field and affects their structure and dynamics, a fact the implications of which have become physically clearer through the study of the structure of the two-neutron halo nucleus ^{11}Li with pair transfer. *Said differently, Cooper pairs are at the basis of (both static and dynamic) spontaneous breaking of gauge invariance, namely, breaking of conservation (of pairs) of particle number. To study the specific consequences of a broken symmetry, one needs a probe which itself breaks the symmetry – in the present case, a probe which changes in two particle number, like a nuclear reaction inducing pair transfer.*

Within the above scenario we apply, in what follows, the formalism worked out in the previous chapters to analyze Cooper pair transfer reactions. In the calculation of the corresponding absolute differential cross sections, use has been made of the computer code COOPER (App. 7.A). A number of examples are considered covering nuclei throughout the mass table, namely, two-particle transfer in ^{10}Be,

---

[1] See Figs. 3.5.7 and 4.4.2.

^{7}Li, ^{11}Li, ^{48}Ca, and ^{206}Pb (systems around closed shells) and on single open-shell superfluid medium heavy nuclei (^{60}Ni and Sn-isotopes), induced both by light and heavy ions.

7.1 The ^{1}H(^{11}Li,^{9}Li)^{3}H Reaction: Evidence for Phonon-Mediated Pairing

We start by discussing the analysis of the two-neutron pickup[2] reaction ^{1}H(^{11}Li,^{9}Li)^{3}H. The results provide evidence for a new mechanism of pairing correlations in nuclei: pygmy dipole resonance (low-energy $E1$-strength) mediated pairing interaction[3], which strongly renormalizes the bare, NN-$^1S_0$ interaction. This is but a particular embodiment of (quasi) phonon-mediated pairing interaction found in nuclei, throughout the mass table.[4] Attention is also paid to the population of the $1/2^-$ first excited state of ^{9}Li lying[5] at 2.69 MeV (see Figs. 2.7.2–2.7.5 and 7.1.1).

The main difference between light halo exotic nuclei and medium heavy nuclei lying along the valley of stability is the role fluctuations play in dressing particles (quasiparticles) and in renormalizing their properties (mass, charge, etc.) and their interactions, like the pairing interaction. In fact, in the case of, for example, Sn isotopes, mean field effects are dominant, while in the case of halo exotic nuclei, renormalization effects can be as large as mean field ones. Concerning the pairing interaction, bare and induced contributions are about equal in the case of Sn-isotopes, while the second one is the overwhelming contribution in the case of ^{11}Li. The collective modes acting as glue of the Cooper pairs are mainly of quadrupole type in Sn and of dipole (pygmy resonance)[6] type in the case of ^{11}Li. In what follows we summarize aspects of structure and reaction which, interweaved, offer insight into this single Cooper pair halo nucleus and allow us to compare theory with experiment in terms of absolute differential cross sections.

7.1.1 Structure

The ground state of ^{11}Li is written as

$$|gs\,(^{11}\text{Li})\rangle = |\widetilde{0}\rangle_\nu \otimes |p_{3/2}(\pi)\rangle, \qquad (7.1.1)$$

where the proton state is assumed to act as a spectator and where the two halo-correlated neutrons are described by the state

$$|0\rangle_\nu = |0\rangle + \alpha|(p_{1/2}, s_{1/2})_{1^-} \otimes 1^-; 0\rangle + \beta|(s_{1/2}, d_{5/2})_{2^+} \otimes 2^+; 0\rangle, \qquad (7.1.2)$$

---

[2] Tanihata et al. (2008).

[3] Barranco et al. (2001); Potel et al. (2010).

[4] See footnote 55 in Chapter 3, and refs. therein.

[5] To assess the direct character of the $1/2^-$ excitation process, the importance of (two-step) inelastic and of knockout channels was considered and found to be small (see Sect. 7.1.2).

[6] Within this context, see Broglia et al. (2019a).

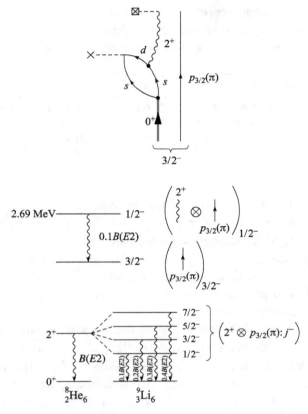

Figure 7.1.1 *Gedankenexperiment,* NFT$_{(s+r)}$ (two-particle transfer γ-decay coincidence) aimed at better individuating the couplings between the elementary modes of excitation, involved in the population of the $1/2^-$, first excited state of ^{9}Li in the ^{1}H(^{11}Li,^{9}Li)^{3}H reaction (Tanihata et al. (2008); Barranco et al. (2001); Potel et al. (2010)). After Potel et al. (2014).

with

$$\alpha = 0.7, \qquad \beta = 0.1, \tag{7.1.3}$$

and

$$|0\rangle = 0.45|s_{1/2}^2(0)\rangle + 0.55|p_{1/2}^2(0)\rangle + 0.04|d_{5/2}^2(0)\rangle, \tag{7.1.4}$$

$|1^-\rangle$ and $|2^+\rangle$ being the (QRPA) states describing the pygmy dipole resonance of ^{11}Li and the low-lying quadrupole vibration of the core ^{9}Li.

We are then in the presence of a paradigmatic nuclear embodiment of Cooper's model which is at the basis of BCS theory: a single weakly bound neutron pair on top of the Fermi surface, namely, the ^{9}Li core. But the analogy goes beyond these aspects and covers also the very nature of the interaction acting between Cooper pair partners (pairing interaction). Due to the high polarizability of the system

Table 7.1.1 *Optical potentials (cf. Tanihata et al. (2008)) used in the calculation of the absolute differential cross sections displayed in Fig. 7.1.2*

					^{11}Li$(p, t)^9$Li							
	V	W	V_{so}	W_d	r_1	a_1	r_2	a_2	r_3	a_3	r_4	a_4
$p, ^{11}$Li	63.62	0.33	5.69	8.9	1.12	0.68	1.12	0.52	0.89	0.59	1.31	0.52
$d, ^{10}$Li	90.76	1.6	3.56	10.58	1.15	0.75	1.35	0.64	0.97	1.01	1.4	0.66
$t, ^9$Li	152.47	12.59	1.9	12.08	1.04	0.72	1.23	0.72	0.53	0.24	1.03	0.83

under study and of the small overlap of halo and core single-particle wavefunctions, most of the Cooper pair correlation energy stems from the exchange of collective vibrations, the role of the strongly screened bare pairing interaction being, in this case, subcritical (see Sect. 3.6). In other words, we are in the presence of a new realization of the Cooper model in which a totally novel Bardeen–Pines–Frölich-like phonon-induced pairing interaction is generated by a self-induced collective vibration of the nuclear medium.

In connection with (7.1.2), it is revealing that the state populated in the inverse kinematics, two-neutron pickup reaction ^{1}H(^{11}Li,^{9}Li)^{3}H is, aside from the ground $|3/2^-$ gs (^{9}Li)$\rangle$ state, is the first excited $|1/2^-, 2.69$ MeV (^{9}Li)$\rangle$ level. The associated absolute differential cross sections calculated making use of the spectroscopic amplitudes displayed in Eqs. (7.1.2)–(7.1.4) and of the optical potentials[7] collected in Table 7.1.1 provide a quantitative account of the experimental data within the NFT scenario (Fig. 7.1.2). It is of note that (**a**) assuming the soft $E1$-mode to be nonoperative, that is, $\alpha = 0$, and normalizing to 1 (7.1.4) (see Fig. 7.1.2, lower right panel labeled $N = 1$), theory overpredicts experiment by a factor of 2; (**b**) the absolute direct two-particle transfer cross section associated with the $1/2^-$ state reflects the value of the amplitude β.

7.1.2 Reaction

In the reaction ^{1}H(^{11}Li,^{9}Li)^{3}H, the $1/2^-$ (2.69 MeV) first excited state of ^{9}Li can, in principle, be populated not only through the direct two-particle transfer reaction channel (sum of successive and simultaneous contributions plus nonorthogonality corrections; see Fig. 7.1.2 and Table 7.1.2, column 2) but also, in principle, through other channels involving breakup and inelastic processes. Of these three channels,

---

[7] The process described by diagram (a) of Fig. 2.7.2 is expected to provide the largest contribution, within the NFT$_{(s+r)}$ framework, to the $(p, ^{11}$Li) optical potential.

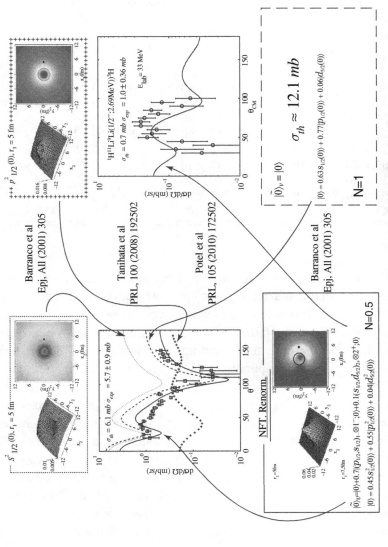

Figure 7.1.2 Absolute, two-nucleon transfer differential cross section associated with the ground state (middle row left) and the first excited state of ^{9}Li (middle row right), excited in the reaction ^{1}H(^{11}Li, ^{9}Li)^{3}H (Tanihata et al., 2008) in comparison with the predicted differential cross sections (Potel et al., 2010) worked out making use of the optical potentials collected in Table 7.1.1 and of the two-nucleon transfer transition amplitudes and Cooper pair wavefunctions, calculated with NFT$_{(s+r)}$. After Broglia et al. (2016). For the color version of this figure, refer to cambridge.org/nuclearcooperpair.

Table 7.1.2 *Probabilities p_l associated with the processes described in the text for each partial wave l. The different channels are labeled by a channel number c. It can be seen that aside from channels 1 and 2 (successive + simultaneous + nonorthogonality), leading to the direct population of the $|^9Li(gs)\rangle$ and $|^9Li(1/2^-; 2.6\,MeV)\rangle$ states, all other channels provide a negligible contribution to the population of the ^9Li first excited state (see also Figs. 2.7.3 (b), (d), 2.7.4 (f), and 2.7.5).*

l \ c	1	2	3	4	5
0	4.35×10^{-3}	1.79×10^{-4}	4.81×10^{-6}	2.90×10^{-11}	3.79×10^{-8}
1	3.50×10^{-3}	9.31×10^{-4}	1.47×10^{-5}	1.87×10^{-9}	1.09×10^{-6}
2	7.50×10^{-4}	8.00×10^{-5}	2.45×10^{-5}	1.25×10^{-8}	1.21×10^{-6}
3	6.12×10^{-4}	9.81×10^{-5}	1.51×10^{-6}	6.50×10^{-10}	2.20×10^{-7}
4	1.10×10^{-4}	1.18×10^{-5}	2.21×10^{-7}	4.80×10^{-11}	1.46×10^{-8}
5	3.65×10^{-5}	2.16×10^{-7}	7.42×10^{-9}	6.69×10^{-13}	9.63×10^{-10}
6	1.35×10^{-5}	6.05×10^{-8}	2.88×10^{-10}	8.04×10^{-15}	1.08×10^{-11}
7	4.93×10^{-6}	7.78×10^{-8}	6.01×10^{-11}	4.05×10^{-16}	5.26×10^{-13}
8	2.43×10^{-6}	2.62×10^{-8}	7.4×10^{-12}	1.26×10^{-17}	9.70×10^{-11}

we choose, in particular, the one in which one of the two neutrons moving in the $1/2^-$ orbital is picked up (the other going into the continuum), together with a neutron from the $p_{3/2}$ orbital. This hole state eventually couples with a new bound $p_{1/2}$ neutron state, resulting from the falling of the continuum $p_{1/2}$ neutron back into the bound orbital, leaving the core in the quadrupole excited state.[8] This is in keeping with the fact that the main amplitude of this vibration is[9] the $X(1p_{3/2}^{-1}, 1p_{1/2})$ (column 3 of Table 7.1.2).

In the second process involving breakup, the proton field (^1H) acting once breaks the halo Cooper pair (assumed again to be in the $|p_{1/2}^2(0)\rangle$ component), forcing one of the $1/2^-$ halo neutrons to populate a $p_{1/2}$ continuum state, and the other $1/2^-$ neutron is assumed to do the same. Acting for a second time, ^1H picks up one of the neutrons moving in the continuum and another one from those occupying the bound $p_{3/2}$ orbital of ^9Li. The resulting hole state can couple, as explained above, to the $p_{1/2}$ bound state into which the continuum one has fallen, leaving again the core in the quadrupole mode of excitation (column 4 Table 7.1.2; see also Fig. 1 (g) of Potel et al. (2010)).

---

[8] See Potel et al. (2010), Fig 1 (f) of this reference.
[9] Barranco et al. (2001).

In the remaining channel considered, the (p, t) process populates the $|gs(^9\text{Li})\rangle$ state, and the final state interaction (FSI) between the outgoing triton and ^9Li results in the inelastic and Coulomb excitation of the $1/2^-$ state (Table 7.1.2, column 5; see also Fig. 1 (h) of Potel et al. (2010)).

Making use of the NFT spectroscopic amplitudes, and of computer codes developed to microscopically take into account the different processes mentioned above, that is, the different reaction channels and continuum states up to 50 MeV of excitation energy, the corresponding transfer amplitudes and associated probabilities p_l were calculated. In Table 7.1.2 the probabilities $p_l = |S_l^{(c)}|^2$ associated with each of the processes mentioned above are displayed, where the amplitude $S_l^{(c)}$ is related to the total cross section associated with each of the channels c by the expression[10]

$$\sigma_c = \frac{\pi}{k^2} \sum_l (2l + 1)|S_l^{(c)}|^2, \tag{7.1.5}$$

k being the wave number of the relative motion between the reacting nuclei.

In keeping with the small values of p_l, the contributions associated with the different reaction channels were calculated making use of second-order perturbation theory. In the case of the $1/2^-$ (2.69 MeV) first excited state of ^9Li, the absolute differential cross section is thus

$$\frac{d\sigma}{d\Omega}(\theta) = \frac{\mu^2}{16\pi^3 \hbar^4} \left| \sum_l (2l + 1) P_l(\theta) \sum_{c=2}^5 T_l^{(c)} \right|^2. \tag{7.1.6}$$

Here μ is the reduced mass and $T_l^{(c)}$ are the transition matrix elements associated with the different channels and for each partial wave of DWBA.[11]

Making use of all the elements discussed above, multistep (simultaneous, nonorthogonality and successive) transfer[12], breakup, and inelastic channels were calculated, and the results are displayed in Figs. 7.1.3 and 7.1.4. Theory provides an overall account of the experimental findings. In particular, in connection with the $1/2^-$ state, this result essentially emerges from multistep two-particle transfer process (Fig. 7.1.4), weighted by the corresponding nuclear structure amplitudes (see Eqs. (7.1.2)–(7.1.4)).

As shown in Figs. 7.1.3 and 7.1.4, the contributions of breakup and inelastic processes to the population of the $1/2^-$ (2.69 MeV) first excited state of ^9Li are negligible as compared to successive two-nucleon transfer. In the case of the breakup channels, this is a consequence of the low bombarding energy of the ^{11}Li beam (inverse kinematics), combined with the small overlap between continuum

[10] Satchler (1980); Landau and Lifshitz (1965).
[11] Satchler (1980).
[12] Bayman and Chen (1982); Igarashi et al. (1991); Bayman and Feng (1973); Broglia and Winther (2004).

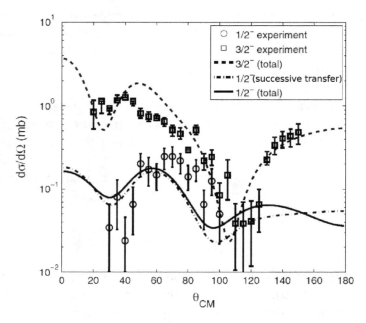

Figure 7.1.3 Experimental (Tanihata et al. (2008)) and theoretical differential cross sections (including successive and simultaneous transfer corrected by nonorthogonality, as well as breakup and inelastic channels; Potel et al. (2010)) associated with the ^{1}H(^{11}Li,^{9}Li)^{3}H reaction populating the ground state (3/2$^-$) and the first excited state (1/2$^-$; 2.69 MeV) of ^{9}Li. Also shown (dash-dotted curve) is the differential cross section associated with this last state, but taking into account only successive transfer. The optical potentials used are from Tanihata et al. (2008); and An and Cai (2006); see Table 7.1.1. The absolute cross section associated with the ground state (3/2$^-$) is predicted to be 6.1 mb (exp: 5.7 ± 0.9 mb), that corresponding to the first excited state (1/2$^-$; 2.69 MeV) being 0.7 mb (exp: 1.0 ± 0.36 mb). After Potel et al. (2010).

(resonant) neutron $p_{1/2}$ wavefunctions and bound state wavefunctions. In the case of FSI processes, it is again a consequence of the relative low bombarding energy. In fact, the adiabaticity parameters ξ_C, ξ_N[13] associated with Coulomb excitation and inelastic excitation in the $t+^9$Li channel are larger than 1, implying an adiabatic cutoff. In other words, the quadrupole mode is only polarized during the reaction but essentially not excited.

Second-order calculations of inelastic, breakup, and final state interaction channels, which in principle can provide alternative routes for the population of the first excited state of ^{9}Li to the direct one predicted by the wavefunction (7.1.2) (β component), lead to absolute cross sections which are smaller by a few orders of magnitude than that shown in Fig. 7.1.3. One can then posit that quadrupole core

---

[13] See Eqs. (IV.12) and (IV.14) of Broglia and Winther (2004).

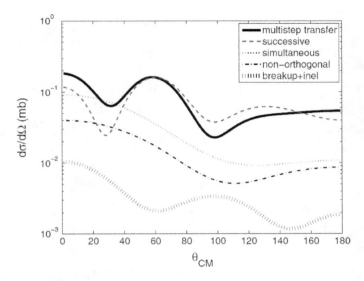

Figure 7.1.4 Successive, simultaneous, and non-orthogonality contributions (prior representation) to the ^{1}H(^{11}Li,^{9}Li)^{3}H differential cross section associated with the population of the $1/2^-$ state of ^{9}Li, displayed in Fig. 7.1.3. Also shown is the sum of the breakup ($c = 3$ and 4) and inelastic ($c = 5$) channel contributions. After Potel et al. (2010).

polarization effects in $|$ gs (^{11}Li)$\rangle$ are essential to account for the observation of the $|1/2^-, 2.69\,\text{MeV}\rangle$ state and provide evidence for phonon-mediated pairing in nuclei.

7.1.3 Further Examples of Pairing Vibrational States

In Fig. 7.1.5, further examples of pairing vibrational states around neutron closed-shell systems are given.[14] The fact that among the (p, t) and (t, p) absolute differential cross sections, one also finds the ^{208}Pb(^{16}O,^{18}O)^{206}Pb(gs) absolute differential cross section is in keeping with the fact that the formalism to treat both light and heavy ion two-nucleon transfer reactions is rather homogeneous and well established.[15] It has thus been implemented in the software COOPER as a standard option (cf. App. 7.A).

7.2 Pairing Rotational Bands

The $|BCS\rangle$ state, which can be viewed as the intrinsic state with deformation α_0 of a pairing rotational band (see Eq. (4.7.26)), is a coherent state (see Eq. (1.4.39)).

[14] For details, see Potel et al. (2013a).
[15] See footnote 31 of Chapter 3 See also Potel et al. (2013a,b).

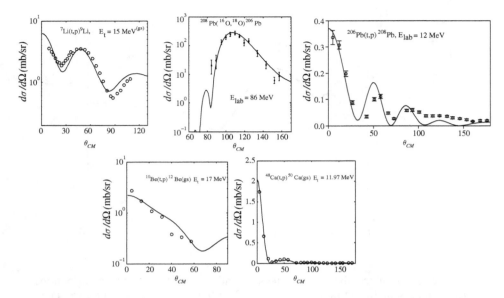

Figure 7.1.5 Absolute two-particle transfer differential cross sections for a number of reactions around a closed shell associated with monopole pair addition and removal modes and populating the ground state of the final system. Making use of spectroscopic amplitudes calculated as described in Sect. 3.5, of global optical parameters and of the code COOPER, absolute differential cross sections for the cases of $N = 6$ (Li,Be), $N = 48$ (Ca), and $N = 126$ (Pb) were calculated. They are displayed in comparison with the experimental data.

One then expects the associated observable ($\sim |\alpha_0|^2$), namely, the absolute two-nucleon transfer differential cross section, to be stable with respect to the models used to calculate the corresponding spectroscopic amplitudes B_ν, that is, $B_\nu \sim U_\nu V_\nu$ ($\alpha_0 = \sum_{\nu>0} U_\nu V_\nu$).

In Table 7.2.1 we collect the two-nucleon spectroscopic amplitudes associated with the reactions $^{A+2}\mathrm{Sn(p,t)}^A$ Sn for A in the interval 112–126. Making use of these results and of global optical parameters (see Table 7.2.2), the absolute differential cross sections $^{A+2}\mathrm{Sn}(p, t)^A\mathrm{Sn(gs)}$ were calculated. They are shown in Fig. 7.2.1 in comparison with the data.

7.2.1 Structure-Reaction: Stability of the Order Parameter α_0

The order parameter or deformation in gauge space can be written as

$$\alpha_0' = \sum_{j_a} \sqrt{\frac{2j_a + 1}{2}} B(j_a^2(0), N \to N + 2). \qquad (7.2.1)$$

Table 7.2.1 *Two-nucleon transfer spectroscopic amplitudes*
$\langle BCS(N)|T_\nu'^\dagger(j_\nu^2(0))|BCS(N+2)\rangle = \sqrt{(2j_\nu+1)/2}\,U_\nu'(N)V_\nu'(N+2)$,
*associated with the reactions connecting the ground states (members of a pairing
rotational band) of two superfluid Sn-nuclei* $^{A+2}Sn(p,t)^A Sn(gs)$. *In the first row,
the target nuclei are indicated, while in the first column, the single-particle
orbitals are listed.*

	^{112}Sn	^{114}Sn	^{116}Sn	^{118}Sn	^{120}Sn	^{122}Sn	^{124}Sn
$1d_{5/2}$	0.664	0.594	0.393	0.471	0.439	0.394	0.352
$0g_{7/2}$	0.958	0.852	0.542	0.255	0.591	0.504	0.439
$2s_{1/2}$	0.446	0.477	0.442	0.487	0.451	0.413	0.364
$1d_{3/2}$	0.542	0.590	0.695	0.706	0.696	0.651	0.582
$0h_{11/2}$	0.686	0.720	1.062	0.969	1.095	1.175	1.222

Table 7.2.2 *Optical potentials used in the calculations of the absolute differential
cross sections displayed in Fig. 7.2.1 (for details, see Potel et al. (2013a)).*

	^{A}Sn$(p,t)^{A-2}$Sn											
	V	W	V_{so}	W_d	r_1	a_1	r_2	a_2	r_3	a_3	r_4	a_4
p, ^{A}Sn	50	5	3	6	1.35	0.65	1.2	0.5	1.25	0.7	1.3	0.6
d, $^{A-1}$Sn	78.53	12	3.62	10.5	1.1	0.6	1.3	0.5	0.97	0.9	1.3	0.61
t, $^{A-2}$Sn	176	20	8	8	1.14	0.6	1.3	0.5	1.1	0.8	1.3	0.6

The quantity

$$B(j_a^2(0), N \to N+2) = \sqrt{\frac{2j_a+1}{2}}U_{j_a}'(N)V_{j_a}'(N+2) \qquad (7.2.2)$$

is the two-nucleon spectroscopic amplitude associated with pair transfer between
members of a pairing rotational band, corresponding to the configuration $j_a^2(0)$.
Thus

$$\alpha_0' = \sum_{j_a,m_a>0} U_{j_a}'V_{j_a}' = \sum_{j_a}\frac{2j_a+1}{2}U_{j_a}'V_{j_a}' = e^{2i\phi}\sum_{j_a}\frac{2j_a+1}{2}U_{j_a}V_{j_a} = e^{2i\phi}\alpha_0 \qquad (7.2.3)$$

defines a privileged orientation in gauge space. *Within a unified description of
structure and reactions, the quantities (7.2.2) are the weighting factors of the successive, simultaneous, and nonorthogonality formfactors (Eq. 7.4.1) involved in
the calculation of the corresponding transfer amplitudes and associated absolute
transfer differential cross sections, making use of global optical parameters.*

Probing Cooper Pairs with Pair Transfer

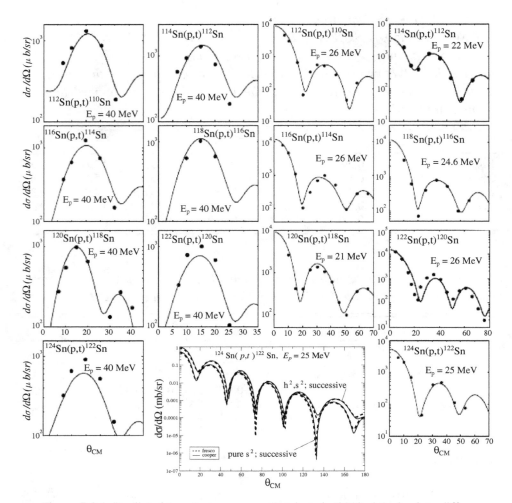

Figure 7.2.1 Predicted (continuous curve; Potel et al. (2013a,b)) absolute differential $^{A+2}$Sn $(p, t)^A$Sn(gs) cross sections for bombarding energies $E_p = 40$ MeV (in the two left columns) and 21 MeV $\leq E_p \leq 26$ MeV (in the two right columns) in comparison with the experimental data (solid dots; Bassani et al. (1965); Guazzoni et al. (1999, 2004, 2006, 2008, 2011, 2012)). In the center of the lowest row, the absolute differential cross section (successive transfer displayed in the angular interval $0° \leq \theta_{CM} \leq 180°$) associated with the process ^{124}Sn$(p, t)^{122}$Sn(f) and calculated with the software COOPER (dashed bold) and FRESCO (thin continuous line) (Thompson (1988)) are displayed. This calculation has been performed both for the $B_{1/2}|s^2_{1/2}(0)\rangle$ contribution alone and for the contribution corresponding to the $(B_{1/2}|s^2_{1/2}(0)\rangle + B_{11/2}|h^2_{11/2}(0)\rangle)$ configuration, where the corresponding B coefficients are the spectroscopic amplitudes reported in the last column of Table 7.2.1. After Potel et al. (2013a).

Consequently, in discussing the properties of the order parameter α_0*, in particular in this section, one has in mind the two-nucleon transfer formfactors.* The
B-coefficients weight the corresponding radial functions, as well as the contributions of energy denominators (Green function) associated with the intermediate
channels.

Empirical renormalization is also operative in connection with these reaction
channels, in terms of selecting the associated *effective Q-values*, and also of the
distorted waves, again by selecting among possible embodiments of the *optical
potentials*. It is to be noted that, within this connection, one is talking about modest changes in the absolute differential cross sections to be used to predict new
experiments, or to relate to the experimental data, as compared with the results
obtained with standard choices (see App. 7.A and Fig. 7.A.1).

7.2.2 A Two-Nucleon Transfer Physical Sum Rule

In what follows we analyze the stability of α_0' within the framework of three
theoretical schemes to calculate the B-amplitudes[16] associated with the reaction
^{120}Sn$(p, t)^{118}$Sn(gs). The first one corresponds to the BCS approximation, making use of a pairing interaction of constant matrix elements. Starting from the
HF solution of a Skyrme interaction, namely, Sly4, the gap and number equations are solved in the pairing approximation with $G = 0.26$ MeV, leading to the
empirical value of the three-point expression of the pairing gap $\Delta^{exp} \approx 1.4$ MeV.
The B-coefficients for the valence orbitals are reported in the fourth column of
Table 7.2.3.

In the second calculational scheme, and making use of the same Skyrme interaction and of the v_{14} Argonne, $^1S_0$ NN-potential and neglecting the influence of the
bare pairing force in the mean field, the HFB equation was solved. As a result, this
step corresponds to an extended BCS calculation over the HF basis, allowing for
the interference between states of equal quantum numbers $a(\equiv lj)$ and both equal
and different numbers of nodes. One includes (N_a) states (for each a) up to ≈ 600
MeV, to properly take into account the repulsive core of v_{14} and be able to accurately calculate Δ^{HFB} (=1.08 MeV). The resulting B-coefficients are displayed in
the third column of Table 7.2.3.

Going beyond mean field, one has to include the particle–vibration coupling
mechanism leading to renormalization phenomena, both in the quasiparticle self-
energy as well as the induced pairing interaction, within the framework of
renormalized NFT ((NFT)$_{ren}$). Propagating the dressing of the quasiparticle states
with the help of the Nambu–Gorkov (NG) equation leads, in the canonical basis,

---

[16] Potel et al. (2017).

Table 7.2.3 *Two-nucleon spectroscopic amplitudes associated with the reaction $^{120}Sn(p, t)^{118}Sn(gs)$. Note the small difference between the values of the fourth column with those reported in the column labeled ^{120}Sn of Table 7.2.1. They are due to a small difference in the value of the pairing coupling constant used here (Potel et al. (2017)) and that employed in Potel et al. (2013b, 2017).*

$a \equiv \{l_j\}$	NFT(NG)	HFBv_{14}	BCS(G)
$d_{5/2}$	0.22	0.29	0.41
$g_{7/2}$	0.46	0.47	0.57
$s_{1/2}$	0.37	0.34	0.41
$d_{3/2}$	0.59	0.60	0.66
$h_{11/2}$	0.95	1.0	1.03

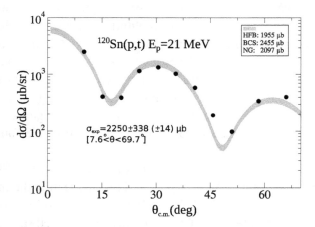

Figure 7.2.2 Absolute differential cross sections associated with the reaction ^{120}Sn(p,t)^{118}Sn(gs) calculated making use of the BCS, HFB, and renormalized NFT(NG) spectroscopic amplitudes and global optical parameters (gray thin area), in comparison with the experimental findings (solid dots; Guazzoni et al. (2008)). After Potel et al. (2017).

to the renormalized spectroscopic amplitudes shown in the second column of Table 7.2.3.

Making use of these *B*-coefficients, together with the global optical parameters reported in Table 7.2.2, the corresponding absolute differential cross sections associated with the reaction $^{120}Sn(p, t)^{118}Sn(gs)$ at 21 MeV of bombarding energy were calculated. They are displayed in Figs. 7.2.2 and 7.2.3 in comparison with the experimental findings. The absolute cross section ratio $|\sigma_i - \sigma_{exp}|/\sigma_{exp}$ is equal

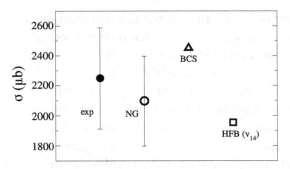

Figure 7.2.3 Integrated absolute cross sections associated with the reaction ^{120}Sn(p,t)^{118}Sn(gs). The error ascribed to the NFT(NG) theoretical result stems from the variation the spin contribution of the different Skyrme interactions induces in the B-coefficients. After Potel et al. (2017).

to 0.09, 0.13, and 0.07 (i = BCS, HFB, NFT(NG)). One can then posit that the relative errors of the associated two-nucleon transfer amplitudes $\alpha_0 (\sim \sqrt{\sigma})$ are[17] 4.5 percent, 6.5 percent, and 3.5 percent. In other words, spontaneous breaking of gauge symmetry, a feature which is embodied in the three descriptions used (BCS, HFB, NFT(NG)), albeit, at very different levels of many-body refinement, seems to give rise to a new emergent property: a physical sum rule resulting from the intertwining of structure both in configuration and in 3D space (spectroscopic amplitudes and radial wavefunction) and reaction aspects of pairing in nuclei, and closely connected with the fact that $|BCS\rangle$ is a coherent state.

7.3 Heavy Ion Reactions between Superfluid Nuclei and Correlation Length

One- and two-neutron transfer reactions have been studied with magnetic and gamma spectrometers in heavy ion reactions between superfluid nuclei[18], namely,

$$^{116}\text{Sn} + ^{60}\text{Ni} \rightarrow \begin{cases} ^{115}\text{Sn} + ^{61}\text{Ni} & (Q_{1n} \approx -1.74 \text{ MeV}) \quad \text{(a)} \\ ^{114}\text{Sn} + ^{62}\text{Ni} & (Q_{2n} \approx 1.307 \text{ MeV}), \quad \text{(b)} \end{cases} \quad (7.3.1)$$

for twelve bombarding energies within the energy range 140.60 MeV$\leq E_{cm} \leq$ 167.95 MeV, that is, from energies above the Coulomb barrier ($E_B = 157.60$ MeV) to well below it. The Cooper pair transfer channel (7.3.1 (b)) is dominated by the ground–ground state transition, while the theoretical calculation of the differential cross section associated with channel (7.3.1 (a)) indicates the incoherent contribution of a number of quasiparticle states of ^{61}Ni lying at energies $\lesssim 2.640$

---

[17] That is, $d\alpha_0' = \frac{1}{2}\sigma^{-1/2}d\sigma$ and $(d\alpha_0'/\alpha_0') = \frac{1}{2}(d\sigma/\sigma)$.
[18] Montanari et al. (2014, 2016).

MeV. This finding is consistent with the fact that the value of the pairing gap of ^{60,62}Ni is of the order of 1.3–1.5 MeV. That is, in the case of the reaction 7.3.1 (a), we are in presence of a S-Q-like transfer (see Fig. 4.6.1).

The analysis of the data associated with the reactions (7.3.1) carried out in Montanari et al. (2014, 2016) makes use of a powerful semiclassical approximation in which also the optical potential was microscopically calculated. The short wavelength of relative motion (de Broglie reduced wavelength $\lambda = 0.36/2\pi$ fm ≈ 0.06 fm) allows to accurately determine the distance of closest approach. Making use of the U, V occupation amplitudes for both Sn and Ni, as well as the optical potential given in Montanari et al. (2014), one has calculated, within the framework of first- and second-order DWBA, the absolute one- and two-nucleon transfer differential cross sections, in the second case, including both successive (dominant channel) and simultaneous transfer, properly corrected by nonorthogonality. Theory is compared with experiment in Fig. 7.3.1. As expected, it provides an overall account of the experimental findings.[19]

As it emerges from direct inspection of Fig. 7.3.1, the distance of closest approach lying within the interval 13.12 fm $\leq D_0 \leq$ 13.49 fm is the largest one for which $d\sigma/d\Omega|_{2n}$ is within a factor of about 0.6 ($\approx (\pi/4)^2$; see Eq. (4.6.20)) of the same order of $d\sigma/d\Omega|_{1n}$. In keeping with Eq. (4.5.12) ($v_F/c \approx 0.27$, $\Delta \approx 1.3-1.5$ MeV) and Eq. (4.3.1), one can posit that the above interval provides a sensible bound to the size of the nuclear Cooper pair correlation length. Already increasing D_0 by 0.6 fm ($D_0 = 14.05$), σ_{1n} becomes a factor 6 larger than σ_{2n}. This is a consequence of forcing Cooper pair partners to tunnel across a barrier of "width" (distance of closest approach) $D \gtrsim \xi$, leading to a strain which plays a role similar to that played by the critical bias $V_{eq} \approx 2\Delta/e$ (i.e., that is giving to the center of mass of the Cooper pairs a critical momentum $q \approx \hbar/\xi$) and resulting in the rupture of the transferred Cooper pair and in the unfreezing of the quasiparticle degrees of freedom. In other words, it results in the transition from the S-S to the S-Q transfer regimes (Fig. 4.6.1). Within this context we refer to Eq. (4.9.6) and Fig. (4.9.2).

7.4 The Structure of "Observable" Cooper Pairs

In his Waynflete lectures on cause and chance, Max Born[20], to whom we owe the statistical interpretation of quantum mechanics[21], states that "quantum mechanics does not describe an objective state in an independent external world, but the aspect of this world gained by considering it from a certain subjective standpoint, or with certain experimental means and arrangements." It is within this context

---

[19] Potel et al. (2021); see also Montanari et al. (2014).
[20] Born (1948).
[21] Born (1964); Pais (1986).

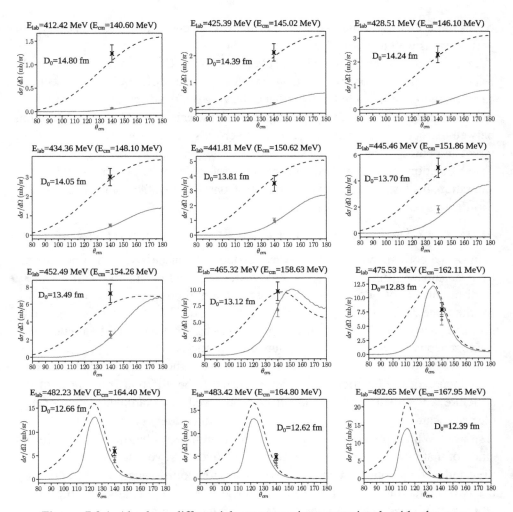

Figure 7.3.1 Absolute differential cross sections associated with the reactions ^{60}Ni+^{116}Sn→^{62}Ni+^{114}Sn (continuous line) and ^{60}Ni+^{116}Sn→^{61}Ni+^{115}Sn (dashed line) in comparison with the experimental data: $d\sigma_{2n}/d\Omega|_{\theta_{cm}=140°}$ (solid dots), $d\sigma_{1n}/d\Omega|_{\theta_{cm}=140°}$ (crosses) (Montanari et al., 2014). For the color version of this figure, refer to cambridge.org/nuclearcooperpair.

that we tried in previous sections to get insight concerning the structure of nuclear Cooper pairs, specifically, in terms of two-nucleon transfer reactions. Being even more *subjective* (concrete), we were interested in shedding light on the structure of one of the 5–6 Cooper pairs participating in the BCS condensate (intrinsic state in gauge space) of the Sn isotopes (ground state rotational band[22]) through pair transfer processes, both in light and in heavy ion reactions. Within this connection,

---

[22] Potel et al. (2013b, 2017).

one is thinking of $^{A+2}$Sn$(p, t)^A$Sn(gs) processes in general and ^{120}Sn$(p, t)^{118}$Sn in particular, as well as ^{116}Sn$+^{60}$Ni$\rightarrow ^{114}$Sn$+^{62}$Ni.

From a strict observational perspective, concerning Cooper pairs, one can only refer to the information two-nucleon transfer absolute differential cross sections carry on these entities. On the other hand, leaving out the discussion regarding the microscopic calculation of the optical potential, the carriers mediating information between structure and absolute differential cross sections, for example, between target and outgoing particle in a standard laboratory setup, are the distorted waves. These functions can be studied independently of the transfer processes under consideration, in elastic scattering experiments. Consequently, the nonlocal, correlated formfactors (see Figs. 4.1.1 and 4.1.2)

$$F(\mathbf{r}_1, \mathbf{r}_2, \mathbf{r}_{Aa}) = F_{succ}(\mathbf{r}_1, \mathbf{r}_2, \mathbf{r}_{Ap}) + F_{sim}(\mathbf{r}_1, \mathbf{r}_2, \mathbf{r}_{Aa}) + F_{NO}(\mathbf{r}_1, \mathbf{r}_2, \mathbf{r}_{Aa}), \quad (7.4.1)$$

sum of the corresponding functions associated with successive and simultaneous transfer and with the nonorthogonality correction, and calculated with a variety of different sets of two-nucleon spectroscopic amplitudes (the same in each case for the three F-functions appearing in the rhs of Eq. (7.4.1)), can be compared at profit to each other. This is in keeping with the fact that they can be related, in a homogeneous fashion, with the absolute cross sections.

It is also of note that the spatial dimensions, structure, nonlocality, and ω-dependence of the function (7.4.1) are expected to be different from those of the structure wavefunction of the Cooper pair, observed, for example, in high-energy electron scattering or (p, pn) reactions – a question closely connected with linear-response. While this concept has been, and continues to be, quite useful in the study of many-body systems, it is a subtle one. In direct two-nucleon transfer reactions induced by both light and heavy (grazing) ion collisions, the contact between the two interacting nuclei is weak. Nonetheless, even a very low (normal) density overlap between target and projectile may induce important modifications in Cooper pairs, in particular, allow pairing correlated partner nucleons to profit from the enlarged volume compared to that available in the target or projectile nucleus. As a consequence the pair can expand, each of the partner nucleons being in a different nucleus, consistent with the fact that successive is the dominant transfer mechanism.

It can be stated that this picture is again an example of the fruitfulness of *linear response* to shed light on subtle questions regarding many-body systems. In the case under discussion, it allows the partners of the nuclear Cooper pair to correlate over dimensions larger than nuclear dimensions and, in so doing, make their intrinsic structure "observable," almost free of the strong pressures of the mean field.[23]

---

[23] Within this context, one can refer to the need to be able to distinguish between right and left weakly coupled (dioxide-layer separated) superconductors, to be able to measure gauge phase difference through the Josephson effect. See also Magierski et al. (2017).

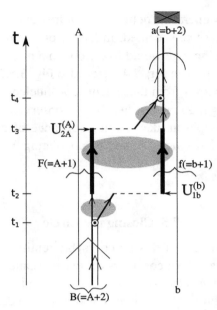

Figure 7.4.1 Diagram describing structure and reaction aspects of the main process through which a Cooper pair (di-neutron) tunnels from target to projectile in the reaction $B(=A+2)+b \rightarrow F(=A+1)+f(=b+1) \rightarrow A+a(=b+2)$. In order that the two-step process takes place, target and projectile have to be in contact, at least in the time interval running between t_2 and t_3 (times at which the mean field potential $U_{1b}^{(b)}$ ($U_{2A}^{(A)}$) acts and induces the transfer of the first (second) nucleon; prior-post representation). During this time, the two systems create, with local regions of ever so low nucleonic presence, a common density over which the nonlocal pairing field can be established and the partners of the transferred Cooper pair can be correlated, even with regions in which the pairing interaction may be zero. Small ellipses (with linear dimensions of the order of the nuclear radius R_0) indicate situations in which the two-neutron correlation is distorted by the external mean field of a single of the systems involved of the reaction, i.e., $B(=A+2)$ in the entrance channel, $a(=b+2)$ in the exit one (see, e.g., Fig. 3.6.2). The large ellipse (with linear dimensions of the order of the correlation length ξ and operative during the interval of time $t_3 - t_2$) indicates the region in which each of the partner nucleons of the Cooper pair are in a different system, correlated over distances of the order of the correlation length ξ. Because successive is the dominant contribution of the two-nucleon transfer process, it is this information that the outgoing particle (nucleus a) of a Cooper pair transfer process brings to the detector, in other words, the "observable" Cooper pair in terms of its specific probe, and the reason why the nucleons are described, in the interval $\Delta t = t_3 - t_1$, in terms of boldface arrowed lines. After Barranco et al. (2019b).

The above discussion is schematically illustrated in Fig. 7.4.1. With it one comes back to the original question (Sect. 2.1): which are the proper variables to be used in an attempt to describe the nuclear system? Elementary modes of excitation is a valid choice. But because these modes are in interaction, the above choice is not sufficient (unique). An operative definition requires that also the specific probe, reaction, or decay process be specified. In fact, if one were to study Cooper pairs through electron scattering, one would likely obtain a picture of the system as that marked by the small ellipses in Fig. 7.4.1 (see also Fig. 3.6.2), thus rather different from the one which emerges from the specific two-nucleon transfer process (large ellipse, correlation length ξ), as testified by the observations made as a function of the width of the Josephson-like junction (distance of closest approach) transiently established in the reaction ^{116}Sn+^{60}Ni populating the **S-S** and **S-Q** channels (Sect. 7.3).

7.5 Closing the Circle

In one of the first references of this monograph,[24] entitled "Quantum Mechanics of Collision Phenomena," Born considers the elastic scattering induced by a static potential, of a beam of N electrons which cross a unit area per unit time, scattered by a static potential.[25] The stationary wavefunction describing the scattering process behaves asymptotically as

$$e^{ikz} + f(\theta, \phi)\frac{e^{kr}}{r}, \qquad \left(k = \frac{mv}{\hbar}\right). \tag{7.5.1}$$

The number of particles scattered into the solid angle $d\Omega = \sin\theta d\theta d\phi$ is given by $N|f(\theta, \phi)|^2 d\Omega$. To connect with Born notation, one has to replace $f(\theta, \phi)$ by Φ_{mn}, where n denotes the initial state plane wave in the z-direction and m the asymptotic final state in which the waves move in the direction fixed by the angles (θ, ϕ). Then Born writes that Φ_{mn} determines the probability for the scattering of the electron from the z- to the $(\theta\phi)$-direction, adding a footnote in proof, as already mentioned, stating that a more precise consideration shows that the probability[26] is proportional to the square of Φ_{mn}.

Within this context, the matrix element between the entrance and exit channel distorted waves of the function $F(\mathbf{r}_1, \mathbf{r}_2, \mathbf{r}_{Ap})$ is proportional to $f(\theta, \phi)$ and thus Φ_{mn}. The function F is not directly measurable, but it is the closest one can come to a theoretical picture connecting the Cooper pair (s+r) to experiment. For superfluid nuclei lying along the stability valley, this construct does not change

---

[24] Born (1926).

[25] In this section we follow closely Pais (1986).

[26] The motion of particles follows probability laws, but the probability itself propagates according to the law of causality. And concerning the distinction between classical and quantal probabilities, he states, "The classical theory introduces the microscopic coordinates which determine the individual processes only to eliminate them because of ignorance by averaging over their values; whereas the new theory get the same results without introducing them at all.... We free the forces of their classical duty of determining directly the motion of particles and allow them instead to determine the probability of states."

much with the theory one uses to calculate the spectroscopic two-nucleon transfer amplitudes, provided they display off-diagonal long-range order (ODLRO)[27], a property closely related to phase coherence between Cooper pairs and deformation in gauge space. This is the reason why the associated absolute cross sections are rather stable concerning the different models used to calculate them, and α_0 provides a natural order parameter to measure deformation in gauge space.

On the other hand, in the case of halo, weakly bound exotic nuclei like ^{11}Li, the sensitivity of the absolute cross sections to the structure models entering in the calculation of $F(\mathbf{r}_1, \mathbf{r}_2, \mathbf{r}_{Aa})$ is expected to be more relevant and directly related to the conspicuous changes which renormalization produces in the radial shape of the single-particle wavefunctions of the halo neutrons forming the Cooper pair. Conversely, that the order parameter of the superfluid phase α_0 is written as a sum of the Cooper pair transfer spectroscopic amplitudes (see Eq. (7.2.1)) testifies to the fact that Cooper pair tunneling is the specific probe of Cooper pairs.

Within this context, self-consistency in (NFT)$_{\text{ren}}$ implies the renormalization both of single-particle energies and occupancies ($\tilde{\epsilon}_j$, $Z_j(\omega)$) and of radial wavefunctions ($\tilde{\phi}(r)^{(i)}$). For each $\tilde{\epsilon}_j$ there can be more than one radial function, depending on whether the nucleon moves around the ground state ($i = $ gs) or around an excited state ($i = $ coll) of the core. In keeping with the fact that $\tilde{\phi}(r)^{(i)}$ enter in the formfactors associated with transfer processes, self-consistency in (NFT)$_{\text{ren}}$ is intimately related to unification of structure and reactions.

7.6 Summary

Nucleons moving in time-reversal states lying close to the Fermi energy and interacting through the short-range $^1S_0$ NN-potential, and the exchange of collective vibrations (long-range, induced pairing interaction), lead to $L = 0, S = 0$ nuclear Cooper pairs. Making use of the two-particle wavefunction describing them, it is found that the partner nucleons become angle correlated and thus come closer to each other, as compared to uncorrelated pairs of nucleons coupled to angular momentum zero.

Cooper pairs were introduced in physics to describe low-temperature superconductivity in metals, and constitute the building blocks of BCS. The BCS ground state wavefunction describes the condensation of these weakly bound (through the exchange of virtual photons, electron plasmons, and lattice phonons), widely extended (dimensions much larger than typical distances between uncorrelated electrons), strongly overlapping (quasi) bosons (pair of electrons moving in time-reversal states ($\nu, \tilde{\nu}$) with momentum and spin ($\mathbf{k} \uparrow, -\mathbf{k} \downarrow$)).

---

[27] See footnote 47 of Chapter 1. See also App. A, Potel et al. (2017), and refs. therein.

In the presence of an external field, for example, that associated with the tunneling interaction of a Josephson junction – involving energies smaller than the transition temperature and thus acting on the Cooper pairs center of mass – all pairs move in an identical (coherent) fashion, leading to a supercurrent across the junction of carriers of charge $2e$, its critical value being approximately equal to the (normal) single-electron current associated with a junction bias equal to the Cooper pair binding energy (2Δ) divided by the electron charge (e), a bias which imparts to the center of mass of the Cooper pairs being transferred a momentum of the order of the inverse of the correlation length. Under such circumstances, the back-and-forth ($\ell = 0$) radial motion of electron partners (intrinsic (v, $\tilde{v}$) motion) cannot follow suit, getting out of step and resulting in the breaking (unbinding) of the pairs (critical current). As a consequence, supercurrent ceases, and normal current of carriers of charge e flows.

The traditional view of nuclear Cooper pairs described in the first paragraph of the present section seems to be partially at odds with that found at the basis of the BCS description of low-temperature superconductivity, in particular concerning insight provided by the specific probe of spontaneous breaking of conservation (of pairs) of particle number, namely, Cooper pair tunneling across a Josephson junction. In each of the two weakly coupled superconductors, pairs of electrons recede from each other as compared to the normal density situation. In this way, the correlated Cooper pair partners lower their relative kinetic energy of confinement, profiting best from the weak, attractive, (retarded) interaction. Being the electron pair phase correlated with a correlation length much larger than the barrier thickness, although they tunnel across the junction one at a time, they do so as a single particle (of mass $2m_e$ and charge $2e$), and the probability of going through the junction is comparable to that of a single electron. It is like interference in optics, with phase coherent wave mixing.

In the transient, time-dependent, Josephson-like junction, established in a heavy ion collision between superfluid nuclei, a Cooper pair tunnels in terms of the successive transfer of two entangled nucleons. At few MeV below the Coulomb barrier, the absorptive component of the optical potential plays essentially no role in the process, and tunneling takes place lossless, free of dissipation. For such bombarding energies for which the distance of closest approach is approximately of the order but smaller than the nuclear Cooper pair mean square radius, each partner nucleon finds itself, in the transfer process, in a different nucleus, at about a correlation length of each other. The associated probability P_2 of pair transfer is, under such circumstances and within a factor of two ($(\pi/4)^2$), equal to the one-neutron transfer probability P_1.

For larger distances of closest approach, larger than the correlation length, the (intrinsic) back-and-forth relative motion characterizing the intrinsic structure of the transferred nuclear Cooper pair gets out of step, resulting in the transition from the superfluid–superfluid (S-S) transfer regime to the superfluid–quasiparticle (S-Q) one, as testified by the fact that P_1 becomes much larger than P_2. Already for distances of closest approach about 0.6 fm larger than the estimated (empirical) nuclear correlation length, the absolute one-nucleon transfer cross section is about an order of magnitude larger than that associated with two-nucleon tunneling. The nuclear (S-S)→(S-Q) (tunneling) phase transition as a function of the distance of closest approach, where the Cooper pair being transferred is probed (stressed) beyond the correlation length, parallels that observed in a Josephson junction, when Cooper pairs are biased to move with a center of mass momentum of the order of the inverse of the correlation length (of note is that ξ and $q(=\hbar/\xi)$ are conjugate variables).

The gained homogeneity between the pictures of Cooper pairs in superconductors and in nuclei is a consequence of fulfilling a basic requirement of quantum mechanics, namely, that to specify the experimental setup in discussing a physical phenomena. In the present case, one has tried to answer the question of what the structure of a nuclear Cooper pair is when probed in terms of its specific probe, namely, a probe which changes particle number in two, comparing theory with experiment in terms of absolute cross sections.

7.A One- and Two-Nucleon Transfer Computer Codes

In this appendix we provide a brief description of the numerical methods implemented in the computer codes ONE and COOPER written to evaluate the absolute differential cross sections associated with one- and two-nucleon transfer reactions.

COOPER[28]: The two-nucleon transfer differential cross section is given by Eq. (6.1.3). The principal task consists in calculating the transfer amplitudes $T^{(1)}(\theta)$, $T_{succ}^{(2)}(\theta)$, and $T_{NO}^{(2)}(\theta)$ described in Eqs. 6.1.4a–6.1.4c by numerically evaluating the corresponding integrals. The dimensionality of the integrals can be reduced by expanding in partial waves (eigenfunctions of the angular momentum operator) the distorted waves and wavefunctions present in the corresponding integrands. The resulting expressions are Eqs. (6.2.36) and (6.2.37) for $T^{(1)}(\theta)$; Eqs. (6.2.128), (6.2.129), and (6.2.130) for $T_{succ}^{(2)}(\theta)$; and Eqs. (6.2.154), (6.2.155), and (6.2.156) for $T_{NO}^{(2)}(\theta)$. The integrals are computed numerically with the method of Gaussian quadratures.

The one-dimensional (radial) functions (6.1.2) appearing in the integrands are defined in a spatial grid up to a given maximum radius r_{max}. The bound state wavefunctions are obtained by numerical integration of the radial Schrödinger equation

---

[28] Potel (2012b).

Table 7.A.1 *Energies (in MeV) of the single-particle orbitals involved in the description of the ground state of* ^{120}*Sn. The second column displays the energies of the single-particle orbitals according to prescription (I). The* $s_{1/2}$ *single-particle state, associated with the* $1/2^+$ *ground state of* ^{119}*Sn, has a binding energy of half the two-neutron separation energy of* ^{120}*Sn,* $S_{2n}(^{120}$*Sn)=15.59 MeV. Calculation (II) (third column) corresponds to binding all the single-particle wavefunctions with half the two-neutron separation energy of* ^{120}*Sn. The last column displays the spectroscopic amplitudes B obtained from a BCS calculation of* ^{120}*Sn (see Table 7.2.3).*

	^{120}Sn (I)	^{120}Sn (II)	B
$d_{5/2}$	−9.49	−7.79	0.41
$g_{7/2}$	−8.89	−7.79	0.57
$s_{1/2}$	−7.79	−7.79	0.41
$d_{3/2}$	−7.29	−7.79	0.66
$h_{11/2}$	−5.89	−7.79	1.03

for a Woods–Saxon potential with a spin-orbit term. Along with the optical potentials governing the relative nucleus–nucleus motion in the initial, intermediate, and final channels, they conform to the basic structure input needed to define the calculation. They are generated by solving a one-body Schrödinger equation for a Woods–Saxon potential of standard geometry, where the radius and diffusivity parameters are given as input. The depth of the potential is adjusted to reproduce the binding energy of the state under consideration, according to the prescription defined below. The resulting potential corresponding to the final (initial) nucleon bound state stands also for the interaction potential featured in the integrand in the prior (post) representation.

The binding energies of the single-particle wavefunctions determine the asymptotic behavior of the transfer formfactors, an essential ingredient for a quantitative account of the two-nucleon transfer cross section.[29] According to the prescription followed in the calculations discussed in the present monograph, the orbital corresponding to the spin and parity of the ground state (quasiparticle) of the $A + 1$ system is bound by half the experimental two-neutron separation energy S_{2n} of the $A + 2$ system (prescription (**I**)). The energies of the remaining single-particle states relative to the $A + 1$ ground state are determined from a structure calculation.

---

[29] For an example with a simple discussion of the problem, see Fig. 6.4.4.

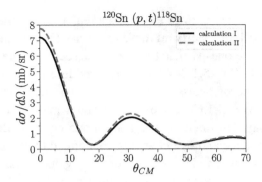

Figure 7.A.1 Numerical computation of the two-neutron transfer cross section corresponding to the reaction ^{120}Sn$(p, t)^{118}$Sn(gs). The two calculations shown in the figure correspond to the two different prescriptions for defining the binding energies of the single-particle orbitals described in the caption of Table 7.A.1. For the color version of this figure, refer to cambridge.org/nuclearcooperpair.

We illustrate this prescription with the reaction ^{120}Sn$(p, t)^{118}$Sn(gs) ($E_p = 21$ MeV; see Table 7.A.1 (second column) and Fig. 7.A.1 (continuous line)). To assess the impact the binding energies of the single-particle states have in the absolute differential cross sections, we also show in Fig. 7.A.1 the result of a calculation in which the single-particle states are all bound by $\frac{S_{2n}(^{120}\text{Sn})}{2}$ (see Table 7.A.1, third column).

In the calculation of the successive transfer amplitude, the energy of the different contributions to the Green function which propagate the intermediate channel $d +$ $(A + 1)$ (Eq. 6.1.11) are taken at the corresponding (experimental) energies of the $(A + 1)$ system. Off-shell effects due to the virtual population of the different $(A + 1)$ intermediate states are thus taken into account.[30]

The distorted waves are obtained by integrating the radial Schrödinger equation with positive energy from $r = 0$ to r_{max} and matching the solution with the corresponding Coulomb wave function at a given $r = r_{match}$, large enough to lie outside of the range of the nuclear interaction. The Woods–Saxon optical potentials used to obtain the distorted waves consist of a real Coulomb term, a real and imaginary

---

[30] It is of note that this approach has to be modified when one deals with systems in which polarization effects play a comparable, let alone larger, role than the mean field in determining the radial formfactors or when the continuum enters the description of the structure of any of the channels (initial, intermediate, final) contributing to the two-nucleon transfer process. An example where both features are present is provided by the reaction ^{1}H$(^{11}$Li,^{9}Li)^{3}H, which virtually populates the unbound nucleus ^{10}Li in the intermediate channel $d+^{10}$Li. In this case, the two-particle self-energy matrix is diagonalized in a spherical spatial box in order to discretize the continuum, making use of techniques similar to those described in Sect. 5.2.3. The resulting formfactor is read by the code as a two-dimensional numerical grid and provides an essential ingredient for the computation of absolute differential cross sections that can be compared with the experimental results (see Fig. 7.1.3).

volume term, an imaginary surface term, and a real and imaginary spin orbit term. The parameters needed to specify all those terms are given as an input.

ONE[31]: To obtain the one-particle transfer cross sections, one has to evaluate the amplitudes defined in Eq. (5.1.26). The corresponding numerical integrations are performed making use of Gaussian quadratures. The radial parts, of the single-particle wavefunctions (5.1.6) and of the partial waves (5.1.7) and (5.1.8) are generated in the way described above. The kinetic energies defining the distorting waves in the entrance and exit channels are defined by the bombarding energy in the center of mass frame and the experimental Q-value of the reaction. The binding energies corresponding to the initial and final single-particle wavefunctions are taken to be the experimental separation energy for each state. It is of note that the difference between the initial and final binding energies should be consistent with the reaction Q-value.

7.B Statistics

Consider two identical particles moving in a one-dimensional harmonic oscillator. It is assumed that one is in the ground state and the other is in the first excited state. According to the superposition principle

$$\Phi(x_1, x_2) = \lambda\phi_1(x_1)\phi_0(x_2) + \mu\phi_0(x_1)\phi_1(x_2), \tag{7.B.1}$$

we calculate the correlation of these particles, that is, the quantity

$$Corr = \frac{\langle x_1 x_2 \rangle - \langle x_1 \rangle \langle x_2 \rangle}{\sqrt{\left(\langle x_1^2 \rangle - \langle x_1 \rangle^2\right)\left(\langle x_2^2 \rangle - \langle x_2 \rangle^2\right)}} \tag{7.B.2}$$

starting with

$$
\begin{aligned}
\langle x_1 x_2 \rangle &= \int dx_1 dx_2 \left(\lambda^*\phi_1^*(x_1)\phi_0^*(x_2) + \mu^*\phi_0^*(x_1)\phi_1^*(x_2)\right) \\
&\quad \times (x_1 x_2)\left(\lambda\phi_1(x_1)\phi_0(x_2) + \mu\phi_0(x_1)\phi_1(x_2)\right) \\
&= |\lambda|^2 \langle\phi_1|x_1|\phi_1\rangle\langle\phi_0|x_2|\phi_0\rangle + \lambda^*\mu\langle\phi_1|x_1|\phi_0\rangle\langle\phi_0|x_2|\phi_1\rangle \\
&\quad + \lambda\mu^*\langle\phi_0|x_1|\phi_1\rangle\langle\phi_1|x_2|\phi_0\rangle + |\mu|^2\langle\phi_0|x_1|\phi_0\rangle\langle\phi_1|x_2|\phi_1\rangle. \tag{7.B.3}
\end{aligned}
$$

In keeping with the fact that

$$\langle\phi_1|x|\phi_1\rangle = \langle\phi_0|x|\phi_0\rangle = 0, \tag{7.B.4}$$

and

$$\langle\phi_0|x|\phi_1\rangle = \langle\phi_1|x|\phi_0\rangle = \sqrt{\frac{\hbar\omega}{2C}}, \tag{7.B.5}$$

---

[31] Potel (2012a).

one obtains

$$\langle x_1 x_2 \rangle = \left(\frac{\hbar \omega}{2C} \right)^2 \Re(\lambda^* \mu) \tag{7.B.6}$$

and

$$\sqrt{} = \left(\frac{\hbar \omega}{2C} \right)^2 \tag{7.B.7}$$

for the denominator of Eq. (7.B.2). From the above results the correlation function between particles 1 and 2 is

$$Corr = 2\Re(\lambda^* \mu) = \begin{cases} 1 & \left(\lambda = +\mu = \frac{1}{\sqrt{2}} \right) \\ -1 & \left(\lambda = -\mu = \frac{1}{\sqrt{2}} \right). \end{cases} \tag{7.B.8}$$

It is of note that, in quantum mechanics, average values imply the mean outcome of a large number of experiments, in this case, of the (simultaneous) measure of the position of the two particles.[32]

7.C Correlation Length and Generalized Quantality Parameter

The correlation length can be defined as[33]

$$\xi = \frac{\hbar v_F}{\pi \Delta} \approx \frac{\hbar^2}{m} \frac{k_F}{\pi \Delta}, \tag{7.C.1}$$

where the Fermi momentum is, in the case of stable nuclei lying along the stability valley,

$$k_F \approx 1.36 \, \text{fm}^{-1}. \tag{7.C.2}$$

Then

$$\xi = 40 \, \text{MeV fm}^2 \times \frac{1.36}{\pi \Delta} \, \text{fm}^{-1} \approx \frac{17}{\Delta} \, \text{fm} \tag{7.C.3}$$

and

$$\xi \approx 14 \, \text{fm}, \quad (\Delta \approx 1.2 \, \text{MeV}). \tag{7.C.4}$$

Thus, the associated (generalized) quantality parameter is, in the present case,

$$q_\xi = \frac{\hbar^2}{2m\xi^2} \frac{1}{2\Delta} \approx 0.04. \tag{7.C.5}$$

---

[32] Basdevant and Dalibard (2005).
[33] See, e.g., Annett (2013), p. 62.

That is, the two partner nucleons are, in the Cooper pair, rigidly correlated with each other.

We now consider $^{11}_{3}\text{Li}_8$ and calculate k_F (neutrons) with the help of the Thomas–Fermi model[34]

$$k_F = \left(3\pi^2 \frac{8}{\frac{4\pi}{3}(4.58)^3}\right)^{1/3} \text{fm}^{-1} \approx \frac{(18\pi)^{1/3}}{4.58} \text{fm}^{-1} \approx 0.8 \, \text{fm}^{-1}. \qquad (7.\text{C}.6)$$

The correlation length can, in the present case, be calculated in terms of the correlation energy ($E_{corr} \approx -0.5$ MeV)

$$\xi \approx \frac{\hbar v_F}{\pi |E_{corr}|} \approx \frac{200 \, \text{MeV} \times \text{fm} \times 0.16}{\pi \times 0.5 \, \text{MeV}} \approx 20 \, \text{fm}, \qquad (7.\text{C}.7)$$

the resulting generalized quantality parameter being

$$q_\xi = \frac{\hbar^2}{2m\xi^2} \frac{1}{|E_{corr}|} \approx 0.1. \qquad (7.\text{C}.8)$$

It is of note that this result is but an alternative embodiment of the relation (4.4.8). One could argue that both (4.4.8) and (7.C.8) (as well as (7.C.5) for stable nuclei) are just a manifestation of (7.B.8). That there is more to it is forcefully expressed by the fact that selecting the pure two-particle configuration $|s^2_{1/2}(0)\rangle$ ($|p^2_{1/2}(0)\rangle$) to describe the halo neutron Cooper pair of ^{11}Li leads to absolute two-particle transfer cross sections which are about one order of magnitude larger (smaller) than the observed cross section (see Fig. 7.1.2). The fact that the NFT result (7.1.2)–(7.1.4) reproduces observations within experimental error underscores the central role played by structure on Cooper pair tunneling, through the emergent property of generalized pairing rigidity.

Summing up, pairing correlations can modify the statistics of the elementary modes from fermionic to (quasi) bosonic ones. This takes place provided[35] that $E_{intr} \ll 2\Delta$, $q \ll \hbar/\xi$, and $D_0 \lesssim \xi$. At the same time, the value of the quantality parameter changes from $q \approx 1$ to $q_\xi \ll 1$, that is, from a regime of delocalized single nucleons to one of strongly overlapping, independent pair motion, each being governed by the same phased wavefunction[36] $(U'_\nu + V'_\nu e^{-2i\phi} P^\dagger_\nu)|0\rangle$ and behaving as a single entity. The operator $P^\dagger_\nu$, being a product of two fermions, does not

---

[34] Quantity which can be related to (v_F/c) according to
$(v_F/c) = \hbar k_F/(mc) = (\hbar c/(mc^2))k_F \approx 0.2(k_F)\text{fm}^{-1})$. In the case in which $k_F \approx 0.8$ fm^{-1} (see Eq. (7.C.6)), $v_F/c \approx 0.16$.

[35] Where $E_{intr}(q)$ stands for the energy (momentum) given to the intrinsic (center of mass) motion of the transferred Cooper pair. The quantity D_0 is the distance of closest approach at which the two-nucleon transfer process takes place, ξ being the correlation length. Within this context, see Sects. 3.4 and 4.7.1. See also footnote 46 of Chapter 3.

[36] Since $P^\dagger_\nu$ commutes for different νs, $|BCS\rangle$ represents uncorrelated occupancy of the various pair states.

fulfill Bose statistics $((P_\nu^\dagger)^2 = 0)$. This property implies the presence of a pairing gap not only for breaking a pair but also for making a pair move differently from the others.[37] As a result, one has (off-diagonal) long-range order in the superfluid nuclear system (ODLRO).[38] This effect leads to generalized gauge rigidity, the detailed renormalizing and dressing mechanisms ultimately deciding on the soundness and applicability of the description under discussion. The fact that in working out the reaction mechanism, one uses, for practical reasons, a single-particle basis (second-order DWBA corrected by nonorthogonality), reconstructing the pair correlations in terms of sums over virtual states is at the basis of the two-neutron transfer physical sum rule discussed in Sect. 7.2.2.

The extension of the validity of the picture of superconductivity in metals based on the motion of $\approx 10^6$ phase-correlated Cooper pairs, down to few (5–8, ^{116}Sn) and eventually one Cooper pair (^{11}Li), seems short of a wonder, in particular in this last case, in which the binding of the Cooper pair is essentially due to a quantum-mechanical many-body retarded interaction resulting from the exchange of a collective vibration. This collective mode at the edge between a soft-$E1$ mode and a Pygmy dipole resonance (see App. 7.D), the associated ZPF affecting, among other things, nuclear diffusivity and the neutron skin.

Within this context, the sense of wonder is common with the emergence of other unexpected quantum materials and macromolecules as a result of interactions resulting from ZPF, like the van der Waals interaction responsible, inter alia, for the solid made out of C_{60} fullerene, as well as associated doped superconductors (Rb_3C_{60}, K_3C_{60}, Rb_2CsC_{60}, etc.)[39], let alone the Casimir interaction and the related (weak and strong)[40] hydrophobic force, which plays an essential role in the folding of the macromolecules responsible for metabolism[41], namely, proteins (see App. 7.E).

7.D Multipole Pairing Vibrations

Although much work has been carried out concerning multipole pairing vibrations[42], that is, modes with transfer quantum numbers $\beta = \pm 2$ and multipolarity

---

[37] See Eqs. (1.4.53) and (1.4.54).

[38] See footnote 47 of Chapter 1.

[39] Hebard et al. (1991); Rosseinsky et al. (1991); Holczer et al. (1991); Fleming et al. (1991); Anderson (1991); Zhou et al. (1992); Hebard (1992); Gunnarsson (1995, 1997, 2004); see also Broglia et al. (2004), and refs. therein.

[40] Chandler (2002).

[41] Dyson (1999).

[42] See Brink and Broglia (2005), Sect. 5.3 and refs. therein. See also Broglia et al. (1974); Ragnarsson and Broglia (1976); Broglia et al. (1971a, 1971b); Bès and Broglia (1971a, 1971b); Flynn et al. (1971); Bès et al. (1972); Broglia (1981); Bohr and Mottelson (1980); Flynn et al. (1972a); Bortignon et al. (1976); see also Kubo et al. (1970), and refs. therein.

and parity λ^π different from 0^+, this remains a chapter essentially missing from the subject of pairing in nuclei, arguably, with the partial exception of quadrupole pairing studied in the multiphonon pairing vibrational spectrum[43] around the closed shell ^{208}Pb, and in connection with strongly populated 0^+ pairing vibrational states in the actinide region[44] and in the quadrupole and hexadecapole pairing vibrations in the multiplet spectrum[45] of ^{209}Bi.

In what follows, we elaborate on the new insight on pairing vibrational modes the studies of two-neutron pickup reactions on ^{11}Li have opened. As already explained, because of the small overlap existing between halo neutrons and core nucleons, both the $^1S_0$, NN-interaction and the symmetry potentials become strongly screened, resulting in a subcritical value of the pairing strength and in a weak repulsion to separate protons from neutrons in the dipole channel. As a result, neither the $J^\pi = 0^+$ correlated neutron state (Cooper pair) nor the $J^\pi = 1^-$ one (vortex-like)[46] pair addition modes are bound (although both qualify to do so) to the $N = 6$ closed-shell system (^{9}Li).

The bare NN-pairing interaction being subcritical, the two neutrons can correlate their motion by exchanging vibrations of the medium in which they propagate, namely, the halo. These modes can hardly be the $\lambda = 2^+, 3^-$, or 5^- surface vibrations found in nuclei lying along the stability valley. This is because the diffusivity of the halo is consistent, and it blurs the very definition of surface. It is of note that those associated with the core (2^+ see Fig. 7.D.1 (c), and eventually also $3^-, 5^-$, etc.) provide some glue, but insufficient to bind the halo dineutron system.

As already mentioned, the next alternative is bootstrapping, a process in which the two partners of the (monopole) Cooper pair exchange pairs of vortices (dipole Cooper pair), as well as one dipole Cooper pair and a quadrupole pair removal mode, while those of the vortex exchange pairs of Cooper pairs (monopole pairing vibrations), but also two dipole pairs, as shown in Figs. 7.D.1 and 7.D.2. In other words, by liaising with each other, the two dineutron contenders at the role of the ^{11}Li ground state settle the issue. As a result, the Cooper pair becomes weakly bound ($S_{2n} = 380$ keV), the vortex state remaining barely unbound, a few hundreds of keV above threshold. There is, in principle, no physical reason why things could not have gone the other way. Within this context we refer to ^{3}He superfluidity, where condensation involves $S = 1$ pairs. It is of note that we are not considering spin degrees of freedom in the present case, at least not dynamic ones.

---

[43] See Flynn et al. (1972a).
[44] Casten et al. (1972); Bès et al. (1972); Ragnarsson and Broglia (1976). It is of note that β-vibrations and monopole pairing vibrations become mixed in quadrupole deformed nuclei.
[45] Bortignon et al. (1976).
[46] One of the effects of a vortex is that it allows rotation of a quantum system to take place about an axis of symmetry.

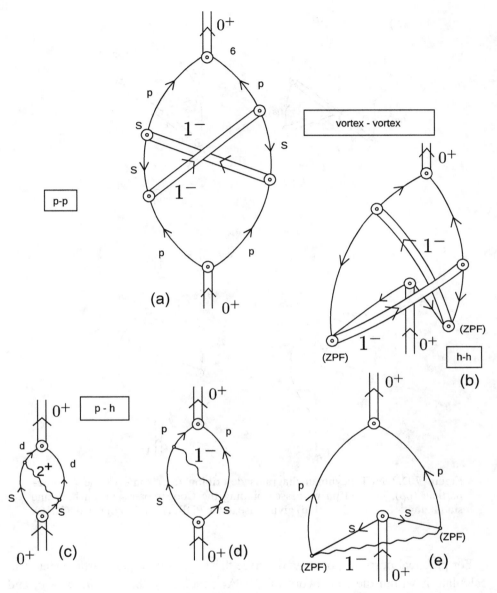

Figure 7.D.1 NFT–Feynman diagrams describing the interweaving between the neutron halo pair addition monopole and dipole modes (double arrowed lines labeled 0^+ and 1^-, respectively). Above, the exchange of dipole modes binding the 0^+ pair addition mode through forward-going particle–particle p-p (h-h) components. Below, the assumption is made that the PDR of ^{11}Li can be viewed as a p-h (two quasiparticle), QRPA mode. After Broglia et al. (2016).

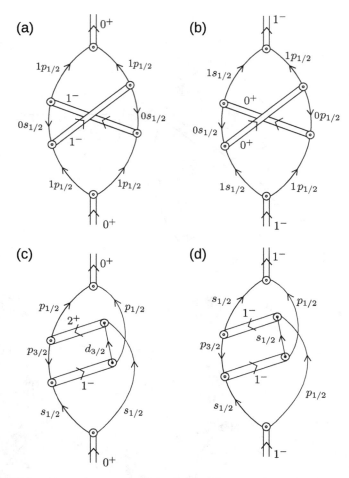

Figure 7.D.2 NFT–Feynman diagrams describing (a,c) some of the particle–particle (pp), hh, and ph processes binding the Cooper pair neutron halo and stabilizing ^{11}Li, as well as (b,d) giving rise to the PDR. After Broglia et al. (2016).

For practical purposes, one can describe the 1^- as a two-quasiparticle state and calculate it within the framework of QRPA[47], including the $p - h$, $p - p$, and $h - h$ channels and adjusting the strength of the dipole–dipole separable interaction to reproduce the experimental findings (Fig. 4.12.1). In this representation, one can refer to it as a Pygmy dipole resonance (PDR).[48] Exchange between the two partners of the Cooper pair (Fig. 7.D.1(d)) leads to essentially the right value of the dineutron binding to the ^{9}Li core. Within this context, one can view the ^{11}Li neutron

[47] Broglia et al. (2019a).
[48] Barranco et al. (2001); Broglia et al. (2019c).

halo as a van der Waals Cooper pair (Fig. 7.D.1 (**e**); see also Sect. 3.7). The trans-formation between this picture and that discussed in connection with (**a**) and (**b**) as well as with Fig. 7.D.2 can be obtained expressing the PDR, QRPA wavefunction in terms of particle creation and destruction operators (Bogoliubov–Valatin trans-formation) as seen from Fig. 7.D.1 (**a**) and (**b**). A vortex–vortex stabilized Cooper pair emerges.

Which of the two pictures is more adequate to describe the dipole-mediated con-densation is an open question, as each reflects important physical properties which characterize the PDR. In any case, both indicate the symbiotic character of the halo Cooper pair addition mode and of the pygmy dipole resonance built on top of and almost degenerate with it. Insight into this question can be obtained by shedding light on the question of whether the velocity field of each of the symbiotic states is more similar to that associated with irrotational or vortex-like flow.[49] Two-nucleon transfer reactions, a specific probe of (multipole) pairing vibrational modes, con-tain many of the answers to the above question (Fig. 7.D.3), to the extent that one is able to calculate absolute differential cross sections within 10 per cent accuracy. In fact, ground state correlations will play a very different role in the absolute value of the ^{9}Li(t,p)^{11}Li (1$^-$) cross section, depending on which picture is correct. In the

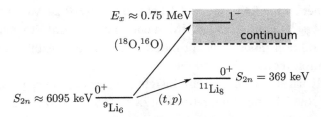

Figure 7.D.3 Schematic representation of levels of ^{11}Li which can be populated in two-nucleon transfer reactions, for example, (t, p), $(^{18}O^{16}O)$, and so on. Indi-cated in keV are the two-neutron separation energies S_{2n}. In labeling the different states, one has not considered the quantum numbers of the $p_{3/2}$ odd proton. After Broglia et al. (2016).

---

[49] See Repko et al. (2013). Within this context, and making use of an analogy, one can mention that a consistent description of the GQR and of the GIQR is obtained assuming that the average eccentricity of neutron orbits is equal to the average eccentricity of the proton orbits (Bès et al. (1975b)), the scenario of neutron skin. The isoscalar quadrupole–quadrupole interaction is attractive. Furthermore, the valence orbitals of nuclei have, as a rule, and aside from intruder states, homogeneous parity. These facts preclude the GQR to play the role of the PDR. In fact, there will always be a low-lying quadrupole vibration closely connected with the aligned coupling scheme and thus with nuclear plasticity. Within this context one can nonetheless posit that the GQR, related to neutron skin, is closely associated with the aligned coupling scheme. Making a parallel, one can posit that the PDR is closely connected with vortical motion. Arguably, support for this picture is provided by the low-lying E1 strength of ^{11}Li. It results from the presence of $s_{1/2}$ and $p_{1/2}$ orbitals almost degenerate and at threshold, leading to a low-lying Cooper pair coupled to angular momentum 1$^-$ (dipole pair addition mode)–the scenario of vortical motion.

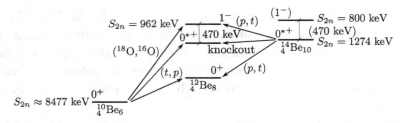

Figure 7.D.4 Levels of ^{12}Be expected to be populated in two-nucleon transfer and knockout processes. S_{2n} are the two-neutron separation energies. After Broglia et al. (2016).

case in which it can be viewed as a vortex (pair addition dipole mode), it will lead to an increase of the two-particle transfer reaction (positive coherence). It will produce the opposite effect if the correct interpretation of the PDR is that of a more $(p - h)$-like excitation.[50] Insight in to the above question may also be obtained by studying the properties of a quantal vortex in a Wigner cell with parameters which approximately reproduce the halo of ^{11}Li. Within this context, and for the sole purpose of providing an analogy, we refer to what has been done in the study of quantal vortices in the environment of neutron stars.[51]

A possible test of the soundness of the physics discussed above concerns the question of whether the first excited, 0^+ halo state (E_x = 2.24 MeV) of ^{12}Be can be viewed as the $|\text{gs}(^{11}\text{Li})\rangle$ in a new environment, in other words, to consider the halo neutron pair addition mode, a novel mode of elementary excitation: the neutron halo pair addition mode, of which the $|1^-(^{12}\text{Be})$; 2.71 MeV$\rangle$ is a fraction of its symbiotic PDR partner. One can gain insight concerning this question, by eventually measuring the E1-branching ratio $|1^- (2.71 \text{ MeV})\rangle \to |0^{+*} (2.24 \text{ MeV})\rangle$) and possibly finding other low-energy E1-transitions populating the 0^* state, as well as through a two-nucleon stripping process and two-nucleon pickup and knockout reactions (Fig. 7.D.4). A resumé of the picture discussed above is given in Fig. 4.12.1.

7.E Vacuum Fluctuations and Interactions: The Casimir Effect

In Fig. 7.E.1 (I) (a), an example is given of zero-point fluctuations (ZPF) of the nuclear vacuum (ground state) in which a surface quantized vibration and an uncorrelated particle-hole mode get virtually excited for a short period of time. Adding a nucleon (odd system; Fig. 7.E.1 (I) (b)) leads, through the particle–vibration

---

[50] Broglia et al. (1971c).
[51] Avogadro et al. (2007, 2008).

coupling mechanism, to processes which contain the effect of the antisymmetry between the single-particle explicitly considered and the particles out of which the vibrations are built (Fig. 7.E.1 (I)(c)). Time ordering gives rise to the graph shown in Fig. 7.E.1 (I)(d). Processes I(c) and I(d), known as correlation (CO) and polarization (PO) contributions to the mass operator, dress the particles and lead to physical nucleons whose properties can be compared with the experimental findings with the help of specific experimental probes, for example, one-particle transfer reactions. Summing up, the processes shown in Fig. 7.E.1 (I) are examples of quantum field theory phenomena. They testify to the fact that the dressing of nucleons is at the basis of the quantal description of the atomic nucleus.

Nuclear superfluidity at large, and its incipience in the single Cooper pair–like case, for example, in ^{11}Li in particular, are among the most quantal of all the phenomena displayed by the nuclear many-body system. Even if the bare $^1S_0$, NN-interaction was not operative, or was rendered subcritical by screening effects, as in the case of ^{11}Li, Cooper binding will still be possible, as a result of the exchange of vibrations between pairs of physical (dressed) nucleons (Fig. 7.E.1 (II) (a), (c)) moving in time-reversal states close to the Fermi energy (Fig. 7.E.1 (II)(b), (d)–(g)). This is a consequence of the ZPF associated with the nuclear vacuum (ground) state. *Having arrived at this point, the reader may find it useful to read again Sect. 1.9, in particular the last paragraphs.*

7.E.1 Measuring QED Vacuum Fluctuations

The many-body character of the attractive interaction between nucleons arising from the nuclear (vacuum) ZPF discussed above, and important to describing phenomena occurring close to the surface of the Fermi sea, has a classical analogue, fittingly, a classical phenomenon which emerges from maritime experience.

At sea, on a windless day in which the water surface resembles a mirror, free-floating ships singly or in groups do not do much; they just stay put. The situation is quite different in a strong swell, still in a windless situation. In this case, single, isolated ships end up lying parallel to the wave crests (see Fig.7.E.2(a)) and start rolling heavily. In the days of the clipper ships, it was believed that under those circumstances, two vessels at close distance attracted each other. This is in keeping with the fact that the rigging of the rolling ships often became entangled, leading to disaster. It was not until quite recently[52] that a quantitative understanding of the phenomenon (based on knowledge of similar quantal effects) was achieved,

---

[52] Boersma (1996).

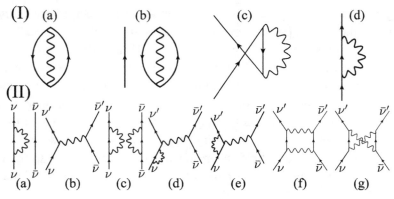

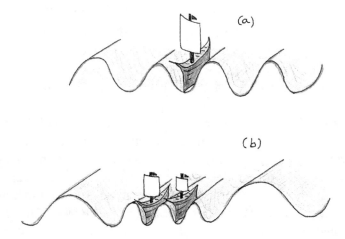

Figure 7.E.1 **(I)** (a) ZPF associated with (particle-hole) surface vibrations; (b) same process that in (a) but in the case of an odd system; (c) the antisymmetrization between the particles considered explicitly and those involved in the vibration; (d) time ordering of (c). Diagrams (c) and (d) lead to the clothing of single-particle motion in lowest order in the particle–vibration coupling vertex. **(II)** A dressed nucleon moving in a state ν in the presence of: (a) a bare nucleon moving in the time reversed state $\bar{\nu}$, (c) another dressed nucleon. Exchange of vibration in (a) leads to (b), NFT lowest-order contribution to the induced pairing interaction, in the particle–vibration coupling vertex. Exchange of vibrations in (c) leads to (d) self-energy, (e) vertex correction of the induced pairing interaction; (f) diagram contributing to the induced pairing interaction. The symmetrization between the bosons displayed in (c) is shown in (g).

Figure 7.E.2 Schematic of the behavior of an isolated ship at sea in a situation of no wind but of strong swell (a), and of two ships close by under similar conditions (b). For the color version of this figure, refer to cambridge.org/nuclearcooperpair.

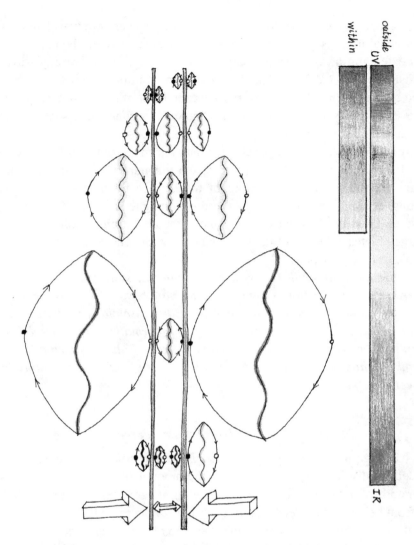

Figure 7.E.3 Casimir effect. Two metallic isolated, conducting plates (gray vertical sections) in vacuum attract each other when they are placed at very small distances (of the order of micron). This is known as the Casimir effect. The origin of such a force can be traced back to quantal zero point fluctuations (ZPF) of the electromagnetic vacuum. In the figure a cartoon of such processes is given. Virtual electrons (e^-, solid dots) and positrons (e^+, open dots) pop up of the vacuum together with a photon, and travel for short distances and times. In their way, some of them hit the plates. Because the range of wavelengths allowed between the plates is smaller than the full spectrum allowed for the photons associated with the electromagnetic ZPF in the right and left unlimited halves, more fermions or bosons will be knocking the plates from outside than from the in between region, thus leading to an imbalance of the "quantal" pressure and consequently to an effective attractive force. For the color version of this figure, refer to cambridge.org/nuclearcooperpair.

providing evidence that the old tale was true. Only waves with wavelength smaller than or equal to the separation of the ships can exist between them. In the region of sea extending away from the ships to the horizon, waves of any wavelength can exist (see Fig. 7.E.2(b)). This fact results in an imbalance between the forces exerted by the internal (between ships) waves in favor of those exerted by the external waves, leading to a net attraction. Quantum mechanically, such an effect is known as the Casimir effect.[53]

Two conducting, neutral plates at very small distances, of the order of the micron ($1\mu = 10^{-6}$ m), attract each other due to the imbalance in electromagnetic field pressure exerted by the bombarding of the surface by photons, electrons, and positrons, arising from the ZPF of the electromagnetic field[54] (see Fig. 7.E.3). It is of note that the Casimir effect, namely, the attraction between two metallic, uncharged plates, which, in Fig. 7.E.3, have been drawn as plane surfaces, can in principle have any shape.

Summing up, in a similar way in which one can state that there are neither bare fermions (e.g., electrons or nucleons) nor bare bosons (or quasi-bosons) (e.g., photons or surface vibrations), one can posit that essentially all bare forces are, with varied degrees of complexity and strength, renormalized by many-body effects. This message comes also from the least suspected of all fundamental forces, the Coulomb interaction, responsible for all of chemistry and most of biophysics (see also Sect. 4.8.3).

7.E.2 The Hydrophobic Force

Water has three noteworthy properties: liquid water is heavier than ice and it has an exceptionally large specific heat[55] – the amount of heat needed to raise the temperature of 1 kg of mass 1 degree Kelvin at room temperature ($\approx$300 K) – aside from being an excellent solvent due to its dipole moment.

All these properties follow from the structure of the water molecule H_2O. This, in turn, is connected with the directional, anisotropic structure of the valence, electronic distribution of O. Oxygen is an open-shell atom[56], having in its ground state only four electrons in the last occupied $2p$-orbital, which can host six ($|gs\rangle = |1s^2\rangle|2s^2\rangle|2p^4\rangle$). It can thus use at profit the electrons of the two H atoms

---

[53] Casimir (1948).

[54] As stated in the last sentences of the caption to Fig. 7.E.3, long wavelengths play the central role in the Casimir effect. A recurring property of the modes renormalizing the bare interaction or the associated collective variables (elementary modes of excitation) found in condensed matter (phonons of much lower frequency (ω_{ip}) than plasmons (ω_{ep}), nuclei (collective quadrupole and soft-E_1 (pygmy) modes ($\hbar\omega_2$, $\hbar\omega_{PDR}$) $\ll \hbar\omega_{GR}$), proteins (in this connection see Micheletti et al. (2004, 2001, 2002); Hamacher (2010); see also footnote 68 of Chapter 1).

[55] Water absorbs at 25°C, 4.18 joules/(g K) of heat when the temperature of 1 g is increased 1°C. For the sake of comparison, it takes $\approx$2.0 joules/(g K) of heat to raise 1 g of acetic acid, acetone, or ethanol 1°C.

[56] See, e.g., Greiner (1998).

to dynamically become a closed-shell system (^{20}Ne noble gas–like electronic configuration; Fig. 7.E.4 (c)).

Hybridization between the four orbitals $|2s\rangle$, $|2p_x\rangle$, $|2p_y\rangle$, and $|2p_z\rangle$ leads to a tetrahedral correlation in which the four orbitals $|i\rangle$ point toward the corners of a tetrahedron with the oxygen atom at the center (Figs 7.E.4 (a) and 7.E.5). Because the electronic distribution has its charge center closer to the oxygen atom than to the two protons of the H atoms, H_2O has a sizable dipole moment[57] ($\approx 0.68ea_0 \approx 0.6$ D, e being the electron charge, a_0 the Bohr radius, and D the Debye unit). Water molecules can form four hydrogen bonds[58] (hb; see Fig. 7.E.4 (b); see also Fig. 7.E.6), a special bond between molecules which is produced in situations when they share a hydrogen nucleus between them. The molecule's two hydrogen atoms form two bonds with neighboring oxygens, while the molecule's two lone pairs interact with neighboring hydrogens.[59]

Water and oil do not mix. The term *hydrophobic* (water-fearing) is commonly used to describe substances that, like oil, do not mix (dissolve) with (in) water. Although it may look as if water repels oil, these two types of molecules actually attract each other, for example, through the van der Waals interaction, but not nearly as strongly as water attracts itself. Mixing enough oil (hydrophobic, nonpolar (NP) molecules) with water leads to a reduction in favorable bonding. Strong mutual attraction between water molecules induces segregation of NP molecules from water and results in an effective NP-NP (hydrophobic) attraction, as observed in surface force measurements.[60] The loss of hydrogen bonds near the two extended hydrophobic surfaces depicted in Fig. 7.E.7 causes water to move away from those surfaces[61], producing thin vapor layers. Fluctuations in the interfaces formed in

---

[57] The dipole moment of a polar molecule is defined as $u = ql$, where l is the distance between the two charges $+q$ and $-q$. Thus, for two electronic charges $q = \pm e$ separated by $l = 0.1$ nm, the dipole moment is $u = 1.602 \times 10^{-19}$ C$\times 10^{-10}$ m $= 4.8$ D, where the Debye $= 1$D $= 3.336 \times 10^{-30}$ Cm is the unit of dipole moment (C stands for coulomb and m for meters). Small polar molecules have moments of the order of 1D (see, e.g., Israelachvili (1985)).

[58] Let us consider a H atom in a covalent bond with oxygen. When a second oxygen atom approaches the H atom, its nucleus, the proton, sees a potential with two minima and tunnels through the corresponding barrier from one minimum to the other. In other words, the effective potential in which the proton moves becomes broader as compared to the single oxygen potential. Thus, the quantum mechanical confinement kinetic energy decreases by roughly a factor of two. This implies that the order of magnitude energy of a hydrogen bond between two oxygen atoms corresponds to the difference in the corresponding ZPF energies, i.e., ≈ 200 meV (≈ 0.38 eV-0.19 eV ≈ 0.2 eV ≈ 4.6 kcal/mol (≈ 8 kT), where kT (≈ 0.6 kcal/mol ≈ 25 meV) is the thermal energy at room temperature (≈ 300 K)). For comparison, a covalent bond corresponds to ≈ 96 kcal/mol ≈ 4 eV, while the van der Waals interaction energy between two H (Eqs. (3.B.18) and (3.B.19)) at a distance of 2.4 Å, that is, of the order of the hydrogen bond length of 1.83 Å, is ≈ 20 meV (see Povh and Rosina (2002)).

[59] A possible pedestrian explanation of this is to impersonate a water molecule. Quoting from Ball (2003): "Your hands are hydrogen atoms, your ankles are the lone pairs of electrons of oxygen. Stand legs apart... Twist 90° at the waist, stretch your arms and you're H_2O. The way that water molecules join up has just one rule: hands can grab ankles, but nothing else. That grasp is an hydrogen bond."

[60] See, e.g., Chandler (2002, 2005); Lum et al. (1999), and refs. therein.

[61] Hydrophobic molecules do not hydrogen bond to water, creating excluded regions where the density of water molecules vanishes. When these units are small enough, water can reorganize near them without losing

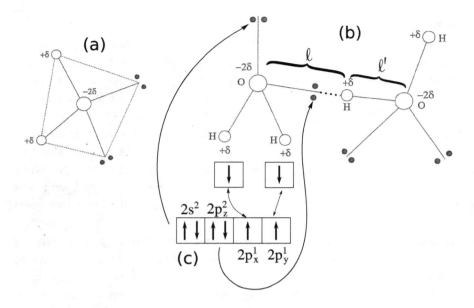

Figure 7.E.4 (a) To understand the behavior of water molecules, one has to realize that in the covalent bonds, where the oxygen and the hydrogen atoms share a couple of electrons, the corresponding electronic cloud is somewhat more concentrated on the oxygen than on the hydrogen. The oxygen acquires a partial charge -2δ while each hydrogen gets a positive one $+\delta$. (b) Such charges give rise to an attractive electric force between close lying molecules, in which the hydrogen atoms of a molecule points to the oxygen atom of the other molecule. One can view this attractive force as a type of chemical bond. It is known as hydrogen bond (dotted line). It is of note that also in this type of bond there is a component which implies the sharing of electrons. Furthermore, the hydrogen atom in a hydrogen bond does not just stick indiscriminantly to the oxygen of another molecule. Being positively charged it goes where the electrons are. So the hydrogen bond is a bond between a hydrogen atom and a lone pair (two solid dots), its length ℓ being 1.83 Å, while $\ell' = 1.04$ Å. This means that a water molecule can form four hydrogen bonds: the molecule's two hydrogens form two bonds with neighboring oxygens, while the molecule's two lone pairs interact with neighboring hydrogens. (c) Schematic electronic structure of H_2O is displayed below (b) with the purpose of indicating the origin of the lone pair of electrons, as well as the role the electrons of the two hydrogen atoms have in closing the $2p$-shell of the oxygen atom.

hydrogen bonds, building an ice-like cage around the NP molecule. The entropic cost of this structural change leads to low solubility for small apolar molecules in water. Said differently, such a force, known as the weak hydrophobic force (WHI), leads only to metastable states of groups of nonpolar molecules (Chandler (2002)). On the other hand, close to a large hydrophobic object (Fig. 7.E.7), the presence of a hydrogen bond network – four bonds per water molecule – cannot persist, being geometrically impossible. In

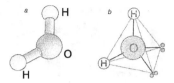

Figure 7.E.5 The water molecule is bent, with the two bonds between oxygen and hydrogen making an angle of 104.5° (**a**). To understand the structure of liquid water, we must also take into account the two "lone pairs" of electrons on the oxygen atom. The hydrogen atoms and the lone pairs sit more or less at the corners of a tetrahedron (**b**).

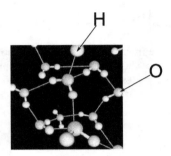

Figure 7.E.6 Stick and ball representation of a network of hydrogen bonds in bulk water. For the color version of this figure, refer to cambridge.org/nuclearcooperpair.

this way can destabilize and expel the remaining liquid contained between these surfaces. The resulting pressure imbalance will cause the surfaces to attract. If the liquid is close to coexistence with the vapor phase, as is the case for water at ambient (biological) conditions[62], this phenomenon occurs also for widely separated surfaces. The similarity with a (generalized) Casimir effect is apparent.

Within this context, and in keeping with the fact that the hydrophobic force plays a central role in the folding and, as a consequence, the biological activity of proteins, one can point out that water at physiological conditions ($T = 300$ K, pH $= 7$, etc.) can be viewed as a vacuum of the macromolecules responsible for metabolism, and thus one of the two origins of life on earth.[63]

this case, water molecules near the hydrophobic cluster (surface) have typically three or fewer hydrogen bonds. The resulting force, known as the strong hydrophobic force (SHI; Chandler (2002)), can be viewed as a bona fide hydrophobic interaction, stabilizing large aggregates of nonpolar molecules.

[62] In particular, at $T \approx 300$ K($\approx 27°$C). This is the reason why proteins unfold at low temperatures, in particular the monomer of the HIV-1-protease homodimer, which denaturizes at 8°C (Rösner et al. (2017)). In general, this temperature lies below the freezing point, and one has to use particular techniques to be able to deal with supercool water (e.g., capilars).

[63] To the question "what is life?" (Schrödinger (1944)), one is forced to answer that life is not one but two things (Dyson (1999)). Which ones? Replication and metabolism. The molecules of DNA and RNA are

liquid   liquid
            vapour

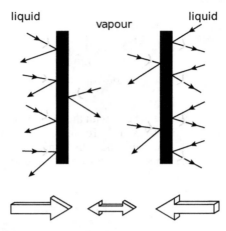

Figure 7.E.7 Schematic representation of water molecules (arrowed lines) impinging from the outside (liquid) and from the interspace (vapor) of two large parallel hydrophobic plates. For the color version of this figure, refer to cambridge.org/nuclearcooperpair.

While the many-body basis of hydrophobicity cannot be denied, the parallel with the zero-body Casimir effect (Fig. 7.E.3) can hardly be avoided, and neither can the nuclear phenomena discussed in Sect. 4.8 in connection with superconductivity in nuclei and in metals (see also Sect. 3.6 concerning pairing in ^{11}Li).

Within this scenario, one can also observe a (generalized) Nambu-tumbling effect, namely, the chain *hydrogen bond→ weak and strong hydrophobicity → polypeptide chain evolution→ proteins (second-order phase transition, spontaneous symmetry breaking in information space[64])→ protein folding (first-order phase transition in conformation space)→ metabolism.* And, at the basis of the first link of the chain, one finds the difference in the ZPF energy which a hydrogen nucleus feels when in the presence of one (O) or two (O$_2$) oxygen nuclei, a quantity closely connected with quantum fluctuation of confinement (Povh and Rosina (2002)).

responsible for the first function (Watson and Crick (1953); Watson (1980)), proteins (i.e., polymers made out of the 20 commonly occurring amino acids in nature) for the second (Sanger (1952)). Because software (replication) is necessary a parasite of hardware (proteins), the becoming of a protein carries, to a large extent, the secret of life (Monod (1970)).

[64] Broglia (2013).

References

E. Abrahams and J. W. F. Woo. Phenomenological theory of the rounding of the resistive transition of superconductors. *Phys. Lett. A*, 27:117, 1968.

Y. Aksyutina, T. Aumann, K. Boretzky, M. Borge, C. Caesar, A. Chatillon, L. Chulkov, D. Cortina-Gil, U. Datta Pramanik, H. Emling, H. Fynbo, H. Geissel, G. Ickert, H. Johansson, B. Jonson, R. Kulessa, C. Langer, T. LeBleis, K. Mahata, G. Münzenberg, T. Nilsson, G. Nyman, R. Palit, S. Paschalis, W. Prokopowicz, R. Reifarth, D. Rossi, A. Richter, K. Riisager, G. Schrieder, H. Simon, K. Sümmerer, O. Tengblad, H. Weick, and M. Zhukov. Momentum profile analysis in one-neutron knockout from Borromean nuclei. *Phys. Lett. B*, 718:1309, 2013.

K. Alder and A. Winther. *Electromagnetic Excitations*. North-Holland, Amsterdam, 1975.

K. Alder, A. Bohr, T. Huus, B. Mottelson, and A. Winther. Study of nuclear structure by electromagnetic excitation with accelerated ions. *Rev. Mod. Phys.*, 28:432, 1956.

K. Allaart, K. Goeke, H. Müther, and A. Faessler. Microscopic, nonharmonic description of rotations and pairing vibrations in deformed pf-shell nuclei. *Phys. Rev. C*, 9:988, 1974.

V. Alva, J. Söding, and A. N. Lupas. A vocabulary of ancient peptides at the origin of folded proteins. *Elife*, 4:e09410, 2015.

V. Ambegaokar. The Green's function method. In R. D. Parks, editor, *Superconductivity, Vol. I*, page 259. Marcel Dekker, New York, 1969.

V. Ambegaokar and A. Baratoff. Tunneling between superconductors. *Phys. Rev. Lett.*, 10:486, 1963.

A. Amelin, M. Gornov, Y. B. Gurov, A. Ilin, P. Morokhov, V. Pechkurov, V. Savelev, F. Sergeev, S. Smirnov, B. Chernyshev, R. Shafigullin, and A. Shishkov. Production of ^{10}Li in absorption of stopped π^- mesons by ^{11}B nuclei. *Sov. J. Nucl. Phys.*, 52:782, 1990.

H. An and C. Cai. Global deuteron optical model potential for the energy range up to 183 MeV. *Phys. Rev. C*, 73:054605, 2006.

P. W. Anderson. New method in the theory of superconductivity. *Phys. Rev.*, 110:985, 1958.

P. W. Anderson and B. T. Matthias, Superconductivity. *Science*, 144:373, 1964a.

P. W. Anderson. Special effects in superconductivity. In E. R. Caianiello, editor, *The Many-Body Problem, Vol. 2*, page 113. Academic Press, New York, 1964b.

P. W. Anderson. How Josephson discovered his effect. *Phys. Today*, 23:23, 1970.

P. W. Anderson. More is different. *Science*, 177:393, 1972.

P. W. Anderson. Uses of solid state analogies in elementary particle theory. In R. Arnowitt and P. Nath, editors, *Gauge Theories and Modern Field Theories: Proceeding of a Conference Held at Northeastern University, Boston*, page 311. MIT Press, Cambridge, MA, 1976.

P. W. Anderson. *Basic Notions of Condensed Matter*. Addison-Wesley, Reading, MA, 1984.

P. W. Anderson. Theories of Fullerene T_c's which will not work. Unpublished manuscript, 1991.

P. W. Anderson. Off-diagonal long-range order and flux quantization. In H. Holden and S. Kjellstrup Ratkje, editors, *The Collected Works of Lars Onsager*, page 729. World Scientific, Singapore, 1996.

P. W. Anderson and J. M. Rowell. Probable observation of the Josephson superconducting tunneling effect. *Phys. Rev. Lett.*, 10:230, 1963.

P. W. Anderson and D. L. Stein. Broken symmetry, emergent properties, dissipative structures, life: Are they related? In P. W. Anderson, editor, *Basic Notions of Condensed Matter*, page 263. Benjamin, Menlo Park, CA, 1984.

P. W. Anderson and D. J. Thouless. Diffuseness of the nuclear surface. *Phys. Lett.*, 1:155, 1962.

J. F. Annett. *Superconductivity, Superfluids and Condensates*. Oxford University Press, Oxford, 2013.

R. J. Ascuitto and N. K. Glendenning. Inelastic processes in particle transfer reactions. *Phys. Rev.*, 181:0 1396, 1969.

R. J. Ascuitto and N. K. Glendenning. Assessment of two-step processes in (p, t) reactions. *Phys. Rev. C*, 2:0 1260, 1970.

R. J. Ascuitto and B. Sørensen. Coupling effects in ^{208}Pb. *Nucl. Phys. A*, 186:0 641, 1972.

R. J. Ascuitto, N. K. Glendenning, and B. Sørensen. Confirmation of strong second order processes in (p, t) reactions on deformed nuclei. *Phys. Lett. B*, 34:0 17,1971.

N. W. Ashcroft and N. D. Mermin. *Solid State Physics*. Holt, Reinhardt and Winston, Hong Kong, 1987.

T. Aumann. Low-energy dipole response of exotic nuclei. *Eur. Phys. J. A*, 55:0 234, 2019.

N. Austern. Direct reactions. In F. Janouch, editor, *Select Topics in Nuclear Theory*, page 17. IAEA, Vienna, 1963.

N. Austern. *Direct Nuclear Reaction Theories*. Interscience monographs and texts in physics and astronomy. Wiley-Interscience, New York, 1970.

S. Austin and G. Bertsch. Halo nuclei. *Sci. Am.*, 272:0 90, 1995.

D. Auton. Direct reactions on ^{10}Be. *Nucl. Phys. A*, 157:0 305, 1970.

A. Avdeenkov and S. Kamerdzhiev. Phonon coupling and the single-particle characteristics of Sn isotopes. In R. A. Broglia and V. Zelevinsky, editors, *50 Years of Nuclear BCS*, page 274. World Scientific, Singapore, 2013.

B. Avez, C. Simenel, and P. Chomaz. Pairing vibrations study with the time-dependent Hartree–Fock–Bogoliubov theory. *Phys. Rev. C*, 78:0 044318, 2008.

P. Avogadro, F. Barranco, R. A. Broglia, and E. Vigezzi. Quantum calculation of vortices in the inner crust of neutron stars. *Phys. Rev. C*, 75:0 012805, 2007.

P. Avogadro, F. Barranco, R. Broglia, and E. Vigezzi. Vortex–nucleus interaction in the inner crust of neutron stars. *Nucl. Phys. A*, 8110 (3):0 378, 2008.

P. Axel. Electric dipole ground-state transition width strength function and 7-MeV photon interactions. *Phys. Rev.*, 126:0 671, 1962.

C. Bachelet, G. Audi, C. Gaulard, C. Guénaut, F. Herfurth, D. Lunney, M. de Saint Simon, and C. Thibault. New binding energy for the two-neutron halo of ^{11}Li. *Phys. Rev. Lett.*, 100:0 182501, 2008.

P. Ball. *H_2O: A Biography of Water*. Phoenix, London, 2003.

J. Bang, S. Ershov, F. Gareev, and G. Kazacha. Discrete expansions of continuum wave functions. *Nucl. Phys. A*, 339:0 89, 1980.

M. Baranger. Recent progress in the understanding of finite nuclei from the two-nucleon interaction. In M. Jean and R. A. Ricci, editors, *Proceedings of the International School of Physics "E. Fermi" Course XL Nuclear Structure and Nuclear Reactions*, page 511. Academic Press, New York, 1969.

C. Barbieri. Role of long-range correlations in the quenching of spectroscopic factors. *Phys. Rev. Lett.*, 103:0 202502, 2009.

C. Barbieri and B. K. Jennings. Nucleon-nucleus optical potential in the particle-hole approach. *Phys. Rev. C*, 72:0 014613, 2005.

J. Bardeen. Tunnelling from a many-particle point of view. *Phys. Rev. Lett.*, 6:0 57, 1961.

J. Bardeen. Tunneling into superconductors. *Phys. Rev. Lett.*, 9:0 147, 1962.

J. Bardeen and D. Pines. Electron–phonon interaction in metals. *Phys. Rev.*, 99:0 1140, 1955.

J. Bardeen, L. N. Cooper, and J. R. Schrieffer. Microscopic theory of superconductivity. *Phys. Rev.*, 106:0 162, 1957a.

J. Bardeen, L. N. Cooper, and J. R. Schrieffer. Theory of superconductivity. *Phys. Rev.*, 108:0 1175, 1957b.

F. Barker and G. Hickey. Ground-state configurations of ^{10}Li and ^{11}Li. *J. Phys. G*, 3:0 L23, 1977.

P. Barnes, E. Romberg, C. Ellegard, R. Casten, O. Hansen, and J. Mulligan. Proton-hole states in ^{209}Bi from the ^{210}Po (t, α) reaction. *Nucl. Phys. A*, 195:0 146, 1972.

S. Baroni, M. Armati, F. Barranco, R. A. Broglia, G. Colò, G. Gori, and E. Vigezzi. Correlation energy contribution to nuclear masses. *J. Phys. G*, 30:0 1353, 2004.

S. Baroni, A. O. Macchiavelli, and A. Schwenk. Partial-wave contributions to pairing in nuclei. *Phys. Rev. C*, 81:0 064308, 2010.

F. Barranco. Estudio de las fluctuaciones y correlaciones en el estado fundamental del nucleo. PhD thesis, University of Sevilla, 1985.

F. Barranco and R. A. Broglia. Correlation between mean square radii and zero-point motions of the surface in the Ca isotopes. *Phys. Lett. B*, 151:0 90, 1985.

F. Barranco and R. A. Broglia. Effect of surface fluctuations on the nuclear density. *Phys. Rev. Lett.*, 59:0 2724, 1987.

F. Barranco, M. Gallardo, and R. A. Broglia. Nuclear field theory of spin dealignment in strongly rotating nuclei and the vacuum polarization induced by pairing vibrations. *Phys. Lett. B*, 198:0 19, 1987.

F. Barranco, R. A. Broglia, and G. F. Bertsch. Exotic radioactivity as a superfluid tunneling phenomenon. *Phys. Rev. Lett.*, 60:0 507, 1988.

F. Barranco, E. Vigezzi, and R. A. Broglia. Calculation of the absolute lifetimes of the variety of decay modes of ^{234}U. *Phys. Rev. C*, 39:0 210, 1989.

F. Barranco, G. F. Bertsch, R. A. Broglia, and E. Vigezzi. Large-amplitude motion in superfluid fermi droplets. *Nucl. Phys. A*, 512:0 253, 1990.

F. Barranco, R. A. Broglia, G. Gori, E. Vigezzi, P. F. Bortignon, and J. Terasaki. Surface vibrations and the pairing interaction in nuclei. *Phys. Rev. Lett.*, 83:2147, 1999.

F. Barranco, P. F. Bortignon, R. A. Broglia, G. Colò, and E. Vigezzi. The halo of the exotic nucleus ^{11}Li: A single Cooper pair. *Eur. Phys. J. A*, 11:385, 2001.

F. Barranco, R. A. Broglia, G. Colò, G. Gori, E. Vigezzi, and P. F. Bortignon. Many-body effects in nuclear structure. *Eur. Phys. J. A*, 21:57, 2004.

F. Barranco, P. F. Bortignon, R. A. Broglia, G. Colò, P. Schuck, E. Vigezzi, and X. Viñas. Pairing matrix elements and pairing gaps with bare, effective, and induced interactions. *Phys. Rev. C*, 72:054314, 2005.

F. Barranco, G. Potel, R. A. Broglia, and E. Vigezzi. Structure and reactions of ^{11}Be: many–body basis for single–neutron halo. *Phys. Rev. Lett.*, 119:082501, 2017.

F. Barranco, R. A. Broglia, G. Potel, and E. Vigezzi. Structure and reactions of N=7 isotones: parity inversion and transfer processes. *EPJ Web Conf.*, 223:01005, 2019a.

F. Barranco, G. Potel, E. Vigezzi, and R. A. Broglia. Radioactive beams and inverse kinematics: Probing the quantal texture of the nuclear vacuum. *Eur. Phys. J. A*, 55:104, 2019b.

F. Barranco, G. Potel, E. Vigezzi, and R. A. Broglia. ^{9}Li(d, p) reaction as a specific probe of ^{10}Li, the paradigm of parity-inverted nuclei around the $N = 6$ closed shell. *Phys. Rev. C*, 101:031305, 2020.

J. L. Basdevant and J. Dalibard. *Quantum Mechanics*. Springer, Berlin, 2005.

G. Bassani, N. M. Hintz, C. D. Kavaloski, J. R. Maxwell, and G. M. Reynolds. (p, t) ground-state $L = 0$ transitions in the even isotopes of Sn and Cd at 40 MeV, $N = 62$ to 74. *Phys. Rev.*, 139:B830, 1965.

B. F. Bayman. Seniority, quasi-particle and collective vibrations. Lecture notes, Palmer Physical Laboratory, Princeton, 1961.

B. F. Bayman. Finite-range calculation of the two-neutron transfer reaction. *Phys. Rev. Lett.*, 25:1768, 1970.

B. F. Bayman. Finite–range calculation of the two-neutron transfer reaction. *Nucl. Phys. A*, 168:1, 1971.

B. F. Bayman and J. Chen. One-step and two-step contributions to two-nucleon transfer reactions. *Phys. Rev. C*, 26:1509, 1982.

B. F. Bayman and C. F. Clement. Sum rules for two-nucleon transfer reactions. *Phys. Rev. Lett.*, 29:1020, 1972.

B. F. Bayman and B. H. Feng. Monte Carlo calculations of two-neutron transfer cross sections. *Nucl. Phys. A*, 205:513, 1973.

B. F. Bayman and A. Kallio. Relative-angular-momentum-zero part of two-nucleon wave functions. *Phys. Rev.*, 156:1121, 1967.

S. Beceiro-Novo, T. Ahn, D. Bazin, and W. Mittig. Active targets for the study of nuclei far from stability. *Prog. Particle Nucl. Phys.*, 84:124, 2015.

M. J. Bechara and O. Dietzsch. States in ^{121}Sn from the ^{120}Sn$(d, p)^{121}$Sn reaction at 17 MeV. *Phys. Rev. C*, 12:90, 1975.

P. A. Beck and H. Claus. Density of states information from low temperature specific heat measurements. *J. Res. National Bureau Standards A*, 74 A:449, 1970.

S. T. Belyaev. Effect of pairing correlations on nuclear properties. *Kgl. Danske Videnskab. Selskab, Mat.-fys. Medd.*, 31:no. 11, 1959.

S. T. Belyaev. Pair correlations in nuclei: Copenhagen 1958. In R. A. Broglia and V. Zelevinski, editors, *50 Years of Nuclear BCS*, page 3. World Scientific, Singapore, 2013.

K. Bennaceur, J. Dobaczewski, and M. Ploszajczak. Pairing anti-halo effect. *Phys. Lett. B*, 496:154, 2000.

F. Bergasa-Caceres and H. A. Rabitz. Interdiction of protein folding for therapeutic drug development in SARS COV-2. *J. Phys. Chem. B*, 2020.

V. Bernard and N. V. Giai. Single-particle and collective nuclear states and the Green's function method. In R. A. Broglia, R. A. Ricci, and C. H. Dasso, editors, *Proceedings of the International School of Physics "Enrico Fermi," Course LXXVII*, page 437. North-Holland, Amsterdam, 1981.

F. Bertrand. Giant multipole resonances – Perspectives after ten years. *Nucl. Phys. A*, 354:129, 1981.

G. Bertsch. Collective motion in Fermi droplets. In R. A. Broglia and J. R. Schrieffer, editors, *International School of Physics "Enrico Fermi" Course CIV, Frontiers and Borderlines in Many-Body Particle Physics*, page 41. North-Holland, Amsterdam, 1988.

G. Bertsch, F. Barranco, and R. A. Broglia. How nuclei change shape. In T. Kuo and I. Speth, editors, *Windsurfing the Fermi Sea*, page 33. Elsevier, New York, 1987.

G. F. Bertsch and R. A. Broglia. Giant resonances in hot nuclei. *Phys. Today*, 39:44, 1986.

G. F. Bertsch and R. A. Broglia. *Oscillations in Finite Quantum Systems*. Cambridge University Press, Cambridge, 2005.

G. F. Bertsch and H. Esbensen. Pair correlations near the neutron drip line. *Ann. Phys.*, 209, 1991.

G. F. Bertsch, R. A. Broglia, and C. Riedel. Qualitative description of nuclear collectivity. *Nucl. Phys. A*, 91:123, 1967.

G. F. Bertsch, P. F. Bortignon, and R. A. Broglia. Damping of nuclear excitations. *Rev. Mod. Phys.*, 55:287, 1983.

G. F. Bertsch, K. Hencken, and H. Esbensen. Nuclear breakup of Borromean nuclei. *Phys. Rev. C*, 57:1366, 1998.

D. R. Bès. The nuclear field theory. In P. Federman, editor, *Notas de Física, Vol 1, n. 1*. Instituto de Física, Universidad Nacional Autónoma de Mexico, 1978.

D. R. Bès. The field treatment of the nuclear spectrum: Historical foundation and two contributions to its ensuing development. *Phys. Scripta*, 91:063010, 2016.

D. R. Bès and R. A. Broglia. Pairing vibrations. *Nucl. Phys.*, 80:289, 1966.

D. R. Bès and R. A. Broglia. Effect of the multipole pairing and particle-hole fields in the particle-vibration coupling of ^{209}Pb. I. *Phys. Rev. C*, 3:2349, 1971a.

D. R. Bès and R. A. Broglia. Effective operators in the analysis of single-nucleon transfer reactions on closed shell nuclei. *Phys. Rev. C*, 3:2389, 1971b.

D. R. Bès and R. A. Broglia. Equivalence between Feynman–Goldstone and particle–phonon diagrams for finite many body systems. In G. Alaga, V. Paar, and L. Sips, editors, *Problems of Vibrational Nuclei: Proceedings of the Topical Conference on Problems of Vibrational Nuclei*, page 1. North-Holland, Amsterdam, 1975.

D. R. Bès and R. A. Broglia. Nuclear superfluidity and field theory of elementary excitations. In A. Bohr and R. A. Broglia, editors, *International School of Physics "Enrico Fermi" Course LXIX, Elementary Modes of Excitation in Nuclei*, page 55. North-Holland, Amsterdam, 1977.

D. R. Bès and J. Kurchan. *The Treatment of Collective Coordinates in Many–Body Systems*. World Scientific, Singapore, 1990.

D. R. Bès and R. A. Sorensen. The pairing–plus–quadrupole model. *Adv. Nucl. Phys.*, 2:129, 1969.

D. R. Bès, R. A. Broglia, and B. Nilsson. Importance of the quadrupole pairing field in the $J^\pi = 0^+$ vibrations of shape deformed nuclei. *Phys. Lett. B*, 40:338, 1972.

D. R. Bès, G. G. Dussel, R. A. Broglia, R. Liotta, and B. R. Mottelson. Nuclear field theory as a method of treating the problem of overcompleteness in descriptions involving elementary modes of both quasi–particles and collective type. *Phys. Lett. B*, 52:253, 1974.

D. R. Bès, R. A. Broglia, G. G. Dussel, and R. Liotta. Simultaneous treatment of surface and pairing nuclear fields. *Phys. Lett. B*, 56:109, 1975a.

D. R. Bès, R. A. Broglia, and B. Nilsson. Microscopic description of isoscalar and isovector giant quadrupole resonances. *Phys. Rep.*, 16, 1975b.

D. R. Bès, R. A. Broglia, G. G. Dussel, R. J. Liotta, and R. P. J. Perazzo. On the many-body foundation of the nuclear field theory. *Nucl. Phys. A*, 260:77, 1976a.

D. R. Bès, R. A. Broglia, G. G. Dussel, R. J. Liotta, and H. M. Sofía. The nuclear field treatment of some exactly soluble models. *Nucl. Phys. A*, 260:1, 1976b.

D. R. Bès, R. A. Broglia, G. G. Dussel, R. J. Liotta, and H. M. Sofía. Application of the nuclear field theory to monopole interactions which include all the vertices of a general force. *Nucl. Phys. A*, 260:27, 1976c.

D. R. Bès, G. G. Dussel, R. P. J. Perazzo, and H. M. Sofía. The renormalization of single-particle states in nuclear field theory. *Nucl. Phys. A*, 293:350, 1977.

D. R. Bès, R. A. Broglia, J. Dudek, W. Nazarewicz, and Z. Szymański. Fluctuation effects in the pairing field of rapidly rotating nuclei. *Ann. Phys.*, 182:237, 1988.

H. A. Bethe. The electromagnetic shift of energy levels. *Phys. Rev.*, 72:339, 1947.

J. H. Bilgram. Dynamics at the solid–liquid transition: Experiments at the freezing point. *Phys. Rep.*, 153:1, 1987.

J. H. Bjerregaard, O. Hansen, O. Nathan, and S. Hinds. States of ^{208}Pb from double triton stripping. *Nucl. Phys.*, 89:337, 1966.

J. B. Bjorken and S. D. Drell. *Relativistic Quantum Mechanics*. Mc Graw-Hill, New York, 1998.

S. Bjørnholm, J. Borggreen, K. Hansen, and J. Pedersen. Electronic shells and supershells in metal clusters. In R. A. Broglia and J. R. Schrieffer, editors, *International School of Physics "Enrico Fermi" Course CXXI, Perspectives in Many-Particle Physics*, page 279. North-Holland, Amsterdam, 1994.

G. Blanchon, A. Bonaccorso, D. M. Brink, and N. V. Mau. ^{10}Li spectrum from ^{11}Li fragmentation. *Nucl. Phys. A*, 791:303, 2007.

S. Boersma. A maritime analogy of the Casimir effect. *Am. J. Phys.*, 64:539, 1996.

S. Bogner, R. Furnstahl, and A. Schwenk. From low-momentum interactions to nuclear structure. *Prog. Particle Nucl. Phys.*, 65:94, 2010.

N. Bogoljubov. On a new method in the theory of superconductivity. *Il Nuovo Cimento*, 7:794, 1958a.

N. Bogoljubov. A new method in the theory of superconductivity. *I. J. Exp. Theor. Phys. USSR*, 34:41, 1958b.

H. G. Bohlen, W. von Oertzen, T. Stolla, R. Kalpakchieva, B. Gebauer, M. Wilpert, T. Wilpert, A. N. Ostrowski, S. M. Grimes, and T. N. Massey. Study of weakly bound and unbound states of exotic nuclei with binary reactions. *Nucl. Phys. A*, 616:254, 1997.

D. Bohm and D. Pines. A collective description of electron interactions. I. Magnetic interactions. *Phys. Rev.*, 82:625, 1951.

D. Bohm and D. Pines. A collective description of electron interactions: III. Coulomb interactions in a degenerate electron gas. *Phys. Rev.*, 92:609, 1953.

A. Bohr. Elementary modes of excitation and their coupling. In *Comptes Rendus du Congrès International de Physique Nucléaire*, vol. 1, page 487. Centre National de la Recherche Scientifique, Paris, 1964.

A. Bohr. *Rotational Motion in Nuclei, in Les Prix Nobel en 1975*. Imprimerie Royale Norstedts Tryckeri, Stockholm, 1976.

A. Bohr and R. A. Broglia. *Elementary modes of excitation in nuclei: Proceedings of the International School of Physics "Enrico Fermi."* North-Holland, Amsterdam, 1977.

A. Bohr and B. R. Mottelson. Submission letter to H. Lipkin Febr. 13th, 1963 for non publication in his jocular but quite influential journal of non-publishable results in nuclear physics concerning the remarkable (coexistence) properties of low-lying 0^+ excited states in ^{16}O, ^{40}Ca, ^{42}Ca, ^{70}Ge, and ^{24}Mg(5.465 MeV) $l = 0$ strength. Unpublished. February 1963.

A. Bohr and B. R. Mottelson. *Nuclear Structure*, Vol. I. Benjamin, New York, 1969.

A. Bohr and B. R. Mottelson. *Nuclear Structure*, Vol. II. Benjamin, New York, 1975.

A. Bohr and B. R. Mottelson. Features of nuclear deformations produced by the alignment of individual particles or pairs. *Phys. Scripta*, 22:468, 1980.

A. Bohr and O. Ulfbeck. Quantal structure of superconductivity gauge angle. In *First Topsøe Summer School on Superconductivity and Workshop on Superconductors*. Roskilde, Denmark Riso/M/2756, 1988.

A. Bohr, B. R. Mottelson, and D. Pines. Possible analogy between the excitation spectra of nuclei and those of the superconducting metallic state. *Phys. Rev.*, 110:936, 1958.

N. Bohr. The quantum postulate and the recent development of atomic theory. *Nature*, 121:580, 1928.

N. Bohr and F. Kalckar. On the transmutation of atomic nuclei by impact of material particles. I. General theoretical remarks. *Mat.-Fys. Medd. Dan. Vidensk. Selsk.*, 14:10, 1937.

N. Bohr and J. A. Wheeler. The mechanism of nuclear fission. *Phys. Rev.*, 56:426, 1939.

M. Born. Zur Quantenmechanik der Stoßvorgänge. *Z. Phys.*, 37:863, 1926.

M. Born. *Natural Philosophy of Cause and Chance*. Oxford University Press, Oxford, 1948.

M. Born. The statistical interpretation of quantum mechanics, in *Nobel Lectures Phys. 1942–1962*, p. 256. Elsevier, New York, 1964.

M. Born. *Atomic Physics*. Blackie, London, 1969.

M. Born and P. Jordan. Zur quantenmechanik. *Z. Phys.*, 34:858, 1925.

M. Born, P. Jordan, and W. Heisenberg. Zur Quantenmechanik II. *Z. Phys.*, 35:557, 1926.

P. F. Bortignon and R. A. Broglia. Role of the nuclear surface in an unified description of the damping of single-particle states and giant resonances. *Nucl. Phys. A*, 371:405, 1981.

P. F. Bortignon and R. A. Broglia. Elastic response of the atomic nucleus in gauge space: Giant pairing vibrations. *Eur. Phys. J. A*, 520 (9):280, 2016.

P. F. Bortignon, R. A. Broglia, D. R. Bès, R. Liotta, and V. Paar. On the role of the pairing modes in the $(h_{9/2} \otimes 3^-)$ multiplet of ^{209}Bi. *Phys. Lett. B*, 64:24, 1976.

P. F. Bortignon, R. A. Broglia, D. R. Bès, and R. Liotta. Nuclear field theory. *Phys. Rep.*, 30:305, 1977.

P. F. Bortignon, R. A. Broglia, and D. R. Bès. On the convergence of nuclear field theory perturbation expansion for strongly anharmonic systems. *Phys. Lett. B*, 76:153, 1978.

P. F. Bortignon, R. A. Broglia, and C. H. Dasso. Quenching of the mass operator associated with collective states in many-body systems. *Nucl. Phys. A*, 398:221, 1983.

P. F. Bortignon, A. Bracco, and R. A. Broglia. *Giant Resonances*. Harwood Academic, Amsterdam, 1998.

M. Brack. The physics of simple metal clusters: Self-consistent jellium model and semiclassical approaches. *Rev. Mod. Phys.*, 65:677, 1993.

D. M. Brink. Some aspects of the interaction of fields with matter, PhD Thesis. Oxford University, 1955 (unpublished).

D. M. Brink. *Semi-classical Methods in Nucleus–Nucleus Scattering*. Cambridge University Press, Cambridge, 1985.

D. M. Brink and R. A. Broglia. *Nuclear Superfluidity*. Cambridge University Press, Cambridge, 2005.

D. M. Brink and G. R. Satchler. *Angular Momentum*. Clarendon Press, Oxford, 1968.

R. Broglia and A. Winther. Semiclassical theory of heavy ion reactions. *Phys. Rep.*, 4:153, 1972.

R. Broglia, R. Liotta, B. Nilsson, and A. Winther. Formfactors for one- and two-nucleon transfer reactions induced by light and heavy ions. *Phys. Rep.*, 29:291, 1977.

R. A. Broglia. Heavy ion reactions www.mi.infn.it/~vigezzi/HIR/HeavyIonReactions.pdf, 1975.

R. A. Broglia. Microscopic structure of the intrinsic state of a deformed nucleus. In F. Iachello, editor, *Interacting Bose–Fermi Systems in Nuclei*, page 95. Plenum Press, New York, 1981.

R. A. Broglia. The surfaces of compact systems: From nuclei to stars. *Surface Sci.*, 500:759, 2002.

R. A. Broglia. From phase transitions in finite systems to protein folding and non–conventional drug design. *La Rivista Nuovo Cimento*, 280 (1):1, 2005.

R. A. Broglia. More is different: 50 years of nuclear BCS. In R. A. Broglia and V. Zelevinsky, editors, *50 Years of Nuclear BCS*, page 642. World Scientific, Singapore, 2013.

R. A. Broglia, editor. *Aage Bohr and the Quantum Finite Many-Body Problem: Selected Papers*. World Scientific, Singapore, 2020.

R. A. Broglia and D. R. Bes. High-lying pairing resonances. *Phys. Lett. B*, 69:129, 1977.

R. A. Broglia and C. Riedel. Pairing vibration and particle-hole states excited in the reaction ^{206}Pb(t, p)^{208}Pb. *Nucl. Phys. A*, 92:145, 1967.

R. A. Broglia and G. Tiana. Reading the three-dimensional structure of a protein from its amino acid sequence. *Prot. Struct. Funct. Desig.*, 45:421, 2001.

R. A. Broglia and A. Winther. On the pairing field in nuclei. *Phys. Lett. B*, 124:11, 1983.

R. A. Broglia and A. Winther. *Heavy Ion Reactions*. Westview Press, Boulder, CO, 2004.

R. A. Broglia and V. Zelevinsky, editors. *50 Years of Nuclear BCS*. World Scientific, Singapore, 2013.

R. A. Broglia, C. Riedel, and B. Sørensen. Two-nucleon transfer and pairing phase transition. *Nucl. Phys. A*, 107:1, 1968.

R. A. Broglia, V. Paar, and D. R. Bes. Diagramatic perturbation treatment of the effective interaction between two-phonon states in closed shell nuclei: The $J^\pi = 0^+$ states in ^{208}Pb. *Phys. Lett. B*, 37:159, 1971a.

R. A. Broglia, V. Paar, and D. R. Bes. Effective interaction between $J^\pi = 2^+$ two–phonon states in ^{208}Pb: Evidence of the isovector quadrupole mode. *Phys. Lett. B*, 37:257, 1971b.

R. A. Broglia, C. Riedel, and T. Udagawa. Coherence properties of two-neutron transfer reactions and their relation to inelastic scattering. *Nucl. Phys. A*, 169:225, 1971c.

R. A. Broglia, C. Riedel, and T. Udagawa. Sum rules and two-particle units in the analysis of two-neutron transfer reactions. *Nucl. Phys. A*, 184:23, 1972.

R. A. Broglia, O. Hansen, and C. Riedel. Two-neutron transfer reactions and the pairing model. *Adv. Nucl. Phys.*, 6:287, 1973. www.mi.infn.it/~vigezzi/BHR/BrogliaHansenRiedel.pdf.

R. A. Broglia, D. R. Bès, and B. S. Nilsson. Strength of the multipole pairing coupling constant. *Phys. Lett. B*, 50:213, 1974.

R. A. Broglia, B. R. Mottelson, D. R. Bès, R. Liotta, and H. M. Sofía. Treatment of the spurious state in nuclear field theory. *Phys. Lett. B*, 64:29, 1976.

R. A. Broglia, G. Pollarolo, and A. Winther. On the absorptive potential in heavy ion scattering. *Nucl. Phys. A*, 361:307, 1981.

R. A. Broglia, T. Døssing, B. Lauritzen, and B. R. Mottelson. Nuclear rotational damping: Finite-system analogue to motional narrowing in nuclear magnetic resonances. *Phys. Rev. Lett.*, 58:326, 1987.

R. A. Broglia, J. Terasaki, and N. Giovanardi. The Anderson–Goldstone–Nambu mode in finite and in infinite systems. *Phys. Rep.*, 335:1, 2000.

R. A. Broglia, G. Coló, G. Onida, and H. E. Roman. *Solid State Physics of Finite Systems: Metal Clusters, Fullerenes, Atomic Wires*. Springer, Berlin 2004.

R. A. Broglia, P. F. Bortignon, F. Barranco, E. Vigezzi, A. Idini, and G. Potel. Unified description of structure and reactions: Implementing the nuclear field theory program. *Phys. Scr.*, 91:063012, 2016.

R. A. Broglia, F. Barranco, A. Idini, G. Potel, and E. Vigezzi. Pygmy resonances: What's in a name? *Phys. Scr.*, 94:114002, 2019a.

R. A. Broglia, F. Barranco, G. Potel, and E. Vigezzi. One- and two-neutron at the drip line. From ^{11}Be to ^{11}Li and back: ^{10}Li and parity inversion. In F. Cerutti, A. Ferrari, T. Kawano, F. Salvat-Pujol, and P. Talou, editors, *Proceedings of the 15th International Conference on Nuclear Reaction Mechanisms*, page 1, https://doi.org/10.23727/CERN-Proceeding-2019-001, 2019b.

R. A. Broglia, F. Barranco, G. Potel, and E. Vigezzi. Characterization of vorticity in pygmy resonances and soft-dipole modes with two-nucleon transfer reactions. *Eur. Phys. J. A*, 55:243, 2019c.

B. A. Brown and W. D. M. Rae. In *NuShell @ MSU*. MSU-NSCL report, 2007.

G. Brown and G. Jacob. Zero-point vibrations and the nuclear surface. *Nucl. Phys.*, 42:177, 1963.

R. Bruce, J. M. Jowett, S. Gilardoni, A. Drees, W. Fischer, S. Tepikian, and S. R. Klein. Observations of beam losses due to bound-free pair production in a heavy-ion collider. *Phys. Rev. Lett.*, 99:144801, 2007.

K. A. Brueckner, A. M. Lockett, and M. Rotenberg. Properties of finite nuclei. *Phys. Rev.*, 121:255, 1961.

M. Buchanan. Wheat from the chaf. *Nat. Phys.*, 11:296, 2015.

J. Burde, E. L. Dines, S. Shih, R. M. Diamond, J. E. Draper, K. H. Lindenberger, C. Schück, and F. S. Stephens. Third discontinuity in the yrast levels of ^{158}Er. *Phys. Rev. Lett.*, 48:530, 1982.

J. Caggiano, D. Bazin, W. Benenson, B. Davids, B. Sherrill, M. Steiner, J. Yurkon, A. Zeller, and B. Blank. Spectroscopy of the ^{10}Li nucleus. *Phys. Rev. C*, 60:064322, 1999.

A. Calci, P. Navrátil, R. Roth, J. Dohet-Eraly, S. Quaglioni, and G. Hupin. Can ab initio theory explain the phenomenon of parity inversion in ^{11}Be? *Phys. Rev. Lett.*, 117:242501, 2016.

A. O. Caldeira and A. J. Leggett. Influence of dissipation on quantum tunneling in macroscopic systems. *Phys. Rev. Lett.*, 46:211, 1981.

A. O. Caldeira and A. J. Leggett. Quantum tunneling in a dissipative system. *Ann. Phys.*, 149:374, 1983.

F. Cappuzzello, D. Carbone, M. Cavallaro, M. Bondí, C. Agodi, F. Azaiez, A. Bonaccorso, A. Cunsolo, L. Fortunato, A. Foti, S. Franchoo, E. Khan, R. Linares, J. Lubian, J. A. Scarpaci, and A. Vitturi. Signatures of the giant pairing vibration in the ^{14}C and ^{15}C atomic nuclei. *Nat. Commun.*, 6:6743, 2015.

F. Cappuzzello, Carbone, D., Cavallaro, M., Spatafora, A., Ferreira, J. L., Agodi, C., Linares, R., and Lubian, J. Confirmation of Giant Pairing Vibration evidence in ^{12,13}C(^{18}O, ^{16}O)^{14,15}C reactions at 275 MeV. *Eur. Phys. J. A*, 57:34, 2021.

J. Casal, M. Gomez-Ramos, and A. Moro. Description of the ^{11}Li(p, d)^{10}Li transfer reaction using structure overlaps from a full three-body model. *Phys. Lett. B*, 767:307, 2017.

H. B. G. Casimir. Hydrophobicity at small and large length scales. *Proc. K. Ned. Akad. Wet.*, 60:793, 1948.

R. F. Casten, E. R. Flynn, J. D. Garrett, O. Hansen, T. J. Mulligan, D. R. Bès, R. A. Broglia, and B. Nilsson. Search for (t, p) transitions to excited 0^+ states in the actinide region. *Phys. Lett. B*, 40:333, 1972.

M. Cavallaro, M. De Napoli, F. Cappuzzello, S. E. A. Orrigo, C. Agodi, M. Bondí, D. Carbone, A. Cunsolo, B. Davids, T. Davinson, A. Foti, N. Galinski, R. Kanungo, H. Lenske, C. Ruiz, and A. Sanetullaev. Investigation of the ^{10}Li shell inversion by neutron continuum transfer reaction. *Phys. Rev. Lett.*, 118:012701, 2017.

M. Cavallaro, F. Cappuzzello, D. Carbone, and C. Agodi. Giant pairing vibrations in light nuclei. *Eur. Phys. J. A*, 550 (12):244, 2019.

E. Chabanat, P. Bonche, P. Haensel, J. Meyer, and R. Schaeffer. A Skyrme parametrization from subnuclear to neutron star densities. *Nucl. Phys. A*, 627:710, 1997.

D. Chandler. Hydrophobicity: Two faces of water. *Nature*, 417:491, 2002.

D. Chandler. Interfaces and the driving force of hydrophobic assembly. *Nature*, 437:640, 2005.

M. Chartier, J. Beene, B. Blank, L. Chen, A. Galonsky, N. Gan, K. Govaert, P. Hansen, J. Kruse, V. Maddalena, M. Thoennessen, and R. Varner. Identification of the ^{10}Li ground state. *Phys. Lett. B*, 510:24, 2001.

B. Chernysev, Y. B. Gurov, L. Y. Korotkova, S. Lapushkin, R. Pritula, and V. Sandukovsky. Study of the level structure of the lithium isotope ^{10}Li in stopped pion absorption. *Int. J. Mod. Phys. E*, 24:1550004, 2015.

R. M. Clark, A. O. Macchiavelli, L. Fortunato, and R. Krücken. Critical-point description of the transition from vibrational to rotational regimes in the pairing phase. *Phys. Rev. Lett.*, 96:032501, 2006.

W. R. Cobb and D. B. Guthe. Angular distribution of protons from the ^{44}Ca(d, p)^{45}Ca reaction. *Phys. Rev.*, 107:181, 1957.

M. H. Cohen, L. M. Falicov, and J. C. Phillips. Superconductive tunneling. *Phys. Rev. Lett.*, 8:316, 1962.

C. Cohen-Tannoudji and D. Guéry-Odelin. *Advances in Atomic Physics*. World Scientific, Singapore, 2011.

G. Colò, N. V. Giai, P. F. Bortignon, and M. R. Quaglia. On dipole compression modes in nuclei. *Phys. Lett. B*, 485:362, 2000.

L. N. Cooper. Bound electron pairs in a degenerate Fermi gas. *Phys. Rev.*, 104:1189, 1956.

L. N. Cooper. Rememberance of superconductivity past. In L. Cooper and D. Feldman, editors, *BCS: 50 Years*, page 3. World Scientific, Singapore, 2011.

C. H. Dasso, H. M. Sofia, and A. Vitturi. Two particle transfer reactions: The search for the giant pairing vibration. *J. Phys. Conf. Ser.*, 580:012018, 2015.

S. Dattagupta. *Relaxation Phenomena in Condensed Matter Physics*. Academic Press, New York, 1987.

J. de Boer. Quantum theory of condensed permanent gases in the law of corresponding states. *Physica*, 14:139, 1948.

J. de Boer. Quantum effects and exchange effects on the thermodynamic properties of liquid helium. *Prog. Low Temp. Phys.*, 2:1, 1957.

J. de Boer and R. J. Lundbeck. Quantum theory of condensed permanent gases III. The equation of state of liquids. *Physica*, 14:520, 1948.

L. de Broglie. Recherches sur la théorie des quanta. *Ann. Phys.*, 10:22, 1925.

R. de Diego, J. M. Arias, J. A. Lay, and A. M. Moro. Continuum-discretized coupled-channels calculations with core excitation. *Phys. Rev. C*, 89:064609, 2014.

P. G. de Gennes. *Superconductivity of Metals and Alloys*. Benjamin, New York, 1966.

P. G. de Gennes. *The Physics of Liquid Crystals*. Clarendon Press, Oxford, 1974.

P. G. de Gennes. *Les objets fragiles*. Plon, Paris, 1994.

W. A. de Heer, W. Knight, M. Chou, and M. L. Cohen. Electronic shell structure and metal clusters. *Solid State Phys.*, 40:93, 1987.

P. Debye. Van der Waals cohesion forces. *Phys. Z.*, 21:178, 1920.

P. Debye. Molecular forces and their electrical interpretation. *Phys. Z.*, 22:302, 1921.

A. Deltuva. Deuteron stripping and pickup involving the halo nuclei ^{11}Be and ^{15}C. *Phys. Rev. C*, 79:054603, 2009.

A. Deltuva. Faddeev-type calculation of three-body nuclear reactions including core excitation. *Phys. Rev. C*, 88:011601, 2013.

P. Descouvemont. Simultaneous study of the ^{11}Li and ^{10}Li nuclei in a microscopic cluster model. *Nucl. Phys. A*, 626:647, 1997.

S. A. Dickey, J. Kraushaar, R. Ristinen, and M. Rumore. The ^{120}Sn$(p, d)^{119}$Sn reaction at 26.3 MeV. *Nucl. Phys. A*, 377:137, 1982.

W. H. Dickhoff and C. Barbieri. Self-consistent Green's function method for nuclei and nuclear matter. *Prog. Particle Nucl. Phys.*, 52:377, 2004.

W. H. Dickhoff and R. J. Charity. Recent developments for the optical model of nuclei. *Prog. Particle Nucl. Phys.*, 105:252, 2019.

W. H. Dickhoff and D. Van Neck. *Many–Body Theory Exposed: Propagator Description of Quantum Mechanics in Many-Body Systems*. World Scientific, Singapore, 2005.

W. H. Dickhoff, R. J. Charity, and M. H. Mahzoon. Novel applications of the dispersive optical model. *J. Phys. G*, 44:033001, 2017.

K. Dietrich. On a nuclear Josephson effect in heavy ion scattering. *Phys. Lett. B*, 320 (6):428, 1970.

K. Dietrich, K. Hara, and F. Weller. Multiple pair transfer in reactions between heavy nuclei. *Phys. Lett. B*, 35:201, 1971.

P. A. M. Dirac. The fundamental equations of quantum mechanics. *Proc. R. Soc. London Ser. A*, 1090 (752):642, 1925.

P. A. M. Dirac. The quantum theory of the electron. *Proc. R. Soc. London Ser. A*, 117 (778):610, 1928a.

P. A. M. Dirac. The quantum theory of the electron. Part II. *Proc. R. Soc. London Ser. A*, 118 (779):351, 1928b.

P. A. M. Dirac. *The Principles of Quantum Mechanics*. Oxford University Press, London, 1930.

F. Dönau, K. Hehl, C. Riedel, R. A. Broglia, and P. Federman. Two-nucleon transfer reaction on oxygen and the nuclear coexistence model. *Nucl. Phys. A*, 101:495, 1967.

F. Dönau, D. Almehed, and R. G. Nazmitdinov. Integral representation of the random-phase approximation correlation energy. *Phys. Rev. Lett.*, 83:280, 1999.

T. Duguet. Bare vs effective pairing forces: A microscopic finite-range interaction for Hartree–Fock–Bogolyubov calculations in coordinate space. *Phys. Rev. C*, 69:054317, 2004.

T. Duguet. Pairing in finite nuclei from low-momentum two- and three-nucleon interactions. In R. A. Broglia and V. Zelevinsky, editors, *50 Years of Nuclear BCS*, page 259. World Scientific, Singapore, 2013.

T. Duguet and G. Hagen. *Ab initio* approach to effective single-particle energies in doubly closed shell nuclei. *Phys. Rev. C*, 85:034330, 2012.

T. Duguet and T. Lesinski. Non-empirical pairing functional. *Eur. Phys. J. Special Topics*, 156:207, 2008.

T. Duguet, T. Lesinski, K. Hebeler, and A. Schwenk. Lowest-order contributions of chiral three-nucleon interactions to pairing properties of nuclear ground states. *Mod. Phys. Lett. A*, 25:1989, 2010.

G. G. Dussel, R. I. Betan, R. J. Liotta, and T. Vertse. Collective excitations in the continuum. *Phys. Rev. C*, 80:064311, 2009.

F. Dyson. *Origins of Life*. Cambridge University Press, Cambridge, 1999.

F. J. Dyson. The radiation theories of Tomonaga, Schwinger, and Feynman. *Phys. Rev.*, 75:486, 1949.

J.-P. Ebran, E. Khan, T. Nikšić, and D. Vretenar. How atomic nuclei cluster. *Nature*, 487:341, 2012.

J.-P. Ebran, E. Khan, T. Nikšić, and D. Vretenar. Localization and clustering in the nuclear Fermi liquid. *Phys. Rev. C*, 87:044307, 2013.

J.-P. Ebran, E. Khan, T. Nikšić, and D. Vretenar. Cluster–liquid transition in finite, saturated Fermionic systems. *Phys. Rev. C*, 89:031303, 2014a.

J.-P. Ebran, E. Khan, T. Niksic, and D. Vretenar. Density functional theory studies of cluster states in nuclei. *Phys. Rev. C*, 90:054329, 2014b.

A. R. Edmonds. *Angular Momentum in Quantum Mechanics*. Princeton University Press, Princeton, NJ, 1960.

J. L. Egido. Projection methods, variational diagonalization of the pairing Hamiltonian and restoration of rotational and gauge invariance. In R. A. Broglia and V. Zelevinski, editors, *50 Years of Nuclear BCS*, page 553. World Scientific, Singapore, 2013.

P. Ehrenfest and J. R. Oppenheimer. Note on the statistics of nuclei. *Phys. Rev.*, 37:333, 1931.

G. M. Eliashberg. Interactions between electrons and lattice vibrations in a superconductor. *Sov. Phys. JETP*, 11:696, 1960.

W. Elsasser. Sur le principe de Pauli dans les noyaux. *J. Phys. Radium*, 4:549, 1933.

H. Esbensen and G. F. Bertsch. Nuclear surface fluctuations: Charge density. *Phys. Rev. C*, 28:355, 1983.

H. Esbensen, G. F. Bertsch, and K. Hencken. Application of contact interactions to Borromean halos. *Phys. Rev. C*, 56:3054, 1997.

P. Federman and I. Talmi. Coexistence of shell model and deformed states in oxygen isotopes. *Phys. Lett.*, 19:490, 1965.

P. Federman and I. Talmi. Nuclear energy levels of calcium isotopes. *Phys. Lett.*, 22:469, 1966.

J. P. Fernández-García, M. A. G. Alvarez, A. M. Moro, and M. Rodriguez-Gallardo. Simultaneous analysis of elastic scattering and transfer/breakup channels for the ^{6}He+^{208}Pb reaction at energies near the Coulomb barrier. *Phys.Lett.*, B693:310, 2010a.

J. P. Fernández-García, M. Rodríguez-Gallardo, M. A. G. Alvarez, and A. M. Moro. Long range effects on the optical model of ^{6}He around the Coulomb barrier. *Nucl. Phys. A*, 840:19, 2010b.

L. Ferreira, R. Liotta, C. H. Dasso, R. A. Broglia, and A. Winther. Spatial correlations of pairing collective states. *Nucl. Phys. A*, 426:276, 1984.

H. Feshbach. Unified theory of nuclear reactions. *Ann. Phys.*, 5:357, 1958.

H. Feshbach. A unified theory of nuclear reactions, II. *Ann. Phys.*, 19:287, 1962.

H. Feshbach. *Theoretical Nuclear Physics: Nuclear Reactions*. Wiley, New York, 1992.

R. P. Feynman. Space-time approach to quantum electrodynamics. *Phys. Rev.*, 76:769, 1949.

R. P. Feynman. The present status of quantum electrodynamics. In R. Stoops, editor, *The Quantum Theory of Fields*. Wiley Interscience, New York, 1961.

R. P. Feynman. *Theory of Fundamental Processes*. Benjamin, Reading, MA, 1975.

R. P. Feynman. *QED: The Strange Theory of Light and Matter*. Princeton University Press, Princeton, NJ, 2006.

R. M. Fleming, A. P. Ramirez, M. J. Rosseinsky, D. W. Murphy, R. C. Haddon, S. M. Zahurak, and A. V. Makhija. Relation of structure and superconducting transition temperatures in A_3C_{60}. *Nature*, 352:787, 1991.

E. R. Flynn, G. Igo, P. D. Barnes, D. Kovar, D. R. Bès, and R. A. Broglia. Effect of the multipole pairing and particle-hole fields in the particle–vibration coupling of ^{209}Pb (II): The ^{207}Pb$(t, p)^{209}$Pb reaction at 20 MeV. *Phys. Rev. C*, 3:2371, 1971.

E. R. Flynn, G. J. Igo, and R. A. Broglia. Three–phonon monopole and quadrupole pairing vibrational states in ^{206}Pb. *Phys. Lett. B*, 41:397, 1972a.

E. R. Flynn, G. J. Igo, R. A. Broglia, S. Landowne, V. Paar, and B. Nilsson. An experimental and theoretical investigation of the structure of ^{210}Pb. *Nucl. Phys. A*, 195:97, 1972b.

S. Fortier, S. Pita, J. Winfield, W. Catford, N. Orr, J. V. de Wiele, Y. Blumenfeld, R. Chapman, S. Chappell, N. Clarke, N. Curtis, M. Freer, S. Galès, K. Jones, H. Langevin-Joliot, H. Laurent, I. Lhenry, J. Maison, P. Roussel-Chomaz, M. Shawcross, M. Smith, K. Spohr, T. Suomijarvi, and A. de Vismes. Core excitation in ^{11}Be$_{gs}$ via the $p(^{11}$Be,^{10}Be)d reaction. *Phys. Lett. B*, 461:22, 1999.

L. Fortunato, W. Von Oertzen, H. Sofia, and A. Vitturi. Enhanced excitation of giant pairing vibrations in heavy-ion reactions induced by weakly bound projectiles. *Eur. Phys. J. A*, 14:37, 2002.

H. Fortune. Structure of exotic light nuclei: Z=2,3,4. *Eur. Phys. J. A*, 54:51, 2018.

H. T. Fortune, G.-B. Liu, and D. E. Alburger. $(sd)^2$ states in ^{12}Be. *Phys. Rev. C*, 50:1355, 1994.

K. Fossez, W. Nazarewicz, Y. Jaganathen, N. Michel, and M. Płoszajczak. Nuclear rotation in the continuum. *Phys. Rev. C*, 93:011305, 2016.

S. Frauendorf. Pairing at high spin. In R. A. Broglia and V. Zelevinski, editors, *50 Years of Nuclear BCS*, page 536. World Scientific, Singapore, 2013.

H. Fröhlich. Interaction of electrons with lattice vibrations. *Procs. R. Soc.*, 215:291, 1952.

R. J. Furnstahl and A. Schwenk. How should one formulate, extract and interpret "non-observables" for nuclei? *J. Phys. G*, 37:064005, 2010.

J. J. Gaardhøje. The dynamics of nuclear structure at high excitation energy. In R. A. Broglia and C. H. Dasso, editors, *Frontiers in Nuclear Dynamics*, page 133. Plenum Press, New York, 1985.

J. J. Gaardhøje, C. Ellegaard, B. Herskind, R. M. Diamond, M. A. Deleplanque, G. Dines, A. O. Macchiavelli, and F. S. Stephens. Gamma decay of isovector giant resonances built on highly excited states in ^{111}Sn. *Phys. Rev. Lett.*, 56:1783, 1986.

J. D. Garrett. The "single-particle" spectrum of states: Correlated or uncorrelated? In R. A. Broglia, G. B. Hagemann, and B. Herskind, editors, *Nuclear Structure 1985*, page 111. North-Holland, Amsterdam, 1985.

J. D. Garrett, G. B. Hagemann, and B. Herskind. Recent nuclear structure studies in rapidly rotating nuclei. *Ann. Rev. Nucl. Particle Sci.*, 36:419, 1986.

E. Garrido, D. Fedorov, and A. Jensen. Structure of exotic light nuclei: Z=2,3,4. *Nucl. Phys. A*, 770:117, 2002.

E. Garrido, D. Fedorov, and A. Jensen. Spin-dependent effective interactions for halo nuclei. *Phys. Rev. C*, 68:014002, 2003.

I. Giaever. Electron tunneling and superconductivity. In P. Norstedt and Söner, editors, *Le Prix Nobel en 1973*, page 84. P. Norstedt and Söner, Stockholm, 1973.

V. L. Ginzburg. Nobel lecture: On superconductivity and superfluidity (what I have and have not managed to do) as well as on the "physical minimum" at the beginning of the XXI century. *Rev. Mod. Phys.*, 76:981, 2004.

V. L. Ginzburg and L. D. Landau. On the theory of superconductivity. *JETP*, 20:1064, 1950.

R. J. Glauber. Coherence and quantum optics. In R. J. Glauber, editor, *International School of Physics "Enrico Fermi" Course XLII on Quantum Optics*, page 15. Academic Press, New York, 1969.

R. J. Glauber. *Quantum Theory of Optical Coherence*. Wiley, Weinheim, 2007.

N. K. Glendenning. Nuclear spectroscopy with two-nucleon transfer reactions. *Phys. Rev.*, 137:B102, 1965.

N. K. Glendenning. *Direct Nuclear Reactions*. World Scientific, Singapore, 2004.

D. Gogny. Lecture notes in physics. In H. Arenhövd and D. Drechsel, editors, *Nuclear Physics with Electromagnetic Interactions*, vol. 108, page 88, Springer, Heidelberg, 1979.

V. I. Goldanskii and A. I. Larkin. An analog of the Josephson effect in nuclear transformations. *Soviet Phys. JETP*, 26:617, 1968.

G. Gori, F. Barranco, E. Vigezzi, and R. A. Broglia. Parity inversion and breakdown of shell closure in Be isotopes. *Phys. Rev. C*, 69:041302, 2004.

L. P. Gor'kov. About the energy spectrum of superconductors. *Soviet Phys. JETP*, 7:505, 1958.

L. P. Gor'kov. Microscopic derivation of the Ginzburg–Landau equations in the theory of superconductivity. *Soviet Phys. JETP*, 9:1364, 1959.

L. P. Gor'kov. Theory of superconductivity: From phenomenology to microscopic theory. In H. Rogallia and P. H. Kes, editors, *100 Years of Superconductivity*, page 72. CRC Press, Boca Raton, FL, 2012.

M. Gornov, Y. Gurov, S. Lapushkin, P. Morokhov, V. Pechkurov, K. Seth, T. Pedlar, J. Wise, and D. Zhao. Spectroscopy of ^{7}, ^{8}He, ^{10}Li, and ^{13}Be nuclei in stopped π^-meson absorption reactions. *Bull. Russ. Acad. Sci.*, 62:1781, 1998.

U. Götz, M. Ichimura, R. A. Broglia, and A. Winther. Reaction mechanism of two-nucleon transfer between heavy ions. *Phys. Rep.*, 16:115, 1975.

W. Greiner. *Quantum Mechanics*. Springer, Berlin, 1998.

H. Grotch. Status of the theory of the hydrogen Lamb shift. *Foundations Phys.*, 240 (2):249, 1994.

P. Guazzoni, M. Jaskola, L. Zetta, A. Covello, A. Gargano, Y. Eisermann, G. Graw, R. Hertenberger, A. Metz, F. Nuoffer, and G. Staudt. Level structure of ^{120}Sn: High resolution (p, t) reaction and shell model description. *Phys. Rev. C*, 60:054603, 1999.

P. Guazzoni, L. Zetta, A. Covello, A. Gargano, G. Graw, R. Hertenberger, H.-F. Wirth, and M. Jaskola. High-resolution study of the ^{116}Sn (p, t) reaction and shell model structure of ^{114}Sn. *Phys. Rev. C*, 69:024619, 2004.

P. Guazzoni, L. Zetta, A. Covello, A. Gargano, B. F. Bayman, G. Graw, R. Hertenberger, H.-F. Wirth, and M. Jaskola. Spectroscopy of ^{110}Sn via the high-resolution ^{112}Sn(p, t) ^{110}Sn reaction. *Phys. Rev. C*, 74:054605, 2006.

P. Guazzoni, L. Zetta, A. Covello, A. Gargano, B. F. Bayman, T. Faestermann, G. Graw, R. Hertenberger, H.-F. Wirth, and M. Jaskola. ^{118}Sn levels studied by the ^{120}Sn(p, t) reaction: High-resolution measurements, shell model, and distorted-wave Born approximation calculations. *Phys. Rev. C*, 78:064608, 2008.

P. Guazzoni, L. Zetta, A. Covello, A. Gargano, B. F. Bayman, T. Faestermann, G. Graw, R. Hertenberger, H.-F. Wirth, and M. Jaskola. High-resolution measurement of the ^{118,124}Sn(p,t) ^{116,122}Sn reactions: Shell-model and microscopic distorted-wave Born approximation calculations. *Phys. Rev. C*, 83:044614, 2011.

P. Guazzoni, L. Zetta, A. Covello, A. Gargano, B. F. Bayman, G. Graw, R. Hertenberger, H. F. Wirth, T. Faestermann, and M. Jaskola. High resolution spectroscopy of ^{112}Sn through the ^{114}Sn(p, t)^{112}Sn reaction. *Phys. Rev. C*, 85:054609, 2012.

O. Gunnarsson. Jahn–Teller effect and on-site interaction for C_{60}^{n-}. *Phys. Rev. B*, 51:3493, 1995.

O. Gunnarsson. Superconductivity in fullerides. *Rev. Mod. Phys.*, 69:575, 1997.

O. Gunnarsson. *Alkali-Doped Fullerides: Narrow-Band Solids with Unusual Properties*. World Scientific, Singapore, 2004.

K. Hagino and H. Sagawa. Pairing correlations in nuclei on the neutron-drip line. *Phys. Rev. C*, 72:044321, 2005.

K. Hagino and H. Sagawa. Dipole excitation and geometry of borromean nuclei. *Phys. Rev. C*, 76:047302, 2007.

K. Hagino, H. Sagawa, J. Carbonell, and P. Schuck. Coexistence of BCS- and BEC-like pair structures in halo nuclei. *Phys. Rev. Lett.*, 99:022506, 2007.

K. Hamacher. Temperature dependence of fluctuations in HIV-1-protease. *Eur. Biophys. J.*, 390 (7):1051, 2010.

I. Hamamoto. Particle-vibration coupling in the nuclei ^{209}Bi and ^{209}Pb. *Nucl. Phys. A*, 126:545, 1969.

I. Hamamoto. Single-particle components of particle-vibrational states in the nuclei ^{209}Pb, ^{209}Bi, ^{207}Pb and ^{207}Tl. *Nucl. Phys. A*, 141:1, 1970a.

I. Hamamoto. Nuclear octupole vibrations in the isotopes of Pb. *Nucl. Phys. A*, 155:362, 1970b.

I. Hamamoto. Particle-vibration coupling: Octupole mode. In A. Bohr and R. A. Broglia, editors, *Elementary Modes of Excitation in Nuclei: Proceedings of the International School of Physics "Enrico Fermi,"* page 252. North-Holland, Amsterdam, 1977.

I. Hamamoto and B. R. Mottelson. Pair correlation in neutron drip line nuclei. *Phys. Rev. C*, 68:034312, 2003.

I. Hamamoto and B. R. Mottelson. Weakly bound $s_{1/2}$ neutrons in the many-body pair correlation of neutron drip line nuclei. *Phys. Rev. C*, 69:064302, 2004.

I. Hamamoto and S. Shimoura. Properties of ^{12}Be and ^{11}Be in terms of single-particle motion in deformed potential. *J. Phys. G*, 34:2715, 2007.

I. Hamamoto and P. Siemens. Large effect of core polarization on the single-particle spectra around ^{208}Pb. *Nucl. Phys. A*, 269:199, 1976.

Y. Han, Y. Shi, and Q. Shen. Deuteron global optical model potential for energies up to 200 MeV. *Phys. Rev. C*, 74:044615, 2006.

O. Hansen. Experimental establishment of the nuclear pairing phases. In R. A. Broglia and V. Zelevinsky, editors, *50 Years of Nuclear BCS*, page 449. World Scientific, Singapore, 2013.

P. G. Hansen and B. Jonson. The neutron halo of extremely neutron-rich nuclei. *Europhys. Lett. (EPL)*, 40 (4):409, 1987.

K. Hara. On the Josephson-current in heavy-ion reactions. *Phys. Lett. B*, 35:198, 1971.

O. Haxel, J. H. D. Jensen, and H. E. Suess. On the "magic numbers" in nuclear structure. *Phys. Rev.*, 75:1766, 1949.

A. F. Hebard. Superconductivity in doped fullerenes. *Phys. Today*, 45:26, 1992.

A. F. Hebard, M. J. Rosseinsky, R. C. Haddon, D. W. Murphy, S. H. Glarum, T. T. M. Palstra, A. P. Ramirez, and A. R. Kortan. Superconductivity at 18 K in potassium-doped C60. *Nature*, 350:600, 1991.

K. Hebeler, T. Duguet, T. Lesinski, and A. Schwenk. Non-empirical pairing energy functional in nuclear matter and finite nuclei. *Phys. Rev. C*, 80:044321, 2009.

P. H. Heenen, V. Hellemans, and R. V. F. Janssens. Pairing correlations at superdeformation. In R. A. Broglia and V. Zelevinski, editors, *50 Years of Nuclear BCS*, page 579. World Scientific, Singapore, 2013.

W. Heisenberg. Über quantentheoretische umdeutung kinematischer und mechanischer beziehungen. *Z. Phys.*, 33:879, 1925.

W. Heisenberg. Über der anschlaulichen inhalt der quantentheoretischen kinematik und mechanik. *Z. Phys.*, 43:172, 1927.

W. Heisenberg. *The Physical Principles of Quantum Theory*. Dover, New York, 1949.

V. Hellemans, A. Pastore, T. Duguet, K. Bennaceur, D. Davesne, J. Meyer, M. Bender, and P.-H. Heenen. Spurious finite-size instabilities in nuclear energy density functionals. *Phys. Rev. C*, 88:064323, 2013.

H. Hergert and R. Roth. Pairing in the framework of the unitary correlation operator method (UCOM): Hartree–Fock–Bogoliubov calculations. *Phys. Rev. C*, 80:024312, 2009.

M. Herzog, O. Civitarese, L. Ferreira, R. Liotta, T. Vertse, and L. Sibanda. Two-particle transfer reactions leading to giant pairing resonances. *Nucl. Phys. A*, 448:441, 1986.

J. Högaasen-Feldman. A study of some approximations of the pairing force. *Nucl. Phys.*, 28:258, 1961.

K. Holczer, O. Klein, S.-M. Huang, R. B. Kaner, K.-J. Fu, R. L. Whetten, and F. Diederich. Alkali-fulleride superconductors: Synthesis, composition, and diamagnetic shielding. *Science*, 252:1154, 1991.

B. R. Holstein. *Weak interaction in nuclei*. Princeton University Press, New Jersey, 1989.

K. Huang. *Statistical Physics and Protein Folding*. World Scientific, Singapore, 2005.

R. Id Betan, R. J. Liotta, N. Sandulescu, and T. Vertse. Two-particle resonant states in a many-body mean field. *Phys. Rev. Lett.*, 89:042501, 2002.

A. Idini. Renormalization effects in nuclei. PhD thesis, University of Milan, http://air.unimi.it/2434/216315, 2013.

A. Idini, F. Barranco, and E. Vigezzi. Quasiparticle renormalization and pairing correlations in spherical superfluid nuclei. *Phys. Rev. C*, 85:014331, 2012.

A. Idini, G. Potel, F. Barranco, E. Vigezzi, and R. A. Broglia. Dual origin of pairing in nuclei. *Phys. Atomic Nuclei*, 79:807, 2014.

A. Idini, G. Potel, F. Barranco, E. Vigezzi, and R. A. Broglia. Interweaving of elementary modes of excitation in superfluid nuclei through particle-vibration coupling: Quantitative account of the variety of nuclear structure observables. *Phys. Rev. C*, 92:031304, 2015.

K. Ieki, D. Sackett, A. Galonsky, C. A. Bertulani, J. J. Kruse, W. G. Lynch, D. J. Morrissey, N. A. Orr, H. Schulz, B. M. Sherrill, A. Sustich, J. A. Winger, F. Deák, A. Horváth, A. Kiss, Z. Seres, J. J. Kolata, R. E. Warner, and D. L. Humphrey. Coulomb dissociation of ^{11}Li. *Phys. Rev. Lett.*, 70:730, 1993.

M. Igarashi, K. Kubo, and K. Yagi. Two-nucleon transfer mechanisms. *Phys. Rep.*, 199:1, 1991.

J. N. Israelachvili. *Intermolecular and Surface Forces*. Academic Press, New York, 1985.

H. Iwasaki, T. Motobayashi, H. Akiyoshi, Y. Ando, N. Fukuda, H. Fujiwara, Z. Fülöp, K. Hahn, Y. Higurashi, M. Hirai, I. Hisanaga, N. Iwasa, T. Kijima, T. Minemura, T. Nakamura, M. Notani, S. Ozawa, H. Sakurai, S. Shimoura, S. Takeuchi, T. Teranishi, Y. Yanagisawa, and M. Ishihara. Quadrupole deformation of ^{12}Be studied by proton inelastic scattering. *Phys. Lett. B*, 481:7, 2000.

D. F. Jackson. *Nuclear Reactions*. Methuen, London, 1970.

B. Jenning. Nonobservability of spectroscopic factors. arXiv: 1102.3721ve [nucl-th], 2011.

H. Jeppesen, A. Moro, U. Bergmann, M. Borge, J. Cederkäll, L. Fraile, H. Fynbo, J. Gómez-Camacho, H. Johansson, B. Jonson, M. Meister, T. Nilsson, G. Nyman, M. Pantea, K. Riisager, A. Richter, G. Schrieder, T. Sieber, O. Tengblad, E. Tengborn, M. Turrión, and F. Wenander. Study of ^{10}Li via the ^{9}Li(^{2}H,p) reaction at REX-ISOLDE. *Phys. Lett. B*, 642:449, 2006.

J. Jeukenne, A. Lejeune, and C. Mahaux. Many-body theory of nuclear matter. *Phys. Rep.*, 250 (2):83, 1976.

B. D. Josephson. Possible new effects in superconductive tunnelling. *Phys. Lett.*, 1:251, 1962.

A. M. Kadin. Spatial structure of the Cooper pair. *J. Superconductivity Novel Magnetism*, 20:285, 2007.

Y. Kanada-En'yo and H. Horiuchi. Structure of excited states of ^{11}Be studied with antisymmetrized molecular dynamics. *Phys. Rev. C*, 66:024305, 2002.

R. Kanungo, A. Sanetullaev, J. Tanaka, S. Ishimoto, G. Hagen, T. Myo, T. Suzuki, C. Andreoiu, P. Bender, A. A. Chen, B. Davids, J. Fallis, J. P. Fortin, N. Galinski, A. T. Gallant, P. E. Garrett, G. Hackman, B. Hadinia, G. Jansen, M. Keefe, R. Krücken, J. Lighthall, E. McNeice, D. Miller, T. Otsuka, J. Purcell, J. S. Randhawa, T. Roger, A. Rojas, H. Savajols, A. Shotter, I. Tanihata, I. J. Thompson, C. Unsworth, P. Voss, and Z. Wang. Evidence of soft dipole resonance in ^{11}Li with isoscalar character. *Phys. Rev. Lett.*, 114:192502, 2015.

K. Kato and K. Ikeda. Analysis of ^{9}Li+n resonances in ^{10}Li by complex scaling method. *Prog. Theor. Phys.*, 89:623, 1993.

N. Keeley, N. Alamanos, and V. Lapoux. Comprehensive analysis method for (d, p) stripping reactions. *Phys. Rev. C*, 69:064604, 2004.

J. B. Ketterson and S. N. Song. *Superconductivity*. Cambridge University Press, Cambridge, 1999.

E. Khan, N. Sandulescu, N. V. Giai, and M. Grasso. Two-neutron transfer in nuclei close to the drip line. *Phys. Rev. C*, 69:014314, 2004.

E. Khan, M. Grasso, and J. Margueron. Constraining the nuclear pairing gap with pairing vibrations. *Phys. Rev. C*, 80:044328, 2009.

V. A. Khodel, A. P. Platonov, and E. E. Saperstein. On the ^{40}Ca-^{48}Ca isotope shift. *J. Phys. G*, 8:967, 1982.

T. Kinoshita, editor. *Quantum Electrodynamics*. World Scientific, Singapore, 1990.

H. Kitagawa and H. Sagawa. Isospin dependence of kinetic energies in neutron-rich nuclei. *Nucl. Phys. A*, 551:16, 1993.

C. Kittel. *Introduction to Solid State Physics*. Wiley, NJ, Hoboken, 1996.

M. Kleber and H. Schmidt. Josephson effect in nuclear reactions. *Z. Phys.*, 245:68, 1971.

T. Kobayashi, S. Shimoura, I. Tanihata, K. Katori, K. Matsuta, T. Minamisono, K. Sugimoto, W. Müller, D. L. Olson, T. J. M. Symons, and H. Wieman. Electromagnetic dissociation and soft giant dipole resonance of the neutron-dripline nucleus ^{11}Li. *Phys. Lett. B*, 232:51, 1989.

T. Kobayashi, K. Yoshida, A. Ozawa, I. Tanihata, A. Korsheninnikov, E. Nikolski, and T. Nakamura. Quasifree nucleon-knockout reactions from neutron-rich nuclei by a proton target: $p(^6\text{He,pn})^5\text{He}$, $p(^{11}\text{Li,pn})^{10}\text{Li}$, $p(^6\text{He,2p})^5\text{H}$, and $p(^{11}\text{Li,2p})^{10}\text{He}$. *Nucl. Phys. A*, 616:223, 1997.

A. J. Koning and J. P. Delaroche. Local and global nucleon optical models from 1 keV to 200 MeV. *Nucl. Phys. A*, 713:231, 2003.

G. J. Kramer, H. P. Blok, and L. Lapikás. A consistence analysis of $(e, e'p)$ and $(d, ^3\text{He})$ experiments. *Nucl. Phys. A*, 679:267, 2001.

A. Krieger, K. Blaum, M. L. Bissell, N. Frömmgen, C. Geppert, M. Hammen, K. Kreim, M. Kowalska, J. Krämer, T. Neff, R. Neugart, G. Neyens, W. Nörtershäuser, C. Novotny, R. Sánchez, and D. T. Yordanov. Nuclear charge radius of ^{12}Be. *Phys. Rev. Lett.*, 108:142501, 2012.

N. M. Kroll and W. E. Lamb. On the self-energy of a bound electron. *Phys. Rev.*, 75:388, 1949.

R. Kryger, A. Azhari, A. Galonsky, J. Kelley, R. Pfaff, E. Ramakrishnan, D. Sackett, B. Sherrill, M. Thoennessen, J. Winger, and S. Yokoyama. ^9Li and n detected in coincidence from a stripping of ^{18}O beam at 80MeV/n on a C target, obtaining relative velocity spectra. *Phys. Rev. C*, 47:R2439, 1993.

K. Kubo, R. A. Broglia, T. Riedel, and C. Udagawa. Determination of the Y_4 deformation parameter from (p, t) reactions. *Phys. Lett. B*, 32:29, 1970.

Y. Kubota, A. Corsi, G. Authelet, H. Baba, C. Caesar, D. Calvet, A. Delbart, M. Dozono, J. Feng, F. Flavigny, J.-M. Gheller, J. Gibelin, A. Giganon, A. Gillibert, K. Hasegawa, T. Isobe, Y. Kanaya, S. Kawakami, D. Kim, Y. Kikuchi, Y. Kiyokawa, M. Kobayashi, N. Kobayashi, T. Kobayashi, Y. Kondo, Z. Korkulu, S. Koyama, V. Lapoux, Y. Maeda, F. M. Marqués, T. Motobayashi, T. Miyazaki, T. Nakamura, N. Nakatsuka, Y. Nishio, A. Obertelli, K. Ogata, A. Ohkura, N. A. Orr, S. Ota, H. Otsu, T. Ozaki, V. Panin, S. Paschalis, E. C. Pollacco, S. Reichert, J.-Y. Roussé, A. T. Saito, S. Sakaguchi, M. Sako, C. Santamaria, M. Sasano, H. Sato, M. Shikata, Y. Shimizu, Y. Shindo, L. Stuhl, T. Sumikama, Y. L. Sun, M. Tabata, Y. Togano, J. Tsubota, Z. H. Yang, J. Yasuda, K. Yoneda, J. Zenihiro, and T. Uesaka. Surface localization of the dineutron in ^{11}Li. *Phys. Rev. Lett.*, 125:252501, 2020.

E. Kwan, C. Wu, N. Summers, G. Hackman, T. Drake, C. Andreoiu, R. Ashley, G. Ball, P. Bender, A. Boston, H. Boston, A. Chester, A. Close, D. Cline, D. Cross, R. Dunlop, A. Finlay, A. Garnsworthy, A. Hayes, A. Laffoley, T. Nano, P. Navrátil, C. Pearson, J. Pore, S. Quaglioni, C. Svensson, K. Starosta, I. Thompson, P. Voss, S. Williams, and Z. Wang. Precision measurement of the electromagnetic dipole strengths in ^{11}Be. *Phys. Lett. B*, 732:210, 2014.

G. A. Lalazissis and P. Ring. Giant resonances with time dependent covariant density functional theory. *Eur. Phys. J. A*, 55:229, 2019.

W. E. Lamb Jr. Fine structure of the hydrogen atom. In *Nobel Lecture Physics 1942–1962*, page 286. Elsevier, Amsterdam, 1964.

W. E. Lamb Jr. and R. C. Retherford. Fine structure of the hydrogen atom by a microwave method. *Phys. Rev.*, 72:241, 1947.

L. Landau. The theory of superfluidity in ^2He. *J. Phys. USSR*, 5:71, 1941.

L. Landau. On analytic properties of vertex parts in quantum field theory. *Nucl. Phys.*, 13:181, 1959.

L. Landau and L. Lifshitz. *Quantum Mechanics*, 3rd ed. Pergamon Press, London, 1965.

W. A. Lanford. Systematics of two-neutron-transfer cross sections near closed shells: A sum-rule analysis of (p, t) strengths on the lead isotopes. *Phys. Rev. C*, 16:988, 1977.

W. A. Lanford and J. B. McGrory. Two-neutron pickup strengths on the even lead isotopes: The transition from single-particle to "collective." *Phys. Lett. B*, 45:238, 1973.

M. Laskin, R. F. Casten, A. O. Macchiavelli, R. M. Clark, and D. Bucurescu. Population of the giant pairing vibration. *Phys. Rev. C*, 93:034321, 2016.

J. A. Lay, A. M. Moro, J. M. Arias, and Y. Kanada-En'yo. Semi-microscopic folding model for the description of two-body halo nuclei. *Phys. Rev. C*, 89:014333, 2014.

J. Le Tourneaux. Effect of the dipole-quadrupole interaction on the width and structure of the giant dipole line in spherical nuclei. *Mat. Fys. Medd. Dan. Vid. Selsk.*, 34, no. 11, 1965.

A. J. Leggett. *The Problems of Physics*. Oxford University Press, Oxford, 1987.

A. J. Leggett. *Quantum Liquids*. Oxford University Press, Oxford, 2006.

T. Lesinski, T. Duguet, K. Bennaceur, and J. Meyer. Non-empirical pairing energy density functional. *Eur. Phys. J. A*, 40:121, 2009.

T. Lesinski, K. Hebeler, T. Duguet, and A. Schwenk. Chiral three-nucleon forces and pairing in nuclei. *J. Phys. G*, 39:015108, 2011.

F. A. Lindemann. The calculation of molecular vibration frequencies. *Phys. Z.*, 11:609, 1910.

J. Lindhard. On the properties of a gas of charged particles. *Kgl. Danske Videnshab. Selskab, Mat. Fys. Medd.*, 28, no. 8, 1954.

E. Lipparini. *Modern Many-Particle Physics: Atomic Gases, Quantum Dots and Quantum Fluids*. World Scientific, Singapore, 2003.

E. Litvinova and H. Wibowo. Finite-temperature relativistic nuclear field theory: An application to the dipole response. *Phys. Rev. Lett.*, 121:082501, 2018.

U. Lombardo, H. J. Schulze, and W. Zuo. Induced pairing interaction in neutron star matter. In R. A. Broglia and V. Zelevinsky, editors, *50 Years of Nuclear BCS*, page 338. World Scientific, Singapore, 2013.

F. London. Zur Theorie der Systematik der Molekularkräfte. *Z. Phys.*, 63:245, 1930.

H. Löwen. Melting, freezing and colloidal suspensions. *Phys. Rep.*, 237:249, 1994.

K. Lum, D. Chandler, and J. D. Weeks. Hydrophobicity at small and large length scales. *J. Phys. Chem. B*, 103:4570, 1999.

S. R. Lundeen and F. M. Pipkin. Measurement of the Lamb shift in hydrogen, $n = 2$. *Phys. Rev. Lett.*, 46:232, 1981.

S. R. Lundeen and F. M. Pipkin. Separated oscillatory field measurement of the Lamb shift in H, $n= 2$. *Metrologia*, 22:9, 1986.

J. E. Lynn. *Theory of Neutron Resonance Reactions*. Oxford University Press, Oxford, 1968.

R. Machleidt, F. Sammarruca, and Y. Song. Nonlocal nature of the nuclear force and its impact on nuclear structure. *Phys. Rev. C*, 53:R1483, 1996.

P. Magierski, K. Sekizawa, and G. Wlazłowski. Novel role of superfluidity in low-energy nuclear reactions. *Phys. Rev. Lett.*, 119:042501, 2017.

C. Mahaux, P. F. Bortignon, R. A. Broglia, and C. H. Dasso. Dynamics of the shell model. *Phys. Rep.*, 120:1–274, 1985.

D. Martin, P. von Neumann-Cosel, A. Tamii, N. Aoi, S. Bassauer, C. A. Bertulani, J. Carter, L. Donaldson, H. Fujita, Y. Fujita, T. Hashimoto, K. Hatanaka, A. Ito, A. Krugmann, B. Liu, Y. Maeda, K. Miki, R. Neveling, N. Pietralla, I. Poltoratska, V. Y. Ponomarev, A. Richter, T. Shima, T. Yamamoto, and M. Zweidinger. Test of the Brink–Axel hypothesis for the pygmy dipole resonance. *Phys. Rev. Lett.*, 119:182503, 2017.

T. P. Martin, U. Näher, and H. Schaber. Metal clusters as giant atomic nuclei. In R. A. Broglia and J. R. Schrieffer, editors, *International School of Physics "Enrico Fermi" Course CXXI, Perspectives in Many-Particle Physics*, page 139. North-Holland, Amsterdam, 1994.

M. Matsuo. Spatial structure of neutron Cooper pair in low density uniform matter. *Phys. Rev. C*, 73:044309, 2006.

M. Matsuo. Spatial structure of Cooper pairs in nuclei. In R. A. Broglia and V. Zelevinsky, editors, *50 Years of Nuclear BCS*, page 66. World Scientific, Singapore, 2013.

M. Matsuo, K. Mizuyama, and Y. Serizawa. Di-neutron correlation and soft dipole excitation in medium mass neutron-rich nuclei near drip line. *Phys. Rev. C*, 71:064326, 2005.

K. Matsuyanagi, N. Hinohora, and K. Sato. BCS-pairing and nuclear vibrations. In R. A. Broglia and V. Zelevinsky, editors, *50 Years of Nuclear BCS*, page 111. World Scientific, Singapore, 2013.

N. V. Mau and J. C. Pacheco. Structure of the ^{11}Li nucleus. *Nucl. Phys. A*, 607:163, 1996.

M. G. Mayer. On closed shells in nuclei. *Phys. Rev.*, 74:235, 1948.

M. G. Mayer. On closed shells in nuclei. II. *Phys. Rev.*, 75:1969, 1949.

M. G. Mayer and J. Jensen. *Elementary Theory of Nuclear Structure*. Wiley, New York 1955.

M. G. Mayer and E. Teller. On the origin of elements. *Phys. Rev.*, 76:1226, 1949.

D. G. McDonald. The Nobel laureate versus the graduate student. *Phys. Today*, 54:46, 2001.

M. H. McFarlane. The reaction matrix in nuclear shell theory. In M. Jean and R. A. Ricci, editors, *Proceedings of the International School of Physics "E. Fermi" Course XL Nuclear Structure and Nuclear Reactions*, page 457. Academic Press, New York, 1969.

J. Mehra. *The Beat of a Different Drum*. Clarendon Press, Oxford, 1996.

U. G. Meißner. Anthropic considerations in nuclear physics. *Sci. Bull.*, 60:43, 2015.

L. Meitner and O. R. Frisch. Products of the fission of the uranium nucleus. *Nature*, 143:239, 1939.

C. Micheletti, J. R. Banavar, and A. Maritan. Conformations of proteins in equilibrium. *Phys. Rev. Lett.*, 87:088102, 2001.

C. Micheletti, F. Cecconi, A. Flammini, and A. Maritan. Crucial stages of protein folding through a solvable model: Predicting target sites for enzyme-inhibiting drugs. *Protein Sci.*, 11:1878, 2002.

C. Micheletti, P. Carloni, and A. Maritan. Accurate and efficient description of protein vibrational dynamics: Comparing molecular dynamics and Gaussian models. *Proteins Struct. Funct. Bioinformatics*, 55:635, 2004.

A. Migdal. Interaction between electrons and lattice vibrations in a normal metal. *Soviet Phys. JETP*, 7:996, 1958.

P. W. Milloni. *The Quantum Vacuum*. Academic Press, New York, 1994.

A. Mitra, A. Kraft, P. Wright, B. Acland, A. Z. Snyder, Z. Rosenthal, L. Czerniewski, A. Bauer, L. Snyder, J. Culver, J. M. Lee, and M. E. Raichle. Spontaneous infra-slow brain activity has unique spatiotemporal dynamics and laminar structure. *Neuron*, 98:297, 2018.

P. Møller, J. R. Nix, W. D. Myers, and W. J. Swiatecki. Nuclear ground-state masses and deformations. *Atomic Data Nucl. Data Tables*, 59:185, 1995.

J. Monod. *Le hasard et la nécéssité*. Seuil, Paris, 1970.

D. Montanari, L. Corradi, S. Szilner, G. Pollarolo, E. Fioretto, G. Montagnoli, F. Scarlassara, A. M. Stefanini, S. Courtin, A. Goasduff, F. Haas, D. Jelavi ć Malenica, C. Michelagnoli, T. Mijatović, N. Soić, C. A. Ur, and M. Varga Pajtler. Neutron pair transfer in ^{60}Ni $+^{116}$Sn far below the Coulomb barrier. *Phys. Rev. Lett.*, 113:052501, 2014.

D. Montanari, L. Corradi, S. Szilner, G. Pollarolo, A. Goasduff, T. Mijatović, D. Bazzacco, B. Birkenbach, A. Bracco, L. Charles, S. Courtin, P. Désesquelles, E. Fioretto, A. Gadea, A. Görgen, A. Gottardo, J. Grebosz, F. Haas, H. Hess, D. Jelavi ć Malenica, A. Jungclaus, M. Karolak, S. Leoni, A. Maj, R. Menegazzo, D. Mengoni, C. Michelagnoli, G. Montagnoli, D. R. Napoli, A. Pullia, F. Recchia, P. Reiter, D. Rosso, M. D. Salsac, F. Scarlassara, P.-A. Söderström, N. Soić, A. M. Stefanini, O. Stezowski, C. Theisen, C. A. Ur, J. J. Valiente-Dobón, and M. Varga Pajtler. Pair neutron transfer in ^{60}Ni + ^{116}Sn probed via γ-particle coincidences. *Phys. Rev. C*, 93:054623, 2016.

A. Moro, J. Casal, and M. Gómez-Ramos. Investigating the ^{10}Li continuum through ^{9}Li$(d, p)^{10}$Li reactions. *Phys. Lett. B*, 793:13, 2019.

M. Moshinsky. Transformation brackets for harmonic oscillator functions. *Nucl. Phys.*, 13:104, 1959.

B. Mottelson. Elementary features of nuclear structure. In H. Nifenecker, J. P. Blaizot, G. F. Bertsch, W. Weise, and F. David, editors, *Trends in Nuclear Physics, 100 Years Later, Les Houches, Session LXVI*, page 25. Elsevier, Amsterdam, 1998.

B. R. Mottelson. Selected topics in the theory of collective phenomena in nuclei. In G. Racah, editor, *International School of Physics "Enrico Fermi" Course XV, Nuclear Spectroscopy*, page 44. Academic Press, New York, 1962.

B. R. Mottelson. Topics in nuclear structure theory. In *Nikko Summer School Lectures*, page 1. Nordita Copenhagen, 1967.

B. R. Mottelson. Properties of individual levels and nuclear models. In J. Sanada, editor, *Proceedings of the International Conference on Nuclear Structure*, page 87. Physical Society of Japan, Tokyo, 1968.

B. R. Mottelson. *Elementary Modes of Excitation in Nuclei, Le Prix Nobel en 1975*. Imprimerie Royale Norstedts Tryckeri, Stockholm, 1976a.

B. R. Mottelson. Atomkernens elementære eksitationer. *Naturens Verden*, 6:169, 1976b.

B. R. Mottelson and S. G. Nilsson. The intrinsic states of odd-A nuclei having ellipsoidal equilibrium shape. *Kgl. Danske Videnskab, Selskab. Mat.-fys. Skrifter*, 1, no. 8, 1959.

J. Mougey, M. Bernheim, A. Boussière, A. Gillebert, Phan Xuan Hô, M. Priou, D. Royer, I. Sick, and G. Wagner. Quasi-free $(e, e'p)$ scattering on ^{12}C, ^{28}Si, ^{40}Ca and ^{58}Ni. *Nucl. Phys. A*, 262:461, 1976.

B. Mouginot, E. Khan, R. Neveling, F. Azaiez, E. Z. Buthelezi, S. V. Förtsch, S. Franchoo, H. Fujita, J. Mabiala, J. P. Mira, P. Papka, A. Ramus, J. A. Scarpaci, F. D. Smit, I. Stefan, J. A. Swartz, and I. Usman. Search for the giant pairing vibration through (p,t) reactions around 50 and 60 MeV. *Phys. Rev. C*, 83:037302, 2011.

T. Myo, Y. Kikuchi, K. Kato, H. Tokim, and K. Ikeda. Systematic study of ^{9,10,11}Li with the tensor and pairing correlations. *Prog. Theor. Phys.*, 110:561, 2008.

T. Nakamura, A. M. Vinodkumar, T. Sugimoto, N. Aoi, H. Baba, D. Bazin, N. Fukuda, T. Gomi, H. Hasegawa, N. Imai, M. Ishihara, T. Kobayashi, Y. Kondo, T. Kubo, M. Miura, T. Motobayashi, H. Otsu, A. Saito, H. Sakurai, S. Shimoura, K. Watanabe, Y. X. Watanabe, T. Yakushiji, Y. Yanagisawa, and K. Yoneda. Observation of strong low-lying E1 strength in the two-neutron halo nucleus ^{11}Li. *Phys. Rev. Lett.*, 96:252502, 2006.

Y. Nambu. Quasi-particles and gauge invariance in the theory of superconductivity. *Phys. Rev.*, 117:648, 1960.

Y. Nambu. Dynamical symmetry breaking. In T. Eguchi and K. Nishigama, editors, *Broken Symmetry: Selected Papers of Y. Nambu (1995)*, page 436. World Scientific, Singapore, 1991.

O. Nathan and S. G. Nilsson. Collective nuclear motion and the unified model. In K. Siegbahn, editor, *Alpha- Beta- and Gamma-Ray Spectroscopy, Vol. 1*, page 601. North-Holland, Amsterdam, 1965.

S. G. Nilsson. Binding states of individual nucleons in strongly deformed nuclei. *Mat. Fys. Medd. Dan. Vid. Selsk.*, 29, no. 16, 1955.

P. J. Nolan and P. J. Twin. Superdeformed shapes at high angular momentum. *Ann. Rev. of Nucl. Particle Sci.*, 38:533, 1988.

W. Nörtershäuser, D. Tiedemann, M. Žáková, Z. Andjelkovic, K. Blaum, M. L. Bissell, R. Cazan, G. W. F. Drake, C. Geppert, M. Kowalska, J. Krämer, A. Krieger, R. Neugart, R. Sánchez, F. Schmidt-Kaler, Z.-C. Yan, D. T. Yordanov, and C. Zimmermann. Nuclear charge radii of ^{7,9,10}Be and the one-neutron halo nucleus ^{11}Be. *Phys. Rev. Lett.*, 102:062503, 2009.

L. Nosanow. On the possible superfluidity of ^{6}He–Its phase diagram and those of ^{6}He-^{4}He and ^{6}He-^{3}He mixtures. *J. Low Temp. Phys.*, 23:605, 1976.

F. Nunes, I. Thompson, and R. Johnson. Core excitation in one neutron halo systems. *Nucl. Phys. A*, 596:171, 1996.

S. E. A. Orrigo and H. Lenske. Pairing resonances and the continuum spectroscopy of ^{10}Li. *Phys. Lett. B*, 677:214, 2009.

T. Otsuka, N. Fukunishi, and H. Sagawa. Structure of exotic neutron-rich nuclei. *Phys. Rev. Lett.*, 70:1385, 1993.

J. M. Pacheco, R. A. Broglia, and B. R. Mottelson. The intrinsic line width of the plasmon resonances in metal microclusters at very low temperatures: Quantal surface fluctuations. *Z. Phys. D*, 21:289, 1991.

A. Pais. *Inward Bound*. Oxford University Press, Oxford, 1986.

A. Pais. *The Genius of Science*. Oxford University Press, Oxford, 2000.

S. S. Pankratov, M. V. Zverev, M. Baldo, U. Lombardo, and E. E. Saperstein. Semi-microscopic model of pairing in nuclei. *Phys. Rev. C*, 84:014321, 2011.

A. Pastore, D. Tarpanov, D. Davesne, and J. Navarro. Spurious finite-size instabilities in nuclear energy density functionals: Spin channel. *Phys. Rev. C*, 92:024305, 2015.

W. Pauli. Zur quantenmechanik ii. *Z. Phys.*, 31:625, 1925.

W. Pauli. Über gasentartung und paramagnetismus. *Z. Physik*, 410 (2):81, 1927.

L. Pauling and E. B. Wilson Jr. *Quantum Mechanics*. Dover, New York, 1963.

J. Pedersen, S. Bjørnholm, J. Borggreen, K. Hansen, T. P. Martin, and H. D. Rasmussen. Observation of quantum supershells in clusters of sodium atoms. *Nature*, 353:733, 1991.

O. Penrose. Bose–Einstein condensation and liquid helium. *Philos. Mag.*, 42:1373, 1951.

O. Penrose and L. Onsager. Bose–Einstein condensation and liquid helium. *Phys. Rev.*, 104:576, 1956.

F. Perey and B. Buck. A non-local potential model for the scattering of neutrons by nuclei . *Nucl. Phys.*, 32:353, 1962.

D. Pines and P. Nozières. *The Theory of Quantum Liquids*. Benjamin, New York, 1966.

W. Pinkston and G. Satchler. Properties of simultaneous and sequential two-nucleon transfer. *Nucl. Phys. A*, 383:61, 1982.

F. M. Pipkin. Lamb shift measurements. In T. Kinoshita, editor, *Quantum Electrodynamics*, page 697. World Scientific, Singapore, 1990.

A. B. Pippard. The historical context of Josephson discovery. In H. Rogalla and P. H. Kes, editors, *100 years of Superconductivity*, page 30. CRC Press, Boca Raton, FL, 2012.

G. Pollarolo, R. A. Broglia, and A. Winther. Calculation of the imaginary part of the heavy ion potential. *Nucl. Phys. A*, 406:369, 1983.

N. Poppelier, A. Wolters, and P. Glaudemans. Properties of exotic light nuclei. *Z. Phys. A*, 346:11, 1993.

G. Potel. ONE: single–particle transfer DWBA code for both light and heavy ions. 2012a.

G. Potel. COOPER: Two-particle transfer in second-order DWBA code for both light and heavy ions. 2012b.

G. Potel, F. Barranco, E. Vigezzi, and R. A. Broglia. Evidence for phonon mediated pairing interaction in the halo of the nucleus ^{11}Li. *Phys. Rev. Lett.*, 105:172502, 2010.

G. Potel, F. Barranco, F. Marini, A. Idini, E. Vigezzi, and R. A. Broglia. Calculation of the transition from pairing vibrational to pairing rotational regimes between magic nuclei ^{100}Sn and ^{132}Sn via two–nucleon transfer reactions. *Phys. Rev. Lett.*, 107:092501, 2011.

G. Potel, A. Idini, F. Barranco, E. Vigezzi, and R. A. Broglia. Cooper pair transfer in nuclei. *Rep. Prog. Phys.*, 76:106301, 2013a.

G. Potel, A. Idini, F. Barranco, E. Vigezzi, and R. A. Broglia. Quantitative study of coherent pairing modes with two–neutron transfer: Sn isotopes. *Phys. Rev. C*, 87:054321, 2013b.

G. Potel, A. Idini, F. Barranco, E. Vigezzi, and R. A. Broglia. Nuclear field theory predictions for ^{11}Li and ^{12}Be: Shedding light on the origin of pairing in nuclei. *Phys. At. Nucl.*, 77:941, 2014.

G. Potel, A. Idini, F. Barranco, E. Vigezzi, and R. A. Broglia. From bare to renormalized order parameter in gauge space: Structure and reactions. *Phys. Rev C*, 96:034606, 2017.

G. Potel, F. Barranco, E. Vigezzi, and R. A. Broglia. Quantum entanglement in nuclear Cooper-pair tunneling with γ rays. *Phys. Rev. C*, 103:L021601, 2021.

B. Povh and M. Rosina. *Scattering and Structures*. Springer, Berlin, 2002.

I. Ragnarsson and R. A. Broglia. Pairing isomers. *Nucl. Phys. A*, 263:315, 1976.

I. Ragnarsson and S. G. Nilsson. *Shapes and Shells in Nuclear Structure*. Cambridge University Press, Cambridge, 2005.

G. H. Rawitscher. The microscopic Feshbach optical potential for a schematic coupled channel example. *Nucl. Phys. A*, 475:519, 1987.

M. Rees. *Just Six Numbers*. Basic Books, New York, 2000.

S. Reich, H. Sofía, and D. Bès. Properties of single-particle states in the ^{208}Pb region. *Nucl. Phys. A*, 233:105, 1974.

P. G. Reinhard and D. Drechsel. Ground state correlations and the nuclear charge distribution. *Z. Phys. A*, 2900 (1):85–91, 1979.

H. Reinhardt. Foundation of nuclear field-theory. *Nucl. Phys. A*, 241:317, 1975.

H. Reinhardt. Functional approach to nuclear field–theory: Schematic model with pairing and particle–hole forces. *Nucl. Phys. A*, 298:60, 1978a.

H. Reinhardt. Nuclear field theory. *Nucl. Phys. A*, 251:77, 1978b.

H. Reinhardt. Application of the nuclear-field theory to superfluid nuclei–quasiparticle–phonon multiplet. *Nucl. Phys. A*, 251:176, 1980.

A. Repko, P. G. Reinhard, V. O. Nesterenko, and J. Kvasil. Toroidal nature of the low-energy E1 mode. *Phys. Rev. C*, 87:04305, 2013.

P. Ring. Berry phase and backbending. In R. A. Broglia and V. Zelevinski, editors, *50 Years of Nuclear BCS*, page 522. World Scientific, Singapore, 2013.

P. Ring and P. Schuck. *The Nuclear Many-Body Problem*. Springer, Berlin, 1980.

R. M. Robledo and G. F. Bertsch. Pairing in finite systems: Beyond the HFB theory. In R. A. Broglia and V. Zelevinski, editors, *50 Years of Nuclear BCS*, page 89. World Scientific, Singapore, 2013.

H. Rogalla and P. H. Kes, editors. *100 Years of Superconductivity*. CRC Press, Boca Raton, FL, 2012.

H. J. Rose and G. A. C. Jones. A new kind of natural radioactivity. *Nature*, 307:245, 1984.

H. I. Rösner, M. Caldarini, A. Prestel, M. A. Vanoni, R. A. Broglia, A. Aliverti, G. Tiana, and B. B. Kragelund. Cold denaturation of the HIV-1 protease monomer. *Biochemistry*, 156:1029, 2017.

M. J. Rosseinsky, A. P. Ramirez, S. H. Glarum, D. W. Murphy, R. C. Haddon, A. F. Hebard, T. T. M. Palstra, A. R. Kortan, S. M. Zahurak, and A. V. Makhija. Superconductivity at 28 K in $Rb_x C_{60}$. *Phys. Rev. Lett.*, 66:2830, 1991.

J. Rotureau, P. Danielewicz, G. Hagen, F. M. Nunes, and T. Papenbrock. Optical potential from first principles. *Phys. Rev. C*, 95:024315, 2017.

D. Sackett, K. Ieki, A. Galonsky, C. A. Bertulani, H. Esbensen, J. J. Kruse, W. G. Lynch, D. J. Morrissey, N. A. Orr, B. M. Sherrill, H. Schulz, A. Sustich, J. A. Winger, F. Deák, A. Horváth, A. Kiss, Z. Seres, J. J. Kolata, R. E. Warner, and D. L. Humphrey. Electromagnetic excitation of ^{11}Li. *Phys. Rev. C*, 48:118, 1993.

H. Sagawa and K. Hagino. Pairing correlations and soft dipole excitations in nuclei on the neutron-drip line. *Phys. Atomic Nuclei*, 70:1321, 2007.

H. Sagawa, B. A. Brown, and H. Esbensen. Parity inversion in the N=7 isotones and the pairing blocking effect. *Phys. Lett. B*, 309:1, 1993.

J. J. Sakurai. *Modern Quantum Mechanics*. Addison-Wesley, Cambridge, MA, 1994.

A. Sanetullaev, R. Kanungo, J. Tanaka, M. Alcorta, C. Andreoiu, P. Bender, A. Chen, G. Christian, B. Davids, J. Fallis, J. Fortin, N. Galinski, A. Gallant, P. Garrett, G. Hackman, B. Hadinia, S. Ishimoto, M. Keefe, R. Krücken, J. Lighthall, E. McNeice, D. Miller, J. Purcell, J. Randhawa, T. Roger, A. Rojas, H. Savajols, A. Shotter, I. Tanihata, I. Thompson, C. Unsworth, P. Voss, and Z. Wang. Investigation of the role of ^{10}Li resonances in the halo structure of ^{11}Li through the ^{11}Li$(p, d)^{10}$Li transfer reaction. *Phys. Lett. B*, 755:481, 2016.

F. Sanger. The arrangement of aminoacids in proteins. *Adv. Protein Chem.*, VII:1, 1952.

P. Santi, J. Kolata, V. Guimaraes, D. Peterson, R. White-Stevens, E. Rischette, D. Bazin, B. Sherrill, A. Navin, P. DeYoung, P. Jolivette, G. Peaslee, and R. Guray. Structure of the ^{10}Li nucleus investigated via the ^{9}Li(d,p)^{10}Li reaction. *Phys Rev. C*, 67:024606, 2003.

E. E. Saperstein and M. Baldo. Microscopic origin of pairing. In R. A. Broglia and V. Zelevinsky, editors, *50 Years of Nuclear BCS*, page 263. World Scientific, Singapore, 2013.

J. R. Sapirstein and D. R. Yennie. Theory of hydrogenic bound states. In T. Kinoshita, editor, *Quantum Electrodynamics*. World Scientific, Singapore, 1990.

R. Sartor and C. Mahaux. Single-particle states in nuclear matter and in finite nuclei. *Phys. Rev. C*, 210 (6):2613, 1980.

G. Satchler. *Introduction to Nuclear Reactions*. Macmillan, New York, 1980.

G. Satchler. *Direct Nuclear Reactions*. Clarendon Press, Oxford, 1983.

M. R. Schafroth. Remarks on the Meissner effect. *Phys. Rev.*, 111:72, 1958.

M. R. Schafroth, S. T. Butler, and J. M. Blatt. Quasichemical equilibrium approach to superconductivity. *Helvetica Phys. Acta*, 30:93, 1957.

J. P. Schiffer, C. R. Hoffman, B. P. Kay, J. A. Clark, C. M. Deibel, S. J. Freeman, A. M. Howard, A. J. Mitchell, P. D. Parker, D. K. Sharp, and J. S. Thomas. Test of sum rules in nucleon transfer reactions. *Phys. Rev. Lett.*, 108:022501, 2012.

A. Schmid. Diamagnetic susceptibility at the transition to the superconducting state. *Phys. Rev.*, 180:527, 1969.

H. Schmidt. Monopole vibrations of closed shell nuclei. *Z. Phys.*, 181:532, 1964.

H. Schmidt. The onset of superconductivity in the time-dependent Ginzburg–Landau theory. *Z. Phys.*, 216:336, 1968.

H. Schmidt. Pairing vibrations in nuclei and solids. In H. Morinaga, editor, *Proceedings of the International School of Physics "Enrico Fermi" Course LIII, Developments and Borderlines on Nuclear Physics*, page 144. Academic Press, New York, 1972.

K. T. Schmitt, K. L. Jones, S. Ahn, D. W. Bardayan, A. Bey, J. C. Blackmon, S. M. Brown, K. Y. Chae, K. A. Chipps, J. A. Cizewski, K. I. Hahn, J. J. Kolata, R. L. Kozub, J. F. Liang, C. Matei, M. Matos, D. Matyas, B. Moazen, C. D. Nesaraja, F. M. Nunes, P. D. O'Malley, S. D. Pain, W. A. Peters, S. T. Pittman, A. Roberts, D. Shapira, J. F. Shriner, M. S. Smith, I. Spassova, D. W. Stracener, N. J. Upadhyay, A. N. Villano, and G. L. Wilson. Reactions of a ^{10}Be beam on proton and deuteron targets. *Phys. Rev. C*, 88:064612, 2013.

J. R. Schrieffer. *Superconductivity*. Benjamin, New York, 1964.

J. R. Schrieffer. Macroscopic quantum phenomena from pairing in superconductors. *Phys. Today*, 26 (7):23, 1973.

E. Schrödinger. Quantisierung als eigenwertproblem. *Ann. Phys.*, 384:273, 1926a.

E. Schrödinger. Der stetige Ubergang von der Mikro-zur Makromechanik. *Nature*, 14:644, 1926b.

E. Schrödinger. *What Is Life? The Physical Aspect of the Living Cell*. Cambridge University Press, Cambridge, 1944.

S. Schweber. *QED*. Princeton University Press, Princeton, NJ, 1994.

J. Schwinger. On quantum-electrodynamics and the magnetic moment of the electron. *Phys. Rev.*, 73:416, 1948.

J. Schwinger. *Quantum Electrodynamics: Selected papers*. Dover, 1958.

J. Schwinger. *Quantum Mechanics*. Springer, Heidelberg, 2001.

A. W. Senior, R. Evans, J. Jumper, J. Kirkpatrick, L. Sifre, T. Green, C. Qin, A. Žídek, A. W. R. Nelson, A. Bridgland, H. Penedones, S. Petersen, K. Simonyan, S. Crossan, P. Kohli, D. T. Jones, D. Silver, K. Kavukcuoglu, and D. Hassabis. Improved protein structure prediction using potentials from deep learnings. *Nature*, 577:706, 2020.

R. F. Service. The game has changed: AI triumphs at solving protein structures. *Science*, 370:1144, 2020.

Y. R. Shimizu. Pairing fluctuations and gauge symmetry restoration in rotating superfluid nuclei. In R. A. Broglia and V. Zelevinsky, editors, *50 Years of Nuclear BCS*, page 567. World Scientific, Singapore, 2013.

Y. R. Shimizu and R. A. Broglia. A comparison of the RPA and number projection approach for calculations of pairing fluctuations in fast rotating nuclei. *Nucl. Phys. A*, 515:38, 1990.

Y. R. Shimizu, J. D. Garrett, R. A. Broglia, M. Gallardo, and E. Vigezzi. Pairing fluctuations in rapidly rotating nuclei. *Rev. Mod. Phys.*, 61:131, 1989.

Y. R. Shimizu, P. Donati, and R. A. Broglia. Response function technique for calculating the random-phase approximation correlation energy. *Phys. Rev. Lett.*, 85:2260, 2000.

S. Shimoura, T. Nakamura, M. Ishihara, N. Inabe, T. Kobayashi, T. Kubo, R. Siemssen, I. Tanihata, and Y. Watanabe. Coulomb dissociation reaction and correlations of two halo neutrons in ^{11}Li. *Phys. Lett. B*, 348:29, 1995.

H. Simon, M. Meister, T. Aumann, M. Borge, L. Chulkov, U. D. Pramanik, T. Elze, H. Emling, C. Forssén, H. Geissel, M. Hellström, B. Jonson, J. Kratz, R. Kulessa, Y. Leifels, K. Markenroth, G. Münzenberg, F. Nickel, T. Nilsson, G. Nyman, A. Richter, K. Riisager, C. Scheidenberger, G. Schrieder, O. Tengblad, and M. Zhukov.

Systematic investigation of the drip–line nuclei ^{11}Li and ^{14}Be and their unbound subsystems ^{10}Li and ^{13}Be. *Nucl. Phys. A*, 791:267, 2007.

J. Smith, T. Baumann, J. Brown, P. DeYoung, N. Frank, J. Hinnefeld, Z. Kohley, B. Luther, B. Marks, A. Spyrou, S. Stephenson, M. Thoennessen, and S. Williams. Spectroscopy of the ^{10}Li nucleus. *Nucl. Phys. A*, 940:235, 2015.

M. Smith, M. Brodeur, T. Brunner, S. Ettenauer, A. Lapierre, R. Ringle, V. L. Ryjkov, F. Ames, P. Bricault, G. W. F. Drake, P. Delheij, D. Lunney, F. Sarazin, and J. Dilling. First penning-trap mass measurement of the exotic halo nucleus ^{11}Li. *Phys. Rev. Lett.*, 101:202501, 2008.

K. A. Snover. Giant resonances in excited nuclei. *Ann. Rev. Nucl. and Particle Science*, 36:545, 1986.

G. S. Stent, editor. *The Double Helix*. W. W. Norton, New York, 1980.

F. H. Stillinger. A topographic view of supercooled liquids and glass formation. *Science*, 237:1935, 1995.

F. H. Stillinger and D. K. Stillinger. Computational study of transition dynamics in 55-atom clusters. *J. Chem. Phys.*, 93:6013, 1990.

I. Talmi and I. Unna. Order of levels in the shell model and spin of Be^{11}. *Phys. Rev. Lett.*, 4:469, 1960.

T. Tamura. Compact reformulation of distorted-wave and coupled-channel born approximations for transfer reactions between nuclei. *Phys. Rep.*, 14:59, 1974.

T. Tamura, D. R. Bès, R. A. Broglia, and S. Landowne. Coupled–channel Born–approximation calculation of two-nucleon transfer reactions in deformed nuclei. *Phys. Rev. Lett.*, 25:1507, 1970.

T. Tamura, T. Udagawa, and C. Mermaz. Direct reaction analyses of heavy-ion induced reactions leading to discrete states. *Phys. Rep.*, 65:345, 1980.

J. Tanaka, R. Kanungo, M. Alcorta, N. Aoi, H. Bidaman, C. Burbadge, G. Christian, S. Cruz, B. Davids, A. Diaz Varela, J. Even, G. Hackman, M. Harakeh, J. Henderson, S. Ishimoto, S. Kaur, M. Keefe, R. Krücken, K. Leach, J. Lighthall, E. Padilla Rodal, J. Randhawa, P. Ruotsalainen, A. Sanetullaev, J. Smith, O. Workman, and I. Tanihata. Halo-induced large enhancement of soft dipole excitation of ^{11}Li observed via proton inelastic scattering. *Phys. Lett. B*, 774:268, 2017.

Y. C. Tang and R. C. Herndon. Form factors of ^{3}H and ^{4}He with repulsive-core potentials. *Phys. Lett.*, 18:42, 1965.

I. Tanihata, M. Alcorta, D. Bandyopadhyay, R. Bieri, L. Buchmann, B. Davids, N. Galinski, D. Howell, W. Mills, S. Mythili, R. Openshaw, E. Padilla-Rodal, G. Ruprecht, G. Sheffer, A. C. Shotter, M. Trinczek, P. Walden, H. Savajols, T. Roger, M. Caamano, W. Mittig, P. Roussel-Chomaz, R. Kanungo, A. Gallant, M. Notani, G. Savard, and I. J. Thompson. Measurement of the two-halo neutron transfer reaction ^{1}H(^{11}Li,^{9}Li)^{3}H at 3A MeV. *Phys. Rev. Lett.*, 100:192502, 2008.

I. Tanihata, H. Savajols, and R. Kanungo. Recent experimental progress in nuclear halo structure studies, *in Particle and Nuclear Physics*, 68:215, 2013.

D. Tarpanov, J. Dobaczewski, J. Toivanen, and B. G. Carlsson. Spectroscopic properties of nuclear Skyrme energy density functionals. *Phys. Rev. Lett.*, 113:252501, 2014.

D. ter Haar. *Men of Physics: L. D. Landau I*. Pergamon Press, Oxford, 1965.

D. ter Haar. *Men of Physics: L. D. Landau II*. Pergamon Press, Oxford, 1969.

D. ter Haar. *Elements of Statistical Mechanics*. Elsevier, 1995a.

D. ter Haar. When is a boson not a boson? *Physics World*, 8:19, 1995b.

J. Terasaki, F. Barranco, R. A. Broglia, E. Vigezzi, and P. F. Bortignon. Solution of the Dyson equation for nucleons in the superfluid phase. *Nucl. Phys. A*, 697:127, 2002.

M. Thoennessen, S. Yokoyama, A. Azhari, T. Baumann, J. A. Brown, A. Galonsky, P. G. Hansen, J. H. Kelley, R. A. Kryger, E. Ramakrishnan, and P. Thirolf. Population of ^{10}Li by fragmentation. *Phys. Rev. C*, 59:111, 1999.

I. J. Thompson and M. V. Zhukov. Effect of a virtual state on the momentum distributions of ^{11}Li. *Phys. Rev. C*, 49:1904, 1994.

I. J. Thompson. Coupled reaction channels calculations in nuclear physics. *Comput. Phys. Reports*, 70 (4):167, 1988.

I. J. Thompson. Reaction mechanism of pair transfer. In R. A. Broglia and V. Zelevinsky, editors, *50 Years of Nuclear BCS*, page 455. World Scientific, Singapore, 2013.

I. J. Thompson and F. M. Nunes. *Nuclear Reactions for Astrophysics*. Cambridge University Press, Cambridge, 2009.

N. K. Timofeyuk and R. C. Johnson. Deuteron stripping and pick-up on halo nuclei. *Phys. Rev. C*, 59:1545, 1999.

M. Tinkham. *Introduction to Superconductivity*. McGraw-Hill, New York, 1980.

W. Tobocman. *Theory of Direct Reactions*. Oxford University Press, Oxford, 1961.

S. Tomonaga. On a relativistically invariant formulation of the quantum theory of wave fields. *Prog. Theor. Phys.*, 1:27, 1946.

M. K. Transtrum, B. B. Machta, K. S. Brown, B. C. Daniels, C. R. Myers, and J. P. Sethna. Perspective: Sloppiness and emergent theories in physics, biology, and beyond. *J. Chem. Physics*, 143:010901, 2015.

J. Ungrin, R. M. Diamond, P. O. Tjøm, and B. Elbek. Inelastic deuteron scattering in the lead region. *Mat. Fys. Medd. Dan. Vid. Selk.*, 38, 1971.

J. Vaagen, B. Nilsson, J. Bang, and R. Ibarra. One- and two-nucleon overlaps generated by a Sturmian method. *Nucl. Phys. A*, 319:0 143, 1979.

J. Valatin. Comments on the theory of superconductivity. *Il Nuovo Cimento*, 7:843, 1958.

V. L. Van der Waerden. *Sources of Quantum Mechanics*. Dover, New York, 1967.

M. van Witsen. Superconductivity in real space. Unpublished manuscript, University of Twente, 2014.

N. L. Vaquero, J. L. Egido, and T. R. Rodríguez. Large-amplitude pairing fluctuations in atomic nuclei. *Phys. Rev. C*, 88:064311, 2013.

N. Vinh Mau. Particle–vibration coupling in one neutron halo nuclei. *Nucl. Phys. A*, 592:33, 1995.

A. Volya, B. Brown, and V. Zelevinsky. Exact solution of the nuclear pairing problem. *Phys. Lett. B*, 509:37, 2001.

W. von Oertzen. Enhanced two-nucleon transfer due to pairing correlations. In R. A. Broglia and V. Zelevinsky, editors, *50 Years of Nuclear BCS*, page 405. World Scientific, Singapore, 2013.

W. von Oertzen and A. Vitturi. Pairing correlations of nucleons and multi-nucleon transfer between heavy nuclei. *Rep. Prog. Phys.*, 64:1247, 2001.

V. V. Vyazovskiy and K. D. Harris. Sleep and the single neuron: The role of global slow oscillations in individual cell rest. *Nat. Revi. Neurosc.*, 14:443, 2013.

J. R. Waldram. *Superconductivity of Metals and Cuprates*. Institute of Physics, Bristol, 1996.

J. C. Ward. An identity in quantum electrodynamics. *Phys. Rev.*, 78:182, 1950.

J. D. Watson. *The Double Helix*. W. W. Norton, New York, 1980.

J. D. Watson and F. H. C. Crick. A structure of the deoxyribose nucleic acid. *Nature*, 171:737, 1953.

L. Weinberg. From BCS to the LHC. In L. Cooper and D. Feldman, editors, *BCS: 50 Years*. World Scientific, Singapore, 2011.

S. Weinberg. Why the renormalizaion group is a good thing. In A. H. Guth, K. Huang, and R. L. Jaffe, editors, *Asymptotic Realms of Physics: Essays in Honor of Francis E. Low*. MIT Press, Cambridge, MA, 1983.

S. Weinberg. *The Quantum Theory of Fields*, Vol. 1. Cambridge University Press, Cambridge, 1996a.

S. Weinberg. *The Quantum Theory of Fields*, Vol. 2. Cambridge University Press, Cambridge, 1996b.

H. Weiss. Semiclassical description of two-nucleon transfer between superfluid nuclei. *Phys. Rev. C*, 19:834, 1979.

V. F. Weisskopf. The formation of superconducting pairs and the nature of superconducting currents. *Contemporary Phys.*, 22:375, 1981.

C. F. Weizsäcker. Zur theorie der kernmassen. *Z. Phys.*, 96:431, 1935.

H. Wibowo and E. Litvinova. Nuclear dipole response in the finite-temperature relativistic time-blocking approximation. *Phys. Rev. C*, 100:024307, 2019.

K. H. Wilcox, R. B. Weisenmiller, G. J. Wozniak, N. A. Jelley, D. Ashery, and J. Cerny. The (^{9}Be, ^{8}B) reaction and the unbound nuclide ^{10}Li. *Phys. Lett. B*, 59:142, 1975.

K. Wimmer, T. Kröll, R. Krücken, V. Bildstein, R. Gernhäuser, B. Bastin, N. Bree, J. Diriken, P. Van Duppen, M. Huyse, N. Patronis, P. Vermaelen, D. Voulot, J. Van de Walle, F. Wenander, L. M. Fraile, R. Chapman, B. Hadinia, R. Orlandi, J. F. Smith, R. Lutter, P. G. Thirolf, M. Labiche, A. Blazhev, M. Kalkühler, P. Reiter, M. Seidlitz, N. Warr, A. O. Macchiavelli, H. B. Jeppesen, E. Fiori, G. Georgiev, G. Schrieder, S. Das Gupta, G. Lo Bianco, S. Nardelli, J. Butterworth, J. Johansen, and K. Riisager. Discovery of the shape coexisting 0^+ state in ^{32}Mg by a two neutron transfer reaction. *Phys. Rev. Lett.*, 105:252501, 2010.

J. Winfield, S. Fortier, W. Catford, S. Pita, N. Orr, J. V. de Wiele, Y. Blumenfeld, R. Chapman, S. Chappell, N. Clarke, N. Curtis, M. Freer, S. Galès, H. Langevin-Joliot, H. Laurent, I. Lhenry, J. Maison, P. Roussel-Chomaz, M. Shawcross, K. Spohr, T. Suomijärvi, and A. de Vismes. Single-neutron transfer from ^{11}Be$_{gs}$ via the (p, d) reaction with a radioactive beam. *Nucl. Phys. A*, 683:48, 2001.

R. B. Wiringa, R. A. Smith, and T. L. Ainsworth. Nucleon–nucleon potentials with and without $\Delta(1232)$ degrees of freedom. *Phys. Rev. C*, 29:1207, 1984.

P. Wölfe. Sound propagation an kinetic coefficients in superfluid ^{3}He. *Prog. Low Temp. Phys.*, 7:191, 1978.

P. G. Wolynes. Moments of excitement. *Science*, 352:150, 2016.

P. G. Wolynes, W. A. Eaton, and A. R. Fehrst. Chemical physics of protein folding. *PNAS*, 109:17770, 2012.

J. Wurzer and H. Hofmann. Microscopic multi-channel calculations for the ^{10}Li system. *Z. Phys. A*, 354:135, 1996.

C. N. Yang. Concept of off-diagonal long-range order and the quantum phases of liquid He and of superconductors. *Rev. Mod. Phys.*, 34:694, 1962.

S. Yoshida. Note on the two-nucleon stripping reaction. *Nucl. Phys.*, 33:685, 1962.

B. M. Young, W. Benenson, J. H. Kelley, N. A. Orr, R. Pfaff, B. M. Sherrill, M. Steiner, M. Thoennessen, J. S. Winfield, J. A. Winger, S. J. Yennello, and A. Zeller. Low-lying structure of ^{10}Li in the reaction ^{11}B(^{7}Li,^{8}B)^{10}Li. *Phys. Rev. C*, 49:279, 1994.

P. G. Young and R. H. Stokes. New states in ^{9}Li from the reaction ^{7}Li(t, p)^{9}Li. *Phys. Rev. C*, 4:1597, 1971.

V. Zelevinsky and A. Volya. Nuclear pairing: New perspectives. *Phys. Atomic Nuclei*, 66:1781, 2003.

V. Zelevinsky and A. Volya. Pairing correlations in nuclei: Old knowledge and new ideas. *Nucl. Phys. A*, 731:299, 2004.

O. Zhou, Q. Zhu, J. E. Fischer, N. Coustel, G. B. M. Vaughan, P. A. Heiney, J. P. McCauley, and A. B. Smith. Compressibility of M_3C_{60} Fullerene superconductors: Relation between T_c and lattice parameter. *Science*, 255:833, 1992.

Y. Zhou, D. Vitkup, and M. Karplus. Native proteins are surface-molten solids: Applications of the Lindemann criterion for the solid versus liquid state. *J. Mol. Biol.*, 285:1371, 1999.

M. Zinser, F. Humbert, T. Nilsson, W. Schwab, T. Blaich, M. J. G. Borge, L. V. Chulkov, H. Eickhoff, T. W. Elze, H. Emling, B. Franzke, H. Freiesleben, H. Geissel, K. Grimm, D. Guillemaud-Mueller, P. G. Hansen, R. Holzmann, H. Irnich, B. Jonson, J. G. Keller, O. Klepper, H. Klingler, J. V. Kratz, R. Kulessa, D. Lambrecht, Y. Leifels, and A. Magel. Study of the unstable nucleus ^{10}Li in stripping reactions of the radioactive projectiles ^{11}Be and ^{11}Li. *Phys. Rev. Lett.*, 75:1719, 1995.

M. Zinser, F. Humbert, T. Nilsson, W. Schwab, H. Simon, T. Aumann, M. J. G. Borge, L. V. Chulkov, J. Cub, T. W. Elze, H. Emling, H. Geissel, D. Guillemaud-Mueller, P. G. Hansen, R. Holzmann, H. Irnich, B. Jonson, J. V. Kratz, R. Kulessa, Y. Leifels, H. Lenske, A. Magel, A. C. Mueller, G. Münzenberg, F. Nickel, G. Nyman, A. Richter, K. Riisager, C. Scheidenberger, G. Schrieder, K. Stelzer, J. Stroth, A. Surowiec, O. Tengblad, E. Wajda, and E. Zude. Invariant-mass spectroscopy of ^{10}Li and ^{11}Li. *Nucl. Phys. A*, 619:151, 1997.

B. Zwieglinski, W. Benenson, R. Robertson, and W. Coker. Study of the ^{10}Be$(d, p)^{11}$Be reaction at 25 MeV. *Nucl. Phys. A*, 315:124, 1979.

Index